S0-DQW-730

Chemistry
A Science for Today

Spencer L. Seager
Weber State College

H. Stephen Stoker
Weber State College

Scott, Foresman and Company
Glenview, Illinois *Brighton, England*

Library of Congress Catalog Card Number: 72-91512
ISBN: 0-673-07808-6

Printed in the United States of America.

Regional offices of Scott, Foresman and Company are located in Dallas, Texas; Glenview, Illinois; Oakland, New Jersey; Palo Alto, California; Tucker, Georgia, and Brighton, England.

Photographs

All photographs by Gene Dekovic.

Preface

The mind is the man, and knowledge mind; a man is but what he knoweth.

—Francis Bacon

What should a person be—what should a person know? These questions provide the basis for many approaches to and disagreements about courses designed for the general education of today's college students. This textbook represents our contribution to the controversy.

We feel that educated people, regardless of their areas of concentration, should know something about themselves and their environment. In addition, they should be able to relate to and function in that environment in a way that is satisfactory (to themselves). Most of us find ourselves in an environment consisting of other living organisms (including many people) as well as nonliving materials and objects. This textbook offers knowledge about the substances of which that environment is composed. We hope to impart to the students some of the feelings of interest, appreciation, and (still) amazement that we, the authors, experience as we learn more about this subject.

With these goals in mind, we have written for students who have little or no formal chemistry background. Although primarily intended for nonmajors who plan to take only one chemistry course, this textbook may also provide a foundation for those who plan to study chemistry further. Enough material exists to permit the instructor and students to exercise a reasonable amount of choice concerning course content.

Although we have followed a fairly traditional order of presentation, we have eliminated some of the traditional topics that our general education students have found to be less interesting, and we have included some high-interest topics that are not traditional—such as environmental chemistry, biochemistry, chemistry of the body, and medicinal chemistry. In a few instances, where the change makes very little difference in the ideas being taught, we have used terms or concepts that are a little less than rigorously correct but are more familiar to the student or more easily understood. The most obvious example is our use throughout the book of *weight* rather than *mass*.

The first nine chapters contain concepts and principles that are needed in order to understand the remaining seventeen chapters. Some of the later chapters are also prerequisite to subsequent ones. For example, Chapters 17 and 18 (organic chemistry) provide a foundation for Chapters 19, 20, 23, and 24; and Chapters 23 and 24 form a foundation for Chapters 25 and 26.

Each chapter is followed by review questions and most also have a list of suggested readings. The questions are designed to draw the students' attention to important and interesting topics and to help them test their understanding. (Answers to these questions are included in the instructor's manual.) The readings will provide an elaboration, expansion, or different perspective on the chapter topics.

We wish the students success and an enjoyable experience as they participate in this study of the substances that make up our world.

S.L.S.
H.S.S.

Contents

1 Chemistry: A Study of Matter

Chemistry is a science in which the substances (matter) making up the universe are studied. In other sciences, such as physics and geology, the same substances are studied but different aspects are emphasized. Chemistry emphasizes (1) the composition and structure of the substances of the universe, (2) the properties of these substances, and (3) the changes they will undergo and the conditions necessary to cause or prevent the changes. In this chapter these chemically important aspects of matter as well as a number of basic terms are discussed.

The Composition and Structure of Matter

Throughout history, human beings have shown a natural curiosity about the composition of matter. According to one early conclusion, based on philosophical speculation, all matter was composed of a small number of fundamental types—earth, air, fire, and water. Modern experimentation has substantiated the idea of a few fundamental types of matter but not the proposed identities.

The **composition** of a substance is known when the identity and amount of each of its components have been determined. The arrangement of the components of a substance defines its **structure.** Knowledge of composition and structure can often be put to practical use. For example, if the components of a natural mineral deposit are known, it is possible to determine what useful products can be obtained from it. Furthermore, the amount of each valuable component must be known before a decision can be made to mine the deposit. Structural information about a substance may provide insight into the reasons for certain behavior and in some cases allow this behavior to be built into other substances. The cleansing action of soap is related to its structure. A knowledge of this structure made possible the manufacture of synthetic detergents having structures and cleaning properties similar to those of soap.

The Properties of Matter

Every day each of us encounters some of the world's many substances. Usually, we find it reasonably easy to differentiate between substances by making use of the **properties** or distinguishing characteristics of matter. Just as we recognize a friend by such characteristics as hair color, walk, shape of nose, and so forth, we recognize each substance by a unique set of properties. These properties can be used in a number of practical ways, such as:

Identifying an unknown substance (law enforcement agencies routinely analyze substances thought to be drugs or poisons).

Distinguishing between different substances (an experienced miner can quickly tell the difference between genuine gold and fool's gold).
Characterizing a newly discovered substance (any new substance must have a unique set of properties different from the set belonging to any known substance).
Predicting the usefulness of a substance for specific applications (a gas known to be poisonous would obviously not be useful as an atmosphere for a space capsule).

The properties of matter are usually classified as physical or chemical. We can observe or measure a **physical property** without changing the fundamental composition of the matter under observation. No original substances are destroyed and no new ones are created. The physical properties of a substance can be described without reference to any other substance. Typical physical properties are hardness, boiling point, melting point, odor, color, and solubility.

A **chemical property** involves changes (or lack of changes) in composition that occur in a substance when it is subjected to different conditions. We cannot measure chemical properties without attempting to change one substance into one or more other substances; that is, an attempt must be made to cause a chemical reaction to take place. For example, attempts to burn a substance in air can produce one of two results—the substance burns or it does not burn. If it burns, one of its chemical properties is flammability in air; if it does not burn, the property is nonflammability in air.

After a reaction, the presence of new materials is indicated by the appearance of new physical and chemical properties. All properties of a sample do not have to change in order to indicate that a chemical change has taken place. A difference in any significant property is generally sufficient.

Changes That Occur in Matter

Changes in matter are common and familiar occurrences. Changes take place, for example, when food is cooked, sugar is dissolved in a cup of coffee, and gasoline is burned in an auto engine.

When we understand these changes, we can, among other things, predict the behavior of substances and choose those most suitable for particular uses. A substance known to react rapidly when heated, for example, should not be stored near a heat source. Similarly, water-soluble substances should not be used in the manufacture of raincoats.

Like properties, changes in matter are classified as physical or chemical. A **physical change** is one that causes no change in composition. A **chemical change,** on the other hand, results in the disappearance of one or more substances, the appearance of one or more new substances, and a change in composition.

The rusting of iron is a chemical change. Hard, shiny iron is converted into a soft, red-colored product with properties obviously different from those of the iron. A physical change takes place when ice is melted to liquid water, since throughout the process the composition of water remains unchanged. This change of water from a solid to a liquid is called a **change of state.** It involves two of the three states of matter: solid, liquid, and gas. Changes of state do not result in composition changes, and are, therefore, classified as physical changes.

Most changes in matter can easily be classified as either physical or chemical. However, some changes—such as the formation of certain solutions—defy simple classification. Common salt dissolves easily in water to form a solution of salt water. The salt can easily be recovered by the physical process of evaporating the water. When gaseous hydrogen chloride is dissolved in water, again a solution results; but in this case the starting materials cannot be recovered by evaporation. The formation of salt water is considered a physical change because the original components can be recovered in an unchanged form. The second solution process presents classification problems because of the possibility that a chemical reaction took place.

The changes involved in the cooking of an egg also present classification problems. The cooked egg contains the same structural units as the uncooked egg. However, some changes in structural arrangement have taken place—so is the change physical or chemical?

In chemistry, as in other sciences, attempts are made to classify information in a way that eliminates exceptions. Obviously, the classification of changes as physical or chemical, using the definitions given earlier, is not completely satisfactory. Other definitions and classification schemes used in science, however, have this same drawback. We shall continue to use the concepts of physical and chemical change because their usefulness far outweighs the problems created by a few exceptions.

Varied techniques are used to bring about physical and chemical changes in matter. Techniques leading to physical changes involve **physical methods,** while those producing chemical changes involve **chemical methods.** The grinding of a solid to a powder is an example of a physical change brought about by a physical method.

Some changes that take place in matter are beneficial because the products are more useful than the starting materials. The metal chemically obtained from an ore is more useful than the ore. Other changes, such as the rusting of iron, are undesirable because the products are not useful. The chemist, by studying the nature of changes in matter, learns how to bring about favorable changes and prevent undesirable ones.

Two hundred years ago most of the familiar materials used by man were only physically changed from the way they occurred in nature. Very few useful materials were produced by chemical changes. Today, many of the commonly used materials are the result of controlled chemical change. Synthetic fibers, plastics, many drugs, and latex paints are examples of chemically produced materials.

Physical vs. Chemical Concepts

Table 1–1 summarizes the physical and chemical terms introduced thus far. It is apparent that the term **physical** is related to the idea that the composition of substances is not changed, while the term **chemical** is associated with changes in composition, or the disappearance of some substances and the appearance of others. The distinction between these two terms is important in the discussions that follow.

Table 1–1. Physical and Chemical Terms

Physical	*Chemical*
Physical property: A property that can be observed without changing the composition of anything present.	**Chemical property:** A property that can be observed only by attempting to change the material into a different substance.
Physical change: A change that does not change the composition of anything present.	**Chemical change:** A change that changes one or more substances into new substances.
Physical method: A technique that can be carried out without changing the composition of anything present.	**Chemical method:** A technique that can be carried out only by changing one or more substances into new substances.

Heterogeneous and Homogeneous Matter

Matter is usually found as complex mixtures of substances. This fact is sometimes quite obvious—as in the case of a T-bone steak (bone, fat, meat, etc.) or concrete (gravel, sand, etc.). Sometimes it is not quite so obvious. Air, for example, is a mixture of at least five pure substances; gasoline is a mixture of ten or more, and steel of at least two. Less commonly, we encounter matter that consists of a single substance. Examples are distilled water, white refined sugar, and copper wire.

An investigation of a number of samples quickly indicates that matter may be classified into two large groups. One group consists of matter with the same observable properties throughout, such as water or sugar. This group will be referred to as **homogeneous.** The second group consists of matter in which different parts have different properties, such as rocky soil or concrete, and is referred to as **heterogeneous.**

More specifically:

Heterogeneous matter consists of two or more phases or regions in which the properties of the matter are different. The phases may or may not be in the same state (solid, liquid, or gas). Concrete is made up of two or more phases in the same solid state. A mixture of sand and water is heterogeneous and contains two phases in two different states (solid and liquid).

Heterogeneous matter contains two or more substances (concrete) *or one substance in two or more states* (ice floating in water).

Heterogeneous matter can theoretically be separated by physical methods into homogeneous components. The term *theoretically* is used because some separations are very difficult to perform—such as the separation of concrete into sand, gravel, and cement.

Similarly, *homogeneous matter consists of only one phase, may contain one substance* (distilled water) *or more than one* (salt dissolved in water), *and is uniform throughout.*

The differences between heterogeneous and homogeneous matter are summarized in Table 1–2.

Table 1–2. **Heterogeneous and Homogeneous Matter**

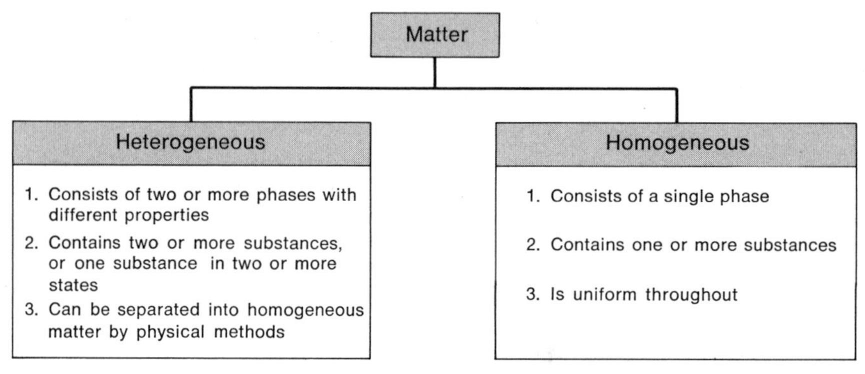

Matter

Heterogeneous	Homogeneous
1. Consists of two or more phases with different properties	1. Consists of a single phase
2. Contains two or more substances, or one substance in two or more states	2. Contains one or more substances
3. Can be separated into homogeneous matter by physical methods	3. Is uniform throughout

Solutions and Pure Substances

Because homogeneous matter may consist of one or more substances, a further classification is useful. Homogeneous matter containing two or more substances is called a **solution** while that containing only one is called a **pure substance.** Common usage of the term *solution* often tends to be limited to liquid mixtures, but in fact it is applicable to all three states of matter—gases, liquids, and solids. Alloys of metals, such as gold and copper, are solid solutions. Salt dissolved in water is a liquid solution, and the air surrounding us is a gaseous solution.

In addition to being homogeneous, solutions are also characterized by the following properties:

Solutions can have variable composition. A saltwater solution may contain $\frac{1}{8}$, $\frac{1}{4}$, or $\frac{1}{2}$ lb of salt in a gallon of water. In all cases, it is a saltwater solution; only the composition has changed.

In most cases all components of a solution except one (the water in the saltwater solutions above) *are present in the smallest unit possible.* This unit, the molecule, is discussed later in this chapter. A sugar cube added to a glass of water does not immediately form a homogeneous solution. The sugar slowly breaks up into the single units mentioned above, until finally no larger units of sugar exist in the mixture. Upon stirring (to hasten the homogenizing processes), the mixture becomes a homogeneous solution. The difference between a heterogeneous mixture and a homogeneous mixture or solution is the size of the participating particles of matter.

Solutions can be separated by physical methods into pure substances. The evaporation of salt water easily separates water as a gas and leaves pure salt behind as a solid.

Homogeneous matter consisting of one pure substance may be classified according to the following properties:

A pure substance has constant composition. Unlike saltwater solutions, two samples of pure water must have the same composition.

A pure substance has a unique set of physical properties not duplicated in any other pure substance.

A pure substance is composed of a unique simplest unit of matter.

A pure substance cannot be separated by physical methods into any simpler pure substance.

Table 1–3 summarizes the classification of homogeneous matter into solutions and pure substances.

Table 1–3. Pure Substances and Solutions

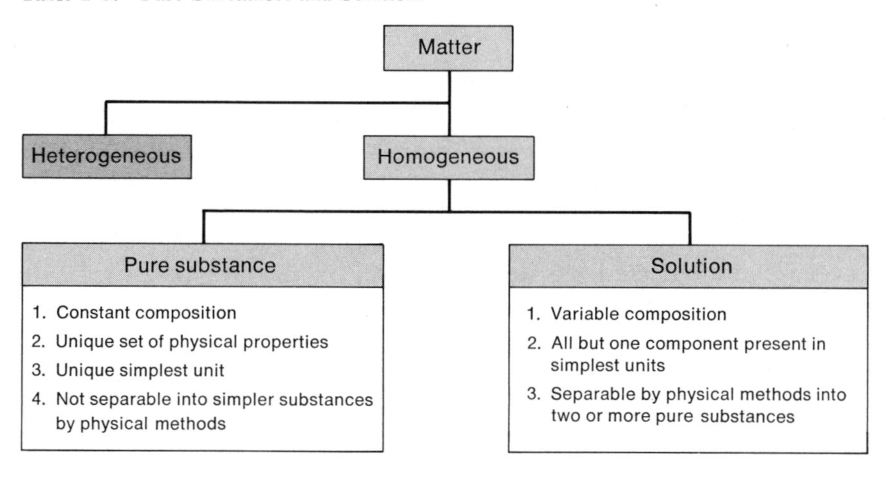

The Meaning of Pure

We have emphasized that mixtures of substances, whether heterogeneous or homogeneous, can theoretically be separated into pure substances by physical methods. Today many sophisticated physical separation methods are known; but after they have been used, two questions still arise: Is the separation complete, and have the expected pure components been produced? A substance is designated **pure** when its properties, under a given set of conditions, cannot be changed by further purification techniques.

Suppose, for example, that powdered iron and sulfur are mixed together. The iron is gray and the sulfur is yellow. We can physically separate the mixture by stirring it with a magnet; the iron clings to the magnet and leaves the sulfur behind. After one stirring, the remaining sulfur is still not pure, as indicated by a gray-yellow color. Repeated stirrings with more powerful magnets would eventually produce bright yellow sulfur whose color would not change with further treatment. Since the property of color does not change upon further purification attempts, the sulfur is assumed to be pure. Such a conclusion, based only on one property, may be in error, and other properties should also be checked. A mixture of powdered sugar and white sand would appear to be a pure substance if only the color were used as a criterion. However, a check of other properties would quickly show the nature of the mixture. Attempts to melt it would show that one part (the sugar) would melt at a lower temperature than the other part (sand). Similar results would be found upon attempting to dissolve the mixture in water; the sugar would dissolve and the sand would not. These examples illustrate the fact that pure substances possess a unique set of physical properties which remain unchanged despite continued attempts at purification.

Pure Substances and Physical Subdivision

Suppose a 1-lb sample of table sugar is divided into $\frac{1}{2}$-lb portions; then each $\frac{1}{2}$ lb is further divided into halves of $\frac{1}{4}$ lb each. How far could such a physical subdivision process be carried? Is there a limit to this subdivision, beyond which our sample would no longer have the properties of sugar? The ancient Greeks wrestled with this same type of problem and at least one of them, Democritus (460–370 B.C.), concluded that a limit to subdivision does exist. Today this conclusion is stated as follows: A limit of physical subdivision does exist for pure substances. The limit is the smallest unit of pure substance capable of a stable independent existence. This smallest unit is called a **molecule.** The useful concept of a molecule will be modified slightly when solids are discussed in Chapter 8, because a single molecule is an unnatural state for certain types of solids.

It must be emphasized that every pure substance has as its smallest, characteristic particle a unique molecule. If two pure substances had the same molecule as a basic unit, both substances would have the same properties;

thus they would have to be classified as the same substance. Also, we must understand the converse: there is only one kind of molecule for each pure substance. Two kinds of molecules would produce properties characteristic of mixtures and could not represent a pure substance.

Pure Substances and Chemical Subdivision

A molecule is the limit of physical subdivision; however, molecules can be broken down chemically into simpler particles if chemical methods are employed. The ultimate limit of chemical subdivision is the unit of matter called the **atom.** The use of chemical methods to divide molecules into atoms causes changes in composition and accompanying changes in properties. As we have discussed before, the sugar molecule is the smallest unit of sugar which demonstrates the properties of sugar. Sugar molecules can be divided into atoms, but the resulting atoms do not show the properties characteristic of sugar.

It is found in practice that chemical treatment of a molecule often results first in a subdivision into smaller molecules, but ultimately the limit of atoms is reached. In some cases, molecules yield only one kind of atom following chemical processes. These molecules are called **homoatomic.** Other molecules, called **heteroatomic,** give two or more kinds of atoms when subdivided chemically. These two kinds of molecules provide the basis for a further classification of pure substances.

Elements and Compounds

Pure substances containing homoatomic molecules are called **elements;** they cannot be broken down into simpler pure substances by chemical processes because only one kind of atom is available. Heteroatomic molecules are contained in pure substances classified as **compounds.** Compounds can be changed chemically into simpler pure substances, each containing one type of atom (elements).

All possible attempts must be made to chemically subdivide a substance into simpler substances before the original substance can be classified as an element. If a sample of pure substance, S, is subjected to a chemical process and produces two new substances, X and Y ($S \rightarrow X + Y$), S would be classified as a compound. However, little could be said about X and Y, since no attempts have been made to chemically change them. If only a few unsuccessful attempts were made to chemically subdivide S, we might correctly call it an element, but until all possible reactions have been tried unsuccessfully, such a classification could be in error. Table 1–4 summarizes the classification of elements and compounds.

Table 1–4. Elements and Compounds

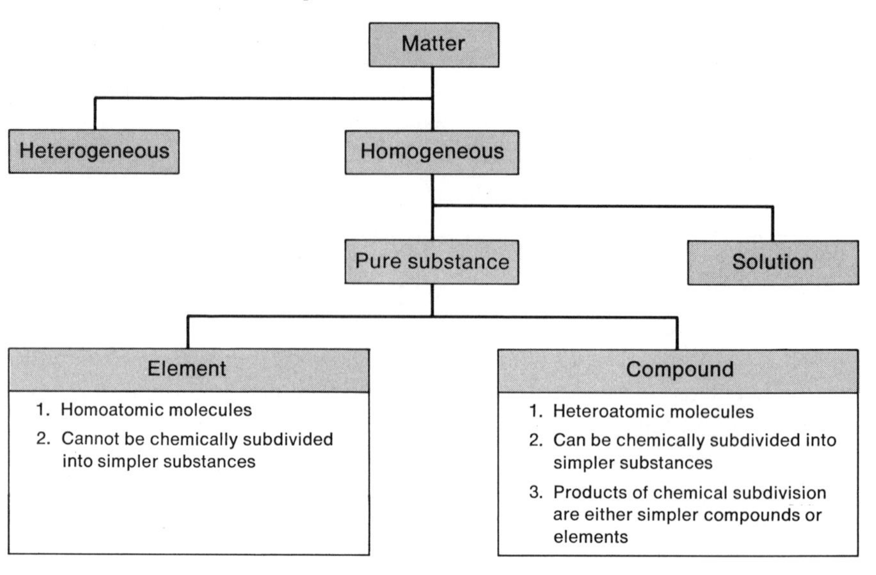

Table 1–5. Atoms, Molecules, and the Classifications of Matter

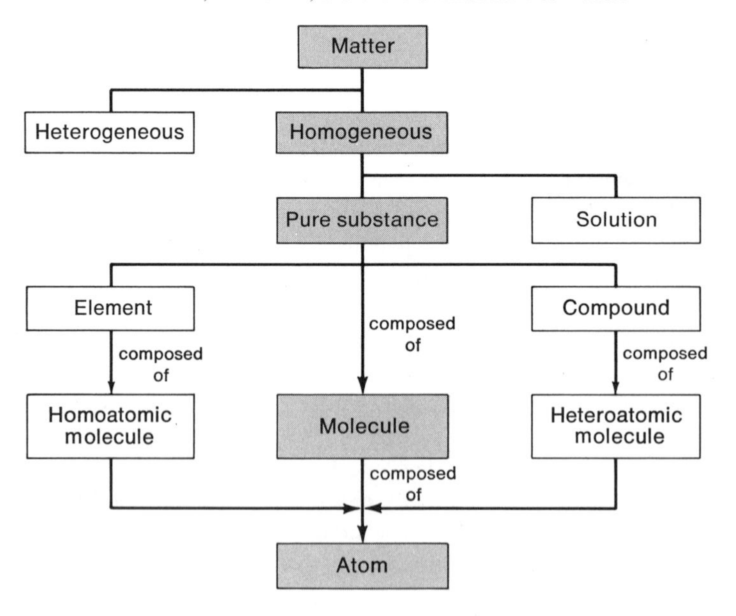

Classifications of Matter

Since atoms represent the limit of chemical subdivision for matter, it is reasonable to consider them to be the units from which matter is made. The atoms combine to form molecules, and each pure substance is made up of one unique type of molecule. The atomic composition of the molecules (hetero- or homoatomic) determines whether the pure substance is a compound or an element. Pure substances (elements or compounds) can be combined chemically to give new pure substances (compounds), or physically to give homogeneous mixtures (solutions) or heterogeneous mixtures. The complete relationships between atoms, molecules, and the classifications of matter are given in Table 1–5.

Review Questions

1. List 3 areas with which the science of chemistry is particularly concerned.
2. How do the terms *composition* and *structure* differ from each other?
3. List three distinguishing characteristics of:
   a) a physical property
   b) a chemical property
4. Classify each of the following properties as chemical or physical:
   a) a substance decomposes on heating
   b) the color of a substance
   c) the melting temperature of a substance
   d) a substance does not react with oxygen
   e) a substance has a high density
   f) a substance is a good reflector of light
   g) a substance is flammable
   h) a substance is very hard
5. Classify each of the following changes as physical or chemical. If the data are not complete enough for you to be sure, indicate the most likely classification.
   a) steam condenses to liquid
   b) a match burns
   c) sugar is dissolved in water
   d) an onion is diced into small pieces
   e) gold is melted
   f) iron rusts
   g) light is emitted by a flashbulb
   h) a table leg is fashioned from a piece of wood
6. Make lists to show what the following pairs have in common and how they differ.
   a) heterogeneous and homogeneous
   b) pure substance and solution
   c) element and compound

7. How is it determined whether or not a substance is pure?
8. A pure substance A is found to change upon heating into two new pure substances, B and C. On the basis of this observation, classify A, B, and C into one of the following categories:
   a) element
   b) compound
   c) insufficient data to allow classification
9. How do molecules of elements and compounds differ from each other?
10. Classify each of the following substances into the most specific category of Table 1–5 that is possible. The category of *compound* is more specific than *pure substance*.
    a) air
    b) copper metal
    c) oil and vinegar salad dressing
    d) pancake syrup
    e) a carrot
    f) table salt
    g) highway asphalt (blacktop)
    h) T-bone steak
11. Indicate whether each of the following statements is true or false. If a statement is false, change it to make it true. Such changes should involve more than merely making the statement the negative of itself.
    a) Some pure substances may be decomposed into simpler pure substances by chemical change.
    b) The unique simplest unit of a pure substance is an atom.
    c) Molecules of pure substances are always heteroatomic.
    d) Solutions can be separated into component substances by physical means.
    e) In order for matter to be homogeneous only one pure substance can be present.
    f) Homogeneous matter can have a variable composition.
    g) The limit of physical subdivision is a molecule.
    h) A compound contains only one kind of molecule.

2 Atoms: The Building Blocks of Matter

At present, 105 different "types" of atoms are known to exist, and all matter is considered to be composed of some combination of them. The situation is somewhat like that of the 26 letters in the alphabet from which all words are derived by various combinations. Because elements are pure substances consisting of only one type of atom (homoatomic molecules), the existence of 105 types of atoms leads to the conclusion that there must be 105 different elements.

Discovery and Abundance of the Elements

The discovery and isolation of the 105 elements have taken place over a period of several centuries. Discovery, for the most part, has occurred since 1700, and in particular during the 1800s. Table 2–1 gives the relationship between time and the number of known elements.

Table 2–1. Number of Elements Discovered During Various Fifty-Year Periods

Time period	*Number of elements discovered during time period*	*Total number of elements known at end of time period*
Ancient–1700	13	13
1701–1750	3	16
1751–1800	18	34
1801–1850	25	59
1851–1900	23	82
1901–1950	16	98
1951–present	7	105

Eighty-eight of the 105 elements occur naturally, and 17 have been synthesized by bombarding atoms of naturally occurring elements with small particles. It is generally accepted by scientists that no more naturally occurring elements will be found, although it is possible that additional ones may be prepared synthetically (see Chapter 16). The last of the naturally occurring elements was discovered in 1925. The elements synthesized by scientists are all radioactive (unstable) and rapidly change into stable elements as the result of radioactive emissions.

Naturally occurring elements are not evenly distributed in the physical world, and the majority are found only in small amounts. Ten of the 105 elements make up over 99 percent of the total weight of material readily available to scientists for examination—the earth's crust, atmosphere, and oceans. Figure 2–1 gives the quantities of the 10 most abundant elements in

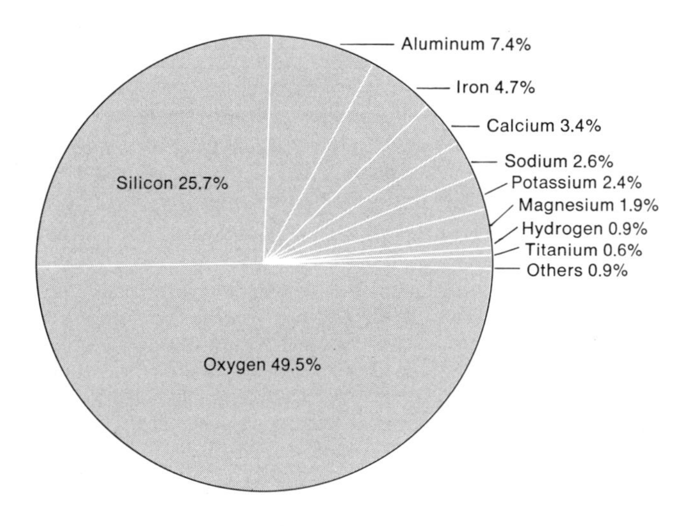

Figure 2–1.
Abundance of the Elements in the Earth's Crust, Water, and Atmosphere

the outer 10 miles of the earth's crust, waters, and atmosphere. Two elements, oxygen and silicon, account for slightly more than three-quarters of the total weight involved.

Names, Symbols, and Formulas: The Language of Chemistry

Each of the elements was initially characterized by its physical and chemical properties. Attempts to communicate in chemical language would be rather cumbersome if, each time a specific element were discussed, it had to be identified by a list of its chemical and physical properties. The assignment of a unique name to each element overcomes this problem, and forms the basis for a nomenclature system that chemists use to communicate with each other.

A great deal of diversity exists among the names of the elements. Some are very simple, such as tin or lead, while others, such as praseodymium or ytterbium, are quite lengthy and even appear somewhat foreign to the English language. This diversity is related to the tradition in chemistry that allows the discoverer to choose the name for a new element.

Historical studies reveal that discoverers, although free to act independently, have derived elemental names from a rather limited number of sources. These sources, along with the number of elements named (in parentheses), are as follows: geography (20), astronomy (8), mythology (8), famous scientists (10), color (11), physical properties other than color (5), chemical properties (4), terms associated with radioactivity (6), aspects of their discovery (8), and alchemical terms (16). Ten elements, known in ancient times, have names of uncertain origin. Table 2–2 gives background information about the origin

Table 2–2. Elemental Name Sources for Selected Elements

Geography	
Germanium	Named after the native country of its discoverer.
Yttrium	Named after the locality (Yttrium, Sweden) where the source material of the element was discovered.
Hafnium	Named after the location of the laboratory where it was first isolated—Copenhagen (Latin name—*Hafnia*).
Astronomy	
Uranium	Named after the planet Uranus, discovered eight years before the element.
Helium	Named from the Greek word *helios* for sun, since it was first observed spectroscopically in the sun's corona during an eclipse.
Selenium	Named from the Greek *selene* for moon. It was chosen because the element had properties very similar to the previously discovered tellurium, which had been named from the Latin *tellus*—earth.
Color	
Chlorine	Name derived from the Greek *chloros*, denoting greenish-yellow, the color of the gas.
Iridium	Name derived from the Greek *iris*, meaning rainbow, alluding to the varying color of the compounds from which it was isolated.

of the names of selected elements in three of the above categories—geography, astronomy, and color. The examples have been chosen to illustrate the many types of reasoning involved in naming an element.

Elemental names are not often written out by chemists, who have found it convenient to set up standard abbreviations. These abbreviations, commonly called elemental symbols, are given for each of the elements in Table 2–3. Thirteen of the elements have symbols consisting of a single capital letter. All of the rest have two-letter symbols in which the first letter is capitalized and the second is always lowercase. The two-letter symbols are often the first two letters of the elemental name, although the choice of the second letter in the symbol may be any letter of the name. Occasionally, the symbol bears no formal resemblance to the English name of the element but is derived from the element's name in another language, usually Latin; the elements with such symbols are indicated in Table 2–3. Many of these elements have been known since biblical times.

Elemental nomenclature has been left largely to the imagination of the discoverers. This is not the case with compounds. Extensive sets of systematic rules have been developed to govern such nomenclature, completely eliminating arbitrariness. This is an absolute necessity, because more than four million compounds are now known, and chaos would result if their nomenclature were left to chance. Much of Chapter 5 is devoted to rules for naming compounds.

Like the names of elements, the names of compounds are usually abbreviated. Compound abbreviations are called formulas. Even though the formal rules for naming compounds will not be presented until later, the meaning

Table 2–3. The Chemical Elements and Their Symbols

Ac	actinium	Ge	germanium	Po	polonium
Ag	silver*	H	hydrogen	Pr	praseodymium
Al	aluminum	Ha	hahnium**	Pt	platinum
Am	americium	He	helium	Pu	plutonium
Ar	argon	Hf	hafnium	Ra	radium
As	arsenic	Hg	mercury*	Rb	rubidium
At	astatine	Ho	holmium	Re	rhenium
Au	gold*	I	iodine	Rf	rutherfordium**
B	boron	In	indium	Rh	rhodium
Ba	barium	Ir	iridium	Rn	radon
Be	beryllium	K	potassium*	S	sulfur
Bi	bismuth	Kr	krypton	Sb	antimony*
Bk	berkelium	La	lanthanum	Sc	scandium
Br	bromine	Li	lithium	Se	selenium
C	carbon	Lu	lutetium	Si	silicon
Ca	calcium	Lr	lawrencium	Sm	samarium
Cd	cadmium	Md	mendelevium	Sn	tin*
Ce	cerium	Mg	magnesium	Sr	strontium
Cf	californium	Mn	manganese	Ta	tantalum
Cl	chlorine	Mo	molybdenum	Tb	terbium
Cm	curium	N	nitrogen	Tc	technetium
Co	cobalt	Na	sodium*	Te	tellurium
Cr	chromium	Nb	niobium	Th	thorium
Cs	cesium	Nd	neodymium	Ti	titanium
Cu	copper*	Ne	neon	Tl	thallium
Dy	dysprosium	Ni	nickel	Tm	thulium
Er	erbium	No	nobelium	U	uranium
Es	einsteinium	Np	neptunium	V	vanadium
Eu	europium	O	oxygen	W	tungsten*
F	fluorine	Os	osmium	Xe	xenon
Fe	iron*	P	phosphorus	Y	yttrium
Fm	fermium	Pa	protactinium	Yb	ytterbium
Fr	francium	Pb	lead*	Zn	zinc
Ga	gallium	Pd	palladium	Zr	zirconium
Gd	gadolinium				

* Elements with symbols not derived from their English names.

** There is disagreement about the names of the two most recently discovered elements. Both American and Russian research groups claim their discovery and consequently have named them. The names given in the table are those suggested by the American group. The Russian suggestions are Kurchatovium (for Rutherfordium) and Bohrium (for Hahnium). The conflicting claims concerning Rutherfordium span the period 1964–1968 and those concerning Hahnium the period 1968–1970. None of the names has been officially accepted.

and significance of a formula can and should be understood at this time. We obtain formulas by writing the symbol of each element in the compound, indicating the number of atoms by a subscript. Table 2–4 gives three typical formulas, along with explanatory comments.

Table 2–4. Formula Writing

Formula	*Comments*
H_2SO_4	This is the formula for a compound whose molecular units contain 7 atoms—2 hydrogen atoms, 1 sulfur atom, and 4 oxygen atoms. Notice that the number of each type of atom is indicated by subscripts following the atomic symbol and that when the subscript is 1 it is not written, but is understood. Verbally this formula is expressed as H-two-S-O-four.
$COCl_2$	This is the formula for a compound whose molecules contain 4 atoms—1 carbon, 1 oxygen, and 2 chlorine.
$CoCl_2$	This is the formula for a compound whose molecules contain 3 atoms—1 cobalt atom and 2 chlorine atoms. In comparing the formula $CoCl_2$ to the previous one, $COCl_2$, we see the importance of the capitalization rules governing the writing of symbols. Co stands for the element cobalt, while CO stands for one atom of carbon and one of oxygen.

Atomic Theory

In our discussion thus far, we have assumed the existence of a simple fundamental unit called the atom. The concept that all matter is made up of individual particles is called the **atomic theory.** In modern terminology the concepts making up the atomic theory may be summarized as follows:

1. All matter is made up of small particles called atoms, of which 105 different "types" are known.
2. The arrangement and relative number of different types of atoms contained in a substance determine its identity.
3. All atoms of a given type are similar to one another and significantly different from all other types.
4. Chemical change is a union, separation, or rearrangement of atoms to give new substances.
5. Only whole atoms can participate in or result from any chemical change, since atoms are considered to be indestructible during such changes.

A large amount of supporting evidence has led to the general acceptance of the atomic theory as fact, although it is still a theory because no one has actually seen an individual atom. As a theory that is used to explain experimental observations, it is subject to revision as observations from, and interpretations of, new experiments become available. Only a completely unforeseen development could cause the abandonment of the basic concepts of the atomic theory, because an enormous amount of scientific evidence has accumulated which can be explained by the theory and therefore supports it.

Chemical Evidence Supporting the Atomic Theory

Many careful observations and measurements have led to several general, summarizing statements concerning chemical changes and compound formation. These statements, commonly referred to as laws, summarize the indirect evidence which supports the atomic theory.

Other theories or explanations have been proposed and provide an explanation for certain of the laws, but only the atomic theory provides a rational explanation for all of the laws. This is the reason for its wide acceptance by scientists.

The type of evidence which supports the atomic theory is illustrated by three laws that are well known to chemists: the law of conservation of matter, the law of constant composition, and the law of multiple proportions.

The Law of Conservation of Matter

When chemical changes are carried out, the total weight of starting materials (reactants) and the total weight of the new materials formed (products) are often measured. Many of these measurements have been made by scientists on a large number of systems. The results, within experimental error, are always the same—the total weight of reactants is equal to the total weight of products. This consistency in observations led to the formulation of the **law of conservation of matter.** A statement of the law is:

In all ordinary chemical changes matter is neither created nor destroyed.

The law can be illustrated by studying the weight relationships in a typical chemical reaction. Calcium and sulfur react to form calcium sulfide according to the following reaction:

$$\mathrm{Ca + S \rightarrow CaS}$$

It has been found that 55.6 grams of calcium will react with 44.4 grams of sulfur to give 100.0 grams of calcium sulfide:

$$55.6\ \mathrm{g\ Ca} + 44.4\ \mathrm{g\ S} \rightarrow 100.0\ \mathrm{g\ CaS}$$

After the reaction is completed, no Ca or S remains. The only substance present is CaS. A comparison of the total weights of reactants and products illustrates the law:

$$\underbrace{55.6\ \mathrm{g} + 44.4\ \mathrm{g}}_{\text{reactants}} = \underbrace{100.0\ \mathrm{g}}_{\text{product}}$$

It has also been found that when 100.0 g of CaS is heated to a high temperature in the absence of air, the CaS decomposes back into Ca and S and produces 55.6 g of Ca and 44.4 g of S:

$$100.0\ \mathrm{g\ CaS} \rightarrow 55.6\ \mathrm{g\ Ca} + 44.4\ \mathrm{g\ S}$$

Once again, the weights of products and reactant are found to be equal.

In terms of the atomic theory the results given above can be explained as follows: (1) The weights of calcium and sulfur used each contain a certain number of indestructible atoms. (2) Since the amounts of calcium and sulfur exactly react with each other, all of each type of atom must be involved. (3) Since every atom initially present is involved in the reaction, the products must contain the same total number of each type of atom as the reactants and must, therefore, have the same weight as the reactants.

As an analogy, suppose a child has a pile of red building blocks and a pile of white building blocks. Once all the red and white blocks have been used to build some object, the weight of the object must be the same as the total weight of red and white blocks that make it up.

The Law of Constant Composition

If a chemist were to analyze pure water obtained from many different sources, he would find that in every case the percentage of oxygen and hydrogen in the compound would be the same, regardless of the amount or source of the samples. (See Table 2–5.)

Table 2–5. Constant Composition of Water

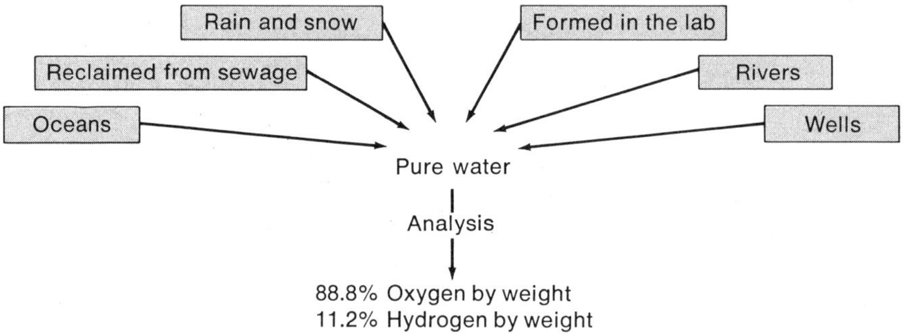

Rain and snow
Formed in the lab
Reclaimed from sewage
Rivers
Oceans
Wells
Pure water
Analysis
88.8% Oxygen by weight
11.2% Hydrogen by weight

This same result has been obtained for all compounds subjected to analysis and has led to the formulation of the **law of constant composition:**

When two or more elements combine to form a compound, the proportion by weight of the elements present is always the same.

In other words, once the composition of a pure substance has been determined, we know that the composition of all other pure samples of the substance will be the same.

An alternate way of stating the law is: A given weight of one element combines with only one definite weight of another element to form a particular compound. This can easily be demonstrated experimentally.

Consider again the reaction between calcium and sulfur to give calcium sulfide. Suppose an attempt is made to combine various weights of sulfur with a fixed weight of calcium. A set of possible experimental data for this attempt is given in Table 2–6. Notice that, regardless of the weight of S present, only a certain amount, 44.4 g, reacts with the 55.6 g of Ca. The excess S is left over in an unreacted form. These data thus illustrate that Ca and S will react in only one fixed weight ratio (55.6:44.4) to form CaS. This fact is consistent with the law of constant composition.

Table 2–6. Data Illustrating the Law of Constant Composition

Weight of Ca used	*Weight of S used*	*Weight of CaS formed*	*Weight of excess unreacted sulfur*
55.6 g	44.4 g	100.0 g	none
55.6 g	50.0 g	100.0 g	5.6 g
55.6 g	100.0 g	100.0 g	55.6 g
55.6 g	200.0 g	100.0 g	155.6 g

These observations are consistent with the atomic theory according to the following line of reasoning: (1) A 55.6-g sample of calcium contains a certain number of Ca atoms. (2) A 44.4-g sample of sulfur contains a number of S atoms just sufficient to react with the number of Ca atoms present. (3) Since the atoms of Ca and S always react in the same way to give calcium sulfide molecules, any amount of sulfur greater than 44.4 g will contain more atoms than needed and will go unreacted.

This is very much like a party attended by men and women. When the dancing starts, each man will take a single partner, and if there are more women than men, the additional women will not be involved in the dancing.

The Law of Multiple Proportions

It has been observed that, under appropriate conditions, two elements can react to form more than just one compound. For example, the elements chlorine and fluorine form three different compounds: ClF, ClF_3, and ClF_5; two different compounds of hydrogen and oxygen are known: H_2O and H_2O_2 (hydrogen peroxide). In such situations it is found experimentally that simple relationships exist between the weights of the elements involved. An analysis of these relationships forms the basis for the **law of multiple proportions:**

When two elements combine to form more than one compound, the various weights of one element that combine with a fixed weight of the second element form ratios of small whole numbers.

As an example of this law consider the previously mentioned series of compounds ClF, ClF_3, and ClF_5. Table 2–7 gives the experimental amounts of fluorine needed to react with 1 gram of chlorine in order to form each of the compounds. The law states that the ratios of the weights of one element (F above) combining with a fixed weight of another element (1.00 g of Cl above) will be small whole numbers. In the above example three ratios are possible, and each is found to consist of small whole numbers.

Table 2–7. Data Illustrating the Law of Multiple Proportions

Formula of compound	*Grams of chlorine (Cl)*	*Grams of fluorine (F)*
ClF	1.00 g	0.536 g
ClF_3	1.00 g	1.608 g
ClF_5	1.00 g	2.680 g

$$\frac{\text{wt. of F in ClF}}{\text{wt. of F in ClF}_3} = \frac{0.536}{1.608} = \frac{1}{3}$$

$$\frac{\text{wt. of F in ClF}}{\text{wt. of F in ClF}_5} = \frac{0.536}{2.680} = \frac{1}{5}$$

$$\frac{\text{wt. of F in ClF}_3}{\text{wt. of F in ClF}_5} = \frac{1.608}{2.680} = \frac{3}{5}$$

The atomic theory easily explains why the ratios always come out as small whole numbers:

1. In the different compounds, ClF, ClF_3, and ClF_5, atoms of F can combine with a given fixed number of Cl atoms.
2. There is no possibility for anything but a whole number of F atoms to combine with the fixed number of Cl atoms, since this would involve fractional F atoms, which do not exist.
3. The ratio of the number of F atoms reacting with the fixed number of Cl atoms in any two reactions will therefore be expressible in small whole numbers.
4. If the number ratio of F atoms is a whole number, then the weight ratio of F will be a whole number since the weight of F is merely the product of the number of F atoms multiplied by the weight of one F atom. The weight of one F atom will appear in both the numerator and denominator of the ratio and thus divide out. For example:

$$\frac{\text{wt. of F in ClF}}{\text{wt. of F in ClF}_3} = \frac{\text{number of F atoms in ClF} \times \cancel{\text{weight of one F atom}}}{\text{number of F atoms in ClF}_3 \times \cancel{\text{weight of one F atom}}}$$

$$= \frac{\text{number of F atoms in ClF}}{\text{number of F atoms in ClF}_3}$$

Subatomic Particles

There is more to the modern concept of the atom than is expressed explicitly by the atomic theory of matter. Extensive experimental evidence indicates that atoms themselves are made up of many smaller particles—more than one hundred different kinds, many discovered in the last decade. This area is one of the modern frontiers of science. No theory yet available can explain all of the experimental observations dealing with subatomic particles. Many relationships among such particles remain to be discovered. An experimenter in this area faces extreme difficulties, because many of the particles are unstable and exist only for a matter of seconds before changing into other particles. Although the picture is still clouded, all current theories concerning subatomic particles do have one common feature—they speak of three "fundamental" particles: the proton, neutron, and electron. Most chemical observations are readily explained in terms of concepts that depend only on these particles—and so only these three will be discussed in detail here.

Among the important physical properties of these particles are the weight and the electrical charge. Values for these two properties, together with comments about each particle, are given in Table 2–8. Notice that exponential notation is used in the table when very large or very small numbers are represented—a practice that will be followed throughout this text. The reader who has questions about the meaning of such numbers is encouraged to turn to Appendix I and review the material before proceeding.

Table 2–8. Properties of Subatomic Particles

Particle	*Property*	*Magnitude*	*Comments*
Electron	Charge	-1	The actual charge on the electron is a negative 4.8×10^{-10} electrostatic units. Since this is the smallest unit of electricity known, it is referred to as a -1.
	Weight	9.1×10^{-28} g	The electron is the lightest of the three fundamental particles.
Proton	Charge	$+1$	The actual charge on the proton is a positive 4.8×10^{-10} electrostatic units. Thus, the charge is equal in magnitude to that of the electron but opposite in sign.
	Weight	1.672×10^{-24} g	The proton is 1837 times heavier than the electron.
Neutron	Charge	0	The neutron is an uncharged particle.
	Weight	1.675×10^{-24} g	The weights of the neutron and proton are practically equal and differ only in the 27th decimal place. The neutron is slightly heavier.

Figure 2–2.
The Nuclear Atom
Electrons are in motion about massive, positively charged nucleus.

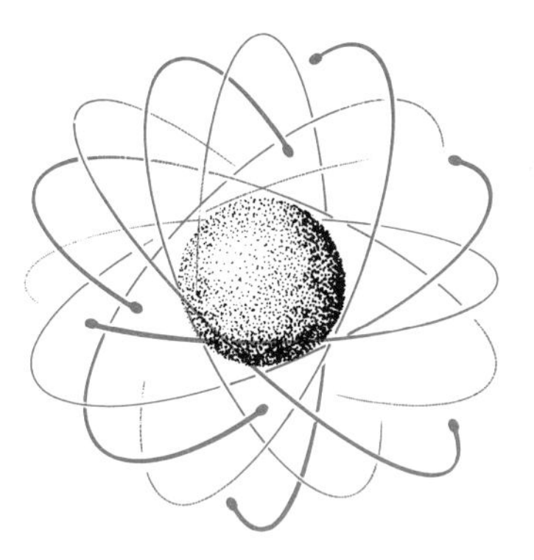

The arrangement of the fundamental subatomic particles in the atom is well known. In the center of the atom is a very small, dense region called **the nucleus.** All of the protons and neutrons of an atom are located in this region, thus giving it the positive charge characteristic of the protons. Collectively, protons and neutrons are often referred to as **nucleons,** indicating their location in the nucleus. The negatively charged electrons are considered to be moving rapidly throughout a relatively large volume surrounding the nucleus. This volume is usually called an **electron cloud.** The motion of the electrons essentially determines the volume of the atom in much the same way as the propeller of an airplane determines an area by its motion. This **nuclear model** of the atom is diagramed in Figure 2–2.

Even though subatomic particles such as protons, neutrons, and electrons exist, the atom remains the fundamental particle of interest in chemistry. This is because subatomic particles do not lead an independent existence for any appreciable length of time. The only way they gain stability is by joining with other particles to form an atom.

Evidence Supporting the Existence and Arrangement of Subatomic Particles

A significant collection of evidence is consistent with and supports the existence, nature, and arrangement of subatomic particles given previously. Two historically important types of experiments illustrate some of the sources of this evidence. Cathode-ray tube experiments resulted in the original concept that the atom contained negatively and positively charged particles. Metal-foil bombardment experiments provided evidence for the existence of a nucleus within the atom.

Cathode-Ray Tube Experiments

Gases at atmospheric pressure are very poor conductors of electricity. In 1859 Julius Plucker, using what came to be known as a cathode-ray tube, discovered that gases under very low pressure are relatively good electrical conductors. His apparatus consisted of a simple glass tube with metal plates sealed into both ends. The tube also had a side arm for attaching a vacuum pump. The metal plates, called **electrodes,** were attached to a high-voltage source. Figure 2–3 is a schematic diagram of such a tube.

It was observed that when the tube was evacuated to a low pressure and a sufficiently high voltage was applied, electricity flowed from one electrode to the other and the residual gas became luminous. After the tube was evacuated to lower pressures, it was found that the luminosity disappeared but the electrical conductance continued—as shown by a glowing of the tube's glass walls. The glowing or fluorescence of the walls was thought to be a result of the bombardment of the tube walls by rays. The name **cathode-**

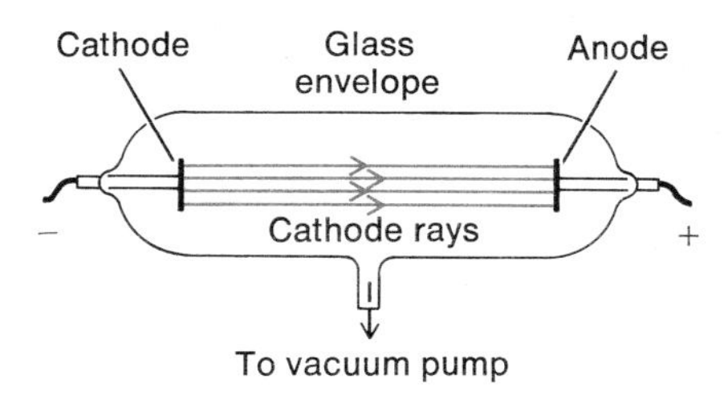

Figure 2–3.
The Cathode–Ray Tube

ray tube comes from the observation that when an obstacle is placed between the negative electrode (cathode) and the opposite glass wall, a sharp shadow is cast on the wall. This indicates that the rays are coming from the cathode.

In the years following their discovery cathode rays were studied intensively by numerous investigators. The following facts were discovered during these studies:

The rays are deflected by a magnet held near the cathode-ray tube (Figure 2–4). The deflection showed that the rays consisted of a stream of particles, and the direction of deflection showed the particles to be negatively charged.

Certain minerals emit intense light when placed between the electrodes. The light is emitted in flashes, which hints that the rays are made up of discrete particles.

The rays deflect in an electrical field in a direction again indicating a negative charge.

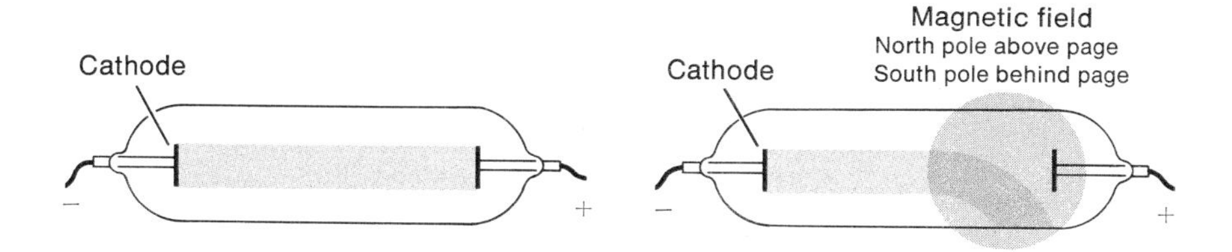

Figure 2–4.
The Deflection of Cathode Rays by a Magnet

In 1897 Sir J. J. Thomson of Cambridge University applied simultaneous electrical and magnetic fields to a cathode-ray tube and was able to determine the charge-to-mass ratio for the particles making up the rays. He found the ratio to be constant, regardless of the metal used in the cathode. The constancy of the ratio led him to conclude that he was dealing with something fundamental to all matter. Further experiments confirmed the fundamental nature of these particles, which today are called electrons.

As soon as it was determined that cathode rays were negatively charged, the search began for positive rays. It was reasoned that when electrons were separated from neutral atoms in the cathode-ray tube, positively charged

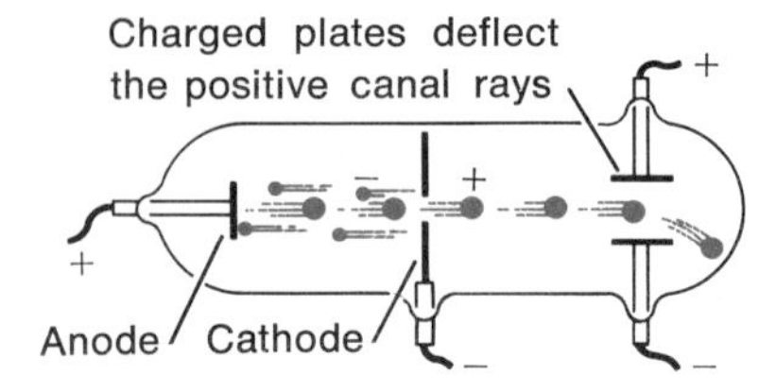

Figure 2–5.
The Goldstein Tube and Canal Rays

residues should be formed. In 1886 a German physicist, Eugene Goldstein, showed that positive particles were also present in the cathode-ray tube. He drilled a hole or canal through the cathode and found that a beam of rays passed through. These rays, called **canal rays,** were found to consist of positive particles (Figure 2–5).

Further research showed that there were many different types of canal rays, as contrasted to only one type of cathode ray, and that canal rays were much heavier than cathode rays. The type of canal ray produced depended upon the gas in the tube. The simplest of these canal rays was eventually identified as the particle now called the proton. Canal rays are now known to be gas atoms which have lost one or more electrons.

On the basis of cathode-ray tube experiments, J. J. Thomson proposed in 1898 that the atom was composed of a sphere of positive electricity containing most of the weight. Small negative electrons were thought to be attached to the surface of the positive sphere. He postulated that a high voltage could pull off the surface electrons to produce cathode rays and leave behind the positive particles found in canal rays. Figure 2–6 shows Thomson's model of the atom along with the mechanism for ray formation.

Thomson's model came to be known as the raisin pudding model, the electrons serving as raisins. The model is now known to be incorrect; but it set the stage for an experiment, commonly called the gold-foil experiment, which led to the nuclear model of the atom.

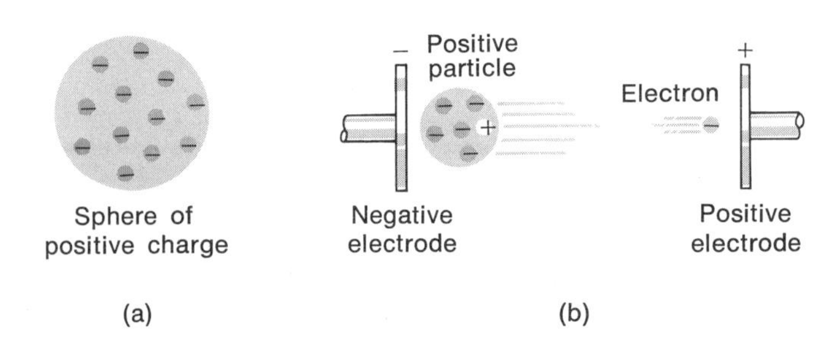

Figure 2–6.

(a) Thomson's Model of the Atom

(b) Formation of Positive Canal Rays and Electrons

Metal-Foil Experiments

In 1911 Ernest Rutherford designed an experiment to test the Thomson model of the atom. In this experiment thin sheets of metal foil were bombarded by alpha (α) particles from a radioactive source. Alpha particles are positively charged particles ejected at high speeds from some radioactive atoms. The phenomenon of radioactivity had been discovered in 1896 and gave further evidence that electrical charges existed within the atom. Gold was chosen as the target metal because it can be easily hammered into very thin sheets. The experimental setup is diagramed in Figure 2–7. Alpha particles do not appreciably penetrate lead, so it was used to produce a narrow alpha-particle beam.

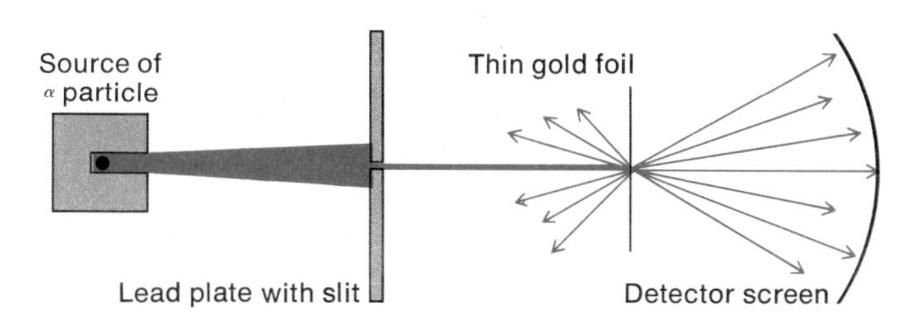

Figure 2–7.
Rutherford's Gold Foil Experiment

Each time an alpha particle hit the detector screen, a flash of light was produced.

Rutherford reasoned that the very energetic alpha particles would pass through the thin gold foil. His reasoning was based on the Thomson model, in which the mass and positive charge of the gold atoms were distributed uniformly throughout each atom. Rutherford assumed the alpha particles would not be appreciably deflected from their path. As each positive particle neared the foil, it would be confronted by a uniform positive charge which would produce no preferred direction of deflection in spite of the mutual repulsion of the positive charges. The lack of deflected alpha particles would support the Thomson model of the atom.

Not all of the observed experimental results were expected. Most of the particles—more than 99 percent—went straight through. A few, however, were appreciably deflected by something that had to be difficult to hit and much heavier than the alpha particles themselves. A very few particles were deflected directly back, almost along the incoming path.

Regarding the last observation, Rutherford commented: "It was quite the most incredible event that has ever happened to me in my life. It was almost as incredible as if you fired a 15-inch shell at a piece of tissue paper and it came back and hit you."

Similar results were obtained when other elements were used as targets. An extensive mathematical analysis of the results (in terms of angles of deflection and so on) led Rutherford to propose the following explanation for his results: (1) A very dense, small nucleus exists in the center of the atom. This nucleus contains most of the weight of the atom and all of the positive charge. (2) Electrons occupy most of the total volume of the atom, are located outside the nucleus, and are in rapid motion about it. (3) When an alpha particle scores a direct hit on a nucleus, it is deflected back along the incoming path. (4) A near miss of a nucleus by an alpha particle results in repulsion and deflection. (5) Most of the alpha particles pass through without any interference, because most of the atomic volume is empty space. (6) Electrons have so little weight that they do not deflect the much larger alpha particles (almost 8000 times heavier than the electron).

Many other experiments have since verified the existence of a nucleus in atoms.

Differences Between Atoms

All atoms are electrically neutral and must therefore contain the same number of protons and electrons. This **atomic number** of an atom is represented by the symbol Z. The nucleus of each atom contains the protons and neutrons, collectively called nucleons. The total number of nucleons in the nucleus is called the **mass number** of the atom, denoted by the symbol A. The nuclear and electronic composition of an atom is uniquely defined by these two numbers. Example 2–1 illustrates this idea.

Example 2–1. A given atom has an atomic number, Z, of 15 and a mass number, A, of 31. How many protons, electrons, and neutrons does the atom contain?

Protons: The number of protons $= Z$; therefore, 15 protons are present. By definition, the atomic number will always equal the number of protons.
Electrons: The number of electrons $= Z$; therefore, 15 electrons are also present. In a neutral atom the atomic number will always equal the number of electrons.
Neutrons: The number of neutrons $= A - Z$. The number of neutrons is 16, $(31 - 15)$. The mass number is equal to the sum of the protons and neutrons, and the atomic number is equal to the number of protons. The difference between the mass and atomic numbers always equals the number of neutrons present in the atom.

Atoms are identified by their chemical properties, which are essentially determined by the number and arrangement of electrons about the nucleus. The protons and neutrons in the nucleus make up only a tiny fraction of the atom's total volume, which is determined by the electron cloud surrounding the nucleus. All neutral atoms containing the same number of electrons will have the same chemical properties and, since identification is based on chemical properties, will be called atoms of the same element. In this way, the atomic number identifies an element and must be unique for each element. It follows, then, that every atom of a specific element must contain the same number of protons.

However, all atoms of an element are not completely identical in subatomic composition. Consider the subatomic particles in the two atoms A and B given below:

ATOM A	ATOM B
$Z = 13$	$Z = 13$
$A = 27$	$A = 28$
number of protons = 13	number of protons = 13
number of electrons = 13	number of electrons = 13
number of neutrons = 14	number of neutrons = 15

These atoms are not identical, since B has one more neutron than A. Nevertheless, they are atoms of the same element, because the atomic number and the number of electrons are the same.

Atoms such as these, with the same atomic number but different mass numbers, are called **isotopes.** Therefore, atoms of the same element may differ from each other, but only in the number of neutrons contained in the nucleus. If two atoms differ in the number of protons or electrons contained, they must have different Z values and must be different elements. The majority of elements in nature exist in two or more isotopic forms. We shall return to this topic later in this chapter.

Isotopes of an element are represented by the same elemental symbol. When it is necessary to distinguish between isotopes, the notation shown in Example 2–2 is used.

Example 2–2. The element chlorine has an atomic number of 17 and exists naturally in two isotopic forms with mass numbers of 35 and 37, respectively. The elemental symbol Cl does not distinguish between them, so some other notation is needed. The notation is

$$^{A}_{Z}\text{symbol}$$

where the atomic number is used as a subscript and the mass number as a superscript. The two isotopes of chlorine given above are represented in this notation by

$$^{35}_{17}\text{Cl} \quad \text{and} \quad ^{37}_{17}\text{Cl}$$

The existence of isotopes clarifies the wording used earlier in stating the postulates of the atomic theory. Postulate 1 states: "All matter is made up of small particles called atoms, of which 105 different 'types' are known." The word *type* was used to indicate a special meaning, which now turns out to involve the existence of isotopes. The word *type* was used to describe elements with the same atomic number.

Postulate 3 states: "All atoms of a given type are similar to one another and significantly different from all other types of atoms." All atoms of an element are similar to each other in chemical properties but may differ by having different numbers of neutrons. All atoms of one element differ significantly from atoms of other elements by having different chemical properties.

The 105 known elements have atomic numbers with consecutive values of 1 to 105. The existence of an element with each of these numbers is an indication of the order existing in nature and the fact that all naturally occurring elements have probably been discovered. Table 2–9 gives the atomic numbers of the known elements. Mass numbers are not usually tabulated in a similar fashion, since, owing to the existence of isotopes, most elements lack a unique mass number. The mass number is unique only when the element is found in a single isotopic form.

Another reason for not listing mass numbers is the possibility that atoms of two different elements may have the same value. For example, the element

Table 2–9. Table of Atomic Weights of the Elements in Order of Increasing Atomic Number.[a,b]

Atomic number	*Name*	*Symbol*	*Atomic weight*	*Atomic number*	*Name*	*Symbol*	*Atomic weight*
1	Hydrogen	H	1.0	54	Xenon	Xe	131.3
2	Helium	He	4.0	55	Cesium	Cs	132.9
3	Lithium	Li	6.9	56	Barium	Ba	137.3
4	Beryllium	Be	9.0	57	Lanthanum	La	138.9
5	Boron	B	10.8	58	Cerium	Ce	140.1
6	Carbon	C	12.0	59	Praseodymium	Pr	140.9
7	Nitrogen	N	14.0	60	Neodymium	Nd	144.2
8	Oxygen	O	16.0	61	Promethium	Pm	(147)
9	Fluorine	F	19.0	62	Samarium	Sm	150.4
10	Neon	Ne	20.2	63	Europium	Eu	152.0
11	Sodium	Na	23.0	64	Gadolinium	Gd	157.3
12	Magnesium	Mg	24.3	65	Terbium	Tb	158.9
13	Aluminum	Al	27.0	66	Dysprosium	Dy	162.5
14	Silicon	Si	28.1	67	Holmium	Ho	164.9
15	Phosphorus	P	31.0	68	Erbium	Er	167.3
16	Sulfur	S	32.1	69	Thulium	Tm	168.9
17	Chlorine	Cl	35.5	70	Ytterbium	Yb	173.0
18	Argon	Ar	39.9	71	Lutetium	Lu	175.0
19	Potassium	K	39.1	72	Hafnium	Hf	178.5
20	Calcium	Ca	40.1	73	Tantalum	Ta	180.9
21	Scandium	Sc	45.0	74	Tungsten	W	183.9
22	Titanium	Ti	47.9	75	Rhenium	Re	186.2
23	Vanadium	V	50.9	76	Osmium	Os	190.2
24	Chromium	Cr	52.0	77	Iridium	Ir	192.2
25	Manganese	Mn	54.9	78	Platinum	Pt	195.1
26	Iron	Fe	55.8	79	Gold	Au	197.0
27	Cobalt	Co	58.9	80	Mercury	Hg	200.6
28	Nickel	Ni	58.7	81	Thallium	Tl	204.4
29	Copper	Cu	63.5	82	Lead	Pb	207.2
30	Zinc	Zn	65.4	83	Bismuth	Bi	209.0
31	Gallium	Ga	69.7	84	Polonium	Po	(210)
32	Germanium	Ge	72.6	85	Astatine	At	(210)
33	Arsenic	As	74.9	86	Radon	Rn	(222)
34	Selenium	Se	79.0	87	Francium	Fr	(223)
35	Bromine	Br	79.9	88	Radium	Ra	(226)
36	Krypton	Kr	83.8	89	Actinium	Ac	(227)
37	Rubidium	Rb	85.5	90	Thorium	Th	232.0
38	Strontium	Sr	87.6	91	Protactinium	Pa	(231)
39	Yttrium	Y	88.9	92	Uranium	U	238.0
40	Zirconium	Zr	91.2	93	Neptunium	Np	(237)
41	Niobium	Nb	92.9	94	Plutonium	Pu	(242)
42	Molybdenum	Mo	95.9	95	Americium	Am	(243)
43	Technetium	Tc	(99)	96	Curium	Cm	(247)
44	Ruthenium	Ru	101.1	97	Berkelium	Bk	(249)
45	Rhodium	Rh	102.9	98	Californium	Cf	(251)
46	Palladium	Pd	106.4	99	Einsteinium	Es	(254)
47	Silver	Ag	107.9	100	Fermium	Fm	(253)
48	Cadmium	Cd	112.4	101	Mendelevium	Md	(256)
49	Indium	In	114.8	102	Nobelium	No	(254)
50	Tin	Sn	118.7	103	Lawrencium	Lr	(257)
51	Antimony	Sb	121.8	104	Rutherfordium	Rf	(261)
52	Tellurium	Te	127.6	105	Hahnium	Ha	(262)
53	Iodine	I	126.9				

[a] Atomic weights are based on carbon-12 and have been rounded to 0.1.

[b] The use of parentheses with atomic weights indicates elements which are relatively unstable. The weight given is usually that of the longest-lived isotope of the element.

indium (Z = 49) exists in two isotopic forms, one of which is $^{115}_{49}In$. Another element, tin (Z = 50), exists in ten isotopic forms, one of which is $^{115}_{50}Sn$.

Table 2–9 does contain another number for each element. This number, the atomic weight, is related to the mass number and is the next topic of this chapter. These atomic weights have been rounded to the nearest 0.1, a value sufficiently accurate for most problems and examples in this book.

Atomic Weights

Atomic weights are numbers developed and used to compare the weights of atoms to each other. A very formal definition of an atomic weight is: *The atomic weight of an element is that number which tells how the relative weight of an average atom of the element compares with the relative weight of an average atom of other elements, with all relative weights based upon the weight of* $^{12}_{6}C$*, which has been arbitrarily set at 12.0000 amu.*

The key to understanding this definition is an understanding of two terms that are used. They are **relative weight** and **average atom.** The meaning of these terms is discussed below.

Relative Weight

The usual standards of weight, such as grams or pounds, are not convenient for use with atoms, because very small numbers are encountered. For example, the weight in pounds of one of the heaviest atoms known ($^{238}_{92}U$) is 8.7×10^{-25} lb (see Appendix I for a review of exponential notation). To avoid the repeated use of such small numbers, scientists have set up a scale of relative weights. On such a scale the zero point can be set at any desired value and small numbers can be avoided. A relative scale provides all of the weight relationships obtainable from an actual weight scale, except the actual weights of atoms cannot be determined. In chemical problems the weight relationships between atoms are as useful as actual weights; consequently, nothing is lost by using a relative scale.

The concept of relative weights can be easily understood by following the process by which they are determined. Example 2–3 illustrates one way of determining relative weights and setting up a relative weight scale.

Example 2–3. Assume a relative weight scale is desired for the hypothetical elements A, B, and C. The following information is known: (I) A and B combine to give a compound with the formula AB, which contains 20 percent A and 80 percent B by weight. (II) A and C combine to give a compound with the formula AC, which contains 25 percent A and 75 percent C by weight.

The information given in (I) indicates that A and B combine in a 1:1 atom ratio, since the formula is AB; and the B atoms contribute 80 percent of the

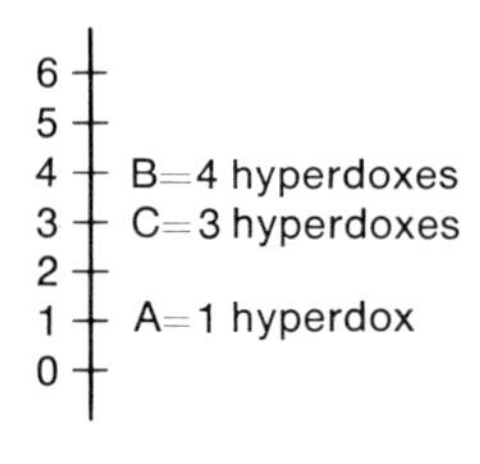

Diagram 1

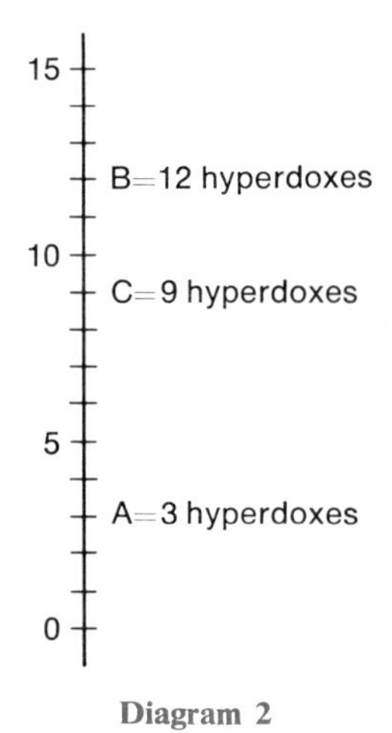

Diagram 2

total compound weight. Since there are equal numbers of atoms, the B atoms must be four times as heavy (80/20 = 4) as the A atoms. Similarly the information in (II) shows that C atoms are three times as heavy (75/25 = 3) as A atoms.

The key pieces of information needed to set up a relative scale are (1) B atoms are four times as heavy as A atoms, (2) C atoms are three times as heavy as A atoms. Because A is the lightest atom, its weight will arbitrarily be assigned a value of one unit. The units can be called anything we wish, and ours will be called the hyperdox. On this basis, one atom of A weighs one hyperdox. This becomes the reference point of a scale on which one atom of B will weigh four hyperdoxes since it is four times as heavy as A. Similarly, one atom of C weighs three hyperdoxes. The relative scale would appear as Diagram 1.

The name used for the weight unit was obviously arbitrary and in no way affects the basic concept. The assignment of one hyperdox for the weight of A was also completely arbitrary. The weight of one atom of A could have just as easily and rationally been chosen to be three hyperdoxes. In this case, the relative scale would appear as Diagram 2. The important point is that the relationships between the weights of A, B, and C are the same, regardless of the units used. On the first scale, for example, B is 1.33 times heavier than C (4/3) and on the second scale B is also 1.33 times heavier than C (12/9). Notice that we did not know or need the actual weights of A, B, and C to set up either scale. On relative scales, one value is arbitrarily assigned, and all others are determined by the weight relationships.

A relative scale of atom weights has been set up in a manner similar to that of the example. The unit is the **atomic mass unit,** abbreviated **amu.** The arbitrary reference point is set by using the weight of a particular isotope of carbon, ${}^{12}_{6}C$. The defined weight of this isotope on the scale is 12.0000 amu. The weights of all other atoms are then determined relative to that of ${}^{12}_{6}C$. For example, if an atom is twice as heavy as a ${}^{12}_{6}C$ atom, its weight is 24.0000 amu on the scale, and if an atom weighs half as much as an atom of ${}^{12}_{6}C$, its scale weight is 6.0000 amu. Experimentally the weights of all atoms have been determined relative to ${}^{12}_{6}C$. Typical values for the weights of selected isotopes on the carbon-12 scale are given in Figure 2–8.

On the basis of the values given in the figure, it is possible to make statements such as: ${}^{238}_{92}U$ is 4.26 times as heavy as ${}^{56}_{20}Fe$ (238.0508 amu/55.9349 amu = 4.26), and ${}^{56}_{26}Fe$ is 2.80 times as heavy as ${}^{20}_{10}Ne$ (55.9349 amu/19.99244 amu = 2.80). Notice that we did not need the actual weights of the atoms in order to make these statements.

Average Atom

The weight of an atom of a specific element may be one of several values because of the existence of isotopes. For example, oxygen atoms can have any one of three weights, since three isotopes exist—${}^{16}_{8}O$, ${}^{17}_{8}O$, and ${}^{18}_{8}O$. Because of this fact, it might seem necessary to specify the isotope of an element used in calculations or discussions involving weights of atoms; however, it is not. In practice, isotopes are seldom mentioned in such calculations and discussions; instead, the atoms of an element are treated as if they all had a single weight. The weight used is an average, relative weight that takes into account the existence of isotopes. This concept reduces the number of weights used from the hundreds characteristic of all isotopes to 105, which represent the different elements. These 105 average weights, called **atomic weights,** are determined by use of the following information: (1) the number of isotopes that exist for each element, (2) the relative weight of these isotopes on the carbon-12 scale, and (3) the percentage abundance of each isotope.

This information, available for each element from experimental studies, is given for selected elements in Table 2–10. The data there illustrate the wide variation in number and abundances of naturally occurring isotopes. A few monoisotopic elements are known, but most elements exist in two or more isotopic forms. The ten naturally occurring isotopes of tin are the maximum number known for any element.

Figure 2–8.
Relative Weights of Atoms on the Carbon–12 Scale

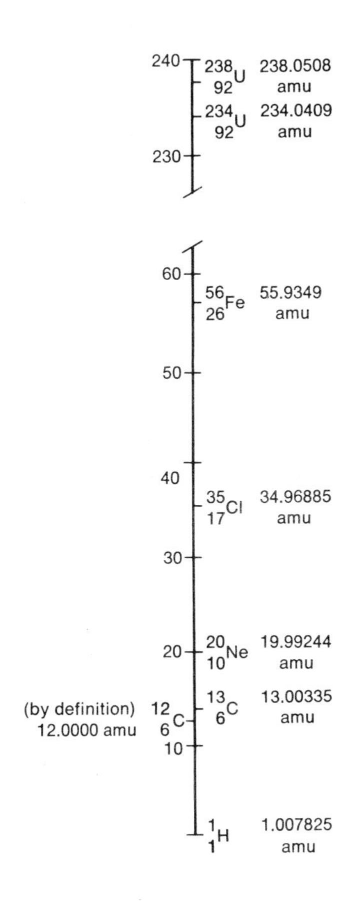

Table 2–10. Percentage Abundances and Weights of Selected Isotopes

Isotopes	*Percentage abundance in nature (%)*	*Weight*
Hydrogen-1	99.985	1.007825 amu
Hydrogen-2	0.015	2.01410 amu
Oxygen-16	99.759	15.99491 amu
Oxygen-17	0.037	16.99914 amu
Oxygen-18	0.204	17.99916 amu
Fluorine-19	100.000	18.9984 amu
Carbon-12	98.89	12.00000 amu
Carbon-13	1.11	13.00335 amu
Chlorine-35	75.53	34.96885 amu
Chlorine-37	24.47	36.96590 amu
Iron-54	5.82	53.9396 amu
Iron-56	91.66	55.9349 amu
Iron-57	2.19	56.9354 amu
Iron-58	0.33	57.9333 amu
Tin-112	0.96	111.9040 amu
Tin-114	0.66	113.9030 amu
Tin-115	0.35	114.9035 amu
Tin-116	14.30	115.9021 amu
Tin-117	7.61	116.9031 amu
Tin-118	24.03	117.9018 amu
Tin-119	8.58	118.9034 amu
Tin-120	32.85	119.9021 amu
Tin-122	4.92	121.9034 amu
Tin-124	5.94	123.9052 amu

The validity of the average-weight concept rests upon two points. First, extensive studies of naturally occurring elements have shown that the percentage abundance of the isotopes for a given element is generally constant. No matter where the element sample is obtained on earth, it generally contains the same percentage of each isotope. Because of these constant isotopic ratios, the weight of an "average atom" does not vary. Second, chemical operations are carried out with very large numbers of atoms. The tiniest piece of matter visible to the eye contains more atoms than can be counted in a lifetime. The numbers are so great that any collection of atoms the chemist uses is very unlikely to contain an isotopic ratio different from that of the naturally occurring element.

Examples 2–4 and 2–5 illustrate the operations used to calculate atomic weights. Example 2–4 is a general calculation of a weighted average, and Example 2–5 calculates a specific atomic weight.

Example 2–4. A weighted average is calculated by using the relative amounts of each item present. Consider the following five numbers: 3, 3, 7, 7, and 10. The average is found by dividing the sum of the numbers by the number of values summed.

$$3 + 3 + 7 + 7 + 10 = 30$$

$$\frac{30}{5} = 6 = \text{average value}$$

The list of numbers could alternatively be expressed as:

40 percent 3s (2 out of 5 numbers, 40 percent, are 3s)
40 percent 7s (2 out of 5 numbers, 40 percent, are 7s)
20 percent 10s (1 out of 5 numbers, 20 percent, are 10s)

To find the average value for data in this form, we multiply each number by its fractional abundance (percentage/100) and then sum the products:

$$\left(\frac{40}{100}\right) \times 3 = (.40) \times 3 = 1.2$$

$$\left(\frac{40}{100}\right) \times 7 = (.40) \times 7 = 2.8$$

$$\left(\frac{20}{100}\right) \times 10 = (.20) \times 10 = \underline{2.0}$$

$$6.0 \quad \text{(same average value as before)}$$

This averaging method is useful for calculating atomic weights, because the fractional abundances of isotopes are readily determined and the number of atoms present is not required.

Example 2–5. Naturally occurring chlorine exists in two isotopic forms—$^{35}_{17}Cl$ and $^{37}_{17}Cl$. The relative weight of $^{35}_{17}Cl$ is 34.97 amu and its abundance is

75.53 percent, while $^{37}_{17}Cl$ has a relative weight of 36.97 amu and a 24.47 percent abundance. What is the atomic weight of chlorine?

The atomic weight can be calculated using the methods of the previous example:

$$\left(\frac{75.53}{100}\right) \times 34.97 = (.7553) \times 34.97 \text{ amu} = 26.41 \text{ amu}$$

$$\left(\frac{24.47}{100}\right) \times 36.92 = (.2447) \times 36.97 \text{ amu} = \underline{9.04 \text{ amu}}$$

$$35.45 \text{ amu} \quad \text{(atomic weight of chlorine)}$$

This calculation involved an element containing just two isotopes. A similar calculation for an element containing three isotopes would be carried out the same way but would have three terms in the sum; an element containing four isotopes, four terms in the sum, and so forth.

The abundances and weights of the individual isotopes in an element are determined by using a device known as a **mass spectrograph.** The important components of this instrument are a source of charged particles of the element under investigation, an aligning system to create a narrow beam of the charged particles, a magnetic field to affect the path of the charged particles, and a means of detecting the charged particles. A schematic diagram of a mass spectrograph is given in Figure 2–9.

Suppose oxygen gas containing all three isotopes is admitted to the instrument. The gaseous molecules are bombarded with an energetic electron beam, producing positively charged oxygen atoms. The aligning system produces a narrow beam of these charged atoms, which then enters the magnetic field. The most massive particles (heaviest isotope) are not deflected by the magnetic field as much as the less massive ones, so the charged atoms are divided into separate beams which strike the photographic plate at different points depending on their weights. The more abundant isotopes will create more intense lines on the plate. Calculations using these experimental data result in numbers such as those of Table 2–10.

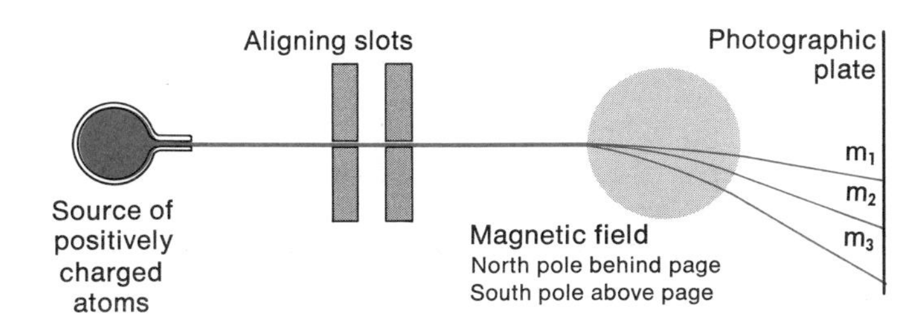

Figure 2–9.
A Schematic Diagram of a Mass Spectrograph

Atoms in Perspective

Atoms are very small, nuclei are even smaller, and subatomic particles are smaller still. Qualitative statements similar to these have been used throughout this chapter. The use of modern instrumentation makes such statements unnecessary by allowing sizes and weights to be determined for atoms and subatomic particles. Some of these measured values, together with comparisons to common objects, are given below to provide some idea of the magnitudes of the quantities involved.

Size of an Atom

The radii of atoms fall within the limits of approximately $(0.3–3.0) \times 10^{-8}$ centimeters. The distance 10^{-8} cm, which is characteristic of atomic size, is extremely small. For example, 1 million fluorine atoms (radius $= 0.72 \times 10^{-8}$ cm) touching each other in a straight line would reach a distance of approximately 6/1000 of an inch. It would take 176 million of these atoms to span 1 inch.

Weight of an Atom

The lightest atom known, ${}^{1}_{1}H$, weighs 1.7×10^{-24} grams. One of the heavier atoms, ${}^{238}_{92}U$, weighs 3.9×10^{-22} g. The characteristic range of weights for atoms is thus seen to be 10^{-24} to 10^{-22} g. Four billion atoms of ${}^{238}_{92}U$ (the same number as the present world population) would weigh 1.6×10^{-13} g. It would take 1.2×10^{24} atoms of ${}^{238}_{92}U$ to weigh a total of 1 pound. This number of atoms is difficult to visualize because it is so large. The following comparison perhaps gives some idea of its magnitude. Assume that each of the 1.2×10^{24} atoms was represented by a dollar. Also assume the 1.2×10^{24} dollars were divided equally among the world's inhabitants (4 billion people). Each person would receive $\$3 \times 10^{14}$, and become a multibillionaire.

Size of the Nucleus

Nuclear radii are of the order of 10^{-13} cm. The nuclear radius of an oxygen atom is estimated to be 2.8×10^{-13} cm. This contrasts to its atomic radius of 0.75×10^{-8} cm. Assuming a spherical nature for both the oxygen atom and its nucleus, and using the radii to calculate their volumes, we find that the volume of the nucleus is only 5×10^{-11} as large as the total volume of the atom. The small size of the nucleus relative to the atom can further be illustrated by means of a model. If the nucleus of the oxygen atom were expanded to the size of a baseball, 7.4 cm in diameter, the atom would be slightly over 1.2 miles across.

Weight of the Nucleus

Most of the weight of an atom is concentrated in the nucleus, since it contains the neutrons and protons, and electrons are only 1/1837 the weight of either of these nuclear particles. The amount of the total weight of an atom contributed by the nucleus varies from 99.95 percent (light atom) to 99.98 percent (heavy atom), and thus the nuclear weights and weights of atoms are essentially the same—approximately 10^{-24} to 10^{-22} g.

Density of the Nucleus

Since essentially all of the weight of an atom is concentrated in the nucleus (an extremely small area—10^{-13} cm radius), nuclear material has an extremely high density (weight per volume). This can best be illustrated by examples. If the head of a straight pin (diameter—0.1 cm) were made of aluminum nuclei (Al atoms stripped of their electrons), it would weigh 2.1×10^5 tons. A copper penny consisting of copper nuclei instead of copper atoms would weigh 1.9×10^8 tons.

Review Questions

1. What percentage of the 105 known elements were known prior to
   a) 1700
   b) 1800
   c) 1900
   d) 1950
2. What are the 5 most abundant elements in the earth's crust, water, and atmosphere?
3. What chemical elements are represented by the following symbols: Np, Se, Ge, O, Es, Br, C, Au?
4. Make a list of the elements whose symbols are not derived from their English names.
5. Distinguish between the terms *symbol* and *formula*.
6. Contrast the meanings of the following:
   a) HF and Hf
   b) $COCl_2$ and $CoCl_2$
   c) SO_2 and SO_3
7. List the five postulates of modern day atomic theory.
8. Consider a reaction in which A and B are reactants, and C and D are products. If 3 g of A completely react with 4 g of B to produce 5 g of C, how many grams of D will be produced? What chemical law did you make use of in determining the answer?
9. Suppose you mixed 39 g of Na with 100 g of Cl. What is the maximum amount of NaCl (NaCl is 39 percent Na and 61 percent Cl, by weight)

that could be formed from the mixture? What chemical law did you make use of in determining the answer?

10. Two compounds containing only sulfur and oxygen were analyzed. An 8.0 g sample of the first compound contained 4.8 g of oxygen and a 6.4 g sample of the second compound contained 3.2 g of oxygen. Show that these compounds obey the law of multiple proportions.
11. It has been found experimentally that two elements, A and B, react to produce a compound or compounds. The combining proportions of the elements are as follows:

Experiment	Grams of A	Grams of B	Grams of compound
1	2.62	3.00	5.62
2	7.86	9.00	16.86
3	1.31	3.00	4.31

    a) Which of the experiments contain(s) data that could be used to illustrate the law of conservation of matter?
    b) Which pair(s) of experiments contain(s) data sufficient to illustrate the law of constant composition?
    c) Which pair(s) of experiments contain(s) data sufficient to illustrate the law of multiple proportions?
    d) If the formula of the compound formed in experiment 1 is AB_2, what are the formulas of the compounds formed in the other two experiments?
12. Contrast the properties of the fundamental subatomic particles
    a) in terms of charge
    b) in terms of weight
13. Describe the arrangement of the fundamental subatomic particles in the atom.
14. Contrast the properties of canal rays with those of cathode rays.
15. How does the Rutherford model of the atom differ from the Thomson model of the atom?
16. How do the actual results of Rutherford's gold-foil experiment differ from his preconceived ideas of what was going to happen?
17. Determine the number of protons, neutrons, and electrons which make up each of the following atoms:
    $^{53}_{24}Cr$, $^{103}_{44}Ru$, $^{256}_{101}Md$, $^{1}_{1}H$, $^{32}_{16}S$
18. For each of the following atoms, determine the symbol (what is X?), name, number of protons, number of neutrons, and number of electrons.
    $^{65}_{29}X$, $^{37}_{17}X$, $^{10}_{5}X$, $^{197}_{79}X$, $^{182}_{74}X$
19. Explain the difference between a mass number and an atomic number.
20. Explain why the atomic number can be used to identify an element.
21. In what ways are isotopes of an element alike? In what ways are they different?
22. Criticize the statement: An element is made up of atoms, all of which are alike.

23. The following symbols represent different atoms: ${}^{50}_{25}X$, ${}^{54}_{30}X$, ${}^{48}_{26}X$, ${}^{49}_{26}X$, ${}^{48}_{24}X$, ${}^{50}_{26}X$, ${}^{49}_{25}X$, ${}^{50}_{30}X$.
    a) Arrange all isotopes into groups.
    b) Arrange all atoms with the same mass numbers into groups.
    c) Arrange all atoms with the same number of neutrons into groups.
24. Write the complete notation (${}^{A}_{Z}$symbol) for an atom of each of the following isotopes:
    a) the nitrogen isotope containing 8 neutrons
    b) the chromium isotope with a mass number of 50
    c) an isotope with 45 protons and 58 neutrons
    d) an isotope with 91 neutrons and a mass number of 155
25. Explain why atoms are considered to be the fundamental building blocks of matter even though smaller particles (subatomic particles) exist.
26. What is meant by the statement: Atomic weights are average weights?
27. What is meant by the statement: Atomic weights are relative weights?
28. Using the data given in Table 2–10, calculate the atomic weight of the following elements. Round all numbers to one decimal place before doing the calculations.
    a) hydrogen
    b) oxygen
    c) iron
    d) fluorine
29. How many times lighter, on the average, is an atom of beryllium than an atom of gold?
30. On a new atomic weight scale the atomic mass unit is defined as one-tenth the weight of a fluorine atom. Only one type of fluorine atom exists in nature—that is, it is monoisotopic. Calculate the atomic weight of gold on this new scale.
31. What is the characteristic size range for atoms in centimeters? What is the characteristic weight range in grams?

Suggestions for Further Reading

Dinga, G. P., "The Elements and the Derivation of Their Names and Symbols," *Chemistry*, Feb. 1968.

Hofstadter, R., "The Atomic Nucleus," *Scientific American*, July 1956.

Nier, A. O. C., "The Mass Spectrometer," *Scientific American*, March 1953.

"The Nature of the Atom," *Life*, May 16, 1949.

Winderleck, R., "History of the Chemical Sign Language," *Journal of Chemical Education*, Feb. 1953.

3 The Electronic Structure of Atoms

On the basis of the information given in the previous chapter, we know the following facts about electrons:

They are one of the three fundamental subatomic particles.
They have very little weight in comparison to the other two fundamental subatomic particles.
They are found outside the nucleus of an atom.
They are moving rapidly about the nucleus in a volume which defines the size of an atom.

Much more must be known about this particle if the chemical behavior of the elements is to be understood, for the number and arrangement of electrons about the nucleus determine the chemical properties of an atom.

The Periodic Law

During the early part of the nineteenth century an abundance of chemical facts became available from detailed studies of the known elements and their properties. With the hope of providing a systematic approach to the study of chemistry, scientists began to look for some order in this increasing amount of chemical information. They were encouraged in their search by the unexplained but well-known fact that certain elements had properties very similar to those of other elements. During the period 1820–1880, numerous attempts were made to find reasons for these similarities and to provide some method for arranging or classifying the elements in an orderly way based on the similarities.

The final development of a classification system based on properties took place in two steps: first, the discovery of an empirical arrangement of the elements which emphasized the similarities in properties, and second, the development of a theory that explained why the empirical arrangement worked. The empirical arrangement used was based on the **periodic law,** which was formulated in 1869. An explanation for this law was not available until 40 years later (1910–1915), when a theory was proposed to explain the electron arrangements around the nucleus. A general statement of the periodic law is:

Elements with similar properties will occur at regular (periodic) intervals in an arrangement based on some fundamental quantity which has a unique and characteristic value for each element.

The choice of the fundamental quantity involved in arranging the elements has changed with time, as shown in Table 3–1.

The use of atomic weights in the original formulation of the law reflected theories prevalent in 1869. According to these theories, the weights of atoms were their most important distinguishing properties—a knowledge of sub-

Table 3–1. Statements of the Periodic Law

Time period	*Statement of the law*
1869–1913	The properties of the elements are periodic functions of their *atomic weights.*
1913–present	The properties of the elements are periodic functions of their *atomic numbers.*

atomic structure was still 30 years away. When the details of subatomic structure were finally discovered, it became obvious that the distinguishing properties of atoms were related not to their weights but to the number and arrangement of their electrons. The periodic law was then changed to include the atomic-number concept.

The periodic law in the original form (using atomic weights) was successful because the arrangement of elements according to increasing atomic weights is very similar to the arrangement in order of increasing atomic numbers. The positions of only six elements differed (Ar-K, Co-Ni, Te-I). These six out-of-place elements were noted, but the problem was ignored in the original periodic arrangement.

It should also be pointed out that the original law was *empirical*—it worked and was useful but no one understood why. The discovery of electrons and the formulation of a theory to describe and predict their arrangement provided a theoretical explanation for the periodic law and removed the empirical label. The results of this theory are given later in this chapter.

The Periodic Table

The periodic table, a tabular representation of the periodic law, constitutes the greatest single aid to the study of chemistry. The elements in a periodic table are arranged according to increasing atomic number in a way that makes the similarities predicted by the periodic law readily apparent. Numerous arrangement schemes have this characteristic, and accordingly many different forms of the periodic table are known. The most common form used today, and the one used throughout this book, is given in Table 3–2.

Each element of the table is represented by a rectangular box in which is given the element's symbol, atomic number (above the symbol), and atomic weight (below the symbol). This representation provides a concise summary of the information given previously in Tables 2–3 and 2–9.

In this form of the periodic table the elements are generally arranged according to increasing atomic number to form rows of different lengths. To reduce the size of the table, an exception to this order is made, involving elements with atomic numbers 58 through 71, and 90 through 103. These elements, found at the bottom of the table, should technically be included

Table 3–2 The Periodic Table of the Elements

	I A	II A	III B	IV B	V B	VI B	VII B	VIII			I B	II B	III A	IV A	V A	VI A	VII A	Rare gases
								1 H 1.01†										2 He 4.00
Period 2	3 Li 6.94	4 Be 9.01											5 B 10.81	6 C 12.01	7 N 14.01	8 O 16.00	9 F 19.00	10 Ne 20.18
Period 3	11 Na 23.00	12 Mg 24.31											13 Al 26.98	14 Si 28.09	15 P 30.97	16 S 32.06	17 Cl 35.45	18 Ar 39.95
Period 4	19 K 39.10	20 Ca 40.08	21 Sc 44.96	22 Ti 47.90	23 V 50.94	24 Cr 52.00	25 Mn 54.94	26 Fe 55.85	27 Co 58.93	28 Ni 58.71	29 Cu 63.55	30 Zn 65.37	31 Ga 69.72	32 Ge 72.59	33 As 74.92	34 Se 78.96	35 Br 79.90	36 Kr 83.80
Period 5	37 Rb 85.47	38 Sr 87.62	39 Y 88.91	40 Zr 91.22	41 Nb 92.91	42 Mo 95.94	43 Tc* 98.91	44 Ru 101.07	45 Rh 102.91	46 Pd 106.4	47 Ag 107.87	48 Cd 112.40	49 In 114.82	50 Sn 118.69	51 Sb 121.75	52 Te 127.60	53 I 126.90	54 Xe 131.30
Period 6	55 Cs 132.91	56 Ba 137.34	57 La 138.91	72 Hf 178.49	73 Ta 180.95	74 W 183.85	75 Re 186.2	76 Os 190.2	77 Ir 192.22	78 Pt 195.09	79 Au 196.97	80 Hg 200.59	81 Tl 204.37	82 Pb 207.2	83 Bi 208.98	84 Po* [210]	85 At* [210]	86 Rn* [222]
Period 7	87 Fr* [223]	88 Ra* 226.02	89 Ac* [227]	104 Rf* [261]	105 Ha* [262]													

Lanthanoids	58 Ce 140.12	59 Pr 140.91	60 Nd 144.24	61 Pm* [147]	62 Sm 150.4	63 Eu 151.96	64 Gd 157.25	65 Tb 158.93	66 Dy 162.50	67 Ho 164.93	68 Er 167.26	69 Tm 168.93	70 Yb 173.04	71 Lu 174.97
Actinoids	90 Th* 232.03	91 Pa* 231.04	92 U* 238.03	93 Np* 237.05	94 Pu* [244]	95 Am* [243]	96 Cm* [247]	97 Bk* [247]	98 Cf* [251]	99 Es* [254]	100 Fm* [257]	101 Md* [258]	102 No* [255]	103 Lr* [256]

† All atomic weights have been rounded to .01
* All isotopes are radioactive
[ ] Indicates mass no. of longest known half-life

Table 3–3. Forms of the Periodic Table (a) Periodic table with elements 58–71 and 90–103 (area in color) in their proper positions. (b) Periodic table (modified to conserve space) with elements 58–71 and 90–103 (in color) placed at the bottom.

														1																	2
3	4																									5	6	7	8	9	10
11	12																									13	14	15	16	17	18
19	20	21															22	23	24	25	26	27	28	29	30	31	32	33	34	35	36
37	38	39															40	41	42	43	44	45	46	47	48	49	50	51	52	53	54
55	56	57	58	59	60	61	62	63	64	65	66	67	68	69	70	71	72	73	74	75	76	77	78	79	80	81	82	83	84	85	86
87	88	89	90	91	92	93	94	95	96	97	98	99	100	101	102	103	104	105													

(a)

						1											2
3	4											5	6	7	8	9	10
11	12											13	14	15	16	17	18
19	20	21	22	23	24	25	26	27	28	29	30	31	32	33	34	35	36
37	38	39	40	41	42	43	44	45	46	47	48	49	50	51	52	53	54
55	56	57	72	73	74	75	76	77	78	79	80	81	82	83	84	85	86
87	88	89	104	105													

58	59	60	61	62	63	64	65	66	67	68	69	70	71
90	91	92	93	94	95	96	97	98	99	100	101	102	103

(b)

in the main body of the table as shown in Table 3–3(a). To save space they are placed in the position shown in Table 3–3(b). This exception presents no problems to the user of the table as long as he is aware of it.

Each vertical column of the periodic table contains elements with similar properties and is called a **group.** For example, elements 8, 16, 34, 52, and 84 have similar properties and constitute a group. The group number is written in Roman numerals at the top of each column. The horizontal rows in the table are called **periods** and are numbered sequentially from the top, as shown in Table 3–2. Each element thus belongs to both a period and a group. Cesium, number 55, belongs to period 6 and group IA. Similarly, cadmium, number 48, belongs to period 5 and group IIB.

The periodic table contains further information about the elements; however, much of it deals with the arrangement of electrons around the nucleus and is obtained from current theories of electronic structure. In addition, these theories provide a justification for the arrangement of elements used in the periodic table.

The Arrangement of Electrons Around the Nucleus

A detailed study of electronic behavior around a nucleus requires the use of a complex mathematical theory called quantum mechanics. The theory itself is beyond the scope of an introductory course, but some of its results are relatively simple and will be used here. These results provide some insight into electronic arrangements around the nucleus and help us understand the basis for the periodic law and table, as well as the basic rules governing compound formation (the topic of Chapter 4).

According to quantum mechanics, an electron can have only one of certain specified energies. In other words, not all energy values are allowed. This condition, called a **quantization** of energy, is somewhat analogous to one encountered when a person climbs a flight of stairs. The energy of a person on the stairs is related to his height above the floor and is obviously quantized, since only certain heights are possible. This situation is shown in Figure 3–1.

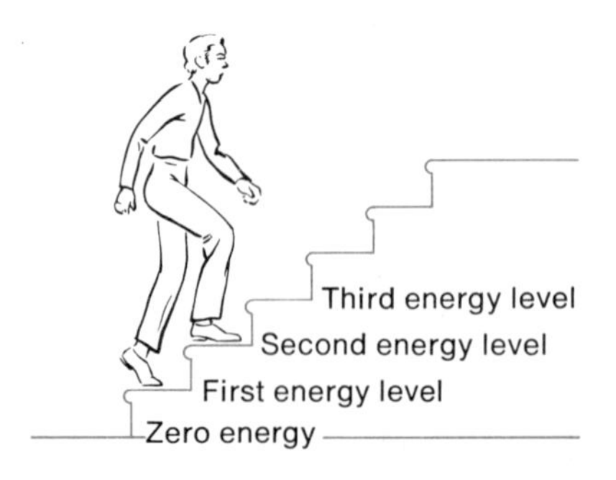

Figure 3–1.
The Quantization of Energy on a Stairway

The energy of an electron determines its behavior about the nucleus. Since only certain energies are allowed, only certain behavioral patterns will be found. The details of electron behavior come from quantum-mechanical calculations, and three terms are used to describe the results of these calculations. These terms—**orbital, subshell,** and **shell**—are discussed below.

Orbital

An orbital is a region of space around the nucleus in which an electron with a specific energy is most likely to be found. Theoretically, an electron can be anywhere around the nucleus at any given time. However, because of energy characteristics, the chances of finding an electron in certain regions is much higher than in other regions. The space extending from the nucleus and including these high-probability regions is the orbital. The following information on orbitals is, again, the result of quantum-mechanical calculations.

Each orbital is a particular region of space around a nucleus in which an electron with a specific energy is most likely to be found.

Each orbital has a definite size and shape, which are related to the energy of the electrons that it can contain.

The orbital volume and distance of extension from the nucleus increase as the energy of contained electrons increases.

Each orbital can contain a maximum of two electrons.

Even though an electron can be at only one point in the orbital at any given time, it is moving very rapidly and "occupies" the entire orbital volume much as a rapidly rotating airplane propeller "occupies" a definite volume.

Each atom has available for electron occupancy a large number of orbitals with various sizes and shapes. They may or may not contain electrons.

Subshell

A subshell is a collection of orbitals that have the same size and similar shapes. Electrons included in such a collection of occupied orbitals necessarily have the same energy, because the energy is related to the orbital sizes and shapes.

Shell

A shell is a collection of subshells, each of which contains orbitals with different shapes but similar sizes. The energies of the electrons included in a shell are not identical but can vary only over a narrow range of values.

The two most important properties of electrons that we will use are (1) the energy and (2) the most probable distance from the nucleus. It is worth pointing out that subshells are collections of electrons with the same energy, and shells are collections of electrons located at similar most-probable distances from the nucleus.

Identification of Electron Arrangements

In order to effectively use the concepts of orbital, subshell, and shell, we must have some way of identifying them and determining the numerical relationships that exist between them. Once again, quantum-mechanical calculations have provided helpful results.

Each shell is identified by a number, n, which is a whole positive integer (1, 2, 3, and so forth). The larger the number, the greater the distance between the nucleus and the most probable position of the contained electrons. Usually, orbitals in a higher-numbered shell contain electrons of higher energy than those in a lower-numbered shell. A few exceptions to this rule are pointed out later.

Figure 3–2.
Shapes of Atomic Orbitals

s orbital

p orbital

d orbital

f orbital

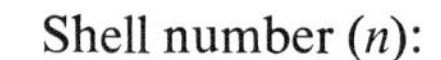

Shell number (n): 1 2 3 4 5 . . .

Distance of electrons from nucleus increases →

Average energy of contained electrons increases →

The number of subshells in a shell is identical to the shell number n. The number of orbitals in each shell is equal to n^2. Because two electrons can occupy an orbital, the maximum number of electrons in a shell must therefore equal $2n^2$. These relationships are given in Table 3–4.

Table 3–4. Numerical Relationships Between Shells, Subshells, Orbitals, and Electrons

Shell number (n)	*Number of subshells* (n)	*Number of orbitals* (n^2)	*Maximum number of electrons* ($2n^2$)
1	1	1	2
2	2	4	8
3	3	9	18
4	4	16	32

Subshells within a shell are denoted by the shell number and the assigned letters *s*, *p*, *d*, and *f*. The energy, complexity, and number of orbitals involved in the subshells increase in the order *s*, *p*, *d*, and *f*.

Subshell letter: *s p d f*

Energy of contained electrons increases →

Complexity of involved orbitals increases →

Number of involved orbitals increases →

The subshell arrangement for the first seven shells is summarized in Table 3–5. In shell number 5 there are five possible subshells but only four are listed.

Similarly, only three of a possible six are listed for shell number 6, and only one of a possible seven for shell 7. The number of subshells listed is sufficient to describe all of the known elements; unneeded ones have been left out.

Table 3–5. Subshell Arrangements in Shells

Shell number (n)	*Subshells*						
1	1*s*						
2	2*s*	2*p*					
3	3*s*	3*p*	3*d*				
4	4*s*	4*p*	4*d*	4*f*			
5	5*s*	5*p*	5*d*	5*f*	—		
6	6*s*	6*p*	6*d*	—	—	—	
7	7*s*	—	—	—	—	—	—

Each subshell contains a definite number of orbitals, which depends only on the type of subshell involved and not on the shell number. Thus, 2*p*, 3*p*, 4*p*, or 5*p* subshells contain the same number of orbitals. The relationships between subshell type and the number of orbitals are given in Table 3–6. All orbitals within a subshell have the same energy and similar shapes, and all are designated by the same notation as that used to designate the subshell. For example, there are three orbitals in a 2*p* subshell, and each one is called a 2*p* orbital.

Table 3–6. Number of Orbitals in Subshells

Type of subshell	*Number of orbitals in the subshell*
s	1
p	3
d	5
f	7

The actual shapes of orbitals are derived from mathematical theory. These shapes are given in Figure 3–2 for typical *s*, *p*, *d*, and *f* orbitals. Notice the increasing complexity, which follows the order *s*, *p*, *d*, and *f*. All *d* and *f* orbitals have shapes related, but not identical, to those given in Figure 3–2.

Some examples will illustrate the information just presented.

Example 3–1. Characterize the similarities and differences between a 3*s* and a 3*p* orbital.

1. Both are in shell number 3 and therefore extend approximately the same distance from the nucleus.
2. They belong to different subshells (*s* and *p*) and therefore have different shapes and energies.
3. An electron in the 3*p* orbital is of higher energy than one in the 3*s*.

Example 3–2. Characterize the similarities and differences between a 3*p* and 6*p* orbital.

1. Both are in *p* subshells and therefore have the same shape.
2. They are in different shells ($n = 3$ and 6) and therefore have different sizes (6*p* is larger).
3. An electron in the 6*p* orbital is of higher energy than one in the 3*p* orbital.

Table 3–7 contains a summary through $n = 4$ of the material needed to designate electron arrangements.

Table 3–7. Relationships Used to Designate Electron Arrangements

Shell number (n)	*Number of subshells*	*Type of subshells*	*Number of orbitals*	*Type of orbital*	*Maximum number of electrons in subshell*
1	1	1*s*	1	1*s*	2
2	2	2*s*	1	2*s*	2
		2*p*	3	2*p*	6
3	3	3*s*	1	3*s*	2
		3*p*	3	3*p*	6
		3*d*	5	3*d*	10
4	4	4*s*	1	4*s*	2
		4*p*	3	4*p*	6
		4*d*	5	4*d*	10
		4*f*	7	4*f*	14

Electron Configurations

Around the nucleus of an atom exists an essentially unlimited number of orbitals which can be occupied by electrons. In spite of this, electrons occupy orbitals in a very predictable pattern which can be determined by applying a simple rule: an electron occupies that orbital in which it achieves the lowest possible energy. Therefore, the arrangement of electrons around a nucleus can be predicted if the relative energy of the various orbitals is known. The following orbitals are arranged in an order of increasing energy.

1*s*, 2*s*, 2*p*, 3*s*, 3*p*, 4*s*, 3*d*, 4*p*, 5*s*, 4*d*, 5*p*, 6*s*, 4*d*, 5*d*, 6*p*, 7*s*, 5*f*, 6*d*

All orbitals within a subshell have identical energies, so this sequence also represents the order of increasing energy for the various subshells. The diagram below will help you remember this order. The order of generally increasing energy is obtained by sequentially following the arrows from top to bottom. Follow each arrow from tail to head, then return to the tail of the next one and continue.

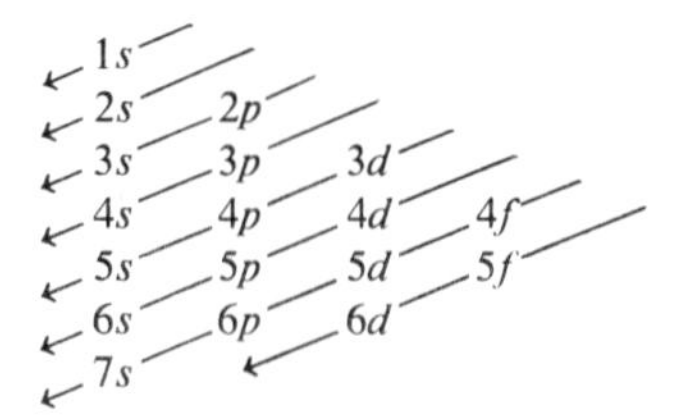

Relative Energies of Subshells or Orbitals

A close look at this sequence reveals that all subshells within a given shell do not necessarily have lower energies than all subshells of higher-numbered shells. For example, the *s* subshell of shell number 4 (4*s* subshell) has a lower energy than the *d* subshell of shell number 3 (3*d* subshell). It is apparent from the order of energies given in this example that this energy overlap of shells is not a rare occurrence.

We now have available the two pieces of information we need in order to predict electron configurations or arrangements around a nucleus. We know the order in which subshells fill and the maximum number of electrons that can occupy each subshell (Table 3–7).

Let's use this information to predict some electron configurations. The element hydrogen ($Z = 1$) contains only 1 electron and is the simplest element known. The electron in hydrogen will be found in the 1*s* subshell, which is the lowest-energy subshell available. Since this subshell can hold a maximum of two electrons, it is half filled in hydrogen atoms. The electronic configuration of hydrogen is written $1s^1$, where the superscript indicates the total number of electrons in the subshell.

Helium ($Z = 2$) has two electrons, both occupying the 1*s* subshell. The electronic configuration of helium is $1s^2$, and the subshell is filled to capacity with the two electrons.

Lithium ($Z = 3$) has three electrons. Two occupy and fill the 1*s* subshell; the third is found in the subshell with the next lowest energy, which is the 2*s*. The lithium electronic configuration is $1s^22s^1$.

Beryllium ($Z = 4$) has the electronic configuration $1s^22s^2$. The fourth electron of beryllium completes the 2*s* subshell, which also has a capacity of two electrons. In boron ($Z = 5$) the 2*p* subshell, with the next lowest energy available, is occupied by a single electron. A *p* subshell can hold a maximum of six electrons, and this subshell is filled by additional electrons added for the elements B, C, N, O, F, and Ne. The electron configurations for these elements are:

B ($Z = 5$): $1s^22s^22p^1$
C ($Z = 6$): $1s^22s^22p^2$
N ($Z = 7$): $1s^22s^22p^3$
O ($Z = 8$): $1s^22s^22p^4$
F ($Z = 9$): $1s^22s^22p^5$
Ne ($Z = 10$): $1s^22s^22p^6$

Examples 3–3 and 3–4 illustrate the extension of this pattern to atoms containing many electrons.

Example 3–3. What is the electronic configuration of iodine ($Z = 53$)? Iodine, with an atomic number of 53, contains 53 electrons distributed among the various subshells. The addition of 53 electrons takes place in the order given below. The bottom line of numbers is a running total of added electrons and is obtained by adding up the superscripts to that point.

Occupied subshells:	$1s^2$ $2s^2$ $2p^6$ $3s^2$ $3p^6$ $4s^2$ $3d^{10}$ $4p^6$ $5s^2$ $4d^{10}$ $5p^5$
Running total of added electrons:	2 4 10 12 18 20 30 36 38 48 53

Notice that the $5p$ subshell contains only five electrons, even though it can hold a maximum of six. The sixth electron would be one more than the 53 needed.

Example 3–4. What is the electronic configuration of hahnium ($Z = 105$)?

$1s^2$ $2s^2$ $2p^6$ $3s^2$ $3p^6$ $4s^2$ $3d^{10}$ $4p^6$ $5s^2$ $4d^{10}$ $5p^6$ $6s^2$ $4f^{14}$ $5d^{10}$ $6p^6$ $7s^2$ $5f^{14}$ $6d^3$
2 4 10 12 18 20 30 36 38 48 54 56 70 80 86 88 102 105

No known element requires more subshells or orbitals than hahnium to describe an electronic configuration, since hahnium has the largest Z of all known elements.

Before we leave the subject of electronic configurations, let us take special note of two points. First, in a few cases the actual electronic configuration of an atom is different from that obtained using the rules given here. This is caused by very small energy differences between some subshells. The exceptions are not important in the uses we shall make of electronic configurations. Second, the material presented concerning orbitals, subshells, and shells is part of the quantum-mechanical theory of electron behavior. These concepts are man-made devices used to describe electron behavior. Their use simplifies the explanation of many facts about the behavior, properties, and reactions of substances.

Electronic Configurations and the Periodic Table

A knowledge of the electronic configurations of the elements provides an explanation for the periodic law and table. We can understand this explanation most easily by considering the electronic configurations of elements which have similar properties. These elements are grouped together as columns in the periodic table (Table 3–2). A typical group, which makes up the first column to the left of the table (group IA), includes the elements $_3$Li, $_{11}$Na, $_{19}$K, $_{37}$Rb, $_{55}$Cs, and $_{87}$Fr. The electronic configurations for the first four of these elements are:

$_3$Li : $1s^2$ $2s^1$
$_{11}$Na: $1s^2$ $2s^2$ $2p^6$ $3s^1$
$_{19}$K : $1s^2$ $2s^2$ $2p^6$ $3s^2$ $3p^6$ $4s^1$
$_{37}$Rb: $1s^2$ $2s^2$ $2p^6$ $3s^2$ $3p^6$ $4s^2$ $3d^{10}$ $4p^6$ $5s^1$

We see that each of these elements has a single electron in the outermost occupied shell (colored). Further, the subshell occupied by the single electron is of the same type s in each element. This similarity in outer-shell electronic arrangements causes the chemical properties of the elements to be similar. It has been found in general that all elements with similar outer-shell electronic configurations have similar chemical properties.

The same type of similarity is found for groups of elements throughout the table. Another example is provided by the elements $_{9}F$, $_{17}Cl$, $_{35}Br$, $_{53}I$, and $_{85}At$, which constitute group VIIA of the periodic table. The electronic configurations for the first four members of this group are:

$_{9}F$: $1s^2\ 2s^2\ 2p^5$
$_{17}Cl$: $1s^2\ 2s^2\ 2p^6\ 3s^2\ 3p^5$
$_{35}Br$: $1s^2\ 2s^2\ 2p^6\ 3s^2\ 3p^6\ 4s^2\ 3d^{10}\ 4p^5$
$_{53}I$: $1s^2\ 2s^2\ 2p^6\ 3s^2\ 3p^6\ 4s^2\ 3d^{10}\ 4p^6\ 5s^2\ 4d^{10}\ 5p^5$

Once again the similarities in electronic configurations are apparent. Therefore, we may say that chemical properties repeat in a periodic manner because electron arrangements repeat in a periodic manner.

The periodic table is more useful when we interpret it in terms of the electronic configurations of the elements that occupy the various regions. Consider the periodic table to be divided into four areas, as shown in Figure 3–3. The electronic configuration of each element differs from that of the preceding element by a single electron. The last electron added to an element is often referred to as the **distinguishing** electron. Elements in the *p* area have a *p* electron as the distinguishing electron, elements in the *s* area have distinguishing electrons of the *s* type, and so forth.

In Figure 3–3, notice that the areas are labeled *s*, *p*, *d*, and *f*, the same as the various subshells. Also notice that the *s* area is two columns wide, the *p* area is six columns wide, the *d* area is ten columns wide, and the *f* area is fourteen columns wide. This correlates directly with the fact that *s*, *p*, *d*, and *f* subshells can contain a maximum of 2, 6, 10, and 14 electrons, respectively.

The position of an element in the periodic table not only indicates the type of subshell containing the distinguishing electron but can also be used to

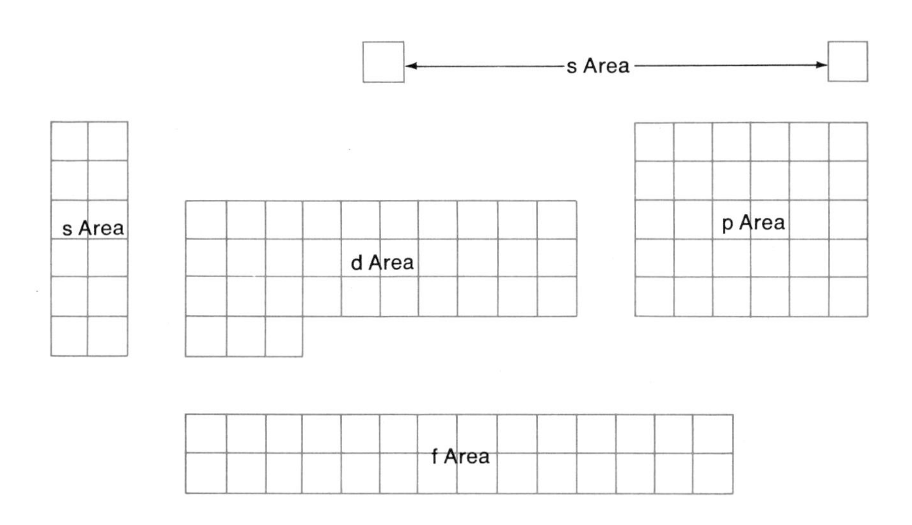

Figure 3–3.
Areas of the Periodic Table

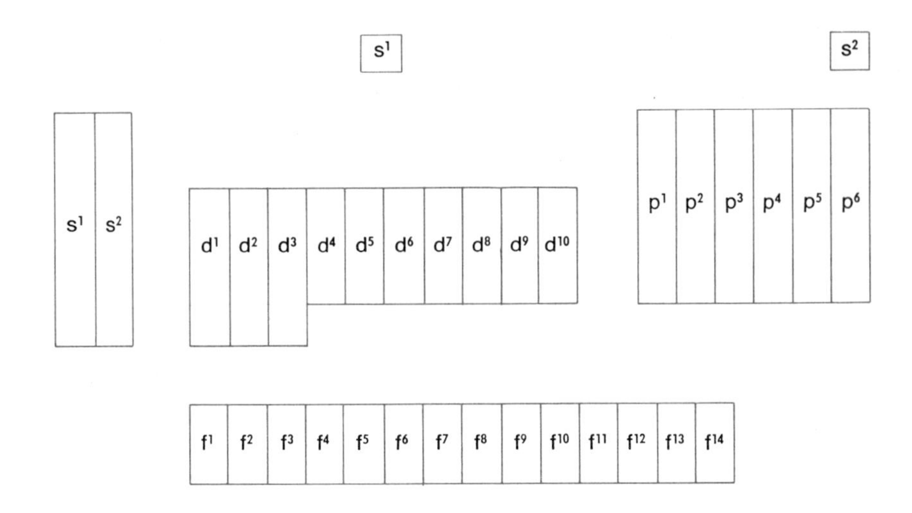

Figure 3–4. Distinguishing Electrons in Various Areas of the Periodic Table

determine the shell involved and the total number of electrons contained in the subshell.

The extent of filling in a subshell is given by the position of an element within an area. For example, all elements in the first column of the *p* area (group IIIA) have an electronic configuration ending in p^1. Elements in the second column of the *p* area (group IVA) end with p^2, and so forth. Similar relationships hold in other areas of the table, as shown in Figure 3–4. This pattern is always followed for elements in the *s* and *p* areas. A few exceptions are found in the *d* and *f* areas, but, again, we will not be concerned with them in this treatment.

The shell occupied by a subshell is related to the period number found in the periodic table. In the *s* and *p* areas, the period number is equal to the shell

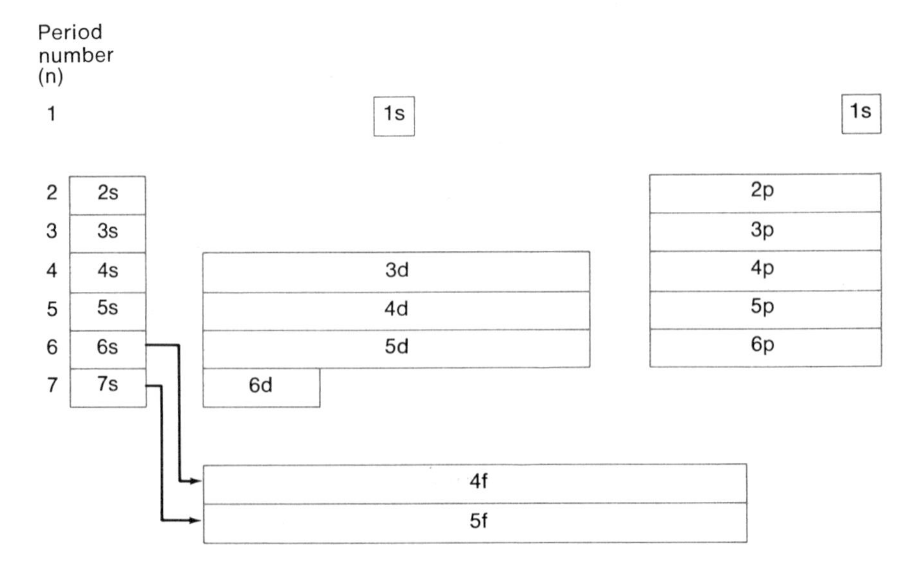

Figure 3–5. The Relationships of Period Numbers to Shell Numbers

number. In the *d* area, the period number minus one is equal to the shell number, and in the *f* area, the period number minus two equals the shell number. Therefore, in period 6 are found the 6*s*, 6*p*, 5*d*, and 4*f* subshells. It must be remembered that the *f* area, even though located at the bottom of the table, correctly belongs in periods 6 and 7. The relationships between period number and shell number are given in Figure 3–5.

The periodic table can also be used to determine the correct ordering of subshells according to increasing energy without the use of the Relative Energies of Subshells diagram given earlier. Merely follow a path of increasing atomic number through the table and note the subshells as they are encountered. The order of subshells determined in this way will be the one of increasing energy shown in our diagram. The results of this operation are given in Table 3–8.

Table 3–8. Subshell Order from the Periodic Table

Atomic number range	*Subshells involved*
1–2	1*s*
3–4	2*s*
5–10	2*p*
11–12	3*s*
13–18	3*p*
19–20	4*s*
21–30	3*d*
31–36	4*p*
37–38	5*s*
39–48	4*d*
49–54	5*p*
55–56	6*s*
57	5*d*
58–70	4*f*
71–80	5*d*
81–86	6*p*
87–88	7*s*
89	6*d*
90–103	5*f*
104–105	6*d*

The following examples illustrate the application of these ideas to actual elements.

Example 3–5. From its position in the periodic table, what can be said about the electronic configuration of $_{33}As$?

1. Since the element is in the *p* area, the distinguishing electron is a *p* electron.
2. Since the element is in the third column of the *p* area, the *p* subshell involved contains three electrons (p^3).
3. Since the element is in period number 4, the *p* subshell involved is the 4*p*, and therefore the outer configuration is $4p^3$.

The complete configuration is

$$1s^2\ 2s^2\ 2p^6\ 3s^2\ 3p^6\ 4s^2\ 3d^{10}\ 4p^3$$

Example 3–6. From its position in the table, what can be said about the electronic configuration of $_{21}Sc$?

1. The element is in the *d* area; therefore the distinguishing electron is a *d* electron.
2. The element is in the first column of the *d* area; therefore the *d* subshell involved contains a single electron (d^1).
3. The element is in period number 4; therefore the *d* subshell involved is a 3*d* subshell (1 less than the period number). The outer configuration is $3d^1$.

The complete configuration is

$$1s^2\ 2s^2\ 2p^6\ 3s^2\ 3p^6\ 4s^2\ 3d^1$$

Common Classifications of the Elements

It is a convenient and common practice to classify the elements of the periodic table into four categories, according to their electronic structures.

Rare Gases

These elements comprise the group found on the extreme right of the periodic table. They are all gases at room temperature and have low natural abundances, hence the group name. With one exception, these elements are characterized electronically by having completely filled *ns* and *np* subshells—ns^2 and np^6. The exception is helium, which has a configuration of $1s^2$. In each rare gas the value of *n* for the filled *s* and *p* subshells is the maximum value found in that element for any electron-containing subshells. This outer electronic structure is found in no other elements; for this reason it is called the **rare-gas configuration.** The eight electrons (two in the *s* subshell and six in the *p* subshell) involved in the rare-gas configuration are the basis for the octet rule of bonding discussed in Chapter 4.

Representative Elements

These elements are found in the *s* and *p* areas of the periodic table but do not include the rare gases. Most of the common elements fall into this category. The last or distinguishing electron of these elements partially or completely fills an *ns* subshell, or partially fills an *np* subshell, where the *n* value is the maximum found in the atom. In the three examples given below, the subshell containing the distinguishing electron is colored.

Li: $1s^2\ 2s^1$ Be: $1s^2\ 2s^2$ F: $1s^2\ 2s^2\ 2p^5$

Transition Elements

The *d* area of the periodic table contains these elements. The common feature of the electronic configurations of these elements is the presence of the distinguishing electron in a *d* subshell with an *n* value one less than the maximum value for electron-containing subshells in the atom. An example of a transition element is vanadium, number 23, with a configuration of

$$1s^2\ 2s^2\ 2p^6\ 3s^2\ 3p^6\ 4s^2\ 3d^3$$

Again, the subshell containing the distinguishing electron is indicated.

Inner-Transition Elements

These elements are found in the *f* area of the periodic table. The characteristic feature of their electronic configurations is the presence of the distinguishing electron in an *f* subshell. The value of *n* for the subshell is two less than the maximum value for *n* in the atom. An example is europium ($Z = 63$) with the following configuration:

$$1s^2\ 2s^2\ 2p^6\ 3s^2\ 3p^6\ 4s^2\ 3d^{10}\ 4p^6\ 5s^2\ 4d^{10}\ 5p^6\ 6s^2\ 4f^7$$

Figure 3–6 contains a periodic table in which these four categories of the elements are indicated.

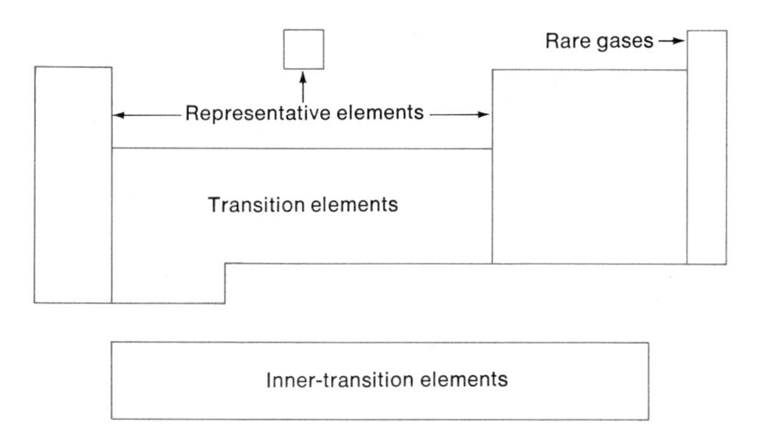

Figure 3–6.
Elemental Classification by Electronic Configurations

Metals, Nonmetals, and Semimetals

In another common type of classification the elements are placed in the categories of metals, nonmetals, and semimetals. Most of the elements are classified as metals, including all transition elements, all inner-transition elements, and the majority of the representative elements. The elements classified as metals have the following characteristic metallic properties:

High thermal conductivity—they transmit heat easily.
High electrical conductivity—they transmit electricity readily.
Ductility—they can be drawn into wires.
Malleability—they can be hammered into sheets.
Metallic luster—they look like metals.

Nonmetals generally have properties the opposite of those given for metals. They often occur as powdery solids or gases under normal conditions.

Semimetals have properties somewhat intermediate between those of metals and nonmetals, and they often exhibit some of the characteristic properties of each type. Because of the existence of these elements, there is no sharp distinction between metals and nonmetals.

Usually a metallic element is characterized electronically by containing one to four electrons in the highest (maximum n) s and p subshells. As we saw before, these subshells together have a capacity for eight electrons. Nonmetals generally contain four or more, but less than eight, electrons in the highest s and p subshells. Consequently, metals are generally located to the left in the periodic table and nonmetals to the right, with the semimetals between them. This is shown in Figure 3–7. Notice the steplike arrangement of the semimetals.

The following generalizations concerning the metallic character of the elements are based on experimental observations:

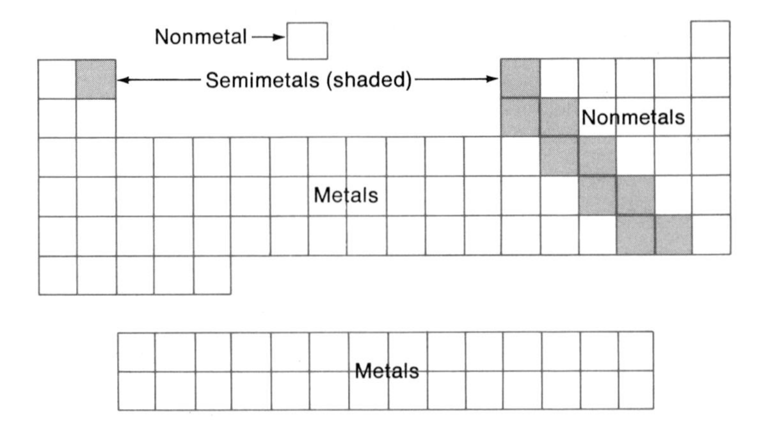

Figure 3–7.
Metals, Semimetals, and Nonmetals

The properties of each succeeding element change gradually from a metallic to a more nonmetallic character as the position of the element changes from left to right along a period of the table. The number of electrons in the highest *s* and *p* subshells also increases in the same direction.

The metallic character of the elements increases from the top to the bottom of a group. Consider group IVA: the top element is nonmetallic carbon while the bottom element is metallic lead. Although both elements have the same number of electrons in the highest *s* and *p* subshells, the subshells are in different shells and are therefore located at different distances from the nucleus. Generally, the further these subshells are from the nucleus, the greater the metallic character of the element.

Review Questions

1. Give a modern statement of the periodic law.
2. What is a periodic table?
3. Give the symbol of the element which occupies each of the following positions in the periodic table:
   a) period 4, group IIIA
   b) period 5, group IVB
   c) group IA, period 2
   d) group IB, period 6
4. From each of the following sets of elements choose the pair that would be expected to have similar properties:
   a) $_{11}Na$, $_{14}Si$, $_{23}V$, $_{55}Cs$
   b) $_{13}Al$, $_{19}K$, $_{32}Ge$, $_{50}Sn$
   c) $_{37}Rb$, $_{38}Sr$, $_{54}Xe$, $_{56}Ba$

5. What does the word *quantization* mean?
6. What property of an electron is quantized?
7. Compare the terms *orbital*, *subshell*, and *shell*.
8. Consider two electron shells—those with $n = 2$ and $n = 4$.
   a) How does the distance between the nucleus and the average most probable position of the contained electrons compare for the two shells?
   b) How does the average energy of the contained electrons compare for the two shells?
   c) How many subshells does each of the shells contain?
   d) How many orbitals does each of the shells contain?
   e) What is the maximum number of electrons each shell may contain?
9. How many orbitals are found in each of the following types of subshells? What is the maximum number of electrons each type of subshell may contain?
   a) *s*
   b) *p*
   c) *d*
   d) *f*
10. What role does electron energy play in the determination of orbital occupancy by electrons?
11. For each of the following sets of orbitals determine the one of lowest energy and the one of highest energy:
    a) 1*s*, 2*s*, 3*s*
    b) 3*s*, 3*p*, 3*d*
    c) 6*s*, 5*d*, 4*f*
    d) 3*d*, 5*p*, 7*s*
12. Write the electronic configuration for each of the following atoms:
    a) ${}_{8}O$
    b) ${}_{35}Br$
    c) ${}_{87}Fr$
    d) ${}_{80}Hg$
    e) ${}_{31}Ga$
13. What is wrong with each of the following attempts to write electron configurations?
    a) $1s^2\ 2s^3$
    b) $1s^2\ 1p^6\ 2s^2\ 2p^6$
    c) $1s^2\ 2s^2\ 2p^8\ 3s^2\ 3p^8$
    d) $1s^2\ 2s^2\ 2p^6\ 3s^2\ 3p^6\ 3d^{10}$
14. For each of the atoms Na, N, Kr, Cu, Ba, Hg, As, Eu
    a) Classify the last electron added (distinguishing electron) as *s*, *p*, *d*, *f*.
    b) Determine the number of electrons in the 4*p* subshell.
    c) Determine the total number of *s* electrons contained in the atom.
    d) Classify the element as a metal, nonmetal, or semimetal.
    e) Classify the element as a rare gas, representative element, transition element, or inner-transition element.

15. Based on the relationship between electronic configurations and the periodic table, indicate:
    a) The symbol of the element containing 82 electrons
    b) The symbol of the element containing only three 5*p* electrons
    c) The symbol of the element containing only one 7*s* electron
    d) The symbol of the element containing a total of 15*p* electrons
    e) The symbol of the element containing only one electron in shell 7
16. Formulate definitions for elements classified as rare gases, representative elements, transition elements, and inner-transition elements in terms of *s*, *p*, *d*, and *f* subshells.
17. How do the successive members of a group in the periodic table differ in electronic structure? How are they alike? How do the successive members of a period differ in electronic structure? How are they alike?
18. In each of the following pairs indicate which element is most metallic:
    a) C, Li
    b) Be, Ca
    c) Al, Pb

4 The Formation of Chemical Compounds

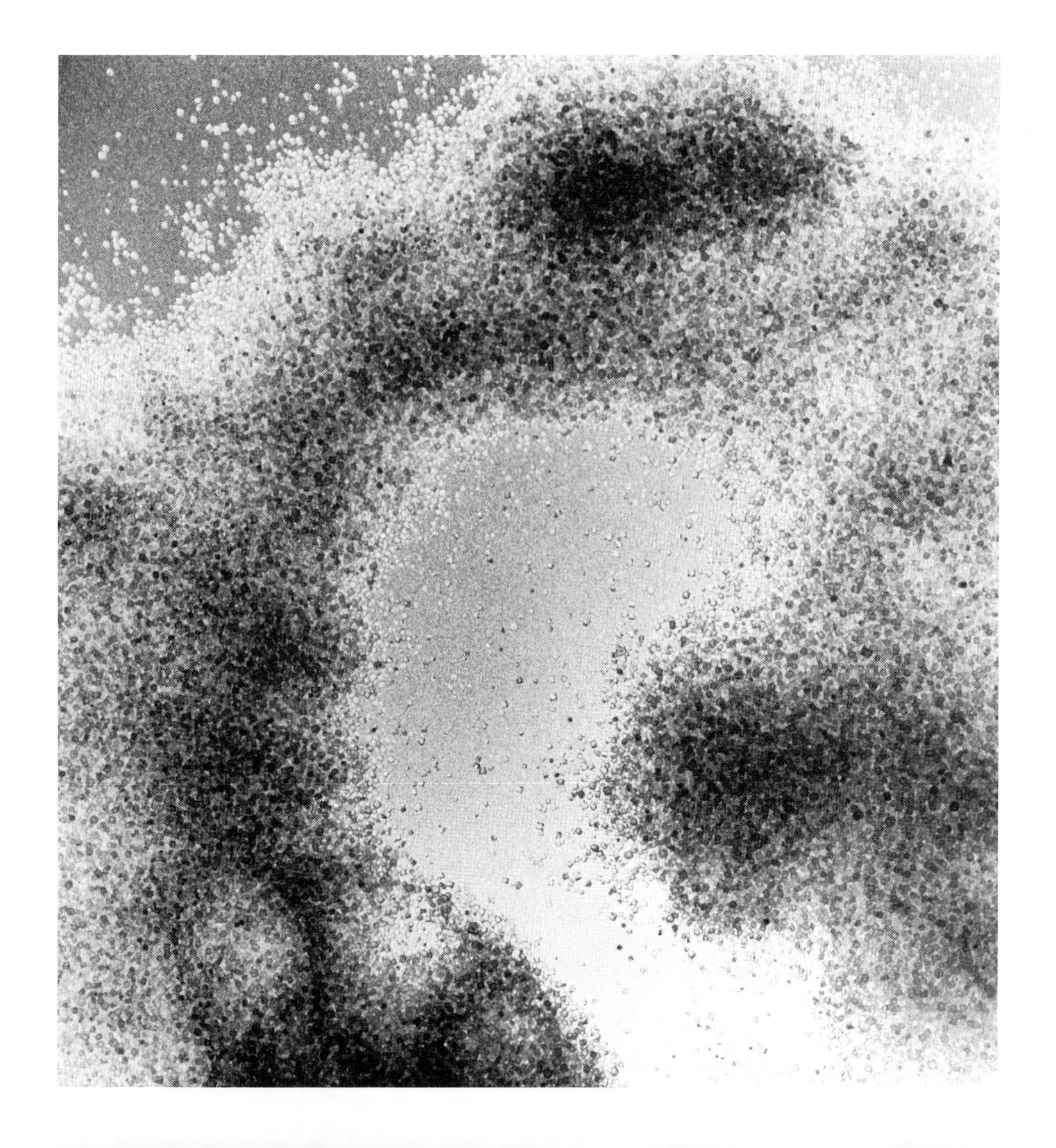

We found in Chapter 1 that pure substances are classified as either element or compound, according to the type of molecules they contain. Elements are composed of homoatomic, and compounds of heteroatomic, molecules. In this chapter we shall discuss the characteristics of elemental atoms that cause them to combine and form molecules.

Chemical Bonding and the Octet Rule

The atoms of most elements do not exist in nature as single discrete particles but are found combined with other atoms. When the combination involves atoms of the same element, homoatomic molecules are formed. For example, oxygen and nitrogen gases exist in nature as the diatomic molecules O_2 and N_2. When the combination involves atoms of two or more different elements, heteroatomic molecules or compounds result.

The atoms involved in these combinations exert attractive forces on each other which bind the group together to form a molecule. These forces are called **chemical bonds** and result, according to current theories, from an interaction of certain electrons contained in the combining atoms. The electrons involved in bonding interactions are called **valence electrons** and include both (1) the electrons of an atom in the *s* and *p* subshells with identical and maximum n values, and (2) the electrons in any partially filled *d* or *f* subshells of an atom. The interaction of the valence electrons of two or more atoms involves the rearrangement of the electrons into a more stable—lower-energy—configuration. It is this decrease in energy that causes the reactions to take place.

The bonding involvement of valence electrons seems reasonable when it is remembered that they will be the first to come into close proximity to each other when atoms collide—an event necessary in combination reactions. Also, they are located farthest from the nucleus and are therefore the most loosely bound and susceptible to rearrangement.

We find evidence supporting the involvement of valence electrons when we consider the formulas of compounds formed when elements in a specific group of the periodic table react with other elements. Table 4–1 contains formulas of this type for sodium (Na), potassium (K), and rubidium (Rb), all found in group IA. The fact that elements in the same group form compounds with identical formulas strongly indicates the involvement of valence electrons. Remember, it is the number and arrangement of valence electrons that elements in a group have in common. In the case of Na, K, and Rb, the valence-electron configurations are $3s^1$, $4s^1$, and $5s^1$, respectively.

A very important development in the formulation of bonding theories was the recognition of what is now considered to be the most stable electronic configuration resulting from valence-electron rearrangements. It was mentioned earlier in this chapter that most elements do not exist in nature as single atoms. The exceptions to this are the rare gases, which make up the

Table 4–1. Formulas of Group IA Compounds

	Formula of compound formed when element reacts with			
Element	*Oxygen*	*Chlorine*	*Nitrogen*	*Hydrogen*
Na	Na_2O	NaCl	Na_3N	NaH
K	K_2O	KCl	K_3N	KH
Rb	Rb_2O	RbCl	Rb_3N	RbH

group of elements located farthest to the right in the periodic table. The elements of this group—He, Ne, Ar, Kr, Xe, and Rn—have only a slight tendency to react chemically and are found in nature as single atoms. No compounds of He, Ne, and Ar are known, and not until 1962 were the few known compounds of Kr, Xe, and Rn first synthesized. The lack of chemical activity for these elements is attributed to their electronic arrangements, which must be very stable.

With the exception of helium, which has a $1s^2$ configuration, each rare gas contains eight electrons in an ns^2np^6 distribution, where n has the maximum value found in the atom. Four of these configurations are given in Table 4–2 with the valence-electron configurations colored. Notice that the valence electrons for Kr are not shown adjacent to each other in Table 4–2. This is caused by the energy overlap between shells mentioned in Chapter 3. The stability of the ns^2np^6 configuration is related to the complete filling of the s and p subshells by eight electrons.

Table 4–2. Electronic Configurations of the Rare Gases

Rare gases	*Electronic configuration*
Helium (He)	$1s^2$
Neon (Ne)	$1s^2\ 2s^2\ 2p^6$
Argon (Ar)	$1s^2\ 2s^2\ 2p^6\ 3s^2\ 3p^6$
Krypton (Kr)	$1s^2\ 2s^2\ 2p^6\ 3s^2\ 3p^6\ 4s^2\ 3d^{10}\ 4p^6$

Current theories indicate that the atoms of many elements which lack this very stable rare-gas configuration tend to attain it in chemical reactions that result in the formation of compounds. This has come to be known as the **octet rule** because of the eight electrons involved in the rare-gas configuration. A formal statement of the octet rule is:

Atoms tend to interact by undergoing electronic rearrangements which produce a rare-gas electronic configuration for each of the atoms involved.

Application of the octet rule to many different systems has shown that it often correctly explains the observed combining ratios of the atoms. For example, it explains why two hydrogen atoms rather than some other number are bonded to one oxygen atom in a water molecule, H_2O.

There are exceptions to the octet rule, but it is still used because of the large amount of information that it is able to correlate. It is particularly effective in explaining compound formation involving only representative elements. Some complications arise with transition and inner-transition elements because of the involvement of *d* and *f* valence electrons in bonding.

Two models, based on the octet rule, are used to describe bond formation. Called the **ionic** and **covalent** models, they represent the limiting extremes of electron transfer and electron sharing. Most real bonds fall somewhere between, but they are described in terms of the model whose characteristics are closest to those of the real bond. The use of only two models presents some difficult boundary-line decisions. However, the problem cannot be alleviated by creating more classifications, which simply create still more boundary-line decisions.

The Ionic-Bond Model

The formation of bonds according to the ionic-bond model involves the complete transfer of one or more electrons from one atom to another during the electron rearrangement that accompanies compound formation. The species that result are no longer neutral atoms; they are charged atoms which have attained a stable rare-gas electronic configuration. These charged atoms are called **ions.** The ions may be either positively or negatively charged, because electrons may have been gained or lost by the neutral atom. When one or more electrons are lost, the atom becomes positively charged, because the number of protons in the nucleus exceeds the remaining number of electrons—the nucleus is undisturbed during the electron-transfer process. When an atom gains electrons, the resulting ion is negatively charged, because the number of electrons exceeds the number of nuclear protons. The number of charges on an ion depends upon the number of electrons transferred. Both positive and negative ions must be produced in any electron-transfer process, because both a donor and acceptor must interact or the reaction cannot occur. The mutual attraction of the resulting positive and negative ions constitutes the force known as an **ionic bond.**

An ion of an element is represented by the elemental symbol and a superscript to show the net positive or negative charge. A positive ion of an element with symbol X and with a net positive charge of 2 would be represented by the symbol X^{2+}, while a negative ion of element Y with a net charge of negative 1 would be written Y^-.

Electron-transfer reactions typically occur most easily between elements of periodic table groups IA or IIA and elements of groups VIA or VIIA. The elements in these groups have electronic configurations most similar to those of the rare gases and can most easily attain rare-gas configurations by electron transfers.

Examples 4–1 and 4–2 illustrate some of the important concepts related to the electron-transfer model.

Example 4–1. What type of ion, positive or negative, will be formed when magnesium ($Z = 12$) becomes involved in an electron transfer process, and what will be the magnitude of the resulting ionic charge?

The rare gas closest to magnesium in the periodic table is neon ($Z = 10$), which is two positions away. The electronic configurations of magnesium and neon are, respectively, $1s^22s^22p^63s^2$ and $1s^22s^22p^6$. Therefore, magnesium differs from neon by only two electrons. The loss of these two electrons would give magnesium a rare-gas configuration and produce an ion represented as Mg^{2+}.

The charge on the magnesium ion is 2+ because of the resulting difference in the proton-electron ratio:

Neutral Mg atom:

12 p (in the nucleus)	=	12^+ charges
12 e (around the nucleus)	=	12^- charges
net charge	=	0

Mg^{2+} ion:

12 p (in the nucleus)	=	12^+ charges
10 e (around the nucleus)	=	10^- charges
net charge	=	2^+

The resulting Mg^{2+} ion and a Ne atom are not identical particles, as shown by a comparison:

	Ne atom	Mg^{2+} ion
Protons (in the nucleus)	10	12
Electrons (around the nucleus)	10	10
Atomic number	10	12
Charge	0	2^+

The only thing they have in common is the number and configuration of their electrons.

Example 4–2. What types of ions will the elements S, Cl, K, and Ca form?

In the periodic table argon ($Z = 18$) is the closest rare gas to each of these elements. Therefore, each element will tend to gain or lose electrons and acquire the argon configuration of $1s^22s^22p^63s^23p^6$.

Sulfur precedes argon in the periodic table by two positions and acquires the desired configuration by gaining two electrons:

$$\text{S } (1s^22s^22p^63s^23p^4) \xrightarrow{\text{gain 2 e}} \text{S}^{2-} \ (1s^22s^22p^63s^23p^6)$$

Chlorine precedes argon by one position and acquires the argon configuration by gaining one electron:

$$\text{Cl } (1s^22s^22p^63s^23p^5) \xrightarrow{\text{gain 1 e}} \text{Cl}^- \ (1s^22s^22p^63s^23p^6)$$

Potassium follows argon by one position and attains the argon configuration by losing one electron:

$$\text{K } (1s^22s^22p^63s^23p^64s^1) \xrightarrow{\text{lose 1 e}} \text{K}^+ \ (1s^22s^22p^63s^23p^6)$$

Calcium follows argon by two positions and by losing two electrons achieves the stable configuration:

$$\text{Ca } (1s^22s^22p^63s^23p^64s^2) \xrightarrow{\text{lose 2 e}} \text{Ca}^{2+} \ (1s^22s^22p^63s^23p^6)$$

A careful analysis of the results of Example 4–2 indicates the existence of a relationship between the number of electrons lost or gained in attaining a rare-gas configuration and the position of an element in the periodic table. The relationship is summarized below.

All group IA elements form +1 ions.
All group IIA elements form +2 ions.
All group VIIA elements form −1 ions.
All group VIA elements form −2 ions.

Theoretically, ions with charges of plus or minus three or four can also be formed, but in most cases the bonding that results is more adequately described by the covalent-bond model discussed later in this chapter. Only a few compounds containing +3 ions from group IIIA elements and −3 ions from group VA elements are known.

We also notice in Example 4–2 that all of the formed ions have the same electronic configuration as argon. The ions are described as being **isoelectronic** with argon and make up an **isoelectronic series.** A similar series exists for each rare gas. Three of these series are given in Table 4–3. Notice that the elements involved are those that closely precede or follow the rare gas in the periodic table.

Table 4–3. Isoelectronic Series of Ions

Helium structure		*Neon structure*		*Argon structure*	
H^-	$1s^2$	N^{3-}	$1s^22s^22p^6$	P^{3-}	$1s^22s^22p^63s^23p^6$
He		O^{2-}		S^{2-}	
Li^+		F^-		Cl^-	
Be^{2+}		Ne		Ar	
		Na^+		K^+	
		Mg^{2+}		Ca^{2+}	
		Al^{3+}			

The combining ratio of positive ions with negative ions in compound formation is determined by the charges on the ions. The total amount of positive and negative charge involved must add up to zero. This is illustrated in Table 4–4, where typical ionic compound formulas are given, together with explanatory comments.

Table 4–4. Ionic Compound Formulas

Compound	*Explanation of combining ratio*
Lithium fluoride, LiF	Li forms a Li^+ ion and F forms an F^- ion. Therefore, a 1:1 ratio provides electrical neutrality.
Magnesium fluoride, MgF_2	Mg forms a Mg^{2+} ion and F forms an F^- ion. Therefore, two F^- ions are needed to balance the charge of a single Mg^{2+} ion.
Potassium oxide, K_2O	K forms a K^+ ion and O forms an O^{2-} ion. Therefore, two K^+ ions are needed to balance a single O^{2-} ion.
Potassium nitride, K_3N	K forms a K^+ ion and N forms an N^{3-} ion. Therefore, three K^+ ions are needed to balance the charge of one N^{3-} ion.
Barium nitride, Ba_3N_2	Ba forms a Ba^{2+} ion and N forms an N^{3-} ion. Three Ba^{2+} ions provide +6 charges, and two N^{3-} ions provide −6 charges. The number 6 is the lowest common multiple of charges possible. Therefore, a 3:2 ratio results.

The Covalent-Bond Model

The theory of ionic bonding explains the formation and stability of many known compounds. However, it cannot explain the formation of molecules such as H_2 or F_2. The atoms in these molecules have identical properties, and there is no reason to suppose that one identical atom would give up electrons to another one.

The bonding in molecules of this type is explained by the concept of electron sharing, which results in the formation of a **covalent bond.** A careful consideration of the bonding in some typical covalent compounds will illustrate the concept. The F_2 molecule is formed from two fluorine atoms which both have the electronic configuration of $1s^22s^22p^5$. Each fluorine atom has seven valence electrons, $2s^22p^5$, and, therefore, needs one more to complete the octet and achieve the rare-gas configuration of neon (Ne, $1s^22s^22p^6$). This is accomplished when the two F atoms share one pair of electrons. The shared pair does a sort of double duty and helps complete the octet of both atoms. This sharing is illustrated diagrammatically by the equation below, where the dots represent valence electrons. The formulas, including the dots, are called **electron-dot formulas.**

$$:\ddot{\underset{..}{F}}\cdot + .\ddot{\underset{..}{F}}: \rightarrow :\ddot{\underset{..}{F}}(:)\ddot{\underset{..}{F}}: \qquad 4\text{–}1$$

shared pair is included in the octet of each atom

The bonding in the H_2 molecule can be explained similarly. Each H atom ($1s^1$) has a single electron, and the sharing of this lone electron results in the helium (He, $1s^2$) rare-gas configuration for each atom.

$$H\cdot + .H \rightarrow H(:)H \qquad 4\text{–}2$$

shared pair

Covalent bonding is also found in molecules composed of nonidentical atoms. The compound ClF contains a covalent bond resulting from the same sort of sharing as that discussed earlier for the F_2 molecule. Both Cl and F have seven valence electrons; the bonding reaction, using electron-dot formulas, is

$$:\ddot{\underset{..}{Cl}}\cdot + .\ddot{\underset{..}{F}}: \rightarrow :\ddot{\underset{..}{Cl}}(:)\ddot{\underset{..}{F}}: \qquad 4\text{–}3$$

shared pair

The covalent bond is not limited to diatomic molecules but occurs also in many polyatomic molecules. A few examples of this type are included in Table 4–5.

From Table 4–5 we see that the C atom has only four valence electrons and needs eight to acquire a rare-gas configuration. The sharing of four pairs of electrons provides the necessary eight. Each H atom has only one electron and needs two in order to have the He configuration. Each atom will therefore share only one pair. The bonding of four H atoms to one carbon atom satisfies the necessary requirements for both elements and forms the methane molecule. It is easy to see why stable compounds with the formulas CH_3 or CH_2 do not exist. In both cases the C atom could not share enough electrons to complete the octet.

In the formation of ammonia we note that, according to Table 4–5, the nitrogen must gain three electrons to complete the octet. This can be accomplished by forming covalent bonds with three H atoms. We also see in this example that only six of the eight electrons in the octet are involved in bonding. There is no requirement that all eight electrons be involved in the bonding of polyatomic molecules.

Oxygen's need for two electrons is satisfied by the sharing of two electron pairs in the formation of both water and oxygen fluoride, as depicted in Table 4–5. When water forms, the two H atoms acquire the He configuration; in the case of OF_2 the F atoms both acquire an octet and a normal rare-gas configuration.

Table 4–5. Covalent Bonding in Some Polyatomic Molecules

Molecule	*Atomic electron-dot formulas*	*Sharing pattern*	*Molecular electron-dot formula*
Methane, CH_4 (each H is bonded to the C)	·C· $(1s^22s^22p^2)$ H· $(1s^1)$	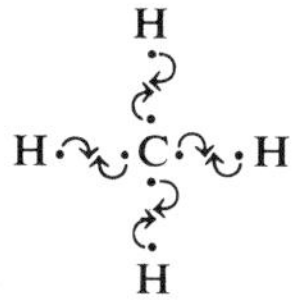 H H·C·H H	H H : C : H H
Ammonia, NH_3 (each H is bonded to the N)	·N· $(1s^22s^22p^3)$ H· $(1s^1)$	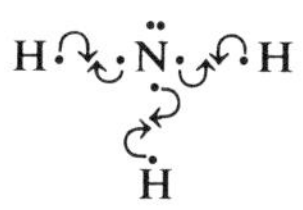 H·N·H H	H : N : H H
Water, H_2O (each H is bonded to the O)	:O· $(1s^22s^22p^4)$ H· $(1s^1)$	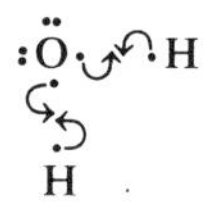 :O·H H	:O : H H
Oxygen fluoride, OF_2 (each F is bonded to the O)	:O· $(1s^22s^22p^4)$:F: $(1s^22s^22p^5)$	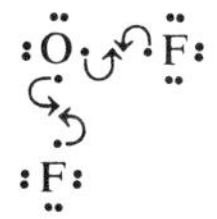 :O·F: :F:	:O : F: :F:

Table 4–5 illustrates some steps that are helpful when covalent-bonding, electron-dot structures are to be written. (1) The number of valence electrons for each element is determined from the elemental electronic configuration. (2) The arrangement of the atoms within the molecule is obtained. Normally this has already been experimentally determined. (3) Enough electrons are shared to provide each element with eight, except in the case of H where two is sufficient.

In all examples of covalent bonding given to this point, only the sharing of a single pair of electrons between atoms has been illustrated. The resulting bond is called a **single covalent bond.** It is sometimes necessary to postulate the sharing of more than one pair of electrons between two atoms in order to provide a complete octet for each atom of a molecule. This type of sharing provides for multiple bonding between atoms. Three examples of molecules involving multiple bonding are given in Table 4–6.

The N_2 molecule contains three shared pairs of electrons called a **triple bond.** Similarly, the carbon-oxygen bonds in CO_2 are called **double bonds** because they contain two shared pairs. The H_2CO molecule contains two single bonds and one double bond. For convenience, covalent bonds are often represented by a dash for each shared pair of electrons. Thus, N_2 is written N≡N, CO_2 as O=C=O, and H_2CO as

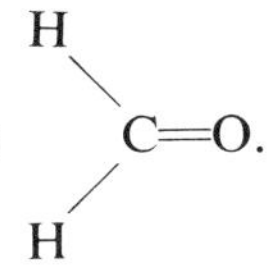

```
H
 \
  C=O
 /
H
```

Notice that only the

Table 4–6. Multiple Covalent Bonding

Molecule	*Atomic electron-dot formula*	*Sharing pattern*	*Molecular electron-dot formula*	*Comments*
Nitrogen gas, N_2	·N̈· ($1s^22s^22p^3$)	:N· ·N:	:N⋮⋮N:	Each N has five electrons and needs three more; therefore three pairs are shared.
Carbon dioxide, CO_2 (each O is bonded to the C)	·Ċ· ($1s^22s^22p^2$) :Ö· ($1s^22s^22p^4$)	:O· ·C· ·O:	O::C::O	Each O must share two pairs and each C four pairs. Therefore, each carbon-oxygen bond involves the sharing of two pairs.
Formaldehyde, H_2CO (each H and the O are bonded to the C)	·Ċ· ($1s^22s^22p^2$) :Ö· ($1s^22s^22p^4$) H· ($1s^1$)	H· C O H·	H C::O H	Both single bonding (carbon-hydrogen) and multiple bonding (carbon-oxygen) are involved.

bonding valence electrons are shown in this representation.

The concept of electron sharing implies that electrons are simultaneously under the influence of two nuclei. A physical model used to visualize this situation makes use of the idea of orbital overlap. Suppose two H atoms are moving toward each other to eventually form H_2. Each atom has a single electron in a spherical 1*s* orbital. As long as the atoms are well separated, the orbitals of the atoms are independent of each other. As the atoms get closer and closer together, the orbitals eventually overlap and create an orbital common to both atoms. When this happens, the electrons move throughout the overlap region between the nuclei and are shared by both nuclei. This process is represented in Figure 4–1. The resulting force (covalent bond) holding the atoms together is related to the fact that both nuclei are attracted toward the electrons and hence toward each other.

Figure 4–1.
Overlap of Orbitals During Electron Sharing

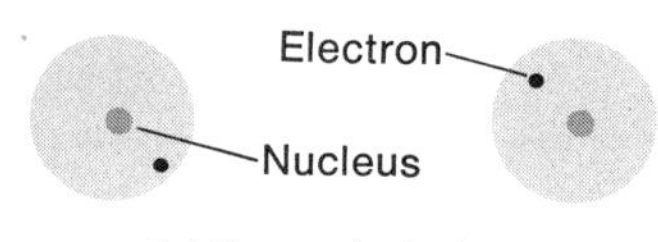

(a) Separated atoms

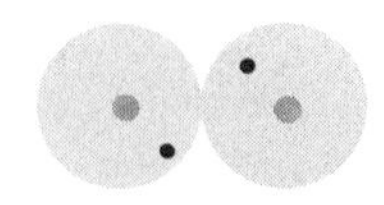

(b) Orbitals touch

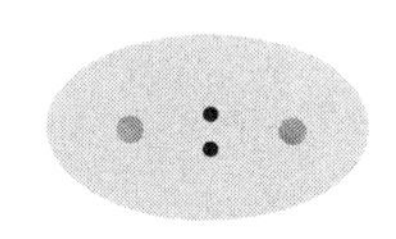

(c) Orbitals overlap--covalent bond is formed

Bond Polarity

The existence of two models for chemical bonding implies that we can explain the properties of every chemical bond by using only one of the models. However, the bonding between atoms in most substances can be classified neither as purely ionic (involving a complete electron transfer) nor as purely covalent (involving a perfectly equal sharing of electrons). Instead, most bonds are described in terms of partial ionic and partial covalent character. This complication can be described in terms of **bond polarity.**

When two identical atoms share a pair of electrons, each exerts the same attraction for the electrons, which are, therefore, equally shared. In this type of bond the electrons spend as much time, on an average basis, associated with one nucleus as they do with the other. The resulting purely covalent bond is called a **nonpolar** covalent bond.

The situation is quite different in covalent bonds between two different atoms. The nuclei of nonidentical atoms have different capacities for attracting the electron pair, and the electrons will spend the largest part of their time more closely associated with the atom showing the strongest attraction. Both atoms were initially uncharged, but the uneven sharing of electrons produces a slight negative charge (δ^-) of less than one unit charge on one bonded atom and an equivalent partial positive charge (δ^+) on the other. A covalent bond of this type, involving unequal electron sharing, is called a **polar** covalent bond.

The inequality in electron sharing increases with an increasing difference between the electron-attracting capacity of the bonded atoms. The largest sharing inequality occurs when the electrons spend 100 percent of their time associated with one atom and 0 percent with the other. This degree of unequal sharing results in the previously described ionic bond.

Electronegativity and Bond Polarity

A number of methods have been devised to estimate the relative electron-attracting abilities of atoms. The results of these methods are contained in the concept of atomic **electronegativity,** which may be defined as

a measurement of the attraction exerted on bonding valence electrons by the nucleus of a bonded atom.

Table 4–7 lists electronegativity values for some of the more common elements (arranged according to their positions in the periodic table).

Table 4–7. Electronegativities for Some Common Elements

		Increasing electronegativity →						
		H 2.1						
Li 1.0	Be 1.5		B 2.0	C 2.5	N 3.0	O 3.5	F 4.0	↑
Na 0.9	Mg 1.2		Al 1.5	Si 1.8	P 2.1	S 2.5	Cl 3.0	
K 0.8	Ca 1.0		Ga 1.6	Ge 1.8	As 2.0	Se 2.4	Br 2.8	*Increasing electronegativity*
Rb 0.8	Sr 1.0		In 1.7	Sn 1.8	Sb 1.9	Te 2.1	I 2.5	
Cs 0.7	Ba 0.9							

Several characteristics of the electronegativity scale and its relationship to the elements are worth noting.

The electronegativity scale is a relative one, calculated relative to a specific reference value. Fluorine, the most electronegative element, has arbitrarily been assigned a value of 4.0 (the maximum on the scale) and serves as the reference element.

The rare gases are not shown in the table. Since electronegativity is defined as the electron-attracting ability of a bonded atom, it would not be reasonable to assign electronegativity values to atoms that do not form compounds (He, Ne, Ar) or do so only rarely (Kr, Xe, Rn).

Electronegativity values increase from left to right across a period of the periodic table.

Electronegativity values increase from bottom to top for a group of the periodic table.

The two trends just mentioned result in high electronegativity values for elements in the upper right area of the periodic table (excluding rare gases) *and low values for elements in the lower left area.* Therefore, metals have somewhat low values.

Table 4–8. Electronegativity Differences and the Nature of Chemical Bonds

Electronegativity difference between combining atoms (ΔEN)	*Amount of ionic character in resulting bonds* (%)
0.2	1
0.4	4
0.6	9
0.8	15
1.2	30
1.6	47
2.0	63
2.4	76
2.8	86
3.2	92

Electronegativity values provide a useful tool for predicting the nature of chemical bonds. A quantitative relationship between electronegativity differences and the nature of bonds between atoms (as percent ionic character) is given in Table 4–8. As expected, a large electronegativity difference, ΔEN, results in a large disparity in electron sharing. Thus, bonds between atoms with small ΔEN values have a large amount of covalent character, and those between atoms showing large ΔEN values have a large amount of ionic character.

It is still convenient to use the words *ionic* and *covalent* to describe bonds, but the terms must be used in a consistent manner. The decision governing the distinction is arbitrary, but we shall use the following rules:

When ΔEN is less than 0.6 (about 10 percent ionic character), the bond will be called **covalent.**

When ΔEN is greater than 2.0 (about 60 percent ionic character), the bond will be called **ionic.**

When ΔEN falls between the two values given above, the bond will be called **polar covalent.**

We see then that the terms *ionic* or *covalent*, when used to describe a particular bond, represent only the predominant character of the bond.

Polarity of Molecules

We can now predict the polarity of bonds within molecules, but it is also important to be able to predict the polar nature of the molecules themselves.

The terms **polar** and **nonpolar,** when applied to molecules, indicate their electrical charge symmetry. When the centers of positive and negative charge in a molecule coincide, the molecule is nonpolar. When the charge centers do not coincide, a polar molecule results.

Both polar and nonpolar molecules carry no net electrical charge—they are neutral. However, in each type the possibility exists for one or more locations in the molecule to carry a slight positive charge while other locations carry slight negative charges because of individual covalent-bond polarizations. The locations of the centers of positive and negative charge within the molecule are determined by the geometrical distribution of these slight positive and negative charges. The positions of the charge centers in diatomic molecules depend completely on the type of bond involved, but this is not true for molecules containing more than two atoms. In the latter, both bond type and molecular geometries are involved. Table 4–9 gives some examples of molecular polarity along with brief explanations for each.

Table 4–9. Polarity of Molecules

Molecular formula	*Geometrical structure*	*Polarity (arrows show direction of electron polarization)*	*Comments*
H_2	H—H linear	H—H nonpolar	Both atoms are identical, therefore no bond polarization occurs and no molecular polarity results.
HCl	H—Cl linear	δ^+ δ^- H—Cl → polar	The two atoms are dissimilar and the bonding electron pair is polarized toward the more electronegative chlorine. The center of negative charge is therefore shifted toward Cl and no longer coincides with the center of positive charge. A polar molecule results.
CO_2	O=C=O linear	δ^- δ^+ δ^- O=C=O ← → nonpolar	The bonding electrons in both carbon-oxygen bonds are polarized toward the more electronegative oxygen. Each oxygen has a slight negative charge but, because of the molecular geometry, the center (average location) of negative charge is on the carbon atom and coincides with the center of positive charge. This symmetrical bond polarization, therefore, results in a nonpolar molecule.
N_2O	N≡N—O linear	δ^+ δ^- N≡N—O → polar	This molecule is linear but still polar. The bonding electrons in the nitrogen-nitrogen bond are not polarized, but those in the nitrogen-oxygen bond are polarized toward the more electronegative oxygen. The resulting charge centers do not coincide, and a polar molecule results.
H_2O	O / \\ H H angular	δ^- O ↗ ↖ H H δ^+ δ^+ polar	The bonding electrons of both hydrogen-oxygen bonds are polarized toward the oxygen. The polarizations, though equal, are unsymmetrical (not in exactly opposite directions) and result in different locations for the positive and negative charge centers. A polar molecule is the result.

Review Questions

1. What is a chemical bond?
2. Define the term *valence electron* and indicate how many valence electrons each of the following atoms contain.
   a) ${}_{5}B$
   b) ${}_{22}Ti$
   c) ${}_{14}Si$
   d) ${}_{35}Br$
   e) ${}_{30}Zn$
3. Define the term *ion* and indicate the type of ion each of the following atoms will form:
   a) ${}_{7}N$
   b) ${}_{13}Al$
   c) ${}_{19}K$
   d) ${}_{35}Br$
   e) ${}_{12}Mg$
4. What is the difference between the following: Mg and Mg^{2+}?
5. What is the *octet rule?*
6. What two bonding models are based on the octet rule and how do they differ from each other?
7. In each of the following sets indicate those ions which are isoelectronic with each other:
   a) Al^{3+}, F^-, Cl^-, Li^+
   b) I^-, Ba^{2+}, Se^{2-}, Cs^+
   c) S^{2-}, P^{3-}, Rb^+, Sr^{2+}
8. Determine the total number of electrons in each of the following ions:
   a) S^{2-}
   b) K^+
   c) Cl^-
   d) Al^{3+}
   e) N^{3-}
9. Write the formula of the compound that would result upon combination of the following pairs of ions:
   a) A^+ and B^-
   b) A^{2+} and B^-
   c) A^+ and B^{2-}
   d) A^{2+} and B^{2-}
   e) A^+ and B^{3-}
   f) A^{3+} and B^-
   g) A^{2+} and B^{3-}
   h) A^{3+} and B^{2-}
   i) A^{3+} and B^{3-}

10. Write the formula of the compound that would result upon ionic combination of the following pairs. Use A B type formulas as in the previous problem.
    a) a group IA element and a group VIA element
    b) a group IIA element and a group VIIA element
    c) a group IIIA element and a group VA element
    d) a group IA element and a group VA element
11. Write formulas for the ionic compounds formed between the following pairs of elements:
    a) Rb and S
    b) Ba and O
    c) Al and F
    d) Mg and N
12. Draw electron dot formulas to illustrate the electrons involved in bonding for the following compounds or elements:
    a) Cl_2
    b) N_2H_2—nitrogens are bonded to each other
    c) CS_2
    d) SCl_2
    e) CCl_4
    f) PBr_3
    g) N_2
13. Write formulas for the covalent compound that would form from the following pairs of nonmetals:
    a) hydrogen and sulfur
    b) phosphorus and chlorine
    c) fluorine and oxygen
    d) hydrogen and fluorine
14. What is the difference between the following bonding notations:
    A—X A═X A≡X
15. Give three examples of molecules which exhibit multiple bonding, and draw their electron dot structures.
16. What is the difference between a polar and a nonpolar bond?
17. Arrange each of the following sets of atoms in order of decreasing electronegativity:
    a) F, I, Rb, Sr
    b) B, Ga, K, Co
    c) F, Cl, S, Se
    d) P, As, N, Bi
18. Classify the bonds in each of the following compounds as covalent, polar covalent, or ionic:
    a) CO
    b) SnO_2
    c) P_2O_5
    d) NI_3
    e) NO_2

19. For each of the following hypothetical molecules indicate whether the bonds are polar or nonpolar, and whether the molecule is polar or nonpolar. Assume A, X, and Y have different electronegativities.
    a) X—A—X
    b) A—X—X
    c) X with bonds to X and X (bent: X at apex, X at lower left, X at lower right)
    d) A with bonds to X and X (bent: A at apex, X at lower left, X at lower right)
    e) Y—A—X

Suggestions for Further Reading

Holliday, L., "Early Views on Forces Between Atoms," *Scientific American*, May 1970.

5 Classification, Nomenclature, and Reactions of Inorganic Compounds

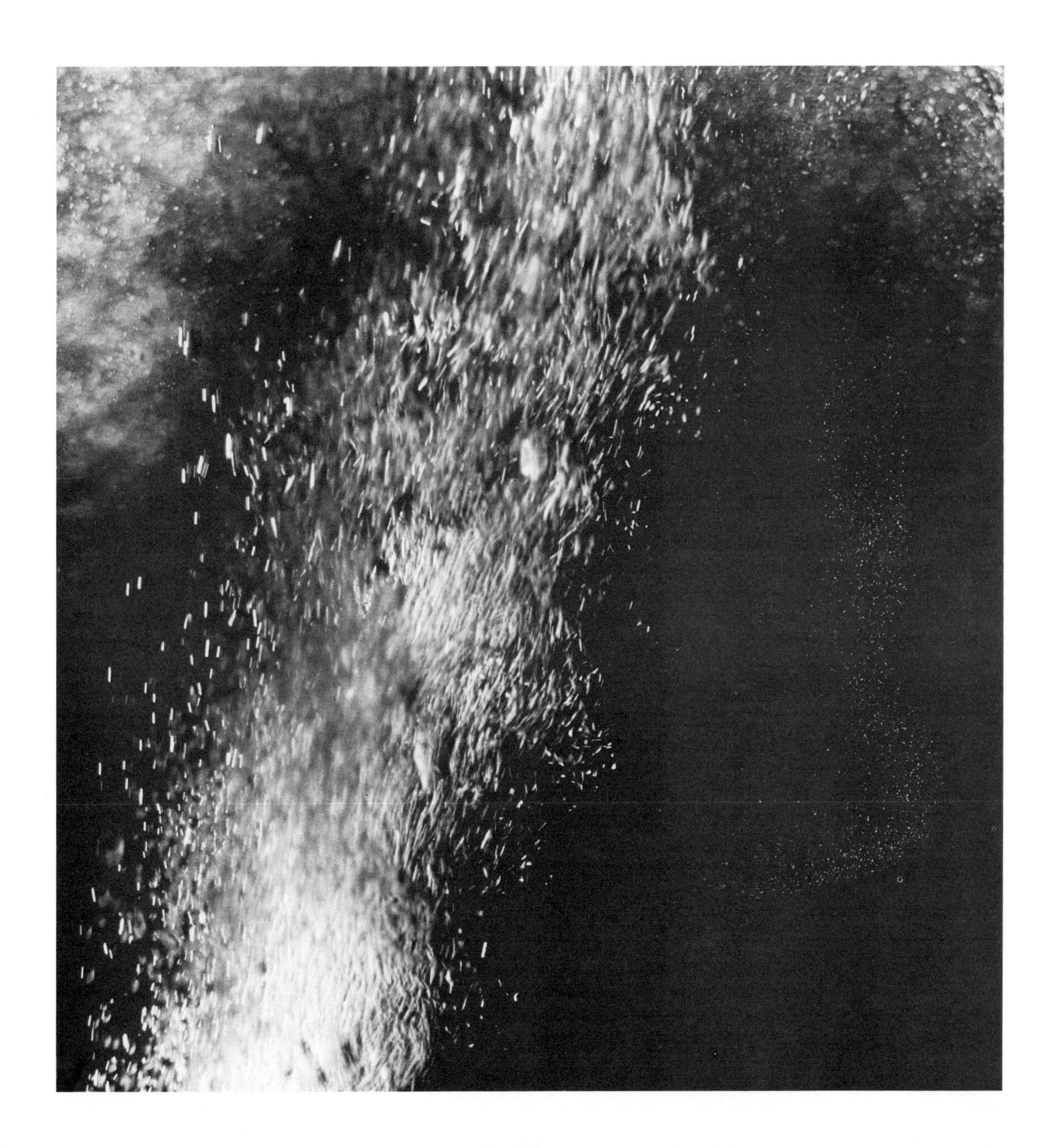

Chemical compounds are generally classified into two broad categories, organic and inorganic. **Organic compounds** are carbon-containing compounds in which carbon atoms are bonded to other carbon atoms. At one time it was thought that only living organisms could produce organic compounds (see Chapter 17). Today this is known to be incorrect, and many organic compounds are produced from nonliving sources.

Inorganic compounds are those made up of the other 104 known elements, plus a few carbon compounds in which no bonds are formed between carbon atoms.

These classifications might appear out of balance, with compounds of a single element comprising an entire classification, and compounds of the remaining 104 elements making up the other. They are justified, however, by the numbers involved. The exact number of organic compounds has not been determined but is estimated to be at least ten times the number of inorganic compounds. Organic compounds are the topics of Chapters 17 through 20, so we shall limit the present discussion to inorganic compounds.

Types of Inorganic Compounds

Every known compound has a unique set of properties by which it can be identified. This does not eliminate the possibility that different compounds may have some properties in common. Many compounds, in fact, are grouped into classes according to their common properties. Three very important classes of this type are *acids*, *bases*, and *salts*. These classes are chemically important because they are involved in a wide variety of reactions.

Acids

Acids may be defined in numerous general and specific ways. We will use a somewhat limited definition that is valid only in water-containing systems. This definition will satisfy our needs, however, because we will deal only with water systems.

We will define an **acid** as any chemical species that reacts with water to produce hydronium ions, H_3O^+. This definition is illustrated by the following reactions:

$$HNO_3 + H_2O \rightarrow H_3O^+ + NO_3^- \qquad 5\text{–}1$$

$$HCl + H_2O \rightarrow H_3O^+ + Cl^- \qquad 5\text{–}2$$

$$HCN + H_2O \rightarrow H_3O^+ + CN^- \qquad 5\text{–}3$$

Notice that the hydronium ion, H_3O^+, is produced in all three reactions. This ion is responsible for the characteristic properties of acidic solutions.

One of these properties is a sour taste, and the word *acid* is derived from the Latin *acidus*, which means *sour*. The hydronium ion is thought to be a hydrogen ion, H^+, to which a water molecule has become attached:

$$H^+ + H_2O \rightarrow H_3O^+ \qquad 5\text{–}4$$

For brevity, the water is sometimes neglected and H_3O^+ is written simply as H^+, even though it does not exist in this form in water solutions.

The acidic strength of a solution increases with increasing H_3O^+ concentration. However, solutions containing identical concentrations of different acids do not always have equal acidities—equal H_3O^+ concentrations. This indicates that all acids do not react with water to the same extent. Those that react to an extent approaching 100 percent—nearly all available acid molecules react—are called **strong acids,** while those that react to an extent less than 100 percent are called **weak acids.** This is illustrated in Table 5–1 for some common acids.

Table 5–1. The Extent of Acid Reactions with Water

Acid name	*Formula*	*Approximate percent of acid molecules reacting with water*	*Strength of acid*
Perchloric acid	$HClO_4$	100	strong
Hydrochloric acid	HCl	100	strong
Hydrobromic acid	HBr	100	strong
Nitric acid	HNO_3	100	strong
Phosphoric acid	H_3PO_4	27	moderately weak
Acetic acid	CH_3COOH	1.3	weak
Carbonic acid	H_2CO_3	0.2	weak
Boric acid	H_3BO_3	0.01	weak

Since, by definition, acids generate H_3O^+ ions during a reaction with water, an acid must contain hydrogen atoms that are loosely bound, and which may be removed from the rest of the acid. However, most of the substances we will call acids consist of predominantly covalently bonded molecules in which hydrogen does not exist as H^+; the H^+ (H_3O^+) forms only during the reaction with water.

Bases

Bases are chemical species that react with water to produce hydroxide ions, OH^-. Unlike acids, many bases are ionic compounds which contain the OH^- ion. The reaction involved when ionic bases form OH^- ions in water solution is simply the dissociation or breaking apart of the compound into its constituent ions, as illustrated in the following reactions:

$$NaOH + H_2O \rightarrow Na^+ + OH^- + H_2O \quad 5\text{–}5$$

$$Mg(OH)_2 + H_2O \rightarrow Mg^{2+} + 2OH^- + H_2O \quad 5\text{–}6$$

One common covalent substance that reacts with water to produce OH^- ions is ammonia, NH_3. The reaction is

$$NH_3 + H_2O \rightarrow NH_4^+ + OH^- \quad 5\text{–}7$$

Soluble ionic compounds dissociate approximately 100 percent into ions upon dissolving in water. Therefore, soluble ionic hydroxides are classified as strong bases. The ion-forming reaction between NH_3 and water occurs to only a slight extent; accordingly, ammonia is classified as a weak base.

Salts

Salts differ from the previous two classes of compounds because they contain no common ion or species that characterizes the group. **Salts** are compounds that contain chemical species held together by ionic bonding. For our purposes the ionic bases already discussed are excluded from this classification.

The ions of salts may be monoatomic or polyatomic. The concept of monoatomic ions was introduced in Chapter 4. **Polyatomic ions** are covalently bonded groups of atoms possessing a net charge. In many chemical reactions they behave in a manner similar to monoatomic ions and retain their identity throughout the reaction. The charges on polyatomic ions may be positive or negative, but the negative is most common. In fact, the ammonium ion, NH_4^+, is the only common polyatomic ion known that has a positive charge. Table 5–2 contains the names and formulas of the more common polyatomic ions.

Table 5–2. Common Polyatomic Ions

Very common		*Common*	
NH_4^+	ammonium	SO_3^{2-}	sulfite
PO_4^{3-}	phosphate	CrO_4^{2-}	chromate
SO_4^{2-}	sulfate	$Cr_2O_7^{2-}$	dichromate
CO_3^{2-}	carbonate	HPO_4^{2-}	monohydrogen phosphate
NO_3^-	nitrate	$H_2PO_4^-$	dihydrogen phosphate
ClO_3^-	chlorate	MnO_4^-	permanganate
OH^-	hydroxide	HCO_3^-	hydrogen carbonate (bicarbonate)
$C_2H_3O_2^-$	acetate	HSO_4^-	hydrogen sulfate (bisulfate)
CN^-	cyanide	NO_2^-	nitrite

Formulas for salts containing polyatomic ions are determined in the same way as those for salts containing only monoatomic ions (see Table 4–4). The basic rule is that the total positive and negative charge must add to zero. In addition, two new conventions are used when necessary: (1) parentheses are used to avoid ambiguity; (2) elemental symbols may appear in more than one place in a formula to preserve the identity of the contained ions. These conventions are illustrated in Examples 5–1 and 5–2.

Example 5–1. Write the formula for the compound that results when the Ba^{2+} and NO_3^- ions combine.

In order to equalize the total positive and negative charge, two nitrate ions (−1 charge) will be needed for each barium ion (+2 charge). Therefore, the formula is $Ba(NO_3)_2$. If parentheses were not used, the formula could appear to be $BaNO_{32}$, which is not intended and which conveys false information. The formula $Ba(NO_3)_2$ indicates that the molecule contains one Ba atom, two N atoms, and six O atoms. However, the formula $BaNO_{32}$ implies that the molecule contains one Ba atom, one N atom, and 32 O atoms.

Example 5–2. What is the formula for the compound that is formed by the combination of NH_4^+ and CN^- ions?

The ions have equal and opposite charges, and will combine in a 1:1 ratio. The formula is written NH_4CN. Notice that the symbol N appears twice. This could be prevented by combining the two N's to give the formula N_2H_4C. This is not done because the resulting formula no longer conveys the existence of both the NH_4^+ and CN^- ions.

The behavior of salts soluble in water is similar to that of strong acids and bases—all molecules in solution dissociate into ions. The terms *strong* and *weak*, as used with acids and bases, are not used with salts because all salts are strong if they dissolve. The solubility of a salt is, therefore, an important factor in the production of ions in solutions. The solubility of salts in water varies, but some rules of thumb are useful for qualitative predictions:

All sodium, potassium, and ammonium salts are water soluble.
All nitrates and acetates are water soluble.
All chlorides, except those of lead, silver, and mercury(I), are water soluble.
Salts not in the three categories above are insoluble or, at best, only slightly soluble.

The meaning of the term **soluble** is somewhat arbitrary. A substance that dissolves to the extent of about 3 g/100 ml of water, at room temperature, is generally considered to be water soluble. The properties affecting the solubility of substances, including salts, are discussed in Chapter 11.

Oxidation Numbers

The assignment of names to existing inorganic compounds can prove difficult unless done systematically. The systematic methods available are based upon the concepts of electronegativity and oxidation numbers. Electronegativity was discussed in Chapter 4, and we shall now discuss oxidation numbers. This concept, in addition to its use in nomenclature, is also useful for classifying reactions into various categories—a later topic of this chapter.

Oxidation numbers are positive or negative numbers that are arbitrarily assigned to the elements contained in molecular formulas. These numbers indicate whether the element has lost or gained electron density, relative to the elemental state, during the formation of the compound. Positive oxidation numbers correspond to a loss in electron density; negative numbers indicate a gain. The assignment of oxidation numbers to individual atoms is based on two arbitrary rules for assignment of shared electrons:

Electrons shared by two dissimilar atoms are arbitrarily assigned to the more electronegative element. For example, the shared pair in HCl is assigned to the Cl, giving it one more than elemental Cl and an oxidation number of -1. At the same time, the H of the HCl molecule has one less electron than elemental H atoms and an oxidation number of $+1$.

Electrons shared by two identical atoms are assigned in equal numbers to each atom. An example is the assignment of the shared pair in an H_2 molecule. One electron is assigned to each H atom, giving each one the same number as found in the elemental state. The oxidation number of each atom is therefore 0.

The arbitrary assignment of electrons according to these rules results in the formation of apparent charges on the atoms, since electrons are either added to or removed from an atom. The apparent charge is the oxidation number (O.N.). The rules for oxidation-number assignment are arbitrary, and the resulting apparent charges do not necessarily exist; nonetheless, they are still very useful.

The results of applying these ideas to elements and compounds are contained in the following rules, which can be used to assign oxidation numbers. Rule 1 applies to uncombined elements; rules 2 through 8 apply to elements in compounds.

1. The O.N. of any element in the free uncombined state is 0.
2. The O.N. of any monoatomic ion is equal to the charge on the ion. Na^+ has an O.N. of $+1$, and O^{2-} has an O.N. of -2.
3. The O.N.'s of group IA and IIA elements are $+1$ and $+2$, respectively.
4. The O.N. of fluorine is always -1. Other members of group VIIA (Cl, Br, and I) commonly exhibit an O.N. of -1, but many also have positive values when they are bonded to more electronegative elements.
5. The O.N. of oxygen is generally -2. Exceptions occur when oxygen is bonded to fluorine (O then has a positive O.N.), and when oxygen is found in compounds containing oxygen-oxygen bonds (peroxides). In peroxides the O.N. for oxygen is -1. Peroxides form only between oxygen and hydrogen (H_2O_2), group IA elements (Na_2O_2, etc.), and group IIA elements (BaO_2, etc.).

6. The O.N. of hydrogen is always $+1$ except in hydrides, where H is bonded to a metal with lower electronegativity. In these cases, H is assigned an O.N. of -1. Examples are NaH, CaH_2, and LiH.
7. The algebraic sum of the O.N.'s of all atoms in a neutral molecule is equal to 0.
8. The algebraic sum of the O.N.'s of all atoms in a polyatomic ion is equal to the charge on the ion.

The use of these rules is illustrated in Examples 5–3, 5–4, and 5–5.

Example 5–3. Calculate the O.N. of Pb in the compound PbO_2.

Oxygen has an O.N. of -2 (rule 5), because PbO_2 is not a peroxide even though two oxygen atoms are present. The O.N. of Pb can be calculated by using rule 7.

$$\begin{aligned}(\text{O.N. Pb}) + 2(\text{O.N. O}) &= 0\\(\text{O.N. Pb}) + 2(-2) &= 0\\(\text{O.N. Pb}) - 4 &= 0\end{aligned}$$

Therefore, O.N. Pb $= +4$.

Example 5–4. Calculate the O.N. of Mn in $KMnO_4$.

K has an O.N. of $+1$ (rule 3). O has an O.N. of -2 (rule 5). The O.N. of Mn can be calculated by using rule 7.

$$\begin{aligned}(\text{O.N. Mn}) + (\text{O.N. K}) + 4(\text{O.N. O}) &= 0\\(\text{O.N. Mn}) + 1 - 8 &= 0\\(\text{O.N. Mn}) &= 8 - 1\\\text{O.N. Mn} &= +7\end{aligned}$$

Example 5–5. What is the O.N. of Cl in the ClO^- ion?

According to rule 8, the sum of the O.N.'s must equal the -1 charge on the ion. Oxygen has an O.N. of -2 (rule 5), and Cl must have an O.N. of $+1$, as calculated below using rule 8.

$$\begin{aligned}(\text{O.N. Cl}) + (\text{O.N. O}) &= -1\\(\text{O.N. Cl}) + (-2) &= -1\\(\text{O.N. Cl}) - 2 &= -1\\\text{O.N. Cl} &= -1 + 2 = +1\end{aligned}$$

Many of the common elements exhibit more than one oxidation number. The specific value shown by an element in a compound can usually be determined by using rule 7 or 8. Table 5–3 contains some commonly occurring oxidation numbers for selected representative and transition elements.

Table 5–3. Common Oxidation Numbers of Selected Elements

Group number	*Elements*	*Common oxidation numbers*
IIIA	B, Al	+3
IVA	C	+4, +2
	Si	+4
VA	N, P, As, Sb	+5, +3, −3
VIA	O	−2
	S	+6, +4, −2
VIIA	F	−1
	Cl, Br, I	−1, +1, +3, +5, +7
Transition elements	Ag	+1
	Cu, Hg	+2, +1
	Zn, Cd	+2
	Ni, Co, Fe	+2, +3
	Cr	+3, +6
	Mn	+2, +7

Nomenclature of Binary Compounds

Binary compounds contain two different elements. Any number of atoms of the two elements may be present in a molecule, so, although diatomic binary compounds are known, all binary compounds are not diatomic.

The formulas for binary compounds are written with the symbol of the least electronegative element first, followed by that of the more electronegative element. Sodium chloride, for example, has a written formula of NaCl instead of ClNa, because Na is less electronegative than Cl.

Nearly all binary compounds fall into one of the following three classes:

Class 1. Those involving a nonmetal, and a metal which exhibits only one oxidation number

Class 2. Those involving a nonmetal, and a metal which can have more than one oxidation number

Class 3. Those involving two nonmetals

Compounds in each of these classes are named in a slightly different way, but the basic rule is the same. Binary compounds of class 1 are named by stating the name of the least electronegative element (metal) followed by the stem of the name of the more electronegative element (nonmetal) with the suffix *-ide*. The general form of the name is:

(name of metal)(stem of nonmetal)(ide)

Metals showing only one oxidation number are those of groups IA and IIA, plus Al, Ga, Zn, Cd, and Ag. Some nomenclature examples of typical class 1 compounds are:

KBr potassium bromide
Al_2O_3 aluminum oxide
Be_3N_2 beryllium nitride
Na_2S sodium sulfide

Table 5–4 contains the name stems for the common nonmetals.

Table 5–4. Name Stems for Some Common Nonmetals

Element name and symbol	*Stem*	*Element name and symbol*	*Stem*
Boron (B)	bor-	Iodine (I)	iod-
Bromine (Br)	brom-	Nitrogen (N)	nitr-
Carbon (C)	carb-	Oxygen (O)	ox-
Chlorine (Cl)	chlor-	Phosphorus (P)	phosph-
Fluorine (F)	fluor-	Silicon (Si)	silic-
Hydrogen (H)	hydr-	Sulfur (S)	sulf- or sulfur-

Class 2 binary compounds are named by using a slight modification of the rule given for class 1. The metals involved in compounds of this class exhibit more than one oxidation number. This oxidation number is indicated in the name by parenthesized Roman numerals. Iron and chlorine, for example, form two different binary compounds, $FeCl_2$ and $FeCl_3$. Obviously, the name *iron chloride* would not distinguish between them. However, the use of Roman numerals makes it possible to give each one a unique name:

$FeCl_2$ iron(II) chloride
$FeCl_3$ iron(III) chloride

The (II) and (III) in these examples are the oxidation numbers of the iron and not the number of chlorine atoms. Other examples are:

ReO_2 rhenium(IV) oxide
Re_2O_5 rhenium(V) oxide
Re_2O_7 rhenium(VII) oxide

Most transition metals and some representative elements display variable oxidation numbers in their compounds.

The class 3 binary compounds, involving two nonmetals, present a similar problem of variable oxidation number. The same solution (Roman numerals) would work, but traditionally these compounds are named another way. Greek prefixes are used to indicate the number of atoms present in a molecule. Table 5–5 gives the Greek prefixes for numbers 1 through 10.

The names of some typical class 3 compounds illustrate the method:

CO carbon monoxide
CO_2 carbon dioxide
NO_2 nitrogen dioxide

Table 5–5. Greek Numerical Prefixes

Number	*Prefix*
1	mono-
2	di-
3	tri-
4	tetra-
5	penta-
6	hexa-
7	hepta-
8	octa-
9	nona-
10	deca-

N_2O_4	dinitrogen tetroxide
N_2O_5	dinitrogen pentoxide
SO_2	sulfur dioxide
SO_3	sulfur trioxide

The prefix *mono* is usually dropped when it appears at the beginning of a name. The final letter *a* that occurs on most of the Greek prefixes is dropped for phonetic reasons when it precedes a vowel—pentoxide is used rather than pentaoxide.

Nomenclature of Ternary Compounds

Ternary compounds contain three different elements; and a substantial number, including most common ones, fall into one of two classes:

Class 1. Those that contain a metal and a polyatomic ion
Class 2. Those that contain hydrogen and a polyatomic ion

Compounds containing a metal and polyatomic ion are named in a manner similar to that already discussed for binary compounds of a metal and nonmetal. The difference is that the name of the polyatomic ion is substituted for the nonmetal-stem-plus *-ide* ending. The general form of the name is:

(name of metal)(name of polyatomic ion)

The names of the common polyatomic ions were given in Table 5–2.

If the metal involved exhibits more than one oxidation number, Roman numerals are used, just as they were in naming binary compounds.

Examples of the use of these rules are:

Na_2SO_4	sodium sulfate
$Ca_3(PO_4)_2$	calcium phosphate
$KMnO_4$	potassium permanganate
$Fe(NO_3)_2$	iron(II) nitrate
$Fe(NO_3)_3$	iron(III) nitrate

Compounds containing hydrogen and a polyatomic ion are not named in the same way, even though it would be simple to do so. For example, HNO_3 could be called hydrogen nitrate; H_2SO_4, hydrogen sulfate. Instead, these compounds are named as acids, because they produce acidic solutions when dissolved in water.

Nomenclature of Acids

The majority of the common inorganic acids either (1) contain hydrogen and a nonmetal—**binary acids,** or (2) contain hydrogen, oxygen, and a nonmetal—**oxyacids.** Many nonmetallic elements can exhibit more than one oxidation number; as a result, series of acids involving particular nonmetals are known. For example, chlorine forms the following acids:

HCl }	binary
$HClO$ $HClO_2$ $HClO_3$ $HClO_4$ }	oxyacids

Similar series of acids are known for other nonmetals, but most are not as extensive as the chlorine series.

The names assigned to acids are related to the position of the acid in a series.

One acid in the series (usually the most common one) is given a name according to the format:

(stem of nonmetal)(suffix *-ic*)(acid)

In the chlorine series, $HClO_3$ is the most common acid and is called *chloric acid* (chlor)(ic)(acid).

The acid in the series containing one less oxygen than the *-ic* acid is named by using the suffix *-ous* instead of *-ic*. In the chlorine series, $HClO_2$ is named *chlorous acid.*

The acid in the series with two less oxygens than the *-ic* acid is named by using the prefix *hypo-* and the suffix *-ous*. On this basis, HClO is named *hypochlorous acid.*

The acid in a series with one more oxygen than the *-ic* acid is named by using the prefix *per-* and the suffix *-ic*. Therefore, $HClO_4$ is called *perchloric acid.*

The acid in a series with no oxygen is denoted by the prefix *hydro-* and the suffix *-ic*. Thus, HCl is named *hydrochloric acid.*

Table 5–6 summarizes these rules.

Table 5–6. Nomenclature Rules for Acids

Type of acid	*Name of acid*
One more oxygen than common acid	per———ic acid
Common oxyacid	———ic acid
One less oxygen than common acid	———ous acid
Two less oxygens than common acid	hypo———ous acid
No oxygen	hydro———ic acid

In order to name an acid it is necessary to know which one in the series is the *-ic* acid. Most *-ic* acids contain either three or four oxygen atoms per molecule, as shown in Table 5–7. On the basis of this information, HNO_2 is named nitrous acid (one less O than HNO_3) and HBrO is named hypobromous acid (two less O's than $HBrO_3$).

Table 5–7. Common -*ic* Acids

Containing three oxygens		*Containing four oxygens*	
$HClO_3$	chloric acid	H_2SO_4	sulfuric acid
$HBrO_3$	bromic acid	H_3PO_4	phosphoric acid
HIO_3	iodic acid	H_3AsO_4	arsenic acid
HNO_3	nitric acid		
H_2CO_3	carbonic acid		

Trivial Names

Names assigned to compounds on the basis of a set of rules are called **systematic names.** They are used not only to identify compounds but also to allow us to write formulas without obtaining additional information.

Prior to the development of systematic rules, compounds were identified by common or trivial names, which were used for identification only and conveyed little or no information about structure. The origins of trivial names were somewhat similar to those of the elements (see Chapter 2) and were often closely related to historical circumstances.

With the advent of systematic nomenclature, most common names were discontinued. A few, however, have persisted and have been officially accepted. A prime example is H_2O, which has a systematic name of dihydrogen monoxide. This name is never used; the compound is known in English as water. It would probably prove quite difficult to get the general population to ask for drinks of dihydrogen monoxide.

Table 5–8 contains officially accepted common names for some important compounds. Compounds containing hydrogen are often among those with accepted common names. The rule of writing the least electronegative element first in the formula is often violated in these compounds, and the "incorrect" formulas have also been accepted and are widely used. For example, the formula for ammonia is written NH_3 rather than H_3N.

Table 5–8. Some Compounds with Officially Accepted Common Names

Compound formula	*Accepted common name*
CH_4	Methane
SiH_4	Silane
SnH_4	Stannane
C_2H_6	Ethane
C_3H_8	Propane
C_6H_6	Benzene
NH_3	Ammonia
PH_3	Phosphine
AsH_3	Arsine
H_2O	Water
H_2S	Hydrogen sulfide
H_2Se	Hydrogen selenide

Chemical Equations

Chemists represent chemical reactions by a shorthand known as chemical equations. These equations describe chemical reactions in terms of formulas instead of words. We have already used a few equations in this text without giving any formal introduction. Because of their importance we shall now discuss the concepts used in the writing of useful chemical equations.

The following format is generally used:

The correct formulas of the **reactants**—substances reacting together—are written on the left.

The correct formulas of the **products**—substances produced in the reaction—are written on the right.
The reactants and products are separated by an arrow pointing toward the products.
Plus (+) signs are used to separate individual reactants and products from each other.

A typical equation written in general form according to this format is:

$$A + B \rightarrow C + D + E \qquad 5\text{–}8$$

We may read this equation as follows: The reactants A and B react to form or yield the products C, D, and E.

To be of maximum use, a chemical equation must be "balanced." A balanced equation obeys the law of conservation of matter. According to this law (Chapter 2), atoms involved in chemical reactions are neither created nor destroyed. This means the same number of atoms of each element must appear on both sides of a balanced equation. Coefficients written to the left of the formulas are used to adjust the number of molecules of each substance to achieve the desired balance of atoms. Examples 5–6 and 5–7 illustrate these ideas.

Example 5–6. Balance the following equation if it is not balanced in the form given:

$$BaCl_2 + Ag_2SO_4 \rightarrow BaSO_4 + AgCl \qquad 5\text{–}9$$

Five different types of atoms are involved—Ba, Cl, Ag, S, and O. A count of the number of each type on each side of the equation gives the following results:

Type of atom	*Number on the left (reactants)*	*Number on the right (products)*
Ba	1	1
Cl	2	1
Ag	2	1
S	1	1
O	4	4

These figures reveal that the equation is not balanced, since the number of Ag and Cl atoms on each side are different—the product side has one less of each than the reactant side. This situation is remedied by writing the coefficient 2 to the left of AgCl on the product side:

$$BaCl_2 + Ag_2SO_4 \rightarrow BaSO_4 + 2AgCl \qquad 5\text{–}10$$

There might be a temptation to achieve the same result by changing the formula AgCl to Ag_2Cl_2. This is not allowed, because both formulas cannot correctly represent the compound AgCl. As a general rule no subscripts in molecular formulas may be changed to balance equations; only coefficients to the left of formulas may be changed.

Example 5–7. Balance the following equation if it is not balanced as given:

$$H_2 + O_2 \rightarrow H_2O \qquad 5\text{–}11$$

The equation is balanced by writing the coefficient 2 before both H_2 and H_2O:

$$2H_2 + O_2 \rightarrow 2H_2O \qquad 5\text{–}12$$

In this form there are four H's and two O's on each side. Other coefficients could also be used to balance the equation. Both of the following forms are balanced:

$$4H_2 + 2O_2 \rightarrow 4H_2O \qquad 5\text{–}13$$

$$H_2 + \tfrac{1}{2}O_2 \rightarrow H_2O \qquad 5\text{–}14$$

In the first of these, the coefficients are double the values necessary, and the lowest possible set of integers is always used to balance equations. The second equation is incorrect because it contains a fractional coefficient; only whole-number coefficients should be used. We can change the equation with fractional coefficients to the correct form by doubling each coefficient.

In the balanced equation, 5–12, hydrogen is written as H_2 instead of H, and oxygen is written as O_2 instead of O. The formulas for elements that are gases at room temperature are written to represent the molecular form in which they actually occur in nature. Both hydrogen and oxygen are composed of diatomic molecules. Many, but not all, elemental gases consist of polyatomic molecules and are always written as such in equations. Table 5–9 lists the correct formulas to use to represent elemental gases in equations. The four elements listed as vapors are not gases at room temperature but vaporize at slightly higher temperatures. The resultant vapors contain molecules with the formulas indicated.

Table 5–9. Molecular Formulas of Gaseous Elements

Element	*Molecular formula*	*Element*	*Molecular formula*
Helium	He	Fluorine	F_2
Neon	Ne	Chlorine	Cl_2
Argon	Ar	Bromine (vapor)	Br_2
Krypton	Kr	Iodine (vapor)	I_2
Xenon	Xe	Oxygen	O_2
Phosphorus (vapor)	P_4	Nitrogen	N_2
Arsenic (vapor)	As_4	Hydrogen	H_2

Most simple equations can be balanced by the inspection method—trial and error—illustrated in the previous examples. Formal mathematical methods are available for balancing more complex equations, but these methods will not be discussed in this text.

Types of Chemical Reactions

Chemical reactions can be grouped into two broad classifications, depending on whether a change in oxidation number occurs for any element involved. Reactions in which oxidation numbers change are called **oxidation-reduction** or **redox** reactions. Those in which no oxidation numbers change are called **metathesis** reactions.

Oxidation-Reduction Reactions

Most reactions are included in the redox category. The changes in oxidation number that occur represent an apparent gain or loss of electrons—remember the definition of oxidation number. A number of terms are used to describe the processes that take place during redox reactions:

The apparent removal of electrons from an atom causes an increase in oxidation number in a process called **oxidation.**

Oxidized atoms are those that have undergone an apparent loss of electrons.

The apparent addition of electrons to an atom causes a decrease in oxidation number in a process called **reduction.**

Reduced atoms are those that have undergone an apparent addition of electrons.

The atom reduced by an apparent gain of electrons is called the **oxidizing agent.**

The atom oxidized by the apparent loss of electrons is called the **reducing agent.**

Table 5–10 summarizes these terms; their use is illustrated in Example 5–8.

Table 5–10. Properties of Oxidizing and Reducing Agents

Oxidizing agent	*Reducing agent*
Gains electrons	Loses electrons
Oxidation number decreases	Oxidation number increases
Becomes reduced	Becomes oxidized

Example 5–8. Discuss the following reaction from the point of view of oxidation and reduction.

$$4Al + 3O_2 \rightarrow 2Al_2O_3 \qquad 5\text{–}15$$

The calculated O.N.'s for the atoms involved in the reaction are:

$$\begin{array}{lccc} & 4Al + & 3O_2 \rightarrow & 2Al_2O_3 \\ \text{O.N.:} & 0 & 0 & +3\ -2 \end{array}$$

The O.N. of Al has increased from 0 to +3, and that of O has decreased from 0 to −2. Therefore, Al has been oxidized and is the reducing agent. Similarly, O has been reduced and is the oxidizing agent.

The fact that electrons are assumed to be exchanged during redox reactions correctly implies that oxidation and reduction reactions must take place in pairs. One cannot take place without the other, or there would be no way for the assumed electron exchange to occur.

Many redox reactions may be classified as either addition, decomposition, or displacement reactions. **Addition reactions** include all those in which substances simply combine to produce larger molecules. The product is always a single substance, and the combining substances may be elements or compounds. A general reaction of this type is:

$$A + B \rightarrow C \qquad 5\text{–}16$$

Specific examples:

$$Zn + S \rightarrow ZnS \qquad 5\text{–}17$$

$$2CO + O_2 \rightarrow 2CO_2 \qquad 5\text{–}18$$

In **decomposition reactions** two or more simple substances are produced by the breaking apart of a more complex substance. These reactions are essentially the reverse of combination reactions. The general form of a decomposition reaction is:

$$A \rightarrow B + C \qquad 5\text{–}19$$

Specific examples:

$$2HI \rightarrow H_2 + I_2 \qquad 5\text{–}20$$

$$2KClO_3 \rightarrow 2KCl + 3O_2 \qquad 5\text{–}21$$

Displacement reactions take place when one element displaces another from its compound. A general representation of a displacement reaction is:

$$A + BX \rightarrow B + AX \qquad 5\text{–}22$$

Specific examples:

$$Zn + H_2SO_4 \rightarrow H_2 + ZnSO_4 \qquad 5\text{–}23$$

$$Cl_2 + Na_2S \rightarrow S + 2NaCl \qquad 5\text{–}24$$

Metathesis Reactions

Metathesis reactions involve no change in oxidation number, and they commonly take place between substances dissolved in water. Many metathesis reactions follow the general form given in equation 5–25 and can be thought of as simply partner swapping:

$$AX + BY \rightarrow BX + AY \qquad 5\text{–}25$$

Specific examples are:

$$HCl + AgNO_3 \rightarrow HNO_3 + AgCl \qquad 5\text{–}26$$

$$Na_2SO_4 + BaCl_2 \rightarrow BaSO_4 + 2NaCl \qquad 5\text{–}27$$

$$HCl + NaOH \rightarrow NaCl + H_2O \qquad 5\text{–}28$$

Equation 5–28 gives an example of a **neutralization reaction,** which may be stated in general form as

$$\text{acid} + \text{base} \rightarrow \text{salt} + H_2O$$

During these reactions the acidic and basic properties of the reactants disappear, and a salt and water are produced. The partner swapping that takes place is more evident when the water is written as HOH instead of H_2O:

$$HBr + KOH \rightarrow KBr + HOH \qquad 5\text{–}29$$

Besides partner swapping, other metathetical reactions may occur, including additions and decompositions. For example, equations 5–30 and 5–31 represent, respectively, a metathetical addition and decomposition:

$$CaO + CO_2 \rightarrow CaCO_3 \qquad 5\text{–}30$$

$$2NaHCO_3 \rightarrow Na_2CO_3 + H_2O + CO_2 \qquad 5\text{–}31$$

Review Questions

1. Distinguish between the three classes of inorganic compounds known as acids, bases, and salts.
2. Write reactions for the ionization of the following acids and bases in water:
   a) HI
   b) H_2SO_4
   c) HClO
   d) LiOH
   e) $Ba(OH)_2$
3. What is a polyatomic ion?
4. Write the formulas for compounds resulting from the combination of each of the following positive ions with each negative. There will be sixteen compounds.
   NH_4^+ Fe^{3+} CN^- PO_4^{3-}
   Ba^{2+} Sn^{4+} SO_4^{2-} NO_3^-
5. Indicate which of the following compounds contain polyatomic ions:
   Be_3N_2 $Ca_3(PO_4)_2$ $Mg(NO_3)_2$ $(NH_4)_2CO_3$ $BaSO_4$
   NH_4I Al_2S_3 AlN Na_3P $GeBr_4$

6. Determine the oxidation numbers of each element in the following compounds:
   a) $NaBr$
   b) Ba_3N_2
   c) NF_3
   d) Na_2O
   e) H_2O
   f) Na_2O_2
   g) PbO_2
   h) NaH
   i) Na_2SO_4
   j) $Pb(NO_3)_2$
7. Name the following binary compounds:

Group 1	Group 2	Group 3
a) $NaCl$	a) $SnCl_4$	a) PCl_3
b) MgF_2	b) $SnCl_2$	b) CO_2
c) Be_3N_2	c) Fe_2O_3	c) CO
d) Li_2O	d) $AuCl_3$	d) OF_2
e) Na_2S	e) FeO	e) N_2O_5

8. Name the following ternary compounds:

Group 1	Group 2
a) Na_2SO_4	a) $Fe(NO_3)_2$
b) K_3PO_4	b) $Fe(NO_3)_3$
c) $LiNO_3$	c) $SnCO_3$
d) $BeCO_3$	d) $Sn(CO_3)_2$
e) $Ca(CN)_2$	e) $Pb_3(PO_4)_2$

9. Name the following acids:
   a) HBr
   b) H_2SO_3
   c) HNO_3
   d) $HClO$
   e) $HClO_4$
   f) H_3PO_3
   g) H_3PO_4
   h) H_2SO_4
   i) HCl
   j) H_2CO_3
10. Balance the following equations:
    a) $C_6H_6 + O_2 \rightarrow CO_2 + H_2O$
    b) $Cu_2O + H_2 \rightarrow Cu + H_2O$
    c) $Ba(NO_3)_2 + Ag_2SO_4 \rightarrow BaSO_4 + AgNO_3$
    d) $Na_2O_2 + H_2O \rightarrow O_2 + NaOH$
    e) $CaC_2 + H_2O \rightarrow C_2H_2 + Ca(OH)_2$
11. Explain what is wrong with the following balanced equations:
    a) $H + 2O \rightarrow H_2O$
    b) $N + 3H \rightarrow NH_3$
12. Classify each of the following reactions as oxidation-reduction (OR) or metathesis (M). If the reactions are metathesis, check (√) those that are neutralization reactions.
    a) $2SO_2 + O_2 \rightarrow 2SO_3$
    b) $2HCl + Zn \rightarrow ZnCl_2 + H_2$
    c) $Ca(OH)_2 + H_2SO_4 \rightarrow CaSO_4 + 2H_2O$
    d) $2Na_3PO_4 + 3Ca(NO_3)_2 \rightarrow Ca_3(PO_4)_2 + 6NaNO_3$
    e) $2H_2O_2 \rightarrow 2H_2O + O_2$
13. For the following oxidation-reduction equations determine (1) what is reduced, (2) what is oxidized, (3) the oxidizing agent, and (4) the reducing agent:
    a) $2NO + O_2 \rightarrow 2NO_2$
    b) $2C_2H_6 + 7O_2 \rightarrow 4CO_2 + 6H_2O$

6 Energy and Entropy

The most casual observer of our surroundings quickly recognizes that many processes are taking place. These processes are both physical and chemical in nature. Objects are observed to fall toward the ground, fuels burn and give off heat, ice melts when heated above 0°C, and the sun warms objects on which its light falls—to mention only a few.

Spontaneous and Nonspontaneous Processes

A close examination reveals that processes can be classified as either **spontaneous**—occurring without external control or stimulus—or **nonspontaneous.** A problem familiar to many of us provides an illustration. Suppose you live on a hill and one morning your car fails to start. Suppose further that another hill stands between you and the nearest service station. The situation is illustrated in Figure 6–1. Because the car is on a hill, you can get in, release the brake, and the car will spontaneously roll down the hill and part way up the next hill. However, unless you get out and push, the car will slow down, stop, and begin rolling backward into the valley. It will not spontaneously roll uphill. If you push on the car, you can move it over the crest of the second hill; it will then spontaneously roll to the service station.

The behavior of the car can be partially explained by the gravitational attraction of the earth, which pulls the car down the hill. A more complete explanation depends upon an understanding of the concept of energy.

Types of Energy

We are familiar with the fact that we can increase the temperature of a substance either by bringing it into contact with another substance that is warmer or by doing work on the first substance. The work may take various forms: liquids can be stirred, solids can be rubbed together (a process often used to warm cold hands), and gases can be compressed.

Careful measurements indicate an equivalence between heat and work. The amount of heat put into or coming out of a substance can be measured, as

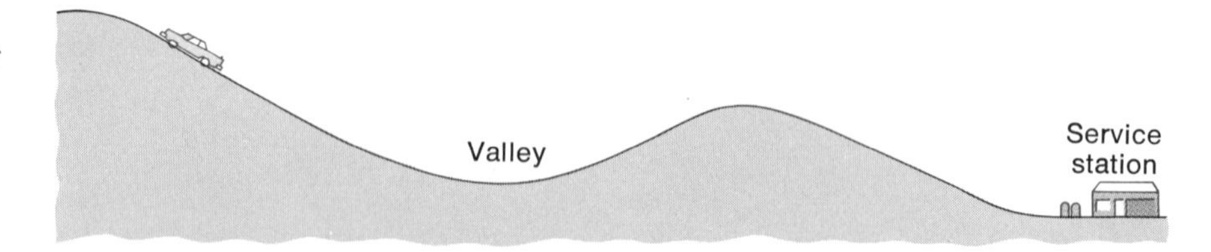

Figure 6–1.
The Problem of the Stalled Car

can the amount of work done on or by a substance. Even though an equivalence exists, different units are often used to express quantities of heat and work. The **calorie** is a commonly used unit of heat, defined as the amount of heat required to raise the temperature of one gram of water by one degree centigrade. Many units of work are used. One example is the **kilowatt-hour** of electrical work. The relationship between the kilowatt-hour and calorie is that one calorie produces the same effect as 1.16×10^{-6} kilowatt-hours. It might be pointed out that the calorie of the weight watcher is actually one kilocalorie, or 1000 calories as defined above. The prefix *kilo* is common in the metric system of measurement. A review of this system is given in Appendix III.

Because of the equivalence between the two, both heat and work are included in the general concept of energy. The total energy content of a substance can thus be changed by the addition or subtraction of either heat or work, and the final energy state is independent of the form involved. So, if one calorie of heat is added to a substance, the change is the same as if 1.16×10^{-6} kilowatt-hours of electrical energy had been added. This is somewhat like adding money to a bank account. Whether you add a single $100 bill, ten $10 checks, or 10,000 pennies to your account, the balance will still increase by 100 dollars.

As an aid to understanding the concept of energy we shall investigate the behavior of systems composed of individual particles. The energy of such systems appears in two forms. The first form is the energy an individual particle possesses by virtue of its motion; this is called **kinetic energy.** As expected, kinetic energy can be converted to work. When a moving hammer comes to rest against a nail, the kinetic energy of the hammer is changed into work that separates the wood and drives in the nail. The kinetic energy of a particle depends on both the weight of the particle and the speed with which it is moving. Mathematically, KE (kinetic energy) $= \frac{1}{2}ws^2$, where w is the weight and s the speed of the particle. Obviously, then, a particle at rest has no kinetic energy; and if a particle in motion is brought to rest, all of the kinetic energy will be converted into other forms of energy or work. In the example of the hammer and nail given above, the kinetic energy of the moving hammer is converted into work (the nail separates the wood fibers) and other forms of energy such as heat and sound.

From the expression $KE = \frac{1}{2}ws^2$, we see that any differences in kinetic energy between particles of the same weight must be caused by differences in their speeds. Similarly, differences in kinetic energy between particles moving at the same speed must be due to weight differences of the particles. These relationships are illustrated by the information in Table 6–1. The numerical values in the table are given without units and are not related to actual weight, speed, or energy units. Notice that doubling the speed of particles with the same weight causes a fourfold increase in the kinetic energy. However, a doubling of particle weight at constant speed only doubles the kinetic energy.

Table 6–1. Kinetic Energy of Particles

Weight	*Speed*	*(Speed)²*	*Kinetic energy*
5	10	100	250
5	20	400	1000
5	20	400	1000
10	20	400	2000

A second form of energy that is important in systems of particles is a consequence of attractive or repulsive forces between the particles. This form is called **potential energy.** A number of these interactions are familiar to us in the case of large particles. Examples include gravitational attractions, the attraction of unlike magnetic poles, the repulsion of like magnetic poles, the attraction of oppositely charged particles, and the repulsion of similarly charged particles. The magnitude of each of these forces depends, in some way, upon the separation distance of the interacting particles. This sort of separation dependence holds also for the forces of attraction and repulsion between atomic-sized particles.

A system of two particles connected by a spring serves to illustrate the potential-energy concept. The results obtained from this model are acceptable for the types of problems we will encounter. Figure 6–2 shows the system in several configurations.

In Figure 6–2(a) the system is in a configuration characterized by no potential energy. The spring is neither stretched nor compressed, and no work can be obtained from the system, since it will not undergo spontaneous change. Work must be done on the system to stretch the spring into the extended configuration of Figure 6–2(b). The work is present in the system as potential energy and can be recovered by allowing the spring to contract. This configuration corresponds to the existence of an attractive force between the particles. The configuration given in Figure 6–2(c) results when work is done on the system to compress the spring. Once again, the resulting potential energy can be converted back into work by allowing the spring to expand to the nonextended configuration. The configuration given in (c) represents the action of a repulsive force between the particles.

Figure 6–3 shows in a general way how the potential energy changes with separation distance when attractive or repulsive forces exist between particles. In Figure 6–3(a) the potential energy decreases as the separation distance decreases, because as the particles are pulled together there is less available energy or ability to do work. In terms of the model, the spring is sponta-

Figure 6–2.
Potential Energy
of Attraction and Repulsion

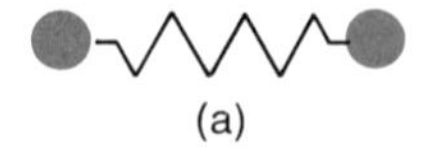

(a)

No extension or compression

(b)

Extended

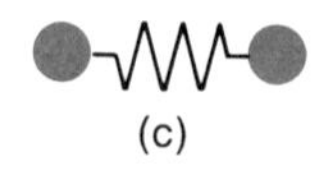

(c)

Compressed

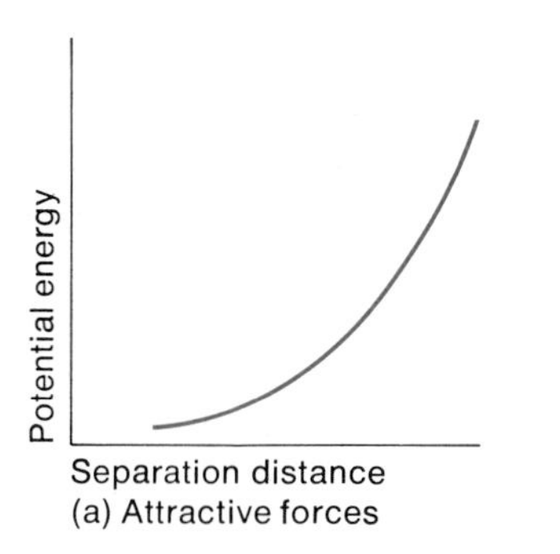

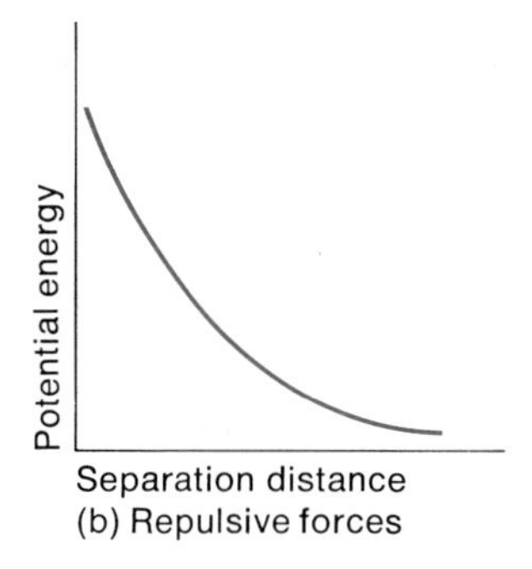

Figure 6–3.
Effect of Separation Distance on Potential Energy

neously contracting back to the nonextended configuration in which it can do no work. In Figure 6–3(b) the potential energy decreases with increasing distance, since, in terms of the model, the compressed spring is spontaneously expanding to the noncompressed configuration from which no work can be obtained.

Energy can be—and often is—converted from one form to another. In the nail-driving example given before, the hammer possesses potential energy when it is in the raised position above the nail. As the hammer falls, the potential energy changes into kinetic energy, which is transferred to the nail. Our model of two particles connected by a spring also illustrates this fact. When the system is set into vibrational motion and the spring is alternately stretched and compressed, the energy of the system continuously changes from one form to another, as summarized in Figure 6–4.

Figure 6–4.
The Interconversion of Kinetic and Potential Energy

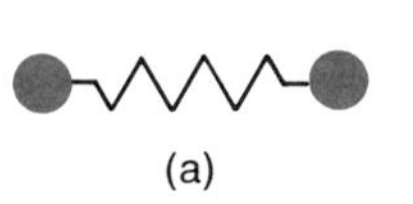

(a)
Particles are motionless and in rest position—no potential or kinetic energy.

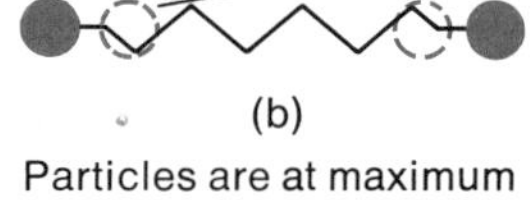

(b)
Particles are at maximum extension and motionless —all energy is potential.

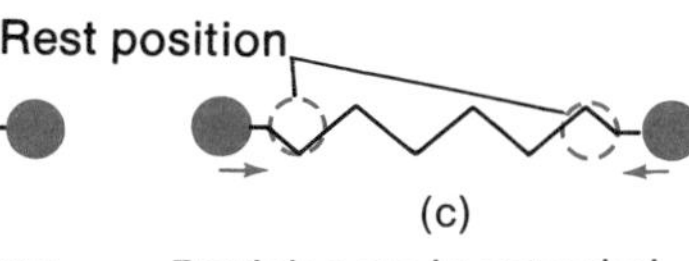

(c)
Particles are in extended position but moving toward each other—energy is part potential (spring is still extended) and part kinetic.

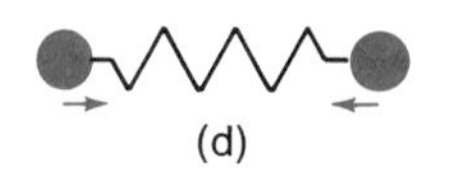

(d)
Particles are in rest position but moving toward each other—energy is all kinetic.

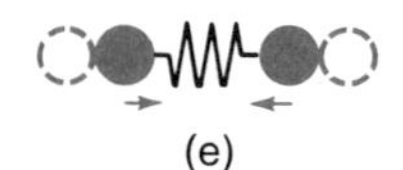

(e)
Particles are moving toward each other and spring is partially compressd—energy is part potential and part kinetic.

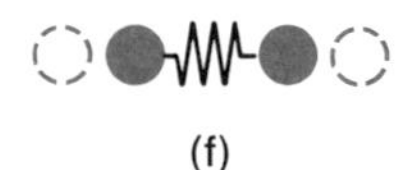

(f)
Particles are motionless, and spring is compressed to its maximum—energy is all potential.

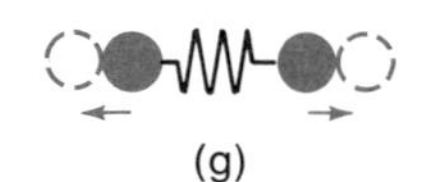

(g)
Particles are moving away from each other but spring is partially compressed—energy is part potential and part kinetic.

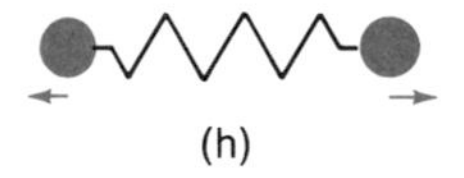

(h)
Particles are in rest position and moving away from each other—all energy is kinetic.

Energy and Spontaneity

We are now prepared to return to the discussion of spontaneous processes. The energy concept will enable us to explain, at least in part, the difference between spontaneous and nonspontaneous processes.

Consider again the car stalled on the hill. The car possessed potential energy because of the gravitational attraction tending to pull it down the hill. The car rolled down the hill spontaneously, and in the process potential energy was changed into kinetic energy. Part of the kinetic energy was converted into heat because of friction between the moving parts of the car and lost to the surroundings. At the bottom of the hill all available potential energy had been changed into kinetic energy and heat. The car then began to roll up the next hill. Kinetic energy was converted back into potential energy and into more heat during this process. Unless work was provided in the form of a push, the car next rolled backward and eventually came to rest in the valley, where it would have no available potential energy since it could no longer roll downhill, and no kinetic energy because it would be motionless. Therefore, the car must have lost energy during the spontaneous move from the side of the hill down into the valley. The lost energy went into the surroundings as heat.

A loss of energy often accompanies spontaneous processes in a system. A piece of wood, once ignited, continues to burn spontaneously and give up energy in the form of heat. Also, it is well known that a hot, molten metal spontaneously cools and solidifies (freezes) at room temperature, and in the process gives up heat to the surroundings.

However, energy losses do not accompany all spontaneous processes. Numerous examples are known in which the energy of a system increases or remains constant during spontaneous processes. Ice, for example, spontaneously melts at 20°C and in the process absorbs energy in the form of heat. A drop of ink placed in a glass of water will, after some time, become uniformly distributed throughout the water even when the system remains undisturbed. No energy change accompanies the dispersion of the ink. These types of spontaneous processes can be explained by the concept of entropy.

Entropy and Spontaneity

Entropy may be defined in a number of ways, but for our purposes it will be defined as a measurement or indication of the randomness or mixed-up character of a system. A high entropy value corresponds to a high degree of randomness or mixed-up character. This concept is illustrated by the various arrangements of colored and white marbles depicted in Figure 6–5. When the 24 marbles are arranged as shown in Figure 6–5(a), the entropy of the system is low—little mixed-up character. A more mixed-up arrangement, representing a higher entropy for the system, is shown in Figure 6–5(b). The still more random arrangement of Figure 6–5(c) represents the highest-entropy state of the three given for the system.

When a spontaneous change occurs in a system and its energy increases, its entropy is found to have increased also. The melting of ice at 20°C is accompanied by an energy increase, as stated before, but in addition the arrangement of water molecules changes from one of order in the solid ice to one of less order and higher entropy, characteristic of liquid water. In a similar way, an ink drop in a glass of water causes the entropy of the system to increase by spontaneously dispersing throughout the water while no change takes place in the energy.

From another but related point of view, a higher entropy for a system represents a statistically more probable arrangement than a lower entropy. Thus, considering entropy changes only, spontaneous processes occur in such a way that arrangements of highest probability result for the components of the system involved. For example, there is a probability that all of the air molecules in the room you now occupy will suddenly collect in one corner, leaving you in dire—but, for you, short-lived—circumstances. It is unlikely that this statement bothers you, because you know from experience that it is a very rare occurrence. In statistical terms, the random distribution (high entropy) of air molecules throughout the room is much more probable than the disastrous (low entropy) collection in the corner.

Figure 6–5.
Entropy, a Measurement of Randomness

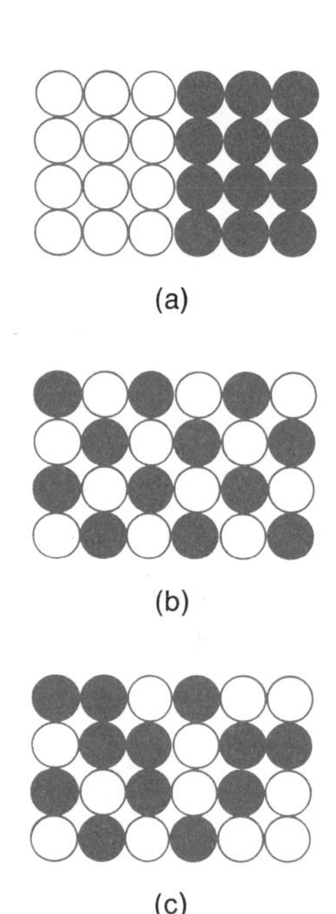

Energy, Entropy, and Spontaneity

We can understand spontaneous processes more completely by considering both energy and entropy changes. A process will be spontaneous if the energy decreases and the entropy increases. An example of such a process is a burning piece of wood. Energy in the form of heat is given up, and the gaseous products of combustion are distributed throughout the surroundings, providing the entropy increase. In a spontaneous process resulting in an energy increase or an entropy decrease, the other term must change sufficiently in the spontaneous direction to compensate for the change in the nonspontaneous direction. Thus, when ice spontaneously melts at 20°C, the increase in entropy is sufficient to compensate for the increase in energy that also takes place. When water freezes at 0°C, the energy loss compensates for the entropy decrease that occurs as the water molecules assume the well-ordered arrangement in ice.

Review Questions

1. What is the difference between a spontaneous and a nonspontaneous process?

2. Which has greater kinetic energy: an airplane that weighs 20 tons and is moving at 600 miles per hour or a freight train that weighs 2000 tons and is moving at 60 miles per hour?
3. Describe the changes in energy that occur as a rubber band is stretched, released, flies across the room, and strikes someone.
4. Discuss the changes in energy that occur as a baseball is pitched and then hit by a bat, and finally comes to rest on the ground.
5. Discuss the concept of entropy from the point of view of the randomness of a system. Use examples in your discussion.
6. Indicate what change takes place in the energy and entropy (increase or decrease) during each of the following processes:
   a) Water freezes
   b) A brick chimney falls apart
   c) A pile of dry leaves is blown by the wind
   d) Two gases react to give a solid, and heat is liberated in the reaction
7. For each of the following combinations of energy and entropy changes indicate whether the process is definitely spontaneous, possibly spontaneous, or definitely nonspontaneous.
   a) Energy increases, entropy increases
   b) Energy increases, entropy decreases
   c) Energy decreases, entropy decreases
   d) Energy decreases, entropy increases
8. In an industrial process, oxygen gas is compressed and stored in steel cylinders. Is the process spontaneous or nonspontaneous? What happens to the entropy of the gas? What happens to the energy of the gas?

Suggestions for Further Reading

Augrist, S. W., "Perpetual Motion Machines," *Scientific American*, Jan. 1968.

7 Reaction Rates and Equilibrium

In Chapter 6 we learned that spontaneous reactions take place when the resulting changes in energy and entropy are favorable. In practice, however, this concept sometimes appears to be invalid. For instance, according to these energy and entropy criteria, carbon in the form of diamond should spontaneously change into graphite—a form hardly suitable for engagement rings. Diamond owners show little concern about this fact because they have not observed this change in their diamonds. The change is taking place, but at a rate too low to detect.

The element barium is very poisonous when it is taken into the body and converted to barium ions (Ba^{2+}). When barium sulfate ($BaSO_4$) dissolves, Ba^{2+} and SO_4^{2-} ions are formed. However, suspensions of solid $BaSO_4$ are administered to patients to facilitate diagnostic X-ray photographs of the stomach and intestinal tract. The patients are not poisoned, because very little of the barium sulfate dissolves in the body. This lack of solubility is an equilibrium property of $BaSO_4$.

Reaction Rates

The rate of a chemical reaction is a measurement of its rapidity or speed, and can be followed by measuring the rate at which reactants disappear or products form. Hydrogen peroxide decomposes slowly according to the following equation:

$$2H_2O_2 \rightarrow 2H_2O + O_2 \qquad 7\text{–}1$$

We can follow the rate of this decomposition by measuring the amount of oxygen gas that bubbles off in a fixed amount of time. Similarly, hydrogen, a colorless gas, and iodine vapor, a violet-colored gas, react to form hydrogen iodide, also a colorless gas. The equation for the reaction is:

$$H_2 + I_2 \rightarrow 2HI \qquad 7\text{–}2$$

We can follow the progress of this reaction by measuring the intensity decrease of the violet-colored iodine as time passes.

Molecular Collisions

It is not unreasonable to assume that, according to equation 7–2, a collision must take place between a hydrogen molecule and an iodine molecule before any hydrogen iodide molecules can be formed. We shall make this assumption in our discussion of reaction rates. In any gas, including those involved in equation 7–2, all the molecules move at extremely high speeds and undergo many collisions each second. The collisions take place between molecules, and between molecules and the walls of the container (this concept is dealt with further in Chapter 8). In a sample of pure hydrogen gas at 0°C and

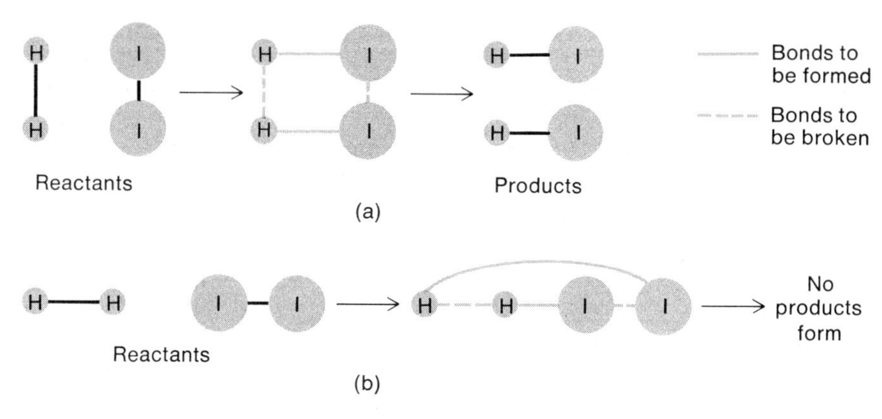

Figure 7–1.
Molecular Orientations During Collisions

normal atmospheric pressure each molecule is involved in 1.6×10^{10} collisions each second.

It immediately becomes obvious that each collision between gaseous reactant molecules does not yield product molecules. For if a reaction took place with each collision, the huge number of collisions would cause all reactions to occur instantaneously.

In order for a collision to be effective and cause a reaction to take place, two conditions must be met. First, the collision must occur between molecules with suitable orientations toward each other, as illustrated in Figure 7–1. In Figure 7–1(a) the colliding molecules have orientations that are conducive to the formation of the two product molecules. In Figure 7–1(b) the orientation at collision is not favorable for product formation. Therefore, one reason that all molecular collisions do not produce reactions is the lack of favorable molecular orientations at the time of collision.

The second condition that must be met involves the kinetic energy of the colliding molecules. As the molecules approach each other, a repulsive potential energy builds up between them because of the mutual repulsion of their electron clouds. The total kinetic energy of the colliding molecules must be large enough to overcome this repulsion; otherwise the molecules will simply rebound before they get close enough to react. At any temperature, some of the molecules in a system have kinetic energies large enough to overcome this repulsion (Chapter 8 says more about this topic). When these molecules collide and have favorable orientations, a reaction takes place and product molecules are formed.

In some cases, a system of molecules contains few with sufficient kinetic energy to overcome the repulsion between molecules, and reactions proceed very slowly. In these instances it is sometimes necessary to provide some **activation energy** to get the reaction started. Once the reaction is started, enough energy is released to activate other molecules and keep the reaction going at a good rate. The striking of a match is a good example. Activation energy is provided by rubbing the head of the match against a rough surface. Once it starts, the match continues to burn.

Figure 7–2.
Activation Energy

E_a = activation energy

E_n = net energy released during reaction

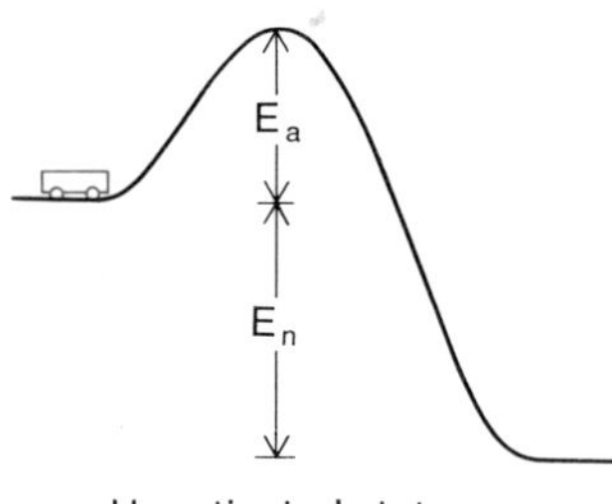

Unactivated state

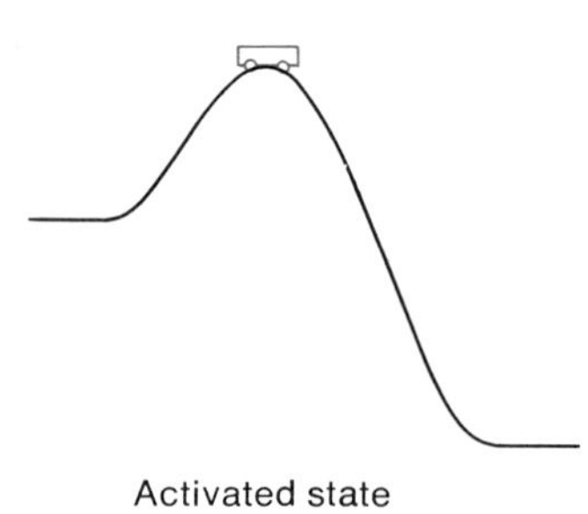

Activated state

Reaction completed

An analogous situation is represented in Figure 7–2, where a cart in a high valley must be pushed up and over a mountain before it can spontaneously roll into a lower valley where it has lower energy. The energy released as the cart rolls down the other side could be transferred, by a system of ropes, to another cart in the high valley, pull it (activate it) over the mountain, and so forth.

Factors Affecting Reaction Rates

Experience has shown that reaction rates are influenced by a number of factors. Four general factors that affect the rates of all reactions are: (1) the nature of the reactants, (2) the concentration of the reactants, (3) the temperature of the reactants, and (4) the presence of catalysts.

The effect of reactant nature can be illustrated by comparing the rates of two reactions. When a solution containing silver ions is mixed with one containing chloride ions, a white solid precipitate of silver chloride is instantly produced. The reaction taking place is:

$$Ag^+ + Cl^- \rightarrow AgCl \qquad 7\text{–}3$$

Because of the attraction between ions with opposite charges, other reactions involving ionic reactants generally go very rapidly as well. On the other hand, reactions involving covalent bonds often proceed quite slowly. The production of water gas, a mixture of CO and H_2, is an example of a covalent-type reaction:

$$H_2O + C \xrightarrow{\text{heat}} CO + H_2 \qquad 7\text{–}4$$

Other structural properties of the reactants may also play a role and must be considered.

We can easily illustrate the effect of reactant concentrations on reaction rates by using the concept of molecular collisions discussed earlier. Suppose a reaction takes place between molecules of compounds A and B, which are mixed in a 1:1 ratio. Effective collisions can take place only between A and B molecules. Collisions involving two molecules of A or two molecules of B will cause no reaction to take place. Statistically, two of every three collisions involving an A molecule will also involve a B molecule and could be effective. When the concentration of B molecules is doubled, the chances of an A molecule's colliding with a B molecule are increased to four out of five collisions, as illustrated in Figure 7–3.

The increased proportion of effective collisions at higher concentrations will produce a larger number of effective collisions in a given amount of time, and thus increase the rate of the reaction. When larger numbers of molecules are used, the results approach more closely those actually observed for simple chemical reactions. Imagine a mixture containing 1000 A and 1000 B molecules instead of the numbers given in Figure 7–3(a). On the average, 1000 out of every 1999 collisions involving a single A molecule will involve a B molecule. This gives essentially a 1:1 ratio of effective to noneffective collisions.

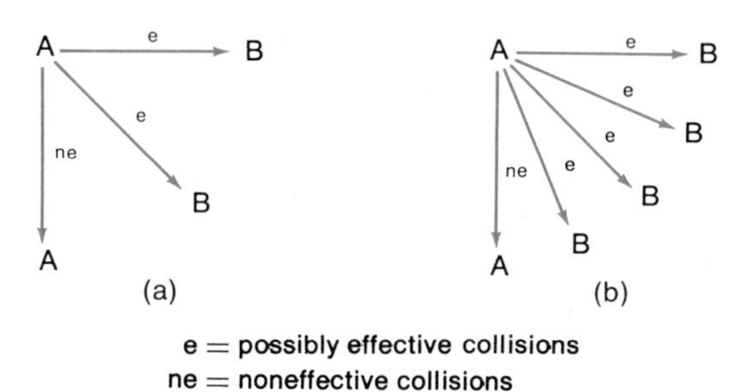

Figure 7–3.
Effect of Concentration on Reaction Rates

When the number of B molecules is doubled to 2000, the ratio becomes 2:1, since 2000 of every 2999 collisions will be effective. The reaction rate should therefore double when the concentration of one reactant is doubled. This result has been observed in numerous chemical reactions.

The concentration of gases is easily related to this approach, but the concentrations of liquids and solids must be looked at a little differently. A large piece of solid contains a large number of molecules, but only those on the surface can come in contact and react with the molecules of another substance. For this reason, the total amount of solid present in a sample is not as important as the surface area of solid in contact with other reactants. The effective concentration of a solid is therefore related to the state of division and surface area. A 100-lb sack of flour is difficult to burn when it is present in a single pile. The same 100 lb, dispersed in the air as a fine dust, burns very rapidly, and a dust explosion results. The effective concentrations of liquids must be thought of in a similar way.

The effect of temperature on reaction rates can also be explained by using the molecular-collision concept. An increase in the temperature of a system shows up as an increase in the average kinetic energy and speed of the molecules in the system (see Chapter 8 for further details). The increased speed of the molecules causes more collisions to take place in a given time period. Also, since the energy of the colliding molecules is greater, a larger fraction of the collisions will be effective from the point of view of activation energy.

As a rough rule of thumb it has been found that for every 10°C increase in temperature, the rate of a chemical reaction doubles. The chemical reactions of cooking take place faster in a pressure cooker because of a higher cooking temperature. The temperature effect on rates is also used when foods are cooled or frozen to slow down the chemical reactions involved in the spoiling of food, souring of milk, and ripening of fruit.

Catalysts are substances with the ability to change the rates of reactions without being used up in the reactions. The term is usually used to describe substances that speed up reactions. However, some catalysts, known as **inhibitors,** have the ability to slow down reactions.

Some catalysts are dispersed throughout a reacting system in the form of individual ions or molecules. Others are solids, usually with large surface areas, on which reactions take place readily.

Catalysts operate by providing a reaction pathway that requires less activation energy than the normal pathway. Our analogy of pushing a cart over a mountain (Figure 7–2) is again useful. If we happen to find a pathway halfway up that allows us to push the cart around the mountain horizontally, less activation energy will be required and the cart will get into the valley much quicker. This situation is depicted in Figure 7–4.

Figure 7–4.
Effect of Catalysts on Activation Energy

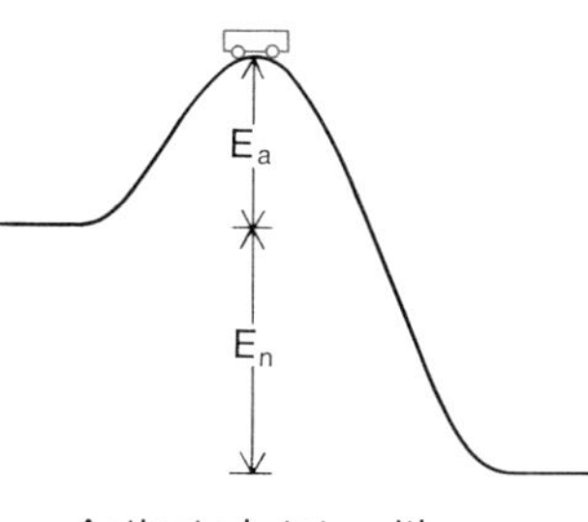

Activated state with no catalyst

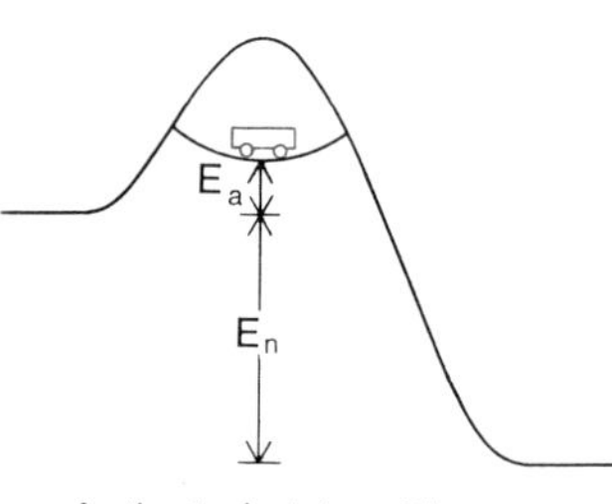

Activated state with catalyst present

According to some theories, a catalyst provides the lower-energy pathway by entering into a reaction and forming an intermediate compound which then breaks up to produce the final products and regenerate the catalyst.

Uncatalyzed:

$$\underset{\text{reactants}}{A + B} \xrightarrow{\text{slow}} \underset{\text{products}}{A{-}B} \qquad 7\text{–}5$$

Catalyzed:

$$\underset{\text{reactants}}{A + B + \text{cat.}} \xrightarrow{\text{fast}} \underset{\text{intermediate compound}}{\overset{A{-}B}{\underset{\text{cat.}}{\backslash\ /}}} \xrightarrow{\text{fast}} \underset{\text{products}}{A{-}B + \text{cat.}} \qquad 7\text{–}6$$

According to another proposed mechanism, solid catalysts provide a surface on which reactant molecules adsorb with a particular orientation. Reactants adsorbed on these surfaces are sufficiently close and favorably oriented toward each other to allow the reaction to take place. The products of the reaction then leave the surface and make it available to catalyze other reactants.

Catalysts do not have the ability to initiate reactions that would not normally occur spontaneously in their absence. Catalysts change reaction rates but have no effect on the energy or entropy factors causing the reaction.

Chemical Equilibrium

In discussing reaction rates to this point, we have looked only at the reactants. In principle, all reactions can be reversed and the products can become the reactants. Suppose equal amounts of H_2 and I_2 are mixed together and allowed to react according to equation 7–2. Initially, no HI is present in the reaction mixture, so the only possible reaction is the one given in equation 7–2. However, after a time some HI molecules are produced and can collide with each other in a way that can cause a reverse reaction to occur:

$$2HI \rightarrow H_2 + I_2 \qquad 7\text{–}7$$

The low concentration of HI will make this reaction slow at first, but as the

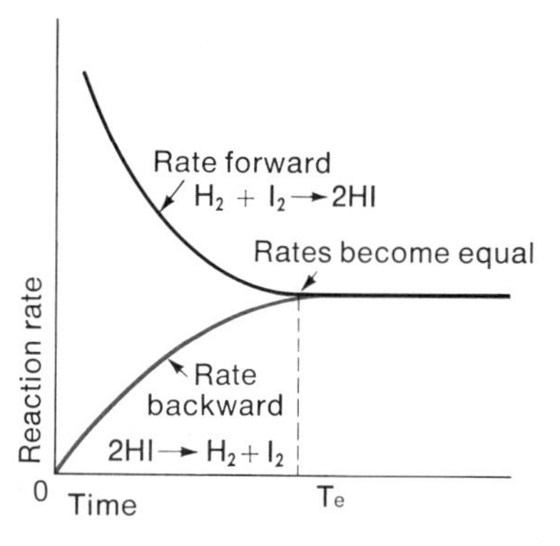

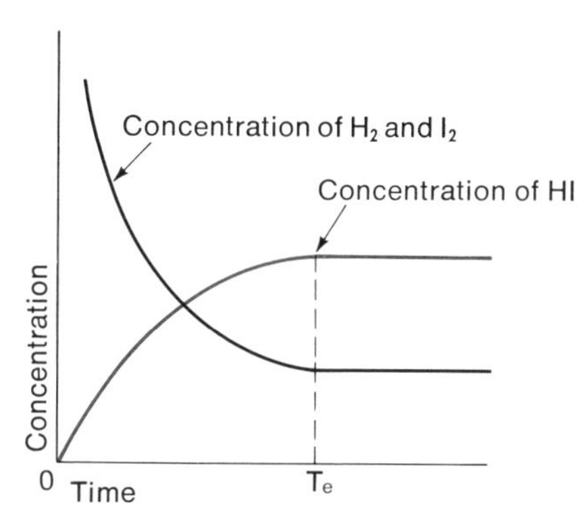

Figure 7–5.
Variation of Reaction Rates and Reactant Concentrations as Equilibrium is Established
T_e is the time needed to reach equilibrium.

concentration increases, so will the rate of the reaction. Eventually the concentrations of H_2, I_2, and HI in the reaction mixture will reach a level at which the rates of the forward (equation 7–2) and reverse (equation 7–7) reactions are equal. From this time on, the concentrations of H_2, I_2, and HI will remain constant, since each substance is produced just as fast as it is used up. When this situation arises, the reaction system is in a state of **equilibrium,** and the concentrations are the **equilibrium concentrations.** The behavior of reaction rates and reactant concentrations for both the forward and reverse reactions is shown graphically in Figure 7–5.

The same equilibrium could have been established just as easily by starting with pure HI and allowing it to change into H_2 and I_2. The final position of equilibrium does not depend on the direction from which equilibrium is approached.

Factors Affecting the Position of Equilibrium

The position of equilibrium for a reaction can be influenced by a number of factors. The nature of the reactants in a system cannot be changed once the system is chosen and cannot be considered a factor in determining the position of the final equilibrium. Similarly, catalysts cannot change the ultimate position of equilibrium but only the time required to reach it.

The concentration of reactants has a profound effect on the position of equilibrium. Suppose an equilibrium mixture of H_2, I_2, and HI is formed. According to the molecular-collision concept, favorable collisions are occurring between H_2 and I_2 molecules to form HI molecules, and favorable collisions are also taking place between HI molecules to form H_2 and I_2 molecules. Now suppose that some additional I_2 is added to the equilibrium mixture. The chances for favorable collisions between H_2 and I_2 molecules are now increased; the rate of formation of HI is increased, and more HI is formed than disappears. The rate of the reaction of HI to give H_2 and I_2 will increase as the concentration of HI increases, and eventually a new equilibrium position will be established. However, the new equilibrium mixture will contain more HI than the original mixture. The original equilibrium has been shifted toward the right according to equation 7–2.

Effects of this type are predicted by a statement known as **Le Chatelier's principle** in honor of its originator:

When a change is made in any factor which determines the position of equilibrium of a system, the system will attempt to minimize or eliminate the change.

Equation 7–2 can be written to represent an equilibrium system as follows:

$$H_2 + I_2 \rightleftharpoons 2HI \qquad 7\text{–}8$$

The double arrows signify the existence of both the forward and reverse reactions. We upset this equilibrium by adding I_2, and, as predicted by Le Chatelier's principle, the equilibrium shifted to the right in an attempt to eliminate the higher concentration of I_2 by combining it with H_2 to form HI.

We can also predict the effect of temperature on equilibrium systems by using Le Chatelier's principle. Temperature has an effect on the position of equilibrium for reactions in which heat is evolved—**exothermic reaction**—or absorbed—**endothermic reaction.** An exothermic reaction is represented by equation 7–9:

$$A + B \rightarrow C + D + \text{heat} \qquad 7\text{–}9$$

The heat in this reaction can be treated as a product. If we add additional heat by raising the temperature of the system, the position of equilibrium shifts to the left in an attempt to use up the added heat. In the system's new equilibrium position, the concentrations of A and B will be higher, those of C and D lower.

Nonchemical Aspects of Equilibrium

The term *equilibrium* is often used to describe nonchemical systems in which no changes are taking place. The stalled car of Chapter 6 is in an equilibrium situation when it comes to rest in the valley between the hills. A man walking up a "down" escalator is in equilibrium, from the point of view of position, when he is walking up at the same rate as the escalator is moving down.

Review Questions

1. Describe how the rate of a reaction can be determined.
2. What factors affect the rate of a reaction?
3. What is an *effective* molecular collision? What evidence indicates that not all molecular collisions are effective?
4. What two conditions must be met in order for a collision between two molecules to be effective?
5. Complete the following table which deals with rates for the hypothetical reaction: $A + B \rightarrow C$

Concentration of A (units)	*Concentration of B (units)*	*Reaction temperature (°C)*	*Time for completion of reaction (seconds)*
5	5	20	30
10	5	20	
	10	20	30
5	5		15

6. What is a catalyst? Describe a proposed mechanism for catalytic behavior.
7. What condition must be met in order for a system to be in equilibrium?
8. The concentrations of all species in an equilibrium mixture are necessarily constant with time. Why is this true?
9. For each of the following, indicate whether the system is at equilibrium, approaching equilibrium, or neither.
   a) Water is being heated by a strong flame.
   b) A man walking up the "down" escalator gets nowhere.
   c) A mixture of ice and water is held at 0°C.
   d) Half a cup of sugar is quickly mixed with half a cup of water.
10. What is meant by the phrase *the position of equilibrium?*
11. What is meant when we say, "the position of equilibrium shifts to the right"?
12. Indicate how each of the following factors affects the position of equilibrium:
    a) The nature of the reactants
    b) The temperature
    c) The presence of a catalyst
    d) The concentration of reactants
13. Consider the equilibrium situation

$$CO_{(g)} + H_2O_{(g)} + \text{heat} \rightleftharpoons CO_{2(g)} + H_{2(g)}$$

For each of the following adjustments of conditions indicate the effect on the position of the equilibrium—shifts left, shifts right, no effect.
   a) Heat the equilibrium mixture
   b) Add CO to the equilibrium mixture
   c) Add H_2 to the equilibrium mixture
   d) Remove CO_2 from the equilibrium mixture
   e) Add a catalyst to the equilibrium mixture
   f) Increase the size of the reaction flask

8 The States of Matter

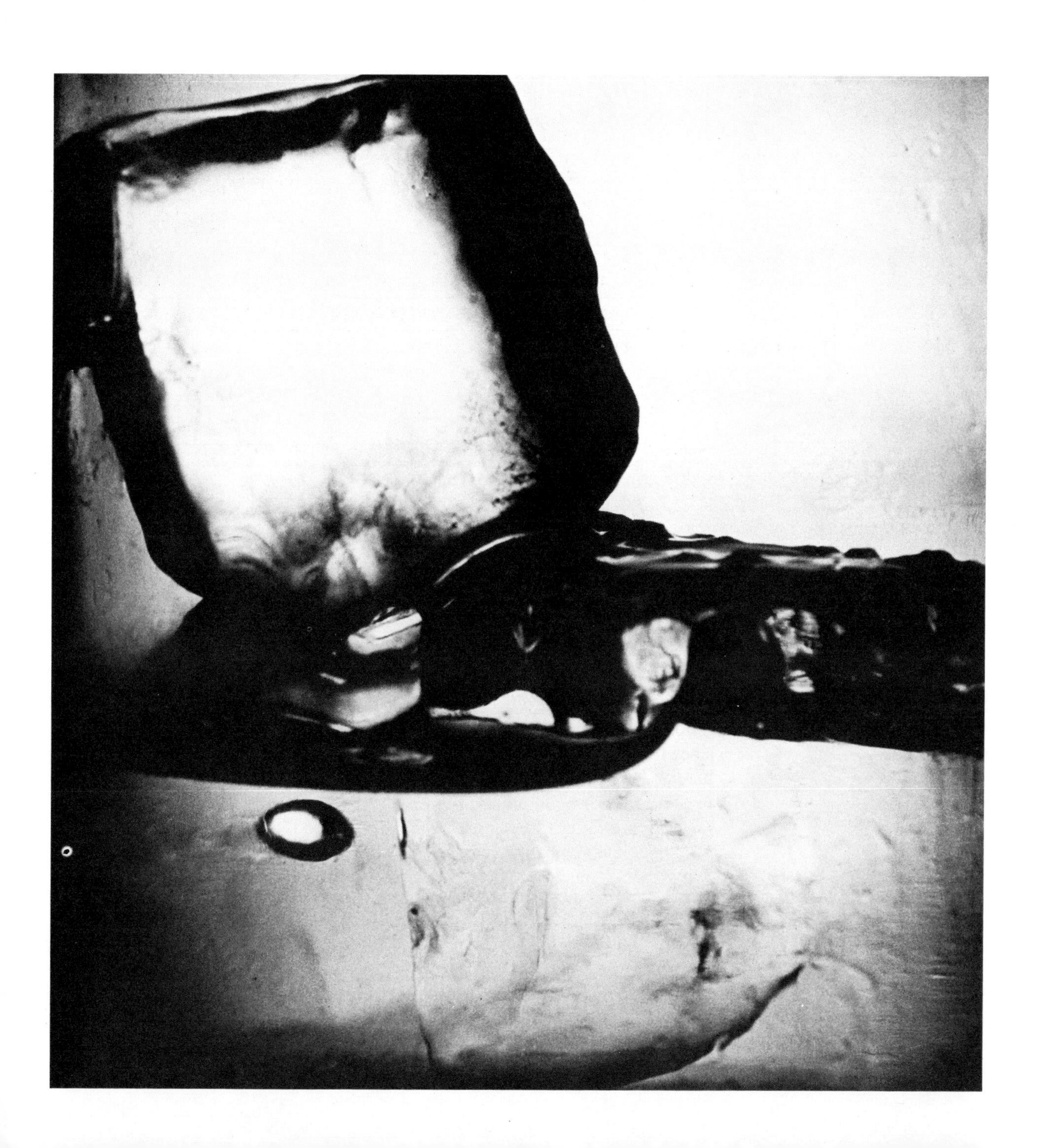

Most of the matter encountered by man exists in one of three common states —solid, liquid, and gaseous. At room temperature (25°C) and normal atmospheric pressure most of the elements are found in the solid state, as shown in Table 8–1. We see that eleven elements are gases, two are liquids (bromine and mercury), the remaining 92 are solids.

Table 8-1. The Common Physical States of the Elements

						1											2
3	4											5	6	7	8	9	10
11	12											13	14	15	16	17	18
19	20	21	22	23	24	25	26	27	28	29	30	31	32	33	34	35	36
37	38	39	40	41	42	43	44	45	46	47	48	49	50	51	52	53	54
55	56	57	72	73	74	75	76	77	78	79	80	81	82	83	84	85	86
87	88	89	104	105													

58	59	60	61	62	63	64	65	66	67	68	69	70	71
90	91	92	93	94	95	96	97	98	99	100	101	102	103

Gases

Liquids

Solids

Water is familiar to everyone in all three of the common states: solid ice, liquid water, and gaseous steam. Gold is usually considered to be a solid, since few people have observed it in the gaseous state. Oxygen, on the other hand, is nearly always thought of as a gas, because few people have seen it in the liquid or solid states. Figure 8–1 shows the solid, liquid, and gaseous temperature ranges for a number of substances. Obviously, these ranges vary widely. Some substances are liquids over a narrow range of 20–30°C (H_2), while others remain liquid over a range of 800°C (MgO). The centigrade temperature scale used in Figure 8–1 is discussed in Appendix III.

In this chapter we will investigate the three common states of matter and attempt to find reasons for the observed behaviors already mentioned. We shall attempt to explain why most elements are solids, why most substances are usually observed in only one or two states, and why some substances remain in one state over wide temperature ranges while others do not.

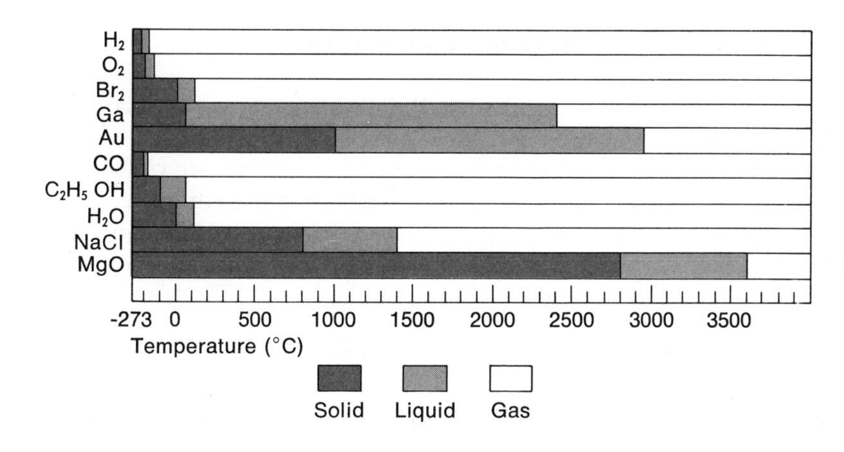

Figure 8–1.
Examples of Solid, Liquid, and Gaseous Temperature Ranges

Observed Properties of Matter

The physical differences between solids, liquids, and gases are easily recognized. The properties that differ are presented here as a starting point for further discussions. Solids are characterized by the following properties:

A high density or large weight per unit volume
A definite shape
A small compressibility (The application of pressure will not cause an appreciable reduction in the volume of a solid.)
A very small degree of thermal expansion (An increase in temperature results in only a very slight increase in volume—about 0.01 percent per degree centigrade.)

Liquids are characterized by:

A high density (usually less than that of the corresponding solid)
No definite shape (A liquid assumes the shape of its container to the degree that the container is filled.)
A small compressibility (usually greater than that of the corresponding solid)
A small degree of thermal expansion (usually greater than that of the corresponding solid—about 0.10 percent per degree centigrade)

The following properties characterize gases:

A low density
An indefinite shape (A gas takes the shape of its container, which it completely fills.)
A large compressibility
An appreciable degree of thermal expansion (When heated at constant pressure, a gas expands about 0.30 percent per degree centigrade.)

The Kinetic Molecular Theory of Matter

The observed behavior and properties of solids, liquids, and gases can be accounted for by using the **kinetic molecular theory of matter.** This theory provides a model for each of the states—and these models are conceptually very useful for predicting and explaining the behavior of matter.

We shall first present the general postulates of the theory that are applicable to all three states of matter; then we shall apply these general postulates to each specific state. The general postulates are:

Matter is ultimately composed of tiny particles (molecules) with definite and characteristic sizes which never change.

The particles are in constant motion and therefore possess kinetic energy.

The particles interact with each other through attractions and repulsions and therefore possess potential energy.

The speed of the particles increases as the temperature is increased. The kinetic energy of the molecules is therefore directly related to the temperature of the system.

The particles in a system move in all possible directions with many different energies (speeds). The average energy of all particles in a system depends upon the temperature, increasing as the temperature increases.

The particles in a system transfer energy from one to another during collisions in which no net energy is lost from the system.

The postulates of this theory provide reasonable explanations for the observed properties of matter. An important factor in these explanations is the relative influence on the system of attractive forces (potential energy), which are often called **cohesive forces,** and the **disruptive forces** of particle motion (kinetic energy). Disruptive forces tend to make the particles of a system increasingly independent of each other; cohesive forces have the opposite effect. The state in which matter is found under a given set of conditions depends upon the relative strengths of the cohesive forces that hold the particles together and the disruptive forces that tend to separate them. Cohesive forces are essentially independent of temperature, since they involve attractions between particles. Disruptive forces increase with temperature, which explains why temperature plays a very important role in determining the states of matter.

The Gaseous State

The gaseous state is one in which the disruptive forces completely dominate cohesive forces, and the molecules of a gas are free to move essentially independently of each other. This causes the molecules, under ordinary pressure, to be relatively far apart—separated by distances many times greater than a molecular diameter. The molecules are in rapid, random motion and move in straight lines between the many collisions which they constantly undergo with each other and the walls of their container. The characteristic properties of gases can readily be explained by using this model:

Low density. The molecules of a gas are widely separated. There are relatively few of them in a given volume, which means little weight per unit volume.

Indefinite shape. The attractive, cohesive forces between molecules have been overcome by kinetic energy, and the molecules are free to travel in all directions. The molecules will, therefore, completely fill the container and assume its shape.

Large compressibility. The molecules in a gas are widely separated, so that a gas in reality is mostly empty space. When pressure is applied to a gas, the molecules are easily pushed closer together, decreasing the amount of empty space and the gas volume. This is represented in Figure 8–2.

Appreciable thermal expansion. The pressure exerted by a gas depends on the frequency (number per unit time) and force of molecular collisions with the container walls. An increase in temperature increases molecular speeds, which in turn increases the frequency and force of the collisions. The frequency can be reduced and the original pressure restored by increasing the distance traveled between collisions. This is accomplished by moving the container walls farther apart—which corresponds to an increased gas volume.

Figure 8–2.
The Compression of a Gas

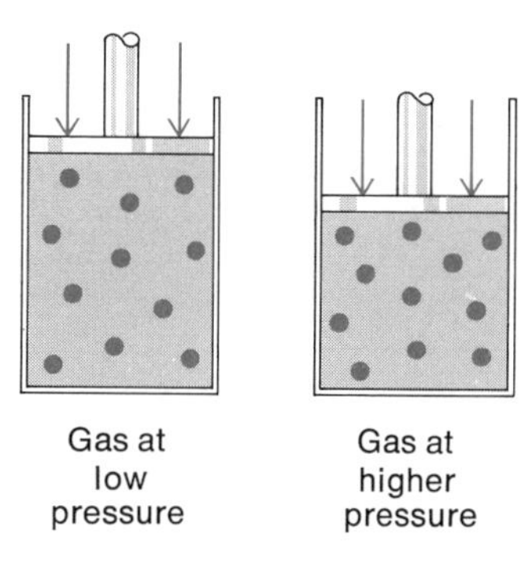

It must be understood that the size of the molecules is not changed during expansion or compression. The molecules are merely moving farther apart or closer together.

The Liquid State

The liquid state consists of particles randomly packed in relatively close proximity to each other. The molecules are in constant, random motion, freely sliding over one another. But they are not able to separate from each other. Liquids, therefore, represent a state in which neither the disruptive nor cohesive forces dominate. The fact that the molecules freely slide over each other indicates the influence of some disruptive forces, but the fact that the molecules do not separate indicates a fairly strong influence from cohesive forces. The characteristic properties of liquids are explained by this model:

High density. The particles in a liquid are not widely separated; they essentially touch each other. There will, therefore, be a large number of molecules per unit volume and a resultant high density.

Indefinite shape. Although not completely independent of each other, the molecules in a liquid are free to move over each other in a random manner, limited only by the container walls.

Small compressibility. Since the molecules in a liquid essentially touch each other, there is very little empty space. Therefore, a pressure increase cannot move the molecules much closer together.

Small thermal expansion. Most of the molecular movement in a liquid is vibrational in nature, because the molecules can move only a short distance before colliding with a neighbor. The increased molecular speed that accompanies a temperature increase results in only increased vibrational amplitudes. The net effect is an increase in the effective volume a molecule "occupies." This causes a slight volume increase in the liquid.

Figure 8–3.
Similarities and Differences in the States of Matter

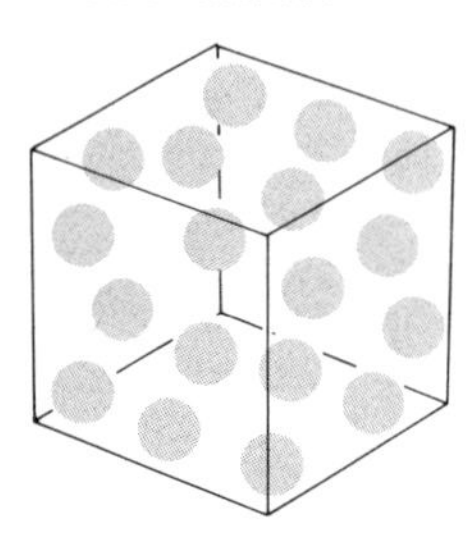

Gaseous state (not to scale)
The particles are far apart and completely fill the container

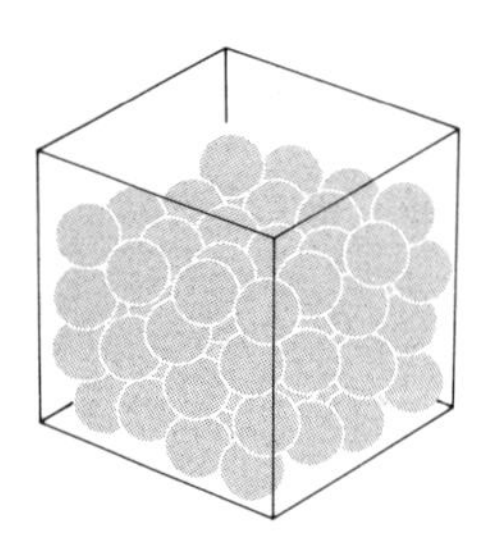

Liquid state
The particles are close together but not held in fixed positions

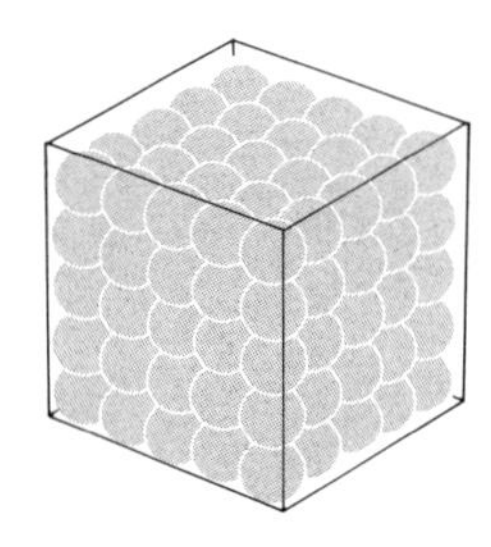

Solid state
The particles are close together and held in fixed positions

From *Chemistry: A Brief Introduction,* Mark M. Jones, et al. (Philadelphia: W. B. Saunders Company, 1969), p. 78.

The Solid State

The solid state is characterized by a dominance of cohesive forces over disruptive forces. Each particle in a solid occupies a fixed position about which it vibrates, owing to disruptive kinetic energy. The particles are drawn close together in a regular pattern by strong cohesive forces. An explanation of the characteristic properties of solids is obtained from this model:

High density. The constituent molecules of solids are located as close together as possible. Therefore, large numbers of molecules are contained in a unit volume, and a high density results.

Definite shape. The strong cohesive forces hold the molecules in essentially fixed positions, resulting in a definite shape for the solid.

Slight compressibility. Since there is very little space between molecules, an increased pressure cannot push them closer together and will have little effect on the solid volume.

Very small thermal expansion. An increased temperature increases the disruptive forces and the vibrational motion of the molecules. Each molecule will vibrate with an increased amplitude and "occupy" a slightly larger effective volume. This results in a slight expansion of the solid. The strong cohesive forces prevent this effect from becoming very large.

A Comparison of Solids, Liquids, and Gases

In some ways the liquid and solid states are very similar to each other and quite different from the gaseous state. This fact is illustrated diagrammatically in Figure 8–3.

The average distance between particles is only slightly different in the solid and liquid states. However, the difference in distance between particles in solids or liquids and those in gases is very pronounced. These distances, at ordinary pressures and temperatures for solids, liquids, and gases, are roughly in the ratios of 1:1.1:10. That is, particles in the liquid state are about 10 percent farther apart than those in the solid state, and gaseous molecules are about ten times as far apart (1000 percent) as those in the solid state.

A further comparison of the three states is given in Table 8–2 in the form of **population densities** (the number of particles per cubic centimeter). We see that in a liquid or solid, between 10^4 and 10^5 times more particles are found in each cubic centimeter than in a gas.

We can put some postulates of the kinetic molecular theory into better perspective by using numbers to illustrate the magnitude of various parameters. For example, we have postulated that the particles which make up a gas are in rapid motion, frequently colliding with each other and the container walls. The validity of these postulates is illustrated by the data in Table 8–3, which were calculated using a mathematical form of the kinetic theory.

Table 8–2. Particle Population Densities for the Three States of Matter

State of matter	*Population density*
Gas	2.7×10^{19}
Liquid	10^{22}–10^{23}
Solid	10^{22}–10^{23}

Table 8–3. Some Numerical Values of Gas Parameters

Gas	*Average speed at 0°C*	*Average distance traveled between collisions at 1 atm of pressure*	*Collisions experienced by 1 molecule in 1 sec at 1 atm of pressure*
Hydrogen	169,000 cm/sec (3,700 mph)	1.12×10^{-5} cm	1.6×10^{10}
Nitrogen	45,400 cm/sec (1,015 mph)	0.60×10^{-5} cm	0.8×10^{10}
Carbon dioxide	36,300 cm/sec (811 mph)	0.40×10^{-5} cm	0.95×10^{10}

Changes in State

Matter can be changed from one state into another by processes such as heating, cooling, or changing of pressure. The processes most often used to bring about such changes in state (phase changes) involve thermal energy. A phase change that accompanies heat input is called endothermic, and one in which heat energy is given up is called exothermic (Chapter 7).

The endothermic changes of state are melting or fusion (solid changes to liquid), sublimation (solid changes to gas), and evaporation or vaporization (liquid changes to gas). The exothermic changes are the reverse of the endothermic changes and include freezing or crystallization (liquid changes to solid), condensation (gas changes to solid), and liquefaction or condensation (gas changes to liquid). These changes are summarized in Figure 8–4.

The temperature at which a change in state occurs, whether exothermic or endothermic, is constant for each pure substance. For example, the temperature at which a solid changes to a liquid, the **melting point,** is identical to the **freezing point,** the temperature at which a liquid changes to a solid.

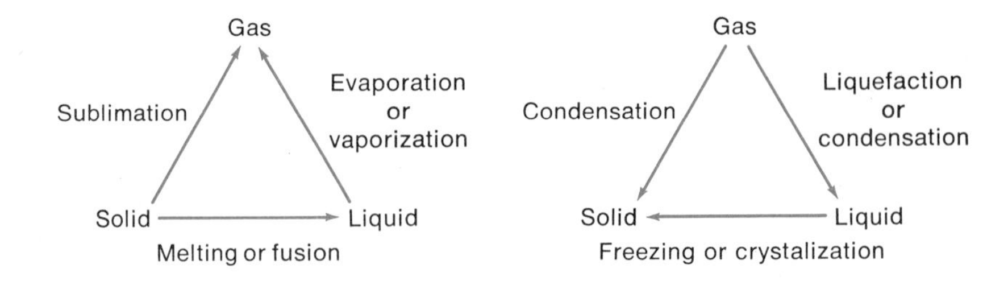

Figure 8–4.
Endothermic and Exothermic Changes of State

Evaporation

Any sample of gas, liquid, or solid contains molecules with a wide variety of kinetic energies. At a specific temperature the average molecular kinetic energy of the system is constant, but many molecules will have kinetic energies larger or smaller than the average. Figure 8–5 shows the distribution of molecular kinetic energies in a system for two temperatures. Kinetic energy is transferred from one molecule to another during collisions. In this way a molecule might have a very high energy at one instant and a very low energy in the next because of collision-induced losses.

Molecules in the liquid state can overcome attractive forces and escape into the vapor state when they momentarily possess very high kinetic energies, are moving in a favorable direction, and are on or near the surface. This evaporation process is shown schematically in Figure 8–6. The name **vapor** is applied to gaseous molecules of a substance which are present at a temperature and pressure that would ordinarily lead us to consider the substance to be a liquid or a solid.

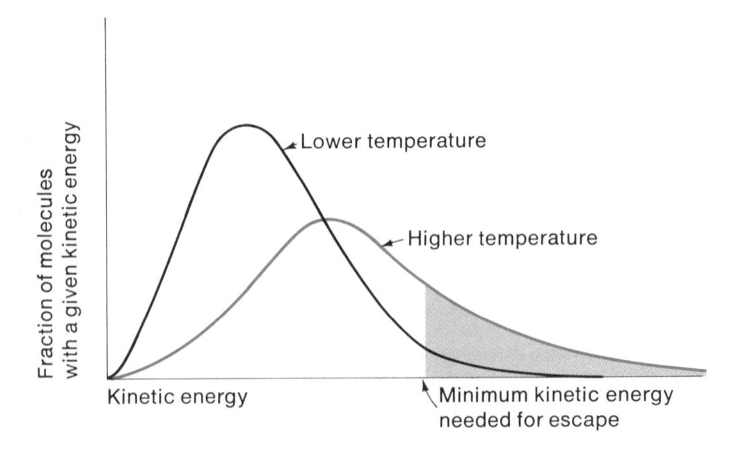

Figure 8–5.
Distribution of Kinetic Energies in a System of Molecules

The rate of evaporation depends upon the amount of liquid surface area and the distribution of kinetic energies among the molecules. If a liquid is heated, or spread out to increase its surface area, it should evaporate faster. We can see the effect of heating by referring to Figure 8–5, where the colored area represents molecular kinetic energies large enough to allow molecular escape from the liquid surface. At the higher temperature a larger fraction of the molecules have the necessary minimum energy, and the rate of evaporation will be increased.

Figure 8–6.
The Evaporation of a Liquid

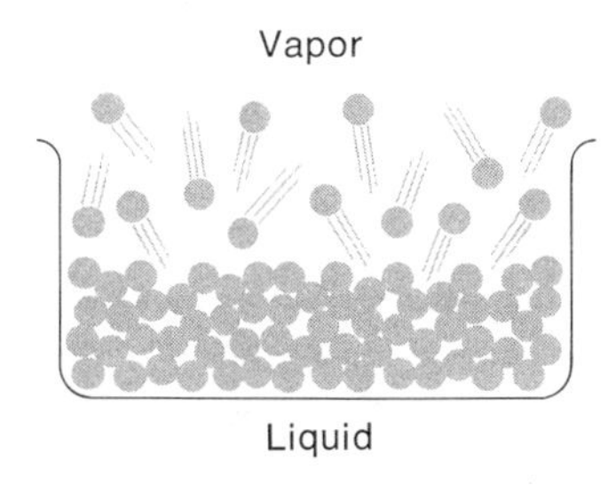

The escape of high-energy molecules from a liquid during evaporation affects the liquid in two ways: the amount of liquid decreases and the liquid temperature is lowered. **Temperature** is a measurement of the average kinetic energy of molecules in a system; as high-energy molecules leave, the average kinetic energy of those remaining is less, and the temperature is lowered. This is the principle behind evaporative cooling processes, such as the one that takes place on the skin.

Water in an open container will completely evaporate if given sufficient time. The process is not slowed by evaporative cooling, because heat flows into the water from the surrounding air and container walls. This maintains the water temperature at or near that of the surroundings, and evaporation goes on at a reasonably constant rate. If the container is constructed to prevent heat flow into or out of the water, a different situation exists. The water temperature drops during evaporation, and at the lower temperature the evaporation rate is decreased. The water level finally stabilizes, and the remaining water is at some temperature below that of the surroundings. The thermos bottle is a commercially constructed container which minimizes heat transfer by using an evacuated space between double walls as insulation. The evaporation of a liquid from a thermos bottle and from an uninsulated container are compared in Figure 8–7.

Figure 8–7.
Evaporation from Insulated and Uninsulated Containers

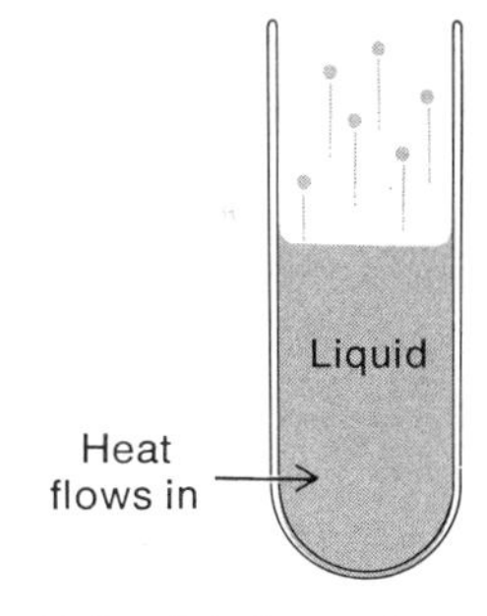

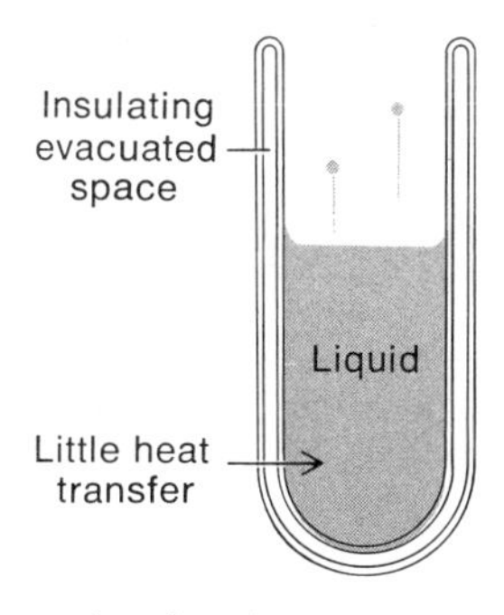

Vapor Pressure

Let us examine the results of an experiment in which a liquid is allowed to evaporate, not into an unlimited volume such as that found above an open container, but into a limited volume created by a closed container. The experimental setup is shown in Figure 8–8.

We find that some evaporation takes place, as indicated by a drop in liquid level. We find also that the liquid level eventually becomes constant. To explain these results we must consider the fate of the molecules involved in the process.

Vapor molecules are unable to move completely away from the vicinity of the liquid as they did in an open container; instead, they are confined to the space immediately above the liquid. Here, the vapor molecules undergo many random collisions with the container walls, with other molecules in the vapor state, and with the liquid surface. Occasionally their random motion carries them directly toward the surface, where they are usually recaptured by the liquid. We therefore see that two processes, evaporation (escape) and conden-

Figure 8–8.
Evaporation of a Liquid in a Closed Container

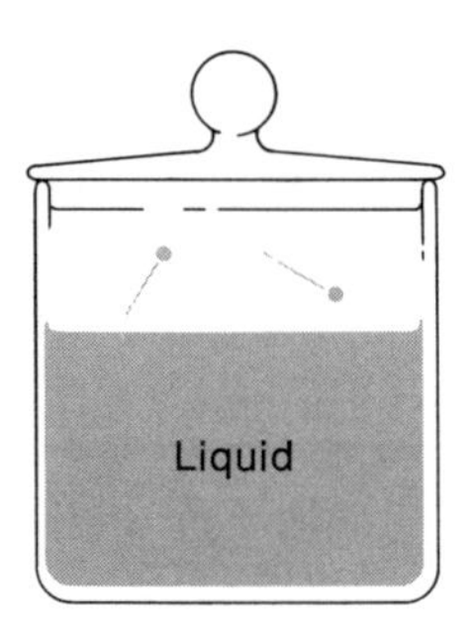

Initially

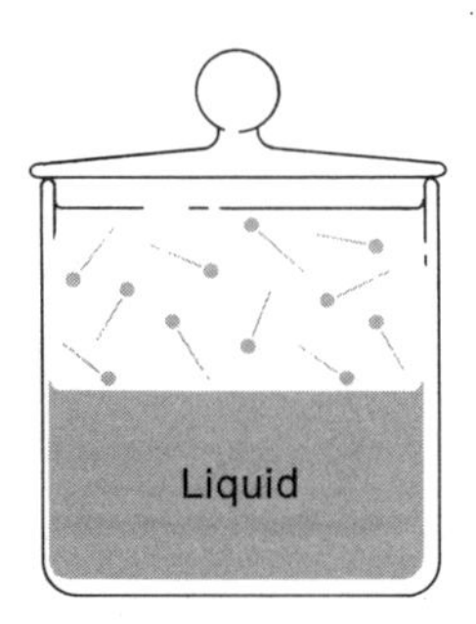

After some time

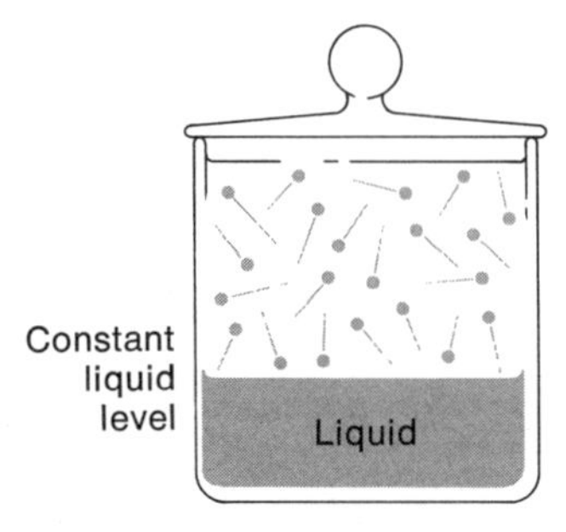

At equilibrium

sation (recapture), actually take place in a closed container of liquid. Initially, the rate of evaporation exceeds the rate of condensation, and the liquid level drops. After a sufficient length of time, however, the rates of the two processes become equal, and an identical number of molecules leave and reenter the liquid per unit time. The system is therefore in equilibrium (Chapter 7) when the liquid level remains constant. The number of molecules in the liquid also remains constant, as does the number in the vapor. We must remember, however, that even in an equilibrium situation a large number of molecules are continually moving back and forth between the liquid and vapor states.

Because the number of vapor molecules is constant at equilibrium, the vapor will exert a constant pressure on the liquid surface and the container walls. This pressure—exerted by a vapor in equilibrium with liquid—is called the **vapor pressure** of the liquid. The value of a vapor pressure depends upon the nature of the liquid (polarity, molecular weight, etc.) and its temperature. The strength of attractive forces is influenced by polar attractions between molecules, and it will obviously influence the vapor pressure by affecting the ability of molecules to escape from the liquid. Vapor pressure depends on temperature because of the effect of temperature on the number of molecules in a system that have sufficient kinetic energy to escape from the liquid (Figure 8–5). Tables 8–4 and 8–5 show the variations of vapor pressure with liquid type and temperature. The vapor pressures are given in terms of the **torr,** a pressure required to support a column of mercury 1 millimeter (mm) in height, and equal to approximately 0.02 lb/in.^{2}. The millimeter, a unit of length in the metric system, is discussed in Appendix III.

Table 8–4. Vapor Pressure of Various Liquids at 25°C

Liquid	*Vapor pressure (torr)*
Mercury	0.0012
Turpentine	4.4
Water	17.5
Bromine	173.0
Ether	442.2

Table 8–5. Vapor Pressure of Water at Various Temperatures

Temperature (°C)	*Vapor pressure (torr)*
0	4.6
20	17.5
40	55.0
60	149.2
80	355.5
100	760.0

Boiling and the Boiling Point

As the temperature of a liquid is increased, the vapor pressure also increases, as shown for water in Table 8–5. If the liquid temperature is raised enough, the vapor pressure of the liquid will reach a value equal to that of the surrounding atmosphere. Up to this point, all vaporization appears to occur on the liquid surface. But when the vapor pressure becomes equal to the atmospheric pressure, vaporization begins to take place beneath the surface as well, and bubbles of vapor form around any speck of dust or on any rough surface of the container. The bubbles of vapor are less dense than the liquid and rise rapidly to the surface, where the vapor escapes. Liquids in which these processes are taking place are said to be **boiling.**

The temperature at which the vapor pressure is equal to the atmospheric pressure is defined as the **boiling point** of a liquid. Atmospheric-pressure variations cause variations in the boiling points of liquids. In order for us to compare boiling points between different liquids, we define a standard or normal boiling point. The **standard boiling point** for any liquid is the temperature at which the liquid boils under a pressure of 1 atmosphere—760 torr or 14.7 lb/in.^{2}.

A liquid boils at temperatures higher than normal when external pressures are greater than 760 torr. Conversely, the boiling point of a liquid will be lower than normal when the external pressure is less than 760 torr. Fluctuations in atmospheric pressure at a specific location usually cause a maximum boiling-point variation of about 2°C for water. However, the variation between locations can be much greater. The average atmospheric pressure at sea level is 760 torr, but at higher elevations the average is somewhat less. At 5000 feet above sea level the average atmospheric pressure is 635 torr; at this pressure, water boils at 95.1°C. At 10,000 feet above sea level the average pressure is only 528 torr, and water boils at 90.1°C.

An increase in pressure above a liquid causes a corresponding increase in the boiling point. This principle is applied in the ordinary pressure cooker used in the kitchen. Table 8–6 gives the boiling temperatures reached by water under normal pressure-cooker conditions. It should be recalled (Chapter 7) that a 10°C increase in temperature approximately doubles the rate of a chemical reaction—in this case the rate of cooking.

Table 8–6. Boiling Point of Water in a Pressure Cooker

Pressure above atmospheric		*Boiling temperature of water*
(lb/in.^{2})	*(torr)*	*(°C)*
5	259	108
10	517	116
15	776	121

Sublimation and Melting

Solids, like liquids, have vapor pressures. The kinetic molecular theory applies to solids in much the same way it does to liquids. Particles in the solid state have a range of energies; and particles on a solid surface can escape into the gas phase if they acquire sufficiently high energies. The vapor pressures of solids are generally much lower than those of liquids because the cohesive forces in solids are not overcome by the opposing disruptive forces to the extent they are in liquids.

As expected, the vapor pressure of a solid increases with increasing temperature. Occasionally the vapor pressure of a solid reaches a value high enough to allow the escape of molecules directly into the vapor state at a significant rate. This direct vaporization of a solid is called **sublimation.** The sublimation of solid carbon dioxide (dry ice) and naphthalene (moth balls) are familiar examples. Sublimation of solid water also takes place. Wet laundry hung out to dry in freezing weather will eventually dry. The water in the clothes freezes and then slowly sublimes at temperatures below the freezing point. This technique, called **freeze drying,** is used to dry biological or other materials that would be damaged by heating.

Although all solids have vapor pressures, most pure substances in the solid state melt, upon heating, before the solid vapor pressure gets high enough to allow appreciable sublimation. Melting involves the breakdown of the relatively rigid solid structure into the mobile liquid state. The collapse of the rigid solid structure takes place at a temperature called the **melting point.** As the temperature of a solid is increased, the motion of the particles about their fixed positions becomes more vigorous and the particles are forced further apart. Eventually, the particles gain sufficient kinetic energy to break down the rigid structure.

Decomposition

Nearly every pure substance found in the gaseous state at ordinary temperatures becomes a liquid and then a solid if the constituent particles are slowed (cooled) enough to allow the cohesive forces, present in all matter, to dominate the disruptive forces. It is not true, however, that all solids can be changed into liquids or that all liquids can be changed into gases by heating. In some cases, the atoms making up the molecules of a solid or liquid may acquire enough energy and vibrate so violently that bonds within the molecules are broken before the solid or liquid can change state. The breaking of bonds in this way changes the substance into other substances in a process called **decomposition.** For this reason, paper and cotton do not melt but discolor, char, and decompose when heated strongly.

Energy Relationships Between States of Matter

The gaseous state of a pure substance is always in a higher-energy condition than the liquid or solid states of the same substance. In addition, the liquid state is always of higher energy than the solid state of the same substance. These statements are true even when the two states are compared at the same temperature.

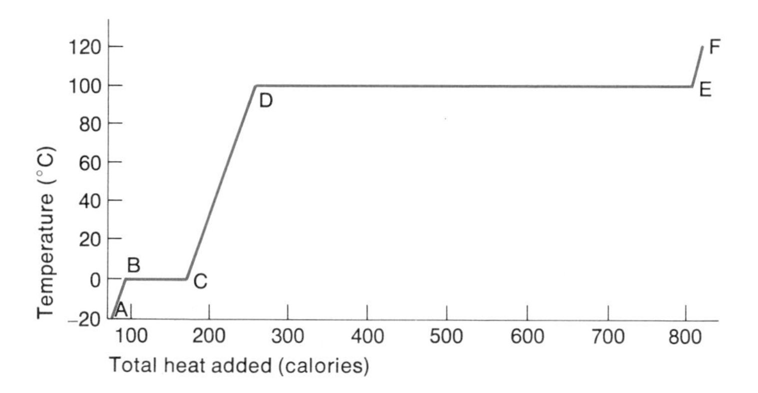

Figure 8–9.
Temperature Behavior of a System During Changes in State

We can see the validity of these statements by following the behavior of a system as it is heated. The system, composed of 1 gram of ice at −20°C, is continually and uniformly heated until the ice is converted into 1 gram of steam at 120°C. The change takes place in several steps. (1) As heat is added, the temperature of the solid ice slowly increases to a value of 0°C, the melting point of ice. (2) Further heating causes no immediate temperature increase, but the ice at 0°C slowly changes into liquid water at 0°C. As long as solid and liquid coexist, the temperature remains at 0°C. (3) After the solid disappears, further heating causes the temperature to increase until a value of 100°C, the boiling point of water, is reached. (4) The addition of more heat does not increase the temperature, but the liquid at 100°C is converted into gaseous steam at 100°C. (5) When the last of the liquid disappears, the temperature again increases as the steam is heated.

The behavior of this system is summarized in Figure 8–9, where the temperature of the system is plotted as a function of the total amount of heat added. The amount of heat added is drawn to scale.

Along line *AB* the solid is heated to its melting point. The increase in temperature indicates that most of the added heat causes an increase in the kinetic energy of the system. Along line *BC* the solid is changed at constant temperature into a liquid. The constant temperature indicates that the added thermal energy increases the potential energy of the system. The amount of heat required to change a solid to a liquid at the melting point is called the **heat of fusion.** At point *B* the system contains only solid and at point *C* only liquid. Between *B* and *C*, both solid and liquid are present. Line *CD* corresponds to heating the liquid to the boiling point. Again, the added heat increases the system's kinetic energy and increases the temperature. Along *DE* the liquid is changing into a gas at the same temperature. The large amount of added heat becomes potential energy in the system as the liquid molecules become widely separated in the gaseous state. This quantity of heat is known as the **heat of vaporization.** At point *D* only liquid is present and at *E* only gas. Between *D* and *E*, both states are found. Line *EF* corresponds to a heating of the gas, and the added energy increases the kinetic energy of the system.

Figure 8–10.
Total Energy of a System During Changes of State

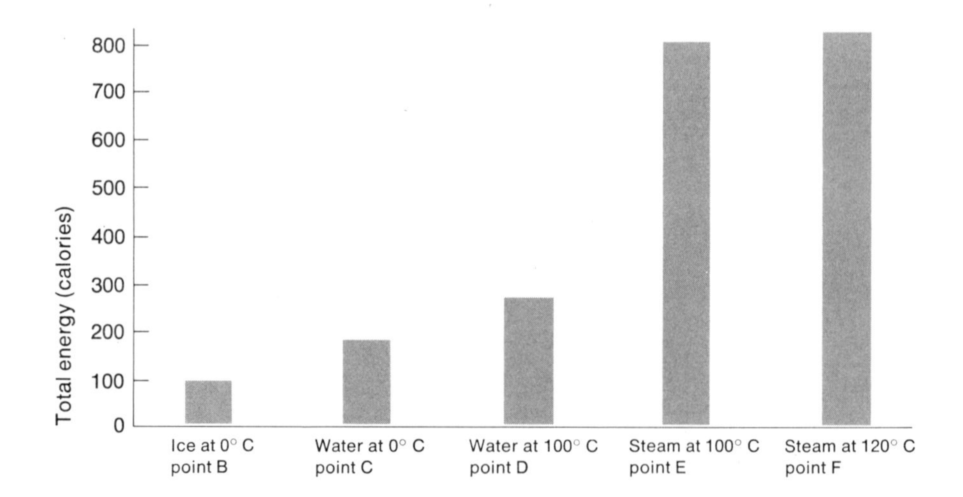

The relative amounts of total energy present in the system at each point in Figure 8–9 can be seen more readily when presented as in Figure 8–10. It is quite obvious that the gaseous state of a substance contains much more energy than the liquid state, which in turn contains more than the solid state. A comparison of the total energy at points *D* (water at 100°C) and *E* (steam at 100°C) reveals why the steam produces much more severe burns. The great energy difference between points *D* and *E* is the result of the very large heat of vaporization for water, which is among the largest known.

Particle Types and the Properties of Matter

The properties of matter depend a great deal upon the nature of the constituent particles. We shall look at the effects of particle type on properties from the point of view of solids, but most of the results can be applied to liquids and gases as well.

The crystalline solid state is characterized by a definite three-dimensional distribution of particles, which occupy essentially fixed positions in space. These three-dimensional arrangements are called **crystal lattices.** The positions occupied by particles in the lattice are called **crystal lattice sites.** Solids can be classified on the basis of the type of particle found at the lattice sites. This classification provides some insight into the reasons why some solids have very low and some very high melting and boiling points. The four common types of particles found in solids are *ions*, *polar molecules*, *nonpolar molecules*, and *atoms*.

In ionic solids the positive and negative ions are arranged so that each one is surrounded by nearest neighbors having the opposite charge. Electrostatic forces of attraction between oppositely charged ions (ionic bonds) are very strong, and as a result ionic solids have very low vapor pressures and high melting points.

The molecules in polar molecular solids are oriented so that the positive end of each molecule is near the negative end of an adjacent molecule in the lattice. The strength of attractive forces between polar molecules is a function of molecular polarity. Some of these forces are fairly strong—but much weaker than ionic forces; others are somewhat weaker. The resulting solids, therefore, have moderate vapor pressures and melting points.

In nonpolar covalent solids the forces between molecules are extremely weak, and the substances have high vapor pressures and low melting points.

Atomic solids contain crystal lattices consisting of atoms bonded to their nearest neighbors by covalent bonds (shared electron pairs). Because of the strong covalent bonds, the crystals of atomic solids are very hard, have very low vapor pressures, and have high melting points. Table 8–7 summarizes the various types of solids and their properties.

Table 8–7. Types of Solids and Their Properties

Solid type	*Particle at the lattice points*	*Forces between particles*	*Properties*	*Examples*
Ionic	Positive and negative ions	Electrostatic attractions	Low V.P., high M.P.	NaCl, CaF_2, MgO
Polar molecular	Polar molecules	Polar attractions	Moderate V.P., moderate M.P.	SO_2, H_2S, HCl
Nonpolar molecular	Nonpolar molecules	Nonpolar molecular attractions	High V.P., low M.P.	I_2, O_2, H_2, CO_2
Atomic	Atoms	Covalent bonds	Very low V.P., very high M.P.	C (diamond), SiC

In Chapter 1 the molecule was defined as the smallest particle of a pure substance capable of a stable existence. We see that this idea must be modified somewhat for ionic and atomic solids.

The formula for an ionic solid—NaCl, for example—does not represent a molecule. It represents the simplest combining ratio for the ions—Na^+ and Cl^-—which occupy the lattice sites. Crystals are the natural stable forms for these compounds, and the term *molecule*, as defined in Chapter 1, is not really applicable. However, the formula is often used as if it did represent a molecule when equations are written and "molecular" weights are calculated (Chapter 9). We shall follow this useful practice throughout this book.

A similar situation exists for atomic solids, in which the formula represents the simplest ratio of the atoms making up the crystal. These formulas will also be used as molecular formulas when equations are written and molecular weights are calculated.

Review Questions

1. What percentage of the known elements are gases at room temperature? What percentage are liquids? Solids? List those that are gases and those that are liquids.
2. Contrast the following properties for the solid, liquid, and gaseous state.
   a) Density
   b) Compressibility
   c) Thermal expansion
3. According to the postulates of the kinetic molecular theory of matter what is the relationship between temperature and kinetic energy?
4. Contrast the effects of cohesive and disruptive forces on a system. What type of energy is related to cohesive forces and to disruptive forces?
5. Distinguish between the gaseous, liquid, and solid states of a substance from the point of view of the kinetic and potential energies of the constituent particles.
6. Explain each of the following observations using the kinetic molecular theory.
   a) Gases have a low density.
   b) The density of the solid and liquid state of a substance are nearly the same.
   c) Solids, liquids, and gases expand upon heating.
   d) A liquid takes the shape, but not the volume of its container.
   e) Solids are practically incompressible.
   f) A gas always exerts a pressure on the object or container with which it is in contact.
7. Contrast the difference in particle separation for solids, liquids, and gases in a semiquantitative manner (using numbers).
8. Contrast the meaning of the terms *exothermic* and *endothermic*. List the terms used to describe changes of state that are exothermic; endothermic.
9. Describe, in terms of the kinetic molecular theory, how it is possible for a liquid to evaporate.
10. In order for a liquid to evaporate at a *constant temperature* heat must be added. Why?
11. Name two factors that affect the vapor pressure. Explain how each factor operates.
12. Why does the boiling point of a liquid vary with pressure?
13. What is the difference in meaning of the terms *boiling point* and *normal boiling point?*
14. Give three examples of sublimation.
15. Criticize the statement: All solids will melt if heated to a high enough temperature.
16. Contrast the terms *heat of fusion* and *heat of vaporization.*
17. Account for the fact that burns caused by steam at 100°C are usually more severe than those from water at the same temperature.

18. Give a brief description of each of the following types of solids by indicating the particles occupying the lattice sites and the type of forces holding them in position.
    a) Atomic solid
    b) Polar molecular solid
    c) Nonpolar molecular solid
    d) Ionic solid
19. Criticize the statement: The smallest unit of sodium chloride (an ionic compound) is an NaCl molecule.

Suggestions for Further Reading

Alder, B., and T. Wainwright, "Molecular Motions," *Scientific American*, Oct. 1959.
Bernal, J. D., "The Structure of Liquids," *Scientific American*, Aug. 1960.
Fergason, J. L., "Liquid Crystals," *Scientific American*, Aug. 1964.
Fullman, R. L., "The Growth of Crystals," *Scientific American*, March 1955.
Lonsdale, K., "Disorder in Solids," *Chemistry*, Dec. 1965.
Scott, N., "The Solid State," *Scientific American*, Sept. 1967.

9 Chemical Calculations

It is often advantageous to know more about the substances involved in chemical reactions than merely their physical and chemical properties. Information of interest might be the combining ratio of reactants, the amounts of products formed, or the formula of a newly prepared compound. This type of information can often be obtained by using calculations based on experimental data or chemical principles.

Molecular Weights

The atomic-weight concept was introduced in Chapter 2 in terms of the relative weights of atoms. Molecular weights are defined by applying the same ideas to molecules. A **molecular weight** is defined as the relative weight of a molecule compared to the weight of a $^{12}_{6}C$ atom, which has been assigned a weight of 12 atomic mass units. The molecular weight of a molecule is easily determined by using the molecular formula and the atomic weights of the constituent elements. The molecular weight is the sum of the atomic weights of all atoms contained in the molecule. Some sample molecular-weight calculations are given in Example 9–1.

Example 9–1. Calculate the molecular weight of each of the following compounds: H_2O, Na_2CO_3, and $Al_2(SO_4)_3$.

H_2O: Each molecule contains two H atoms and one O atom with atomic weights of 1.0 and 16.0 amu, respectively. The atomic weights are obtained from Table 2–9.

$$\begin{aligned} 2\text{ H atoms} &= (2)(1.0\text{ amu}) = 2.0\text{ amu} \\ 1\text{ O atom} &= (1)(16.0\text{ amu}) = \underline{16.0\text{ amu}} \\ \text{Molecular weight} &= 18.0\text{ amu} \end{aligned}$$

Na_2CO_3: Each molecule contains two Na atoms, one C atom, and three O atoms. The atomic weights of Na, C, and O are, respectively, 23.0, 12.0, and 16.0 amu.

$$\begin{aligned} 2\text{ Na atoms} &= (2)(23.0\text{ amu}) = 46.0\text{ amu} \\ 1\text{ C atom} &= (1)(12.0\text{ amu}) = 12.0\text{ amu} \\ 3\text{ O atoms} &= (3)(16.0\text{ amu}) = \underline{48.0\text{ amu}} \\ \text{Molecular weight} &= 106.0\text{ amu} \end{aligned}$$

$Al_2(SO_4)_3$: Each molecule contains two Al atoms, three S atoms, and twelve O atoms. Note that the subscript 3 in the formula affects the number of both the S and O atoms. The atomic weights of Al, S, and O are 27.0, 32.1, and 16.0 amu, respectively.

$$\begin{aligned} 2\text{ Al atoms} &= (2)(27.0\text{ amu}) = 54.0\text{ amu} \\ 3\text{ S atoms} &= (3)(32.1\text{ amu}) = 96.3\text{ amu} \\ 12\text{ O atoms} &= (12)(16.0\text{ amu}) = \underline{192.0\text{ amu}} \\ \text{Molecular weight} &= 342.3\text{ amu} \end{aligned}$$

In most problems the atomic weights can be rounded off to one decimal place, as we have done in these sample calculations.

Counting Particles by Weighing

Atoms are so small that it is impossible to count them one-by-one, and yet it is often necessary to determine their number. Chemists are able to count particles by weighing substances and, therefore, get around the problem of counting them individually. This concept is fundamental to an understanding of chemical calculations and should be well understood.

The process of counting by weighing will be illustrated by comparing atoms of the elements hydrogen and fluorine. The choice of elements is arbitrary; any two could be used.

Suppose we begin with one atom of each element. The F atom weighs 19.0 amu, its atomic weight, and the H atom weighs 1.0 amu, its atomic weight. The F atom is obviously 19.0 times heavier than the H atom. Now suppose we make the same sort of comparison for ten atoms of each element. The ten F atoms will weigh a total of 190.0 amu (10×19.0 amu) and the ten H atoms will weigh a total of 10.0 amu (10×1.0 amu). Once again, the total weight of the ten F atoms is 19.0 times the total weight of the ten H atoms. Let us continue our analysis, this time using Y atoms of both F and H, where Y is any number. The Y atoms of F weigh $Y \times 19.0$ amu, and the Y H atoms weigh $Y \times 1.0$ amu. We notice that once again the total weight of the F atoms is 19.0 times the total weight of the H atoms. The ratio of the total weight of F to the total weight of H for each situation treated above is summarized as follows.

One atom of each:

$$\frac{\text{total weight of F}}{\text{total weight of H}} = \frac{(1)(19.0\text{ amu})}{(1)(1.0\text{ amu})} = \frac{19.0}{1.0}$$

Ten atoms of each:

$$\frac{\text{total weight of F}}{\text{total weight of H}} = \frac{(10)(19.0\text{ amu})}{(10)(1.0\text{ amu})} = \frac{190.0}{10.0} = \frac{19.0}{1.0}$$

Y atoms of each:

$$\frac{\text{total weight of F}}{\text{total weight of H}} = \frac{(Y)(19.0\text{ amu})}{(Y)(1.0\text{ amu})} = \frac{19.0Y}{1.0Y} = \frac{19.0}{1.0}$$

In each case the F/H weight ratio is identical (19.0/1.0), and we can generalize this observation by the following statement: the ratio of the total weights of equal numbers of F and H atoms will always be 19.0/1.0. Another useful generalization results if this statement is turned around: samples of F and H found to have a F/H weight ratio of 19.0/1.0 must contain identical numbers of atoms. This latter statement is true regardless of the weight units used. For example, 19.0 lb of F contains the same number of atoms as 1.0 lb of H, and 19.0 tons of F contains the same number of atoms as 1.0 ton of H.

Note that the weight ratio of significance is the ratio of the relative weights or atomic weights of the two elements. This will be true regardless of the elements involved in the calculations. In general, samples of two or more elements found to have weight ratios equal to their atomic weight ratios must contain identical numbers of atoms. Therefore, the following samples will contain the same number of atoms:

19.0 g of F	(atomic weight of F = 19.0 amu)
1.0 g of H	(atomic weight of H = 1.0 amu)
16.0 g of O	(atomic weight of O = 16.0 amu)
197.0 g of Au	(atomic weight of Au = 197.0 amu)

If we substitute molecular weights in place of atomic weights, these same results can be applied to molecules. The substitution of molecular weights for atomic weights is valid because both are based on the same ${}^{12}_{6}C$ standard. Therefore, the following samples all contain the same number of molecules:

18.0 g of H_2O	(molecular weight of H_2O = 18.0 amu)
44.0 g of CO_2	(molecular weight of CO_2 = 44.0 amu)
64.1 g of SO_2	(molecular weight of SO_2 = 64.1 amu)

We can now make the following generalization, which is applicable to both elements and compounds: samples of two or more pure substances found to have weight ratios equal to the ratios of their atomic or molecular weights must contain identical numbers of particles (atoms or molecules). The following samples, therefore, contain the same number of particles:

19.0 g of F	(atomic weight = 19.0 amu)
16.0 g of O	(atomic weight = 16.0 amu)
18.0 g of H_2O	(molecular weight = 18.0 amu)
44.0 g of CO_2	(molecular weight = 44.0 amu)

We see that, in order to obtain samples of elements or compounds containing equal numbers of particles, we merely have to weigh out quantities (in any units) numerically equal to the atomic or molecular weights.

The Mole Concept

Even the smallest detectable weight of any pure substance contains an extremely large number of constituent particles—atoms or molecules. Chemists use the term **mole** to specify a chemically significant large number, which is equal to 6.02×10^{23}. To a chemist, *one mole* always means 6.02×10^{23} (atoms, molecules, or any other object), just as to most people one dozen means 12 (eggs, donuts, or any other object).

The number 6.02×10^{23} is chosen as a fundamental quantity because it represents the experimentally determined number of atoms or molecules contained in a sample of a pure substance with a weight in grams numerically equal to the atomic or molecular weight of the substance. The mole is therefore closely related to the process of counting by weighing already discussed. In terms of this definition of the mole, the following relationships hold:

$$
\begin{aligned}
19.0\text{ g of F (atomic wt.} = 19.0\text{ amu)} &= 1\text{ mole of F atoms} \\
&= 6.02 \times 10^{23}\text{ F atoms} \\
16.0\text{ g of O (atomic wt.} = 16.0\text{ amu)} &= 1\text{ mole of O atoms} \\
&= 6.02 \times 10^{23}\text{ O atoms} \\
18.0\text{ g of } H_2O\text{ (molecular wt.} = 18.0\text{ amu)} &= 1\text{ mole of } H_2O\text{ molecules} \\
&= 6.02 \times 10^{23}\ H_2O\text{ molecules} \\
44.0\text{ g of } CO_2\text{ (molecular wt.} = 44.0\text{ amu)} &= 1\text{ mole of } CO_2\text{ molecules} \\
&= 6.02 \times 10^{23}\ CO_2\text{ molecules}
\end{aligned}
$$

The number of objects in one mole, 6.02×10^{23}, is commonly called **Avogadro's number** in honor of Amedeo Avogadro, an Italian physicist who developed concepts that led to its determination.

This definition of the word mole has complicated your vocabulary. You must now remember three meanings: a burrowing animal, a skin blemish, and a large number. Chemical calculations, however, have been simplified, as we shall see.

The Mole and Chemical Formulas

The formula for any substance contains the symbol of each element present, and indicates the number of atoms of each by subscripts. The ratios of the number of different atoms present may be determined from a formula. A formula also provides the number of moles of each atom contained in one mole of a substance.

This last statement is verified by the following line of reasoning: water is a compound with a molecular formula of H_2O; this formula indicates that H and O atoms are present in the molecule in a 2:1 ratio; therefore, one H_2O molecule contains two H atoms and one O atom.

The following statements also are true:

Two H_2O molecules contain four H atoms and two O atoms.
Ten H_2O molecules contain twenty H atoms and ten O atoms.
6.02×10^{23} H_2O molecules contain 12.04×10^{23} H atoms and 6.02×10^{23} O atoms.

This last statement is very significant because it incorporates the mole concept into our line of reasoning. We see that it can be changed to read:

One mole of H_2O molecules contains two moles of H atoms and one mole of O atoms.

This is true because, according to the mole concept, 6.02×10^{23} H_2O molecules equals one mole of H_2O molecules, 12.04×10^{23} H atoms equals two moles of H atoms, and 6.02×10^{23} O atoms equals one mole of O atoms.

The following examples illustrate the relationships between the mole concept and chemical formulas.

Example 9–2. How many ears, tails, and legs are contained in one mole of normal rabbits?

This example, although nonchemical, is helpful in illustrating the relationships that exist between the individual parts of a formula and the formula as a whole. An analogy exists between rabbits and chemical formulas. The elements of a formula are related to the formula as a whole, just as the parts of a rabbit are related to the rabbit as a whole.

Ears: Each rabbit has two ears, and a 2:1 ratio exists between the number of ears and rabbits. Therefore, one mole of rabbits contains two moles of ears.

Tails: Similarly, a 1:1 ratio exists between the number of tails and rabbits, and one mole of rabbits contains one mole of tails.

Legs: The four legs of each rabbit produce a 4:1 ratio between legs and rabbits, and one mole of rabbits contains four moles of legs.

Example 9–3. How many moles of atoms of each constituent element are contained in one mole of a compound with the molecular formula $C_6H_5NO_2$?

Just as each rabbit has two ears, each $C_6H_5NO_2$ molecule contains five H atoms, two O atoms, and so forth. Therefore:

One mole of $C_6H_5NO_2$ contains six moles of C atoms.
One mole of $C_6H_5NO_2$ contains five moles of H atoms.
One mole of $C_6H_5NO_2$ contains one mole of N atoms.
One mole of $C_6H_5NO_2$ contains two moles of O atoms.

Example 9–4. How many moles of O atoms are contained in one mole of oxygen molecules (O_2)?

A single oxygen molecule (O_2) contains two O atoms. Therefore, one mole of O_2 molecules contains two moles of O atoms.

The Mole and Chemical Equations

The coefficients in a balanced equation can be used to tell us two things—the number of individual particles involved and the number of moles of individual particles involved.

The following statements are consistent with the equation

$$2H_2 + O_2 \rightarrow 2H_2O$$

1. Two H_2 molecules + one O_2 molecule → two H_2O molecules (a 2:1:2 ratio).

2. Four H_2 molecules + two O_2 molecules → four H_2O molecules (a 2:1:2 ratio).
3. One hundred H_2 molecules + 50 O_2 molecules → 100 H_2O molecules (a 2:1:2 ratio).
4. 12.04×10^{23} H_2 molecules + 6.02×10^{23} O_2 molecules → 12.04×10^{23} H_2O molecules (a 2:1:2 ratio).
5. According to the mole definition, 12.04×10^{23} is equal to two moles, and 6.02×10^{23} is equal to one mole. Therefore, two moles of H_2 molecules + one mole of O_2 molecules → two moles of H_2O molecules (a 2:1:2 ratio).

The coefficients in statements 1 and 5 are the same as those in the original equation. Therefore, the original equation can be used to determine the number of particles or moles involved in the reaction.

Grams, Particles, and Moles

We are now prepared to perform some chemical calculations. The important quantities used in these calculations are: (1) the number of particles involved, (2) the number of moles involved, and (3) the number of grams (weight) involved. As we have seen, these three quantities are related by the following:

Avogadro's number—provides a relationship between the number of particles of a substance and the number of moles of the same substance.

Molecular or atomic weight—provides a relationship between the number of grams of a substance and the number of moles of the same substance.

Formula of a compound or coefficients in a chemical equation—these terms relate the number of moles of one substance to the number of moles of another substance involved in the same reaction.

The important quantities and the relationships between them are summarized in Figure 9–1. The arrows between boxes indicate that the direct relationship noted exists between the two quantities. This diagram can be used to set up the solutions to many problems.

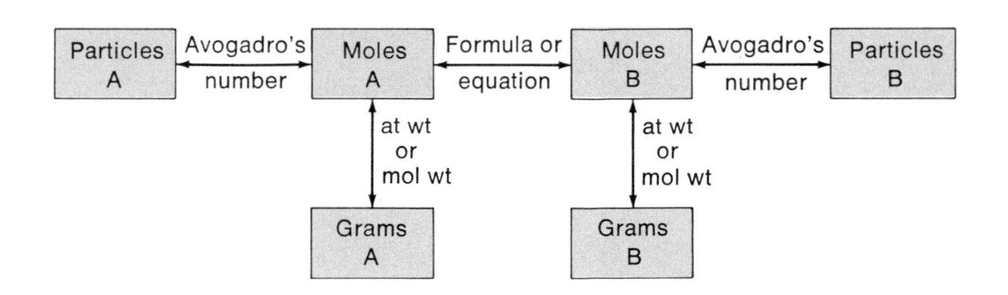

Figure 9–1. Relationships Useful for Problem Solving

Problem Solving

The factor method is used for problem solving in this text. This method is very useful where a known quantity must be converted into a desired or unknown quantity—an operation that many chemical problems require. The following steps, when followed, make this form of problem solving very systematic.

1. Write down the known quantity. Include both the numerical value and the units or name of the quantity.
2. Leave some working space, and set the known quantity equal to the units of the desired unknown quantity.
3. Multiply the known quantity by one or more equivalence factors. These are fractions in which the numerator and denominator are equivalent—for example, 12 inches/1 foot or 1 foot/12 inches (since the reciprocal of an equivalence factor is also an equivalence factor).
4. Arrange the equivalence factors so the units of the known quantity divide out and the units of the unknown quantity are produced.
5. After the desired units are obtained, carry out the indicated numercial calculations to produce the final answer.

These steps are demonstrated in Example 9–5.

Example 9–5. (a) Calculate the number of eggs contained in a full 20-dozen egg container.

step 1. 20 doz eggs

step 2. $20 \text{ doz eggs} \qquad = \qquad \text{eggs}$

step 3. $$20 \text{ doz eggs} \times \frac{12 \text{ eggs}}{1 \text{ doz eggs}} = \text{eggs}$$

step 4. $$20 \cancel{\text{doz eggs}} \times \frac{12 \text{ eggs}}{1 \cancel{\text{doz eggs}}} = \text{eggs}$$

step 5. $$\frac{(20)(12)}{(1)} \text{ eggs} = 240 \text{ eggs}$$

(b) How many inches are in one mile?

step 1. 1 mile

step 2. $1 \text{ mile} \qquad = \qquad \text{inches}$

step 3. $$1 \text{ mile} \times \frac{5280 \text{ feet}}{1 \text{ mile}} \times \frac{12 \text{ inches}}{1 \text{ foot}} = \text{inches}$$

step 4. $$1 \cancel{\text{mile}} \times \frac{5280 \cancel{\text{feet}}}{1 \cancel{\text{mile}}} \times \frac{12 \text{ inches}}{1 \cancel{\text{foot}}} = \text{inches}$$

step 5. $$\frac{(1)(5280)(12)}{(1) \quad (1)} \text{ inches} = 73{,}360 \text{ inches}$$

(c) How many atoms are contained in two moles of copper?

step 1. 2 moles Cu

step 2. 2 moles Cu = atoms Cu

step 3. $2 \text{ moles Cu} \times \frac{6.02 \times 10^{23} \text{ atoms Cu}}{1 \text{ mole Cu}} = \quad \text{atoms Cu}$

step 4. $2 \,\cancel{\text{moles Cu}} \times \frac{6.02 \times 10^{23} \text{ atoms Cu}}{1 \,\cancel{\text{mole Cu}}} = \quad \text{atoms Cu}$

step 5. $\frac{(2)(6.02)(10^{23})}{(1)} \text{ atoms Cu} = 12.04 \times 10^{23} \text{ atoms Cu}$

We shall now use the factor method to solve some typical chemical problems in Examples 9–6 through 9–9. The appropriate equivalence factors and their sequence can be determined for each problem by referring to Figure 9–1. The figure provides the correct pathway to follow in order to solve for the unknown quantity by using the data or known quantity.

Example 9–6. The hallucinogenic drug LSD has the formula $C_{24}H_{30}N_3O$. How many molecules of LSD would be contained in 0.10 gram?

In this problem we have 0.10 gram of LSD as the given quantity and the number of LSD molecules as the desired unknown quantity. Therefore, this is a "grams A → particles A" problem, and we solve it by following the pathway obtained from Figure 9–1:

$$\text{grams A} \xrightarrow{\text{molecular weight}} \text{moles A} \xrightarrow{\text{Avogadro's number}} \text{particles A}$$

where the source of each equivalence factor in the pathway is indicated over each arrow.

step 1. 0.10 g LSD

step 2. 0.10 g LSD = molecules LSD

step 3. The equivalence factor needed to change grams of LSD into moles of LSD comes from the molecular weight, determined to be 376.0 amu, and the mole definition. The factor is

$$\frac{1 \text{ mole LSD}}{376.0 \text{ g LSD}}$$

The equivalence factor needed to change moles of LSD into molecules of LSD is obtained from Avogadro's number and the mole definition. The factor is

$$\frac{6.02 \times 10^{23} \text{ molecules LSD}}{1 \text{ mole LSD}}$$

The use of these two factors produces the following:

$$0.10 \text{ g LSD} \times \frac{1 \text{ mole LSD}}{376 \text{ g LSD}} \times \frac{6.02 \times 10^{23} \text{ molecules LSD}}{1 \text{ mole LSD}}$$

$$= \quad \text{molecules LSD}$$

step 4. $$0.10\ \cancel{\text{g LSD}} \times \frac{1\ \cancel{\text{mole LSD}}}{376\ \cancel{\text{g LSD}}} \times \frac{6.02 \times 10^{23}\ \text{molecules LSD}}{1\ \cancel{\text{mole LSD}}} = \quad \text{molecules LSD}$$

step 5. $$\frac{(0.10)(1)(6.02)(10^{23})}{(376)\ (1)}\ \text{molecules LSD} = 1.6 \times 10^{20}\ \text{molecules LSD}$$

(A review of the multiplication and division of exponential numbers is given in Appendix II.)

Example 9–7. What is the weight in grams of a single aspirin molecule which has the formula $C_9H_8O_4$?

The setup for this problem is the reverse of that for the preceding one, since we are given the number of particles and asked to find the number of grams of the same substance. The pathway obtained from Figure 9–1 is therefore:

$$\text{particles A} \xrightarrow{\text{Avogadro's number}} \text{moles A} \xrightarrow{\text{molecular weight}} \text{grams A}$$

step 1. 1 molecule $C_9H_8O_4$

step 2. 1 molecule $C_9H_8O_4$ $\quad = \quad$ g $C_9H_8O_4$

step 3. $$1\ \text{molecule}\ C_9H_8O_4 \times \frac{1\ \text{mole}\ C_9H_8O_4}{6.02 \times 10^{23}\ \text{molecules}\ C_9H_8O_4} \times \frac{180\ \text{g}\ C_9H_8O_4}{1\ \text{mole}\ C_9H_8O_4} = \quad \text{g}\ C_9H_8O_4$$

step 4. $$1\ \cancel{\text{molecule}\ C_9H_8O_4} \times \frac{1\ \cancel{\text{mole}\ C_9H_8O_4}}{6.02 \times 10^{23}\ \cancel{\text{molecules}\ C_9H_8O_4}} \times \frac{180\ \text{g}\ C_9H_8O_4}{1\ \cancel{\text{mole}\ C_9H_8O_4}} = \quad \text{g}\ C_9H_8O_4$$

step 5. $$\frac{(1)(1)(180)}{(6.02)(10^{23})(1)}\ \text{g}\ C_9H_8O_4 = 3.0 \times 10^{-22}\ \text{g}\ C_9H_8O_4$$

Example 9–8. Propane, C_3H_8, is a fuel gas which burns in oxygen according to the following balanced equation:

$$C_3H_8 + 5O_2 \rightarrow 4H_2O + 3CO_2$$

How many grams of O_2 are needed to completely react with 10.0 grams of propane?

There is an important difference between this problem and the preceding two; here we are dealing with two substances, oxygen and propane. The given quantity is grams of propane (substance A), and we are asked to find the grams of oxygen (substance B). The appropriate pathway according to

Figure 9–1 is:

$$\text{grams A} \xrightarrow{\text{mol. wt.}} \text{moles A} \xrightarrow{\text{equation}} \text{moles B} \xrightarrow{\text{mol. wt.}} \text{grams B}$$

step 1.

$$10.0 \text{ g } C_3H_8$$

step 2.

$$10.0 \text{ g } C_3H_8 \qquad = \qquad \text{g } O_2$$

step 3.

$$10.0 \text{ g } C_3H_8 \times \frac{1 \text{ mole } C_3H_8}{44.0 \text{ g } C_3H_8} \times \frac{5 \text{ moles } O_2}{1 \text{ mole } C_3H_8} \times \frac{32.0 \text{ g } O_2}{1 \text{ mole } O_2} = \qquad \text{g } O_2$$

step 4.

$$10.0 \cancel{\text{ g } C_3H_8} \times \frac{1 \cancel{\text{ mole } C_3H_8}}{44.0 \cancel{\text{ g } C_3H_8}} \times \frac{5 \cancel{\text{ moles } O_2}}{1 \cancel{\text{ mole } C_3H_8}} \times \frac{32.0 \text{ g } O_2}{1 \cancel{\text{ mole } O_2}} = \qquad \text{g } O_2$$

step 5.

$$\frac{(10.0)(1)(5)(32.0)}{(44.0)(1)(1)} \text{ g } O_2 = 36.4 \text{ g } O_2$$

Example 9–9. How many grams of nitrogen are contained in 100.0 grams of TNT, $C_7H_5O_6N_3$?

This problem also involves the conversion "grams A → grams B," and the following pathway will be used:

$$\text{grams A} \xrightarrow{\text{mol. wt.}} \text{moles A} \xrightarrow{\text{formula}} \text{moles B} \xrightarrow{\text{mol. wt.}} \text{grams B}$$

step 1.

$$100.0 \text{ g TNT}$$

step 2.

$$100.0 \text{ g TNT} \qquad = \qquad \text{g N}$$

step 3.

$$100.0 \text{ g TNT} \times \frac{1 \text{ mole TNT}}{227.0 \text{ g TNT}} \times \frac{3 \text{ moles N}}{1 \text{ mole TNT}} \times \frac{14.0 \text{ g N}}{1 \text{ mole N}} = \qquad \text{g N}$$

step 4.

$$100.0 \cancel{\text{ g TNT}} \times \frac{1 \cancel{\text{ mole TNT}}}{227.0 \cancel{\text{ g TNT}}} \times \frac{3 \cancel{\text{ moles N}}}{1 \cancel{\text{ mole TNT}}} \times \frac{14.0 \text{ g N}}{1 \cancel{\text{ mole N}}} = \qquad \text{g N}$$

step 5.

$$\frac{(100.0)(1)(3)(14.0)}{(227.0)(1)(1)} \text{ g N} = 18.5 \text{ g N}$$

The Determination of Chemical Formulas

As we have seen, chemical formulas provide a great deal of useful information about the substances they represent. The formula for a compound can be calculated from experimentally determined information. Two types of formulas can be obtained, depending on the amount of information available. An **empirical formula,** representing the correct and simplest combining ratio of the atoms involved, is obtainable from percentage composition data alone. A **molecular formula,** which gives the actual combining ratio of the atoms, is obtainable only if both composition data and molecular-weight information are available.

The actual molecular formula will always be the same as, or some multiple of, the empirical formula. For example, the compound benzene has a molecular formula of C_6H_6, which gives a simplest combining ratio of 1:1 and an empirical formula of CH. The actual formula is a multiple of six times the empirical formula. Water, on the other hand, has an actual formula of H_2O, which gives a simplest combining ratio of 2:1 and an empirical formula of H_2O, identical to the molecular formula. Molecular weights must be known if a molecular formula is to be obtained from composition data. These ideas are illustrated in Examples 9–10 and 9–11.

Example 9–10. Hydrazine, a compound of nitrogen and hydrogen, is a component of some rocket fuels. It has been found to have the following composition: 87.5 percent N and 12.5 percent H. Other experiments have shown the molecular weight to be 32.0 amu. Calculate the empirical and molecular formulas of hydrazine.

We have already seen that the subscripts in a molecular formula give either the ratio of atoms in one molecule or the ratio of moles of atoms in one mole of the substance. The simplest combining ratios of the atoms (and the subscripts of the formula) will therefore be the same as the molar combining ratios. We will obtain the formula by changing the composition data into molar combining ratios. The percentage compositions can be converted into weights by assuming a sample size of 100 grams. The weight of each element becomes the same as the percentage composition, and can be converted into moles by methods already discussed.

$$87.5\,\cancel{\text{g N}} \times \frac{1 \text{ mole N atoms}}{14.0\,\cancel{\text{g N}}} = 6.25 \text{ moles N atoms}$$

$$12.5\,\cancel{\text{g H}} \times \frac{1 \text{ mole H atoms}}{1\,\cancel{\text{g H}}} = 12.5 \text{ moles H atoms}$$

The molar combining ratio of 6.25:12.5 provides an empirical formula of $N_{6.25}H_{12.5}$. We prefer subscripts that are whole numbers. The combining ratio can be changed into whole numbers by dividing both numbers by the smallest of the two numbers.

$$\frac{6.25}{6.25} = 1, \qquad \frac{12.5}{6.25} = 2$$

This ratio of 1:2 gives an empirical formula of NH_2.

The molecular weight of the empirical formula is 16.0 amu, which is just half the experimental value. Therefore, the actual formula must contain twice as many atoms but in the same 1:2 ratio. The molecular formula is therefore N_2H_4.

Example 9–11. An analysis of automobile exhaust showed that the primary lead-containing component has a composition of 64.2 percent Pb, 11.0 percent Cl, and 24.8 percent Br. Calculate the empirical formula of the compound.

We shall convert the percentages to weights by assuming we have a 100-g sample of compound; then we shall convert the weights into moles.

$$64.2\ \cancel{\text{g Pb}} \times \frac{1\ \text{mole Pb}}{207.2\ \cancel{\text{g Pb}}} = 0.31\ \text{moles Pb}$$

$$11.0\ \cancel{\text{g Cl}} \times \frac{1\ \text{mole Cl}}{35.5\ \cancel{\text{g Cl}}} = 0.31\ \text{moles Cl}$$

$$24.8\ \cancel{\text{g Br}} \times \frac{1\ \text{mole Br}}{79.9\ \cancel{\text{g Br}}} = 0.31\ \text{moles Br}$$

We change the moles into whole numbers by dividing through by the smallest.

$$\text{Pb:}\ \frac{0.31}{0.31} = 1, \qquad \text{Cl:}\ \frac{0.31}{0.31} = 1, \qquad \text{Br:}\ \frac{0.31}{0.31} = 1$$

The empirical formula is PbClBr. The actual formula cannot be determined without molecular-weight information.

Occasionally, whole numbers are not obtained when the fractional numbers of moles are divided by their smallest member. Instead, common fractions such as 1.5 or 1.33 result. When this happens, the fractions can be converted into whole numbers by multiplication—$1.5 \times 2 = 3.0$ and $1.33 \times 3 = 3.99 \cong 4.0$.

Review Questions

1. Calculate the molecular weight for each of the following:
   a) CO_2
   b) $MgCO_3$
   c) $(NH_4)_2SO_4$
   d) C_2H_5OH
   e) $Fe_3(PO_4)_2$
2. What is Avogadro's number? What is its numerical value? What is a mole?
3. Calculate the weight (in grams) of one mole of each of the following:
   a) SO_2
   b) N_2H_4
   c) XeF_6

4. How many A atoms does each of the following represent?
   a) A_8
   b) $2A_4$
   c) $4A_2$
   d) 8A
5. How many discrete particles does each of the following represent?
   a) A_8
   b) $2A_4$
   c) $4A_2$
   d) 8A
6. How many nitrogen atoms are there in each of the following:
   a) 1 mole of N_2
   b) $\frac{1}{2}$ mole of NO
   c) 2 moles of N_2H_4
   d) 14 grams of N_2
   e) 60 grams of NO
   f) 32 grams of N_2H_4
7. How many oxygen molecules are there in 20 moles of O_2? How many oxygen atoms?
8. How many grams will each of the following weigh?
   a) 6.02×10^{23} O_2 molecules
   b) 3.01×10^{23} PF_6 molecules
   c) $\frac{1}{2}$ mole of SO_2 molecules
   d) 6.02×10^{23} gold atoms
   e) one carbon atom
9. Consider the contents of the following five flasks, each of which contains a total of 28 grams of material:

   Flask A—28 g N_2
   Flask B—28 g CO
   Flask C—28 g Si
   Flask D—9 g Be + 19 g F_2 (assume they do not react)
   Flask E—20 g Ne + 8 g H_2 (assume they do not react)

   a) Which flasks contain the same number of atoms?
   b) Which flasks contain the same number of moles of particles?
   c) Which flasks contain the same number of grams?
10. How many moles of fluorine, F_2, are required to form 3 moles of xenon hexafluoride, XeF_6, according to the equation $Xe + 3F_2 \rightarrow XeF_6$?
11. According to the equation $2Fe_2O_3 + 3C \rightarrow 4Fe + 3CO_2$, how many atoms of C will be needed to produce 6 moles of Fe?
12. According to the equation $2H_2SO_4 + Cu \rightarrow SO_2 + 2H_2O + CuSO_4$, how many grams of $CuSO_4$ will be produced when 10^{24} molecules of SO_2 are generated?
13. If 138 grams of sodium react completely with water as given by the equation $2Na + 2H_2O \rightarrow 2NaOH + H_2$, how many moles of hydrogen gas will be liberated?

14. How many grams of MnO_2 are required to produce 142 grams of Cl_2 according to the following reaction? $MnO_2 + 4HCl \rightarrow MnCl_2 + Cl_2 + 2H_2O$
15. Carbon monoxide may be reacted to give CO_2 according to the following equation: $2CO + O_2 \rightarrow 2CO_2$. Calculate the weight of O_2 needed to completely react with 120 grams of CO.
16. What is the molecular formula for each of the following compounds:
    a) Empirical formula = CH_2, molecular weight = 28
    b) Empirical formula = HO, molecular weight = 34
    c) Empirical formula = $PNCl_2$, molecular weight = 348
17. Explain why C_6H_8 could not possibly be the molecular formula for a compound whose empirical formula is C_3H_2.
18. Find the empirical formula of a compound consisting of 2.90 grams Na, 4.05 grams S, and 3.04 grams O.
19. Find the empirical formulas from the given data in each case:
    a) 45.5 percent Ni, 54.5 percent Cl
    b) 52.83 percent Pt, 16.79 percent P, 30.40 percent O
    c) 87.6 percent Zn, 12.5 percent N
20. Calculate the percentage composition (the percentage of each element) of the following:
    a) H_2O
    b) $CaCO_3$
    c) NH_4NO_3
    d) $CuCO_3 \cdot Cu(OH)_2$ (include Cu and O from both sources in your calculation)

Suggestions for Further Reading

Benson, S. W., *Chemical Calculations*, 3rd ed., John Wiley and Sons Inc., New York, 1971 (paperback).

Johnson, M. D., *Problem Solving and Chemical Calculations*, Harcourt Brace Jovanovich, Inc., New York, 1972 (paperback).

Kieffer, W. F., *The Mole Concept in Chemistry*, Reinhold Publishing Corp., New York, 1962 (paperback).

Pierce, C., and R. N. Smith, *General Chemistry Workbook*, 4th ed., W. H. Freeman Co., San Francisco, 1971 (paperback).

10 Water: An Abundant and Vital Compound

Water, a compound formed by the reaction of two simple gases, is the most abundant and essential compound with which man deals. This substance covers approximately 75 percent of the earth's surface and constitutes about 70 percent of the weight of each human body.

Unusual Properties of Water

Much of the usefulness of water results from some rather unusual properties that are not expected on the basis of periodic relationships. These properties include the boiling point, heat of vaporization, freezing point, heat of fusion, and the temperature dependence of density.

These unusual properties become apparent when a comparison is made between water, the hydride of oxygen, and the hydrides of the other group VIA elements. The attractive forces between molecules generally increase with molecular size in a series of such compounds. On this basis, more energy should be required to separate the molecules of higher-molecular-weight compounds from each other during change-of-state processes such as fusion or evaporation. Thus, water, the smallest molecule in the series H_2O, H_2S, H_2Se, and H_2Te, should undergo changes in state at the lowest temperature.

In Figure 10–1 the boiling points of group VIA hydrides are plotted as a function of molecular weight. According to this plot (which is not to scale), water should have a boiling point of −90°C, as obtained by an extrapolation of the line connecting the three heavier compounds. The actual boiling point of water, under a pressure of one atmosphere, is 100°C—a difference of 190°C from the expected value.

Water's heat of vaporization, 540 calories per gram, is also higher than

Figure 10–1.
Boiling Points
of the Group VI A Hydrides

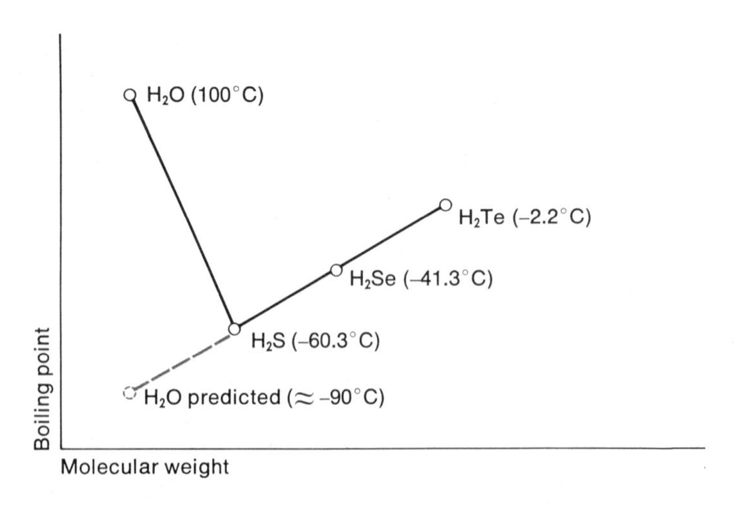

expected. It is more than four times as large as the heat of vaporization for H_2S, the next least heavy of the compounds found in Figure 10–1.

A higher-than-expected freezing point is also characteristic of water, and a plot of freezing point versus molecular weight has the same general shape as Figure 10–1. This high freezing point is accompanied by an unusually large heat of fusion of 80 calories per gram.

A very striking and important behavior of water is the variation of its density with temperature. The **density** of a substance is defined as the weight per unit volume (weight/volume); for our purposes it will be given in units of grams/ml. This means that the numerical value for a density is merely the weight in grams of a single milliliter of the substance in question. For most materials the volume of a given weight decreases as the temperature falls, owing to the decreasing kinetic energy of the particles making up the material. Since the volume of a fixed weight decreases, the weight/volume ratio increases, and so an increase in density normally accompanies a decrease in temperature. Table 10–1 contains density and temperature data for water.

Table 10–1. Water Density as a Function of Temperature

Temperature (°C)	*Water density (g/ml)*
100	0.9586
80	0.9719
60	0.9833
40	0.9923
20	0.9982
10	0.9997
5	0.9999
4 (actually 3.98)	1.0000
2	0.9999
0 (liquid H_2O)	0.9998
0 (solid H_2O)	0.9170

It is easily seen from Table 10–1 that water behaves normally, and the density increases, down to a temperature of 3.98°C. However, below 3.98°C water becomes less dense than at 3.98°C. At the freezing point of water, 0°C, a marked decrease in density occurs as the liquid water is transformed into solid ice.

Hydrogen Bonding in Water

All of the unexpected properties of water mentioned previously can be explained when the structure and characteristics of the water molecule are understood. Water is an angular molecule, owing to the geometry of the *p* atomic orbitals of oxygen that are involved in the bonding. The molecule can be represented as

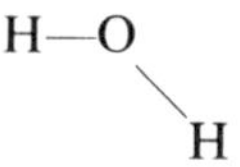

The oxygen atom is much more electronegative than the hydrogen atoms, and the resulting attraction of the bonding electrons toward the oxygen causes the bonds to become polarized. The polarized bonds and angular nature of the molecule produce a polar molecule which can be represented as

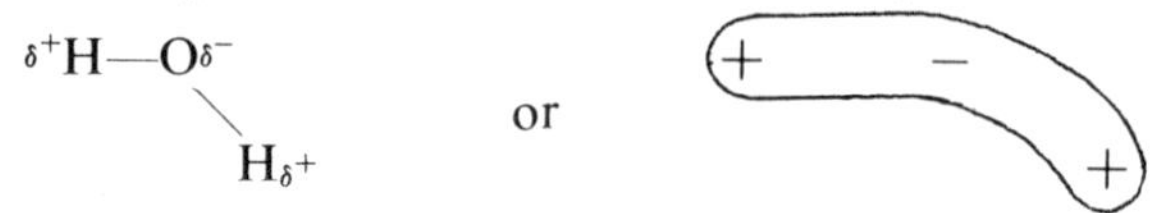

The bonding electrons of the water molecule are the only electrons around the hydrogen nuclei. The polarization of these electrons toward the oxygen atom leaves essentially bare protons at the positive ends of the resulting polar molecule. These positive ends possess a high charge density because of their small size and lack of electronic charge influence. These ends are very strongly attracted toward the negative oxygen ends of other water molecules, which results in a bonding force between the molecules.

Because of the electronegativity differences, the small size of the proton ends of molecules, and the high charge density, the resulting bond is much stronger than the usual polar bond. This relatively strong bond is called a **hydrogen bond,** because it appears that hydrogen atoms are, to some extent, bonding oxygen atoms together in the resulting network of molecules. This bonding network for water is shown in Figure 10–2, where the dotted lines are hydrogen bonds. Hydrogen bonding in the other group VIA hydrides is not appreciable because of the much lower electronegativity differences between the elements and hydrogen, and the resulting smaller amount of bond polarization.

The presence of hydrogen bonding between water molecules accounts for the unexpected properties mentioned earlier. The higher-than-expected boiling and freezing points result because the forces of hydrogen bonding are able to keep the molecules in the more condensed state in spite of high kinetic energies. The molecular vibrations in the solid can be more energetic in the presence of hydrogen bonding than would otherwise be possible without forming a liquid. Similarly, the molecules in the liquid, when constrained by hydrogen bonds, can move at higher speeds without forming vapor. The large heats of fusion and vaporization that accompany phase changes are reflections of the large amount of potential energy that is accumulated as the strongly attracting molecules are separated from each other.

The peculiar density behavior can be explained by visualizing the events that occur on a molecular scale as water is cooled to a temperature below its freezing point. At some higher temperature, such as 80°C, the kinetic energy of the molecules as they move over and past one another in the liquid state is

Figure 10–2.
Hydrogen Bonding in Water

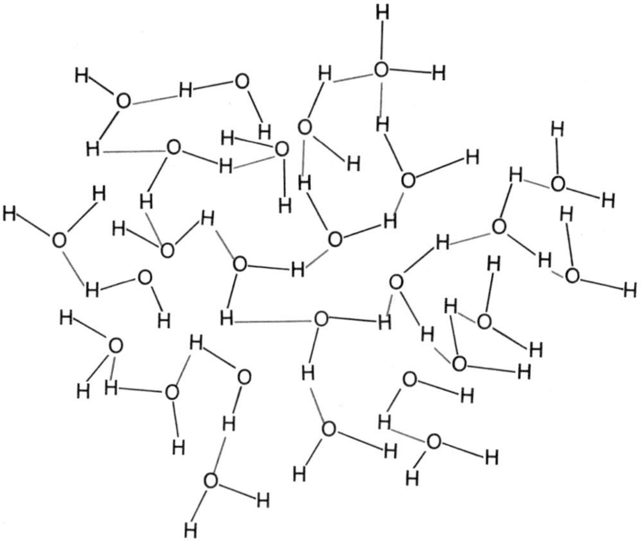

Figure 10–3.
The Structure of Ice

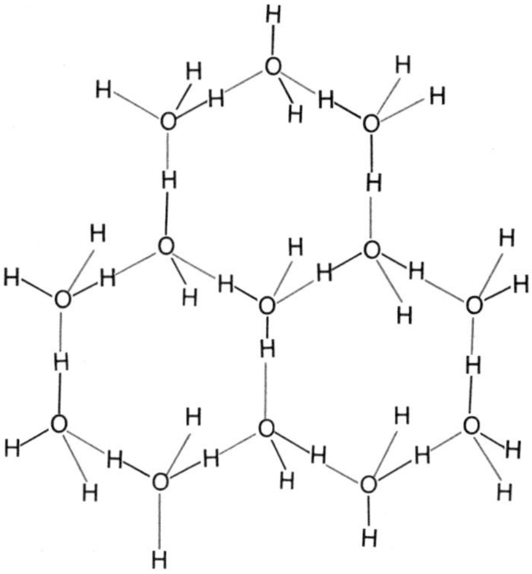

sufficiently large to prevent hydrogen bonding from having much of an orientation effect on the molecules. As the temperature is lowered, the decrease in kinetic energy allows the molecules to move closer together, but still the hydrogen bonds have little orientation effect on the molecules. This expected behavior continues until the temperature reaches 3.98°C. At this temperature the average distance between molecules is as small as it ever gets. When the temperature is lowered further, the kinetic energy of the moving molecules becomes too low to prevent hydrogen bonding from orienting the water molecules into a hydrogen-bonded structure. This orientation effect becomes more pronounced as the temperature drops lower; and the density decreases because the oriented water has a more open structure—more empty space between molecules—than less oriented water. This process continues down to the freezing point, at which temperature the hydrogen bonds orient the molecules to the maximum degree, and the solid crystal lattice of ice is formed. This solid lattice is an extremely open type of structure, which results in the very low density of the solid form (see Figure 10–3).

The Results of Water's Unusual Properties

The abundance of water accounts in part for its widespread use as a coolant. However, the large heats of vaporization and fusion are also important factors. These properties make water an effective coolant for many natural and man-made processes, because changes of state cause large amounts of energy to be absorbed.

Large bodies of water exert a temperature-moderating effect on their surroundings, partially as a result of water's large heats of vaporization and fusion. In the heat of a summer day large amounts of water evaporate, and in the process energy is absorbed from the surroundings, so that their temperature is lowered. In the cool evening some of this water vapor condenses back to the liquid state, releasing heat, which raises the temperature of the surroundings. In this way the temperature variation between night and day is reduced. In the winter a similar process occurs; water freezes on cold days and releases energy or heat to the surroundings. When spring arrives, the melting ice absorbs energy, and the temperature variation between winter and spring is reduced. This last effect is also the basis for the old practice of placing containers of water in the unheated root cellars of farms when freezing temperatures were expected. The freezing water liberated enough heat energy to keep the temperature of the air higher than the freezing point of the fruits and vegetables stored in the cellar.

The heat of vaporization is involved in the cooling of the human body. The coolant in this case is perspiration, which evaporates, absorbs heat in the process, and lowers the skin temperature. For similar reasons a swimmer upon emerging from the water on a warm but windy day will feel cold and will shiver as excess water rapidly evaporates from his body.

Water at its boiling point is still an effective coolant, because 540 calories of heat are absorbed for every gram of water vaporized. This explains why a car can generally be driven without damaging the engine in spite of a boiling radiator.

The peculiar density behavior of water is very important, especially in regions which have freezing temperatures during part of the year. If a container is filled with water and sealed, the expansion of the water upon freezing will break the container. This is the reason automobile radiators must be protected from freezing in the winter. Any water trapped inside the engine or radiator can, upon freezing, generate sufficient force to burst even the steel or copper parts.

Of greater importance are the effects of density change with temperature in a lake or other open body of water. In the fall, surface water is cooled through contact with cold air. The water becomes more dense than the warmer water underneath and sinks. In this way, cool water is circulated from the top of the lake to the bottom until the entire lake has reached the temperature of maximum density, 4°C. During the circulation process, oxygen and nutrients are distributed throughout the water. Upon further cooling of the water, a new behavior pattern emerges. Surface water cooled below 4°C becomes less dense than the water underneath and remains on the surface. In this way a thin layer of surface water is cooled to the freezing point and changed to ice, which floats because of its still lower density. Even in the coldest winters, lakes will usually not freeze to a depth of more than a few feet, because the ice forms an insulating layer over the water.

These cooling, circulation, and freezing processes contribute to the temperature-moderating effects mentioned earlier, and also allow aquatic life to survive through the winter. The aquatic organisms use the oxygen that was distributed to the lower depths during the circulation of the water. If water behaved like most substances, freezing would occur from the bottom up, and most aquatic life would be destroyed. In fact, many lakes would remain frozen all year long except for a few feet on the surface.

For a few weeks in the early spring, circulation again occurs during the melting and warming of the surface water. This process quickly stops, however, as the surface waters are warmed above 4°C. Above this temperature they become less dense and form a layer over the colder water of higher density. This causes a permanent thermal stratification that persists through the hot summer months. In the fall, circulation again begins and the cycle is repeated.

The peculiar density behavior of water also makes possible the sport of ice skating. As noted in Chapter 8, most substances can exist in more than one of the physical states: solid, liquid, or gas. When pressure is placed on a substance in one of these states, the substance will tend to revert to the state of higher density. For example, when a gas is placed under pressure, it tends to become more like a liquid; if the temperature is low enough, it may even liquefy. This occurs because liquids are more dense than gases. In the case of ice, the liquid is more dense than the solid, and pressure placed upon the solid will cause it to become liquid if the temperature is high enough. Ice skating is made possible by the liquid layer of water which forms under the skate blade. The pressure on the ice under the blade is quite high, because all the skater's weight is concentrated on a small area. Presumably, it can become too cold for skating. If the temperature is low enough, water will not form under the skate, skating becomes difficult, and the ice is said to be "slow." Under these circumstances a heavy person has an advantage, since the pressure under the skate will be higher and can, perhaps, cause the liquid layer to form even at low temperatures.

Natural Waters

All water, as it occurs in nature, is impure in a chemical sense. The impurities include suspended matter, microbiological organisms, and dissolved minerals and gases. The composition of the dissolved material varies; some typical constituents are listed in Table 10–2.

The diversity of dissolved substances illustrates the broad solvent properties of water. These properties are discussed in more detail in Chapter 11. Water has the ability to dissolve a large variety of materials because of the highly polar nature of its molecules. Its good solvent properties are both beneficial and detrimental from the viewpoint of man. They are beneficial because they make water a useful medium for carrying nutrients and oxygen

Table 10–2. Dissolved Substances Found in Natural Waters

Gases	*Positive ions*	*Negative ions*
N_2	Na^+	Cl^-
O_2	K^+	F^-
CO_2	H^+	OH^-
SO_2	NH_4^+	NO_3^-
H_2S	Ca^{2+}	HS^-
NH_3	Mg^{2+}	HCO_3^-
CH_4	Fe^{2+}	CO_3^{2-}
	Zn^{2+}	SO_4^{2-}

throughout the body, and they make water a useful material in washing processes. They are detrimental because materials tend to dissolve which are difficult or expensive to remove when the water is needed for specific purposes.

Natural waters are classified into three categories on the basis of their dissolved mineral content: fresh water, inland brackish water, and sea water. Water containing less then 0.1 percent total dissolved solids is generally considered to be fresh, but the U.S. Public Health Service has set 0.05 percent as the recommended upper limit for drinking water. The inland underground and surface supply of brackish water contains dissolved solids in the range of 0.1 to 3.5 percent, with an average of about 0.6 percent. Sea water normally contains 3.5 percent total dissolved solids, but bodies of water such as the Great Salt Lake and the Dead Sea contain much higher levels—up to 25 percent. Ninety-seven percent of the earth's total water supply is ocean water. The inland brackish supplies make up about 2.5 percent, leaving only 0.5 percent as usable fresh water.

Because of the impurities present, nearly all water intended for human use must undergo some kind of treatment. The extent of purification necessary, and the methods used, depend upon the ultimate use of the water. For water to be used for cooking or drinking, the treatment is aimed primarily at disease prevention. If the water is to be used for cleaning or industrial purposes, the treatment must accomplish the removal of particular mineral substances.

The Purification of Water

The treatment of fresh water for public use involves processes designed to remove or destroy suspended particles, disease-causing agents, and objectionable colors and odors. Dissolved minerals are not usually present in sufficient quantities to require their removal for health or esthetic reasons.

The usual steps in the treatment sequence are sedimentation, filtration, and chemical disinfection.

Finely divided suspended solids are removed by treating the water with substances that form a **flocculating agent,** composed of a gelatinous solid that is highly adsorbent toward small particles and some microorganisms. This flocculating agent settles to the bottom of sedimentation tanks and carries most suspended matter with it. Two common flocculating agents are aluminum hydroxide and iron(II) hydroxide, which are formed by reactions between the corresponding sulfates and slaked lime, $Ca(OH)_2$.

$$3Ca(OH)_2 + Al_2(SO_4)_3 \rightarrow 2Al(OH)_3 + 3CaSO_4 \qquad 10\text{–}1$$

$$Ca(OH)_2 + FeSO_4 \rightarrow Fe(OH)_2 + CaSO_4 \qquad 10\text{–}2$$

The by-product, $CaSO_4$, is a crystalline precipitate that also settles out. After the flocculation treatment, the water is passed through filtering beds of gravel and sand, where any remaining flocculating agent, suspended particles, and some bacteria are removed.

In some installations the water is sprayed into the air to remove odors—a process called **aeration.** In another method, finely divided carbon is used to adsorb and remove odor-causing substances.

As a final precaution, a small amount of disinfectant is usually added. Chlorine gas is the most widely used, because it is inexpensive and quite effective in small quantities. It behaves as an oxidizing agent and converts microorganisms into such simple compounds as CO_2 and H_2O. The Cl_2 gas is reduced to Cl^- ions in the process.

Hard Water

The term **hard water** is used to describe water that contains more than 0.005 percent dissolved minerals. According to this definition, most natural waters are hard; and conventional purification steps do not alter the situation, since dissolved minerals are not removed. Such naturally hard water is fit to drink, but the high mineral content creates problems when the water is used for other purposes.

Two very noticeable effects of hardness are the formation of a sticky, curdy precipitate (scum) with soap; and the deposition of a hard scale in steam boilers, tea kettles, and hot water pipes. The scum, among other things, produces a ring on bathtubs and imparts a dull, gray appearance to washed clothes. The hard scaly deposit inside boilers and pipes forms an effective insulating layer, and heat is not transferred efficiently to the water inside.

Both of these effects are related to the insolubility of certain magnesium and calcium salts. Magnesium and calcium ions are two of the positive ions present in largest amounts in hard water. The other prevalent ion, Na^+, does not cause problems, since all common sodium salts are soluble.

Commercial soaps are, for the most part, soluble sodium salts of weak organic acids (Chapter 18). Sodium stearate, $NaC_{18}H_{35}O_2$, is the most common soap used. It dissolves in water and forms the sodium ion and stearate ion, $C_{18}H_{35}O_2^-$. The stearate or a similar ion is the active cleaning agent in a soap. It exerts a cleansing action by allowing nonpolar soiling agents such as oils or greases to become suspended in water. The negative ionic charge is localized on an oxygen atom found on one end of the long, chainlike ion. The other end of the ion is uncharged, nonpolar, and similar in character to the soiling agents. The nonpolar end dissolves in the soiling agents, and the charged end dissolves in water. A link is thus formed between the two that allows the soiling agents to be lifted from the soiled item and suspended in the wash water, as illustrated in Figure 10–4.

Hard water causes problems because salts such as $Ca(C_{18}H_{35}O_2)_2$ or $Mg(C_{18}H_{35}O_2)_2$ are not as soluble in water as the sodium salts. When a soap is dissolved in hard water, the negative stearate ion reacts with the hard-water ions, and the resulting insoluble salts precipitate out of solution. In this way the stearate ion is removed from solution and can no longer function as a cleaning agent. The precipitated solid remains in the clothes and causes them to look dirty and gray.

One answer is the use of synthetic detergents. These cleaning agents are structurally similar to soaps but form more soluble salts with Ca^{2+} and Mg^{2+}. These substances are also discussed further in Chapter 18.

Boiler scale consists primarily of calcium carbonate, calcium sulfate, and magnesium hydroxide. Calcium sulfate is produced because some hydrated forms of $CaSO_4$ are less soluble in hot water than in cold water. The $CaCO_3$ results from the decomposition of soluble $Ca(HCO_3)_2$ according to the following reaction:

$$Ca(HCO_3)_2 \xrightarrow{\text{heat}} CaCO_3 + H_2O + CO_2 \qquad 10\text{–}3$$

The production of CO_2 in this reaction increases the *p*H (see Chapter 11)—produces a higher concentration of OH^- ions—which in turn causes the precipitation of $Mg(OH)_2$. Thus, all of the reactions leading to boiler-scale formation result from an increase in temperature.

Figure 10–4.
The Cleansing Action of Soap

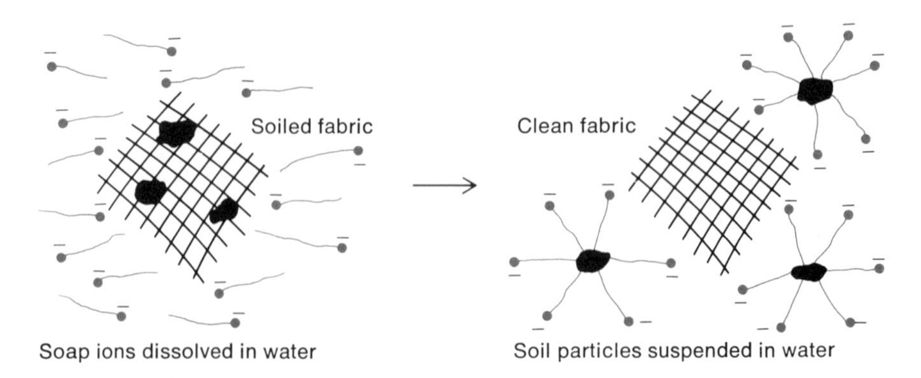

Water Softening

Several methods are available for removing objectionable minerals from water. The removal of these ions is referred to as **softening** the water.

There are two types of hardness, **temporary** and **permanent.** Temporary hardness is caused by the presence of dissolved $Ca(HCO_3)_2$ and $Mg(HCO_3)_2$. Permanent hardness is characterized by the presence of dissolved $CaCl_2$, $MgCl_2$, $CaSO_4$, and $MgSO_4$. Temporary hardness can be eliminated by boiling the water. This causes the bicarbonate ion to decompose into a carbonate ion. The Ca^{2+} and Mg^{2+} ions react with the CO_3^{2-} and precipitate out of solution. The reaction involved is given in equation 10–3. Permanently hard water cannot be softened by boiling; the SO_4^{2-} and Cl^- ions are unaffected by heat. It can, however, be softened by the following methods: precipitation, sequestration, ion exchange, and distillation.

Both Ca^{2+} and Mg^{2+} can be removed by **precipitation.** The carbonates and phosphates of both ions are insoluble, and the addition of soluble carbonates and phosphates to hard water causes the insoluble salts to form and come out of solution. Both Na_2CO_3 (washing soda) and Na_3PO_4 have been used for this purpose.

$$Ca^{2+} + CO_3^{2-} \rightarrow CaCO_3 \qquad 10\text{–}4$$

$$3Mg^{2+} + 2PO_4^{3-} \rightarrow Mg_3(PO_4)_2 \qquad 10\text{–}5$$

A disadvantage of this method is the formation of an insoluble product that might show up as a scum.

Sequestration involves the formation of soluble complexes containing the offending ions. The ions remain in solution as complexes but can no longer form precipitates with soaps. A commonly used sequestering agent is sodium polyphosphate $(NaPO_3)_n$, which is marketed under the trade name Calgon. The sequestering reaction is

$$Ca^{2+} + (NaPO_3)_n \rightarrow \underset{\text{soluble complex}}{[Ca(NaPO_3)_n]^{2+}} \qquad 10\text{–}6$$

Many commercial water-softening products contain large amounts of sequestering agents. One function of the phosphates found in some detergents is that of sequestering. The sequestering method has one advantage over the precipitation method: no scum can be produced.

The most effective method for softening water in the home utilizes a process called **ion exchange,** in which the offending ions are removed and some non-offending ions (usually Na^+) are substituted. This process is carried out by percolating the hard water through a container filled with a finely divided substance capable of performing the ion exchange. Naturally occurring minerals with this ability are zeolites, complex sodium aluminum silicates with formulas such as $Na_2Al_2Si_4O_{12}$. The reactions that occur can be represented as follows, where Z stands for the complex negative ion of the zeolite:

$$Ca^{2+} + 2NaZ \rightleftharpoons CaZ_2 + 2Na^+ \qquad 10\text{–}7$$

or

$$Mg^{2+} + 2NaZ \rightleftharpoons MgZ_2 + 2Na^+ \qquad 10\text{–}8$$

Notice that the offending ions remain on the zeolite and are replaced by an equivalent amount of sodium ion. After sufficient use, the zeolite becomes "saturated" with hard-water ions and can no longer soften water. The resin can be regenerated by washing it with a concentrated salt (NaCl) solution. This treatment shifts the equilibrium of equations 10–7 and 10–8 to the left and converts most of the resin back into NaZ. Man-made ion-exchange resins are available which work the same way as the zeolites and which are used in some commercial water softeners. The operation of an ion-exchange type water softener is illustrated in Figure 10–5.

Extremely pure water can be obtained for use in the laboratory by using mixtures of ion-exchange resins which substitute H^+ for metal ions and OH^- for negative ions. The H^+ and OH^- get together and form water, so the net effect is total removal of all ions. The reactions, where R and R′ represent the negative ions of the resins, are

$$Ca^{2+} + 2HR \rightleftharpoons CaR_2 + 2H^+ \qquad 10\text{–}9$$

$$SO_4^{2-} + 2R'OH \rightleftharpoons R'_2SO_4 + 2OH^- \qquad 10\text{–}10$$

$$2H^+ + 2OH^- \rightarrow 2H_2O$$

The water resulting from this treatment is called **deionized** water. The resins involved can be regenerated if they are used in separate containers.

Figure 10–5.
Ion Exchange Water Softener

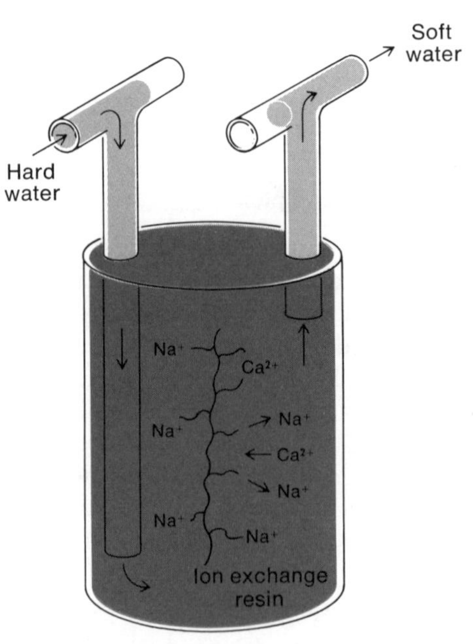

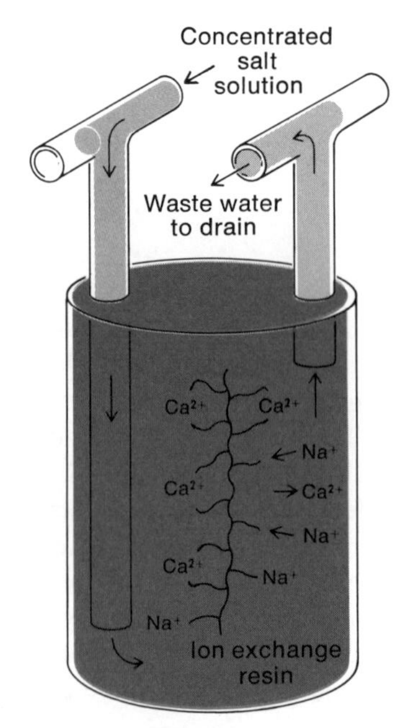

Pure water, primarily for laboratory use, can be prepared by **distillation.** In this process water is boiled and the steam, which contains no ions, is condensed into ion-free water. At present, distillation is too expensive for large-scale use.

Fresh Water from Sea Water

The demand for fresh water in the United States has increased nearly tenfold since 1900—40 billion gallons per day (bgd) in 1900 compared to over 400 bgd in 1970. This demand is expected to climb to about 500 bgd by 1980. A seemingly obvious answer to the impending fresh-water supply problem is to reclaim some of the huge quantities of water found in the sea.

The development of economical, large-scale processes for producing fresh water from sea water has been the subject of intensive investigation. At present, desalted sea water costs at least 85 cents per 1000 gallons when produced in the most efficient desalting plants. By contrast, the cost of producing municipal water from conventional fresh-water sources is 30–35 cents per 1000 gallons. The cost of desalting water is a function of plant size, energy costs, and the salt concentration in the water. Figure 10–6 gives estimated

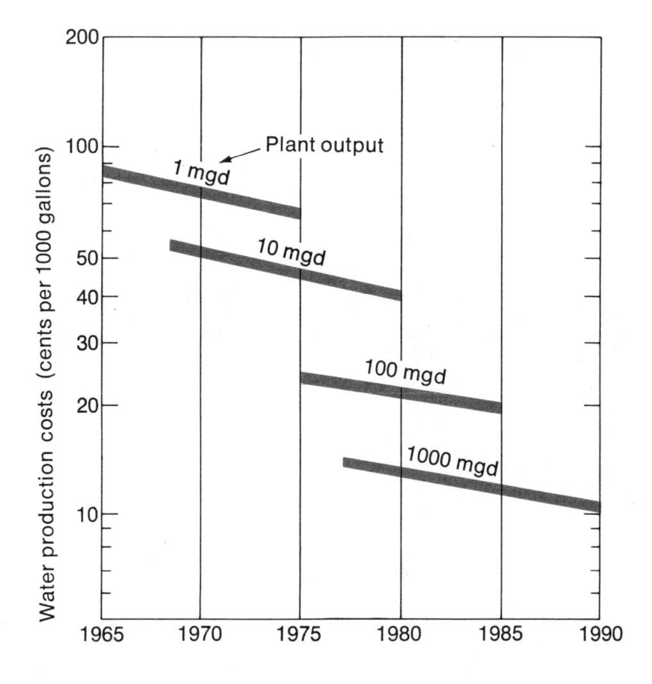

Courtesy of the Office of Saline Water.

Figure 10–6.
Production Costs for Desalting Sea Water

production costs for various plant sizes based on current technology and anticipated advances in such technology. The largest plant operating during 1970 in the United States produced 2.6 million gallons per day (mgd).

According to a 1970 worldwide inventory, there were 680 desalting plants in operation or under construction with capacities of 25 thousand gpd or greater. This amounts to a worldwide capacity of 247 million gpd. The Middle East leads the world in desalting capacity with 74 plants producing a total of 62.9 mgd. The United States and its territories are second with 53.4 mgd coming from 322 plants.

About 98 percent of the total world's desalted water is produced by distillation processes. The remaining 2 percent comes from the use of crystallization and membrane systems.

The various types of distillation processes are designed to boil the sea water and condense the resulting vapor into pure fresh water. As easy as it may appear, the process of distillation presents many difficult technological problems. As sea water is heated to about 72°C, some dissolved salt ($CaSO_4$) begins to precipitate out of solution and form a scale on all heat-transferring surfaces. The scale acts as an insulator and reduces the efficiency of the heat-transfer operation. Techniques have been developed to control scale formation up to a temperature of about 140°C, and research is continuing in an attempt to find control methods that are effective at still higher temperatures.

Another problem associated with all distillation cycles is the corrosive effect of hot sea water on exposed metal surfaces. Metals are available that are capable of withstanding such corrosion, but their high costs prevent their use in economical processes. Such obstacles as these must, however, be overcome before economical multimillion-gpd plants become a reality.

Figure 10–7.
Electrodialysis of Salt Water

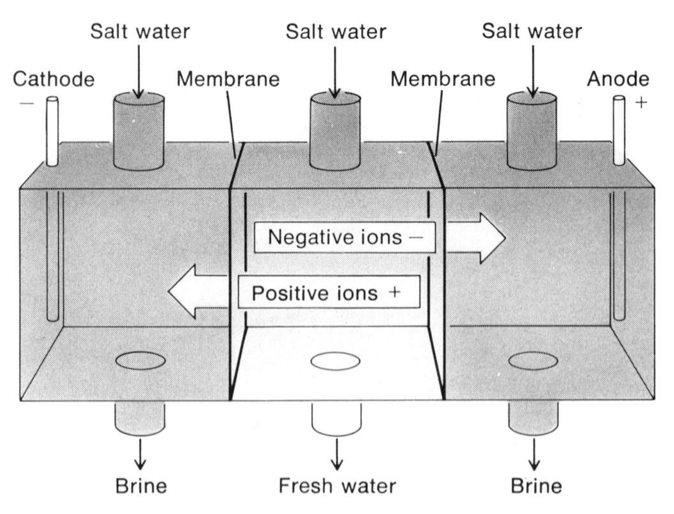

Redrawn from Richard H. Wagner, *Environment and Man* (New York: W. W. Norton and Company, Inc., 1971), p. 102.

Other processes being explored in smaller plants involve freezing and the use of membranes. Freezing processes take advantage of the fact that when salt water is frozen, crystals of pure water are formed. Freezing processes show some promise because freezing requires less energy than evaporation, and corrosion and scaling are less of a problem at lower temperatures. The freezing may be accomplished by the flash evaporation of sea water at a low pressure of 3 torr, or by the vaporization of a refrigerant in contact with the sea water. A major problem with freezing processes involves the lack of efficient methods for separating the ice crystals from the brine.

Electrodialysis is the most advanced of the membrane processes. Electrodialysis removes salts from water by the application of an electric current. The dialysis cell contains two different types of ion-selective membranes. One allows the passage of negative ions and one the passage of positive ions. An electric current provides the driving force that moves the ions through the membranes. The compartment between the membranes becomes depleted of salt, as shown in Figure 10–7.

Review Questions

1. List five properties of water that are not "normal" on the basis of the periodic law. What process in water causes these anomalies?
2. What is meant by the term *hydrogen bonding?*
3. Account for the abnormally high boiling point of water.
4. Most substances contract on changing from the liquid to the solid state. Water expands. Account for this "abnormal" behavior of water.
5. Discuss the importance of the "abnormal" density behavior of water as the temperature changes.
6. How many grams of water at 100°C could be vaporized by 1000 calories of heat?
7. Discuss the basis for classification of natural waters into categories.
8. Explain how aluminum sulfate and calcium hydroxide are used to help purify water.
9. What metal ions are commonly found in natural waters? Which of these are responsible for the hardness of water?
10. List two common problems which arise from hard water use.
11. From the points of view of dissolved substances and properties, compare hard water and water softened by reaction with the following substances:
    a) A zeolite-like resin
    b) Washing soda
    c) A sequestering or complexing agent (Calgon)
12. Contrast the terms *water softening* and *water demineralizing*.
13. Distinguish between temporary and permanent hardness of water.
14. What is the difference between distilled water and deionized water?

15. How does the current price for producing municipal water from sea water compare with that for producing it from conventional fresh water sources? What are the 1985 price estimates for desalting sea water?
16. What is the principal process in use today for desalting sea water? What problems are associated with this process?

Suggestions for Further Reading

Buswell, A. M., and W. H. Rodebush, "Water," *Scientific American*, April 1956.

Chalmers, B., "How Water Freezes," *Scientific American*, Feb. 1959.

Gillam, W. S., and J. W. McCutchan, "Demineralization of Saline Waters," *Science*, Oct. 13, 1961.

Penman, H. L., "The Water Cycle," *Scientific American*, Sept. 1970.

Revelle, R., "Water," *Scientific American*, Sept. 1963.

Runnels, L. K., "Ice," *Scientific American*, Dec. 1966.

Snyder, A. E., "Desalting Water by Freezing," *Scientific American*, Dec. 1962.

11 The Chemistry of Solutions

Solutions represent one of the most widely used forms of chemicals. A solution consists of one or more substances uniformly and homogeneously dissolved in another substance. The substance present in the largest amount is called the **solvent** and any other components are referred to as **solutes.**

Types of Solutions

Solutions representing all three states of matter—solid, liquid, and gas—are known. The type of solution depends upon the final state after solutes and solvent have been combined. The final state is often the same as the original state of the solvent. The solute is usually no longer in its original form after dissolving. For example, sugar is no longer a solid after it has been dissolved in water; it is in the form of separate molecules which are free to move about like the solvent water molecules. Table 11–1 contains examples of the various combinations of solvent and solute found in solutions. The original states of the solvent and solute are given in parentheses.

Table 11–1. Examples of Solution Types

Solution	*Solution state*	*Solvent*	*Solute*
Salt water	Liquid	Water (liquid)	Sodium chloride (solid)
Alcoholic beverage	Liquid	Water (liquid)	Alcohol (liquid)
Carbonated water	Liquid	Water (liquid)	Carbon dioxide (gas)
Gold alloy (jewelry)	Solid	Gold (solid)	Copper (solid)
Gold amalgam	Solid	Gold (solid)	Mercury (liquid)
Hydrogen in palladium	Solid	Palladium (solid)	Hydrogen (gas)
Air	Gaseous	Nitrogen (gas)	Oxygen (gas)
Humid oxygen	Gaseous	Oxygen (gas)	Water (liquid)
Camphor in nitrogen	Gaseous	Nitrogen (gas)	Camphor (solid)

Concentration of Solutions

The amount of solute dissolved in a specified quantity of solvent is called the **concentration** of a solution. A number of different units are used to express solution concentrations. Two commonly used units are **weight percent,** the number of grams of solute in 100 grams of solution, and **molarity,** the number of moles of solute in one liter of solution.

In general, the amount of solute that can dissolve in a specified amount of solvent is limited. A solution is **saturated** when the maximum amount of solute is dissolved in a specific quantity of solvent. The amount of solute needed to produce a saturated solution is referred to as the **solubility** of the solute. For example, at 20°C exactly 36.0 grams of sodium chloride will dissolve in 100 grams of water. The resulting saturated solution has a concentration given by the following calculation:

$$\frac{36 \text{ g solute}}{136 \text{ g solution}} \times 100 = 26.4 \text{ percent sodium chloride}$$

By definition the solubility of sodium chloride at 20°C is 36.0 g/100 g H_2O.

The concentration of a solution in moles of solute per liter of solution can be calculated in a similar way. Suppose a solution is made by placing 4.0 grams of sodium hydroxide into a container and adding enough water to make the final volume of the resulting solution exactly 1.0 liter. The molarity of the solution is given by the following calculation, in which the molecular weight of NaOH is 40.0 amu:

$$\frac{4.0 \text{ g NaOH} \times \dfrac{1 \text{ mole NaOH}}{40.0 \text{ g NaOH}}}{1 \text{ liter of solution}} = \frac{0.1 \text{ mole NaOH}}{1 \text{ liter solution}} = 0.1 \text{ molar NaOH}$$

The *p*H of Solutions

We saw in Chapter 5 that acids and bases react with water to produce H_3O^+ and OH^- ions, respectively. The molar concentrations of these ions are often expressed by a quantity known as the *p*H.

In any water solution the following relationship always exists between the molar concentrations of H_3O^+ and OH^-:

$$(\text{molar conc. } H_3O^+)(\text{molar conc. } OH^-) = 10^{-14}$$

or, in abbreviated form,

$$[H_3O^+][OH^-] = 10^{-14} \qquad 11\text{–}1$$

where [ ] denotes molar concentrations. On the basis of this relationship, the concentration of either H_3O^+ or OH^- can be calculated if the concentration of the other ion is known. For example, the 0.1 molar NaOH solution described above has a concentration of OH^- (and Na^+) equal to 0.1 molar, because NaOH is a strong base that completely dissociates into ions according to the reaction

$$NaOH \rightarrow Na^+ + OH^- \qquad 11\text{–}2$$

The concentration of H_3O^+ in this solution is determined by the following calculation:

$$[H_3O^+][OH^-] = 10^{-14}, \qquad [H_3O^+] = \frac{10^{-14}}{[OH^-]}$$

and

$$[H_3O^+] = \frac{10^{-14}}{10^{-1}} = 10^{-13} \text{ moles/liter}$$

The pH of a solution is equal to the negative value of the exponent on 10 used to express the molar concentration of H_3O^+. Accordingly, the above solution, in which $[H_3O^+] = 10^{-13}$, has a pH of $-(-13)$ or 13. A similar quantity, the pOH, is equal to the negative of the exponent used to express the OH^- concentration. The pOH of the solution in which $[OH^-] = 10^{-1}$ is therefore $-(-1)$ or 1. When these pH and pOH definitions are applied to equation 11–1, the following relationship is obtained:

$$pH + pOH = 14 \qquad 11\text{–}3$$

In pure water the concentrations of H_3O^+ and OH^- are identical and therefore equal to 10^{-7} molar (equation 11–1). The resulting pH value is 7. Solutions in which the pH is the same as water are called **neutral.** Those with a pH lower than 7—a higher H_3O^+ concentration than water—are **acidic,** and those with a pH higher than 7—a lower H_3O^+ concentration than water—are called **basic** or **alkaline.** Table 11–2 summarizes this discussion.

Factors Affecting Solubility

The solubilities of various solutes in different solvents at the same temperature vary widely. Sugar is very soluble in water but only slightly soluble in gasoline, while iodine is only slightly soluble in water but very soluble in alcohol.

We can understand the factors contributing to this variation by considering the solution process. A solute's going into solution can be thought of in terms of a series of steps. The solute molecules must first be separated from one another, then the solvent molecules must be separated to provide space for the solute molecules, and then the solvent molecules will close in around the dissolved solute molecules. On the basis of these steps, it is not surprising that solubility depends upon the nature of the solvent and solute. The separation processes will require more or less energy depending upon the strength of the forces holding the respective particles together. The attraction and surrounding of the solute molecules by solvent molecules will liberate energy in amounts also dependent upon the strength of the attractive forces involved.

Both an increase in entropy and decrease in energy enhance the solution process. However, if a given solute is dissolved, the entropy change will be essentially the same regardless of the solvent, since it appears as an increase in the disorder of the solute as it goes from the nondissolved to the dissolved state. This increase in disorder will generally be independent of the solvent. Entropy considerations will generally be important only when the energy effects become quite small.

The energy changes accompanying the solution process depend upon three different particle interactions: solvent-solvent attractions, solute-solute at-

Table 11-2. Relationships Between *p*H and *p*OH

	$[H_3O^+]$	$[OH^-]$	*p*H	*p*OH	*Examples* (solids are dissolved in H_2O)
Increasing acidity ↑	10^0	10^{-14}	0	14	HCl (1 mole/liter)
	10^{-1}	10^{-13}	1	13	
					Gastric juice
	10^{-2}	10^{-12}	2	12	
					Lemon juice
	10^{-3}	10^{-11}	3	11	
					Vinegar, carbonated drink
					Aspirin
	10^{-4}	10^{-10}	4	10	Orange juice
					Apple juice
	10^{-5}	10^{-9}	5	9	Black coffee
	10^{-6}	10^{-8}	6	8	Normal urine (average value)
					Milk, liquid dishwashing detergent
Neutral	10^{-7}	10^{-7}	7	7	Saliva, pure water
					Blood
Increasing alkalinity ↓	10^{-8}	10^{-6}	8	6	Soap (not synthetic detergent)
					Baking soda
					Phosphate-containing detergent
	10^{-9}	10^{-5}	9	5	
					Milk of magnesia
					Powdered household cleanser
	10^{-10}	10^{-4}	10	4	
					Phosphate-free detergent
	10^{-11}	10^{-3}	11	3	Household ammonia
					Liquid household cleaner
	10^{-12}	10^{-2}	12	2	
	10^{-13}	10^{-1}	13	1	NaOH (0.1 mole/liter)
	10^{-14}	10^0	14	0	NaOH (1 mole/liter)

tractions, and solvent-solute attractions. The first two act against the solution process while the third enhances it.

The relative strengths of the various attractive forces involved in interactions between polar and nonpolar molecules is as follows: polar-polar, very strong; polar-nonpolar, moderately weak; and nonpolar-nonpolar, very weak. Table 11–3 summarizes the effects of these interactions for various combinations of solvent and solute molecules.

A general observation concerning solubilities is contained in the statement "likes dissolve likes"; in other words, polar solutes are more soluble in polar solvents and nonpolar solutes are more soluble in nonpolar solvents. This behavior is consistent with the information of Table 11–3. In the first situation, involving a polar solvent and a polar solute, the interactions appear to be strongly for and against solubility. In this case, the solubility is due either to a stronger solvent-solute interaction, or to the entropy effect mentioned earlier, or both. In the second and third situations, both involving polar-nonpolar interactions, it is obvious that low solubility would result because of the strong influence opposing solubility and the weak influence favoring it.

Table 11–3. Effects of Solvent-Solute Interactions on the Solution Process

Type of solution	Interactions		
	Solvent-solvent	*Solute-solute*	*Solvent-solute*
Polar solvent, polar solute	Strong (opposes solution formation)	Strong (opposes solution formation)	Strong (favors solution formation)
Polar solvent, nonpolar solute	Strong (opposes solution formation)	Very weak (opposes solution formation)	Mod. weak (favors solution formation)
Nonpolar solvent, polar solute	Very weak (opposes solution formation)	Strong (opposes solution formation)	Mod. weak (favors solution formation)
Nonpolar solvent, nonpolar solute	Very weak (opposes solution formation)	Very weak (opposes solution formation)	Very weak (favors solution formation)

Heat of Solution

When a solute dissolves in a solvent, energy in the form of heat is usually absorbed or evolved. This energy, which is more often evolved, represents the net energy change that accompanies the solution steps mentioned before: (1) separation of solute molecules from each other, (2) separation of solvent molecules from each other, and (3) solvation—the surrounding of solute molecules by solvent molecules. The term *heat of solution* is used to denote this net energy change, which depends upon the concentration of the solution. Heats of solution are called positive when energy is absorbed during the solution process and the solution is found to cool. This type of behavior represents an **endothermic** reaction. A negative heat of solution occurs when heat is evolved during the solution process and the solution is found to warm up. The reaction is **exothermic.**

We can understand these ideas by following the processes which take place when an ionic substance dissolves in water, as illustrated in Figure 11–1. The first step is the separation of the solute ions from each other. Energy, E_1, must be added to pull apart the strongly attracting ions [Figure 11–1(a)]. The water molecules must be separated from one another to provide space for the ions [Figure 11–1(b)]. This also requires an input of energy, E_2, since the polar water molecules are attracted to one another through hydrogen bonds. The last step involves the solvation or surrounding of the ions by water molecules [Figure 11–1(c)]. Since the polar water molecules are attracted to the charged ions, the energy, E_3, is given off.

When the sum $E_1 + E_2$ is larger than E_3, the net energy will be absorbed and the reaction will be endothermic. Under these conditions the resulting entropy increase causes the solution to form. When $E_1 + E_2$ is less than E_3, the net energy will be evolved and the reaction will be exothermic. Therefore, both an energy decrease and entropy increase cause the solution to form.

Because the molecules in a gas are separated by large distances, little energy

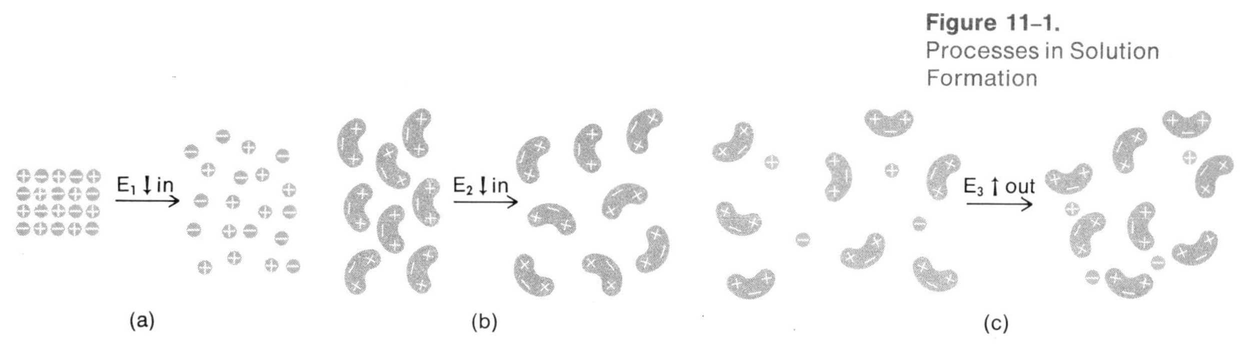

Figure 11–1.
Processes in Solution Formation

(E_1) is needed to further separate them during solution formation. The solvation energy (E_3) is usually larger than the separation energy for water (E_2); consequently, the process of solution formation for gases in water is usually exothermic.

Solutions and Equilibrium

An equilibrium situation exists in every saturated solution that is in contact with undissolved solute. Even though the concentration of solute does not change with time, solute particles are continually being exchanged between the solution and solute in contact with the solution. This is illustrated in Figure 11–2 for a solution containing a solid polar solute dissolved in water.

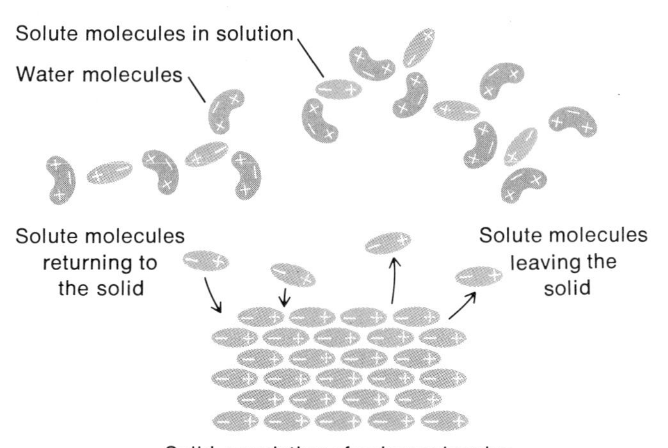

Figure 11–2.
Equilibrium in Solution Processes

According to Le Chatelier's principle (Chapter 7), any equilibrium system will, when placed under stress, change its position of equilibrium in a direction that tends to relieve the stress. Solution formations involving heat can be represented as:

$$\text{Exothermic:} \quad \text{solute} + \text{solvent} \rightleftharpoons \text{sat. soln.} + \text{heat} \qquad 11\text{–}3$$

$$\text{Endothermic:} \quad \text{solute} + \text{solvent} + \text{heat} \rightleftharpoons \text{sat. soln.} \qquad 11\text{–}4$$

The addition of heat, by increasing the temperature, constitutes a stress on the equilibrium represented by both equations. In each case a reaction will occur which tends to use up the added heat. In each of the two systems this reaction will be away from the heat. Therefore, in exothermic reactions the solute is less soluble at higher temperatures because the reaction shifts toward the left. In endothermic reactions the reverse is true: the shift is toward the right, and solute solubility increases with increasing temperature.

The solution process for most gases in water is exothermic and the solubility increases with decreasing temperature. This fact becomes quite apparent when we compare the results of opening a cold and a warm carbonated beverage.

The pressure on a solution may also influence the solubility of the solute. This effect is very minimal when the solute is a liquid or a solid, but it becomes quite appreciable when the solute is a gas. It has been found that the amount of gas dissolved in a solution at constant temperature is directly proportional to the pressure of the gas above the solution. This means a doubling of gas pressure above the solution will cause a doubling in the amount of dissolved gas.

This effect is used to help provide the "fizz" of carbonated beverages. The beverage is saturated with carbon dioxide gas under pressure. When the bottle is opened, the pressure is relieved and the gas, now less soluble, comes out of solution as fine bubbles.

A similar, though less enjoyable, effect takes place in the bloodstream of deep-sea divers. The air pumped to them under pressure is more soluble in the blood than at normal atmospheric pressure. If the diver is brought to normal pressure too quickly, the extra gases dissolved in the blood cannot be removed rapidly enough and bubbles of gas are formed in the bloodstream and joints. The resulting affliction, the bends, is painful and dangerous. The air supplied to deep divers usually contains a mixture of oxygen and helium rather than the oxygen and nitrogen of natural air; helium is less soluble in the body than nitrogen and the chances of the bends occurring are diminished.

Rate of Solution Formation

Care must be taken to avoid confusing the solubility of a solute with the rate at which it dissolves. It is true that some very soluble solutes dissolve rapidly, but others do not. A number of factors contribute to the rate at which a solute will dissolve:

Particle size of solute—smaller particles provide more surface area for solvent attack and therefore dissolve more rapidly than large particles.

Solvent temperature—at high temperatures solvent molecules are moving faster and have more frequent collisions with the solute.

Agitation or stirring of solution—stirring removes locally saturated, or at least concentrated, solution from the vicinity of the solute and allows less concentrated solution to take its place.

Bulk concentration of solution—it is more difficult for solute particles to find positions in the solution when the concentration of solute already in solution is high.

Water occupies a unique role in solution chemistry. It is the solvent for all solutions involved in the makeup of the human body. It is the solvent in which most waste products of the body are dissolved and eliminated. It is the solvent for other plant and animal circulatory systems. In short, all life on this planet is based upon a water solvent system. This ability of water to dissolve all the substances necessary for living systems—ionic compounds, polar molecules, and some nonpolar molecules—originates with its structure as an angular, polar, hydrogen-bonding molecule (Chapter 10).

Figure 11–3.
Effect of Solute on Solvent Vapor Pressure

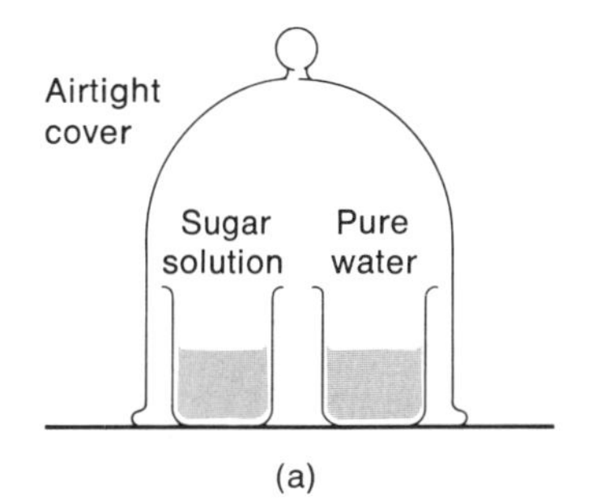

(a)

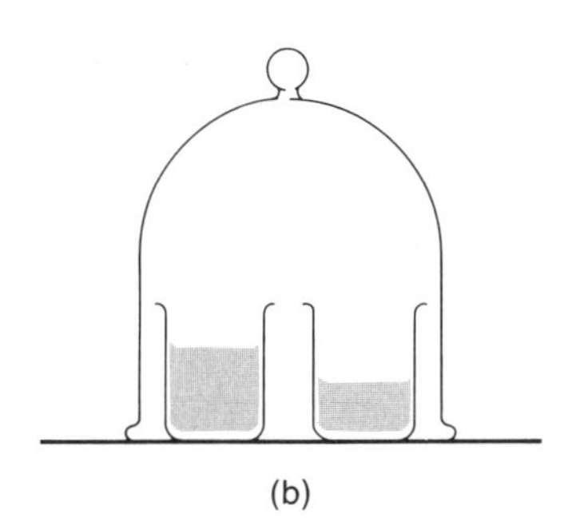

(b)

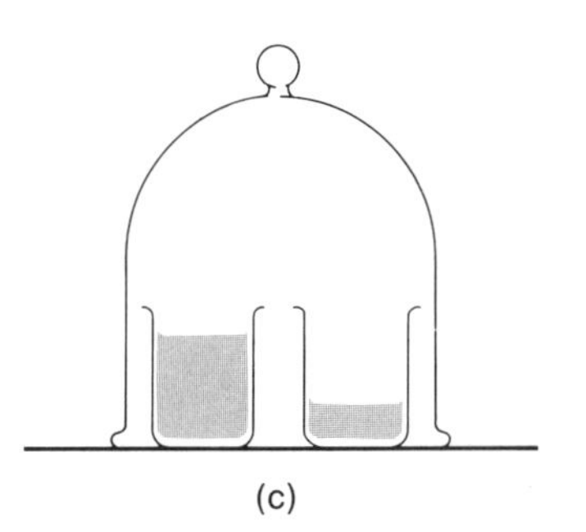

(c)

Colligative Properties of Solutions

The chemical properties of solutions depend primarily on the chemical identity of the solutes and solvents. A few properties of solutions, however, depend only on the amount of solute present and are generally the same regardless of the solute involved. These properties, collectively called **colligative properties,** include the solution vapor pressure, boiling point, freezing point, and osmotic pressure.

Vapor Pressure

An experiment illustrating the effect of a solute on the vapor pressure of a solvent is represented in Figure 11–3. In this experiment, a container of pure water and one of a solution of sugar in water are placed under an airtight cover. The contents of the two containers are isolated from one another except through the air surrounding them. The initial situation is shown in Figure 11–3(a), the situation after some time in (b), and that after a longer time—several weeks—in (c).

Observe that as time passes, the level of pure water drops while the level of solution rises. Since the only connection between the two is through the gaseous or vapor state, it is obvious that water molecules are leaving the pure water surface and going into the solution. Water is being transferred from the pure solvent to the solution through the vapor phase. The transfer occurs because the vapor pressure of water under the cover is uniform and equal to that of the pure water sample, but the equilibrium vapor pressure of the water in the sugar solution is lower than that of pure water. Therefore, water molecules leave the vapor under the cover (higher pressure) and enter the sugar solution (lower pressure). The vapor pressure of water under the cover is not reduced by this process, because water molecules leave the pure water sample and replace those lost to the sugar solution. The net effect is the observed transfer of water from the pure sample to the solution.

Direct experimental measurements have verified that the vapor pressure of water above a sugar solution is lower than the vapor pressure of pure water. The sugar molecules interfere with the escape of water molecules from the solution. It is found as a general rule that the vapor pressure of a solvent is lowered when a solute is added.

Boiling Point and Freezing Point

The **boiling point** of a liquid is the temperature at which the liquid vapor pressure is equal to atmospheric pressure (Chapter 8). When the vapor pressure of a solvent is lowered by the addition of solute, it becomes necessary to increase the temperature to a level above the normal boiling point in order to increase the vapor pressure up to the value of one atmosphere. The net effect is an increase in the boiling point of the solution. This is illustrated in Figure 11–4(a).

The **freezing point** of a liquid is the temperature at which the liquid and solid have the same vapor pressure. Since the vapor pressure of the liquid solvent in a solution is lower than pure solvent, no solid will form until a temperature is reached at which the solid vapor pressure is equal to this lower value. The freezing temperature is thus found to be lower than normal, as illustrated in Figure 11–4(b).

Figure 11–4.
Effect of Solute on Boiling and Freezing Points of Solvent

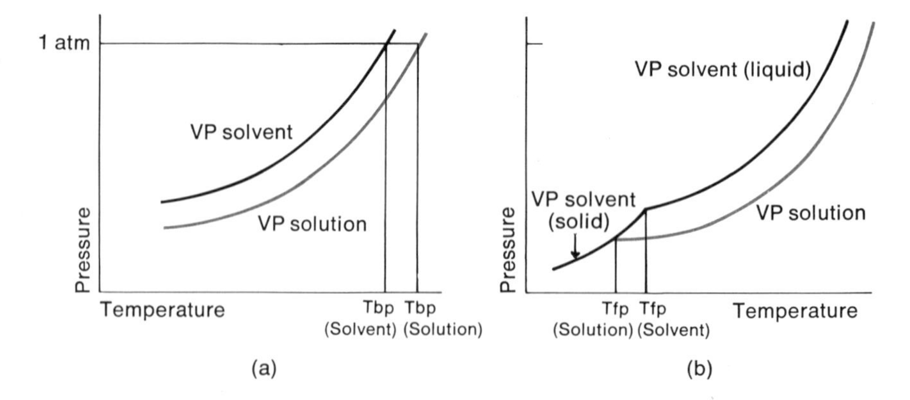

A number of applications are made of this lowering of vapor pressure. Most have to do with the freezing-point depression. Homemade ice cream is a frozen mixture of cream and other materials which has a freezing point lower than the normal freezing point of water. We freeze the mixture by surrounding it with another mixture of ice, water, and salt. The salt dissolves in the water, producing a solution with a vapor pressure lower than that of ice. The ice is then at a higher temperature than that representing an equilibrium with the salt solution—the ice is at a temperature higher than the melting point. The ice melts to establish an equilibrium, and in the process heat is drawn from the ice cream mixture, which freezes.

In climates where temperatures drop below 0°C in the winter, it is necessary to protect water-cooled engines from freezing. This is done by adding some solute to the cooling water of the engine. Ethylene glycol, CH_2OHCH_2OH, is commonly used. The vapor pressure and freezing point of the resulting solution are lower than those of water, and freezing does not take place until temperatures much lower than 0°C are reached.

Also in winter, salt is spread on sidewalks to melt ice or prevent its formation. This technique works like the one for freezing ice cream.

The crop of the fruit farmer remains a few degrees safer in freezing weather, owing to the lower freezing point of the solutions contained in the blossoms or fruit of the trees. The freezing point of these solutions is about 28°F (or −2.3°C) rather than 32°F (0°C). Smudge pots and other devices are used to keep orchard temperatures above these critical values.

The property of boiling-point elevation is used by candy makers to ensure a good product. In most candy-making procedures sugar and other materials are dissolved in water and cooked at the boiling point of the resulting solution. As the boiling proceeds, water leaves the solution and the concentration of sugar increases. It is necessary to stop boiling the solution when the concentration of sugar reaches a specific value. The boiling point of the solution depends upon the sugar concentration, and increases as water is lost from the solution. When the sugar reaches the proper concentration, as indicated by a thermometer reading, the cooking is stopped. Perfect candy results if the cook has remembered to also correct the boiling point for changes in atmospheric pressure (Chapter 8).

Figure 11–5.
Diffusion Eliminates Concentration Gradients

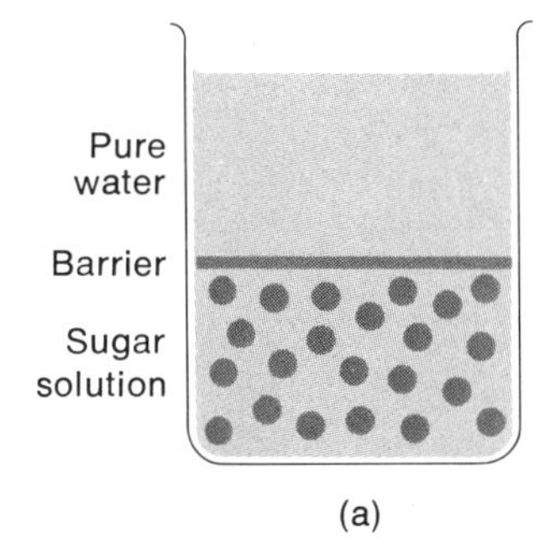

(a)

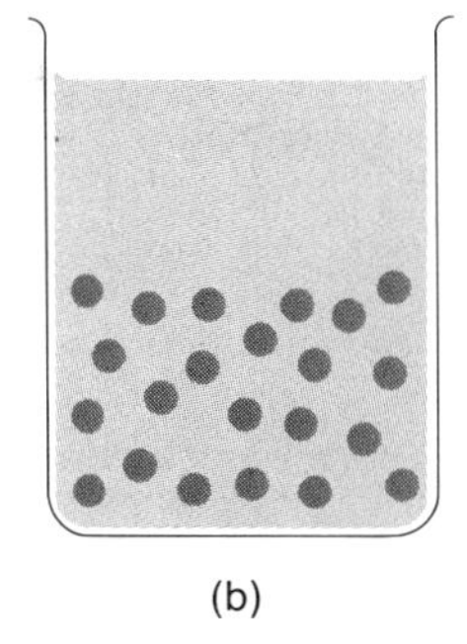

(b)

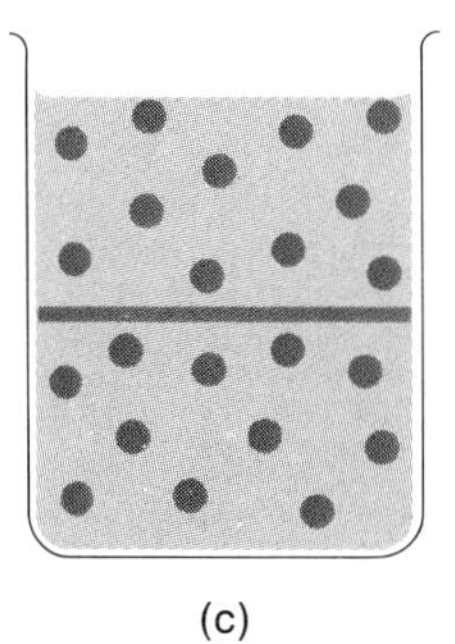

(c)

Osmotic Pressure

Osmotic pressure, another important colligative property, can be illustrated by a hypothetical experiment. Figure 11–5 shows the setup. Sketch (a) gives the equipment and initial conditions. A solution of pure water and sugar is separated from pure water by a dividing barrier. The more dense sugar solution is on the bottom to prevent mixing because of density differences, and the two solutions are at the same temperature to prevent mixing by thermal convection currents. In (b) the barrier has been removed but the liquids have not been stirred. In (c) the final state of the system is shown after an appreciable time has passed and the barrier has been replaced. The results of this experiment would have been predicted by each of us. Of course, sugar molecules moved throughout the entire system. We would have been very surprised to find all of the sugar molecules still on one side when the barrier was replaced. The process took place spontaneously, not because of an energy change—the solutions were at the same temperature—but because of the increase in entropy that resulted upon mixing.

Now consider the experiment shown in Figure 11–6, in which two solutions similar to those above are separated by a membrane. This membrane is of a type referred to as **semipermeable,** which means that is has pores large enough to allow some types of molecules to pass through but small enough to prevent others. The membrane in this case allows water molecules to pass through but not sugar molecules.

In this experiment we have created a concentration gradient but we have prevented a mechanism for its removal in the same way as the experiment of Figure 11–5. Because the membrane will not allow sugar molecules to move and increase the sugar concentration of the pure water, the water moves through in the opposite direction and lowers the sugar concentration so that it is closer to that of pure water. The entropy of the system is increased as the sugar molecules are given a greater volume in which to move. Because of the movement of the water, a hydrostatic pressure, *h*, is created across the membrane. This pressure, coupled with the actual dilution that takes place, creates a chemical condition identical to that existing when the concentration of sugar is the same on both sides of the membrane. When this condition is achieved, the flow of water stops. The pressure difference across the membrane necessary to accomplish this is called the **osmotic pressure,** and the movement of water molecules through the membrane is called **osmosis.**

All plants and animals contain membranes which are semipermeable to different substances, and the osmosis process thus plays an important role in various physiological processes. Salt water poured over a plant will kill it because the solution outside the root membranes is more concentrated in salts than the solutions inside. Water flows out of the roots in an attempt to dilute the concentrated solution outside, and the plant is dehydrated and

Figure 11–6.
The Osmosis Process

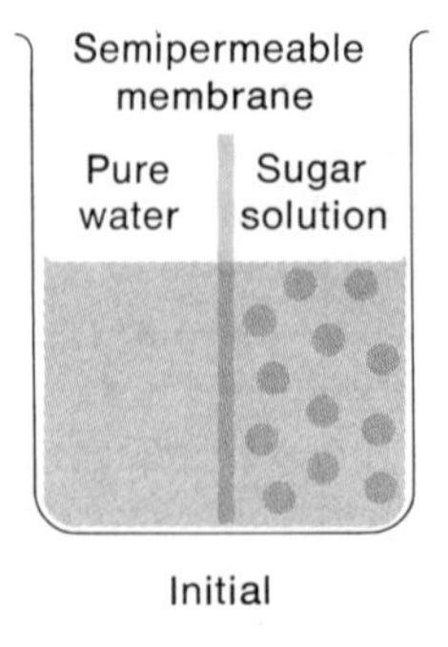

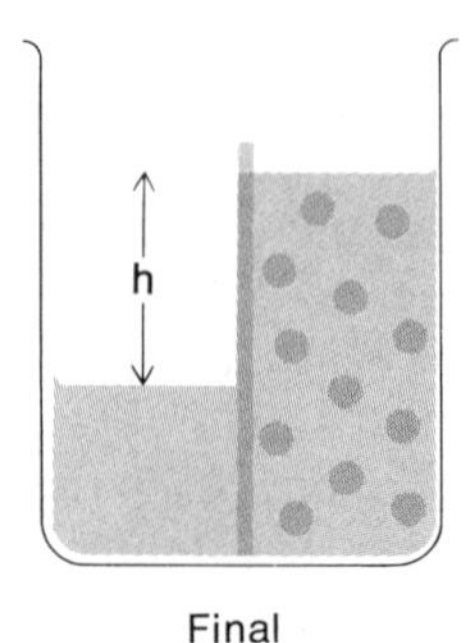

dies. This same process occurs in the stomach and intestines of animals. When salt water is taken into the stomach and is more concentrated than the solutions on the other side (within the stomach membrane), water flows out of the membrane into the stomach, and the tissues are dehydrated. Therefore, the drinking of sea water will cause more thirst rather than less, since the body will lose water rather than absorb it. A cut finger stings when placed in salt water because of the water flowing through the exposed cell walls into the salt water. A person swimming in fresh water often notices an uncomfortable sensation in the eyes caused by water passing into the eye through its membrane surface. The water goes into the eye because the solute concentrations inside the eye are greater than those of the pure water outside. Medical solutions used for injection into the bloodstream are made 0.9 percent salt (NaCl), the same concentration as that of blood. These so-called **isotonic solutions** thus create no concentration gradients between the blood and other body fluids, and prevent osmotic movements of water which might prove unfortunate for the patient.

Review Questions

1. What is the difference in meaning of the terms *solvent* and *solute?*
2. List two commonly used concentration units and define each one.
3. Calculate the concentrations of the following solutions:
   a) The weight percent of a solution containing 200 g of H_2O and 26.0 g of NH_4Cl
   b) The weight percent of a solution if 70 g of it leaves a residue of 11.8 g of salt upon evaporation
   c) The molarity of a solution in which 17.0 g of $AgNO_3$ are dissolved in enough water to give 500 ml of solution
4. How many grams of a salt should be added to 50 grams of water to give a 30 percent solution of the salt?
5. Calculate the *p*H and *p*OH of solutions having the following indicated $[H^+]$ or $[OH^-]$:
   a) $[H^+] = 10^{-9}$
   b) $[OH^-] = 10^{-5}$
   c) $[H^+] = 10^{-4}$
   d) $[OH^-] = 10^{-2}$
6. Calculate the $[H^+]$ and $[OH^-]$ for solutions with one of the following *p*H's or *p*OH's:
   a) $pOH = 10$
   b) $pOH = 3$
   c) $pH = 6$
   d) $pH = 12$

7. Calculate the ratio of $\frac{[H^+]}{[OH^-]}$ for each of the following pH values:
   a) pH = 7
   b) pH = 10
   c) pH = 3
   d) pH = 1
8. List the types of particle interactions upon which the solution process is dependent. What determines the strength of each type of interaction?
9. How will an increase in temperature affect the amount of solute that dissolves to form a saturated solution when
   a) The heat of solution is positive
   b) The heat of solution is negative
10. List four factors that affect the rate of solution of a solute. Indicate why each has the effect that it does.
11. What is a colligative property of a solution? List four colligative properties.
12. Account for the fact that under identical conditions sugar water boils at a higher temperature than pure water.
13. What is the purpose of the salt water-ice mixture used to make homemade ice cream?
14. List three examples of situations where either osmosis or osmotic pressure is an important factor.

12 Electrochemistry

Electrochemistry is the branch of chemistry concerned with the interactions of matter and the particular form of energy called electricity. Electrochemical processes or the results of electrochemical processes are routinely used by most people in the course of their daily living. When we recall the electrical nature of matter, it isn't surprising that the energy involved in chemical reactions might appear in the form of electricity. In fact, electrochemical applications date back to the late eighteenth century, when Alessandro Volta obtained a continuous current of electricity from a "pile" consisting of alternate sheets of silver and copper separated by paper soaked in salt water. Other investigators used similar voltaic piles as sources of electricity for their electrochemical experiments.

Today, two types of applications are usually made. Elements are electrically obtained from their compounds by electrolysis and electroplating, and electricity is produced from chemical reactions in electric cells, batteries, and fuel cells.

Electrolysis Processes

We shall first consider **electrolysis** processes, in which an electric current is passed through either a solution or melt of a substance in order to produce desired products. Figure 12–1 shows the apparatus required. The essential components are the electrodes, the melt or solution (electrolyte), and a source of direct current.

The electric current moves through the external circuit (electrodes and wire) in the form of electrons, which travel by a displacement or bumping process. As an electron is pushed into one end of a wire, it bumps a loosely held electron down the wire a short distance. This second electron bumps another and so on, until an electron is displaced out of the other end of the wire. The passage of a current in this fashion, **metallic conduction,** depends upon the presence of loosely held electrons in the metal. When an electron reaches the internal part of the circuit (melt or solution), another mechanism for transporting electricity takes over. Free electrons are not found in solutions or melts, and the movement of electricity between the electrodes is accomplished by the movement of charged ions. The negatively charged ions move in the same direction as the electrons of the external circuit, while the positively charged ions move in an opposite direction. The process is known as **ionic conduction.**

Figure 12–1.
Electrolysis Apparatus

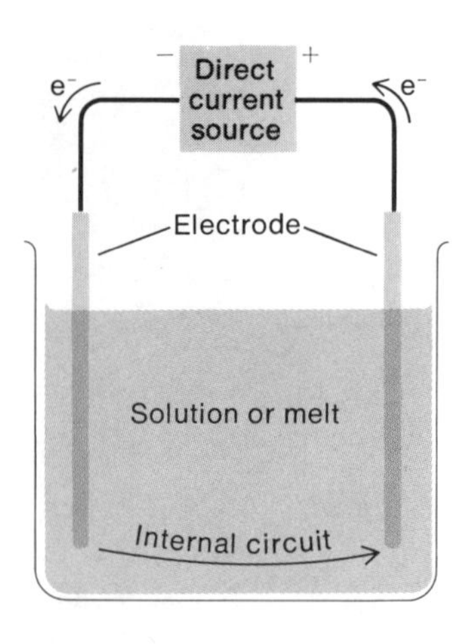

Let's consider two specific electrolysis processes. In the first, sodium metal and chlorine gas are produced from molten sodium chloride. The apparatus needed is diagrammatically represented in Figure 12–2. The arrows on the ions represent the direction of their movement when the current is flowing.

As soon as the current begins to flow, it is carried through the melt by the ions. The positive sodium ions (cations) move toward the negatively charged electrode (the cathode), and the negative chloride ions (anions) move toward the positively charged electrode (the anode). If nothing else happened, each electrode would quickly become covered with ions (polarized), and the current would cease to flow. Experience shows that the current does continue to flow and something else happens at each electrode. A silver-colored molten metal (Na) floats to the surface around the cathode, and an unpleasant-smelling gas (Cl_2) bubbles to the surface near the anode.

At the cathode electrons are transferred to the sodium cations, converting them into neutral atoms of sodium metal, which collect together and float to the surface of the melt. At the anode the chloride ions each give up their extra electron and become neutral chlorine atoms, which pair together to form molecules and bubble off as chlorine gas. The electrons given up at the anode are pumped around the circuit by the direct-current source and are made available at the cathode. The process taking place at the cathode (acceptance of electrons by ions) is a reduction, and the anode reaction (donation of electrons by ions) is an oxidation.

The reactions involved at each electrode are given below. Since each reaction represents only one-half of the total process going on in the electrolysis, they are called **half reactions.** The sum of the half reactions, adjusted to involve the same number of electrons, gives the overall electrolysis reaction.

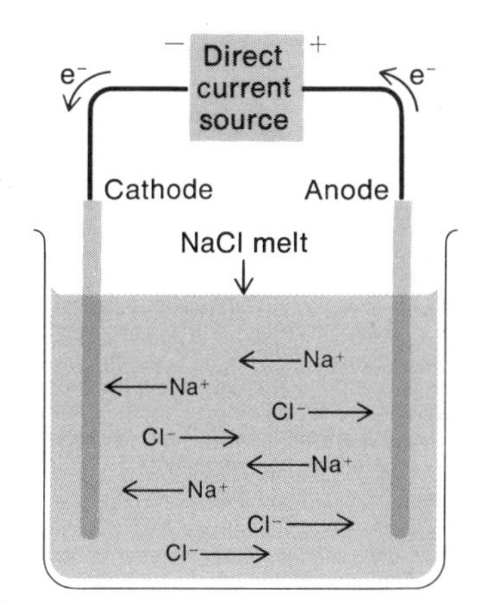

Figure 12–2.
The Electrolytic Production of Sodium and Chlorine

Cathode reaction (reduction):

$$Na^+ + 1e^- \rightarrow Na \qquad 12\text{–}1$$

Anode reaction (oxidation):

$$2Cl^- \rightarrow 2e^- + Cl_2 \qquad 12\text{–}2$$

Overall reaction (the sum of adjusted half reactions):

$$2Na^+ + 2Cl^- \rightarrow 2Na + Cl_2 \qquad 12\text{–}3$$

The second electrolysis process we want to consider is carried out in the same way, except that a concentrated brine is used as the electrolyte. The electrolysis of such a strong solution of salt in water again results in the formation of chlorine gas at the anode. However, hydrogen gas, not sodium metal, is produced at the cathode. When both water and Na^+ ions are present in an electrolysis cell, water is reduced more easily than the Na^+ ions, and the characteristic product, H_2, is formed. In addition, hydroxide ions are generated in the solution around the cathode. These, together with the available sodium ions, constitute a solution of sodium hydroxide from which solid NaOH can be obtained by evaporating away the excess water. The electrode reactions that occur in this process are:

Cathode reaction (reduction):

$$2H_2O + 2e^- \rightarrow H_2 + 2OH^- \qquad 12\text{–}4$$

Anode reaction (oxidation):

$$2Cl^- \rightarrow 2e^- + Cl_2 \qquad 12\text{–}5$$

Overall reaction:

$$2H_2O + 2Cl^- \rightarrow H_2 + Cl_2 + 2OH^- \qquad 12\text{–}6$$

Electrolysis processes are used in industry to prepare a number of useful products, including all of the metallic aluminum and sodium and nearly all of the magnesium produced annually. Chlorine gas is produced almost exclusively by the electrolysis of either molten or aqueous sodium chloride (equations 12–2 and 12–5); and practically all commercial NaOH is a by-product of the latter process. All elemental fluorine, the strongest elemental oxidizing agent known, must be prepared by electrolysis. Table 12–1 contains the reactions for the preparation of all these substances except as given in equations 12–1 through 12–6.

Table 12–1. Production of Elements by Electrolysis

Product	*Raw material*	*Reactions*	
Hydrogen gas	Water	$2H_2O + 2e^- \rightarrow H_2 + 2OH^-$	(cathode)
Oxygen gas	Water	$2H_2O \rightarrow 4e^- + O_2 + 4H^+$	(anode)
		$6H_2O \rightarrow 2H_2 + O_2 + 4OH^- + 4H^+$ ($4OH^- + 4H^+ \rightarrow 4H_2O$)	
		$2H_2O \rightarrow 2H_2 + O_2$	(overall)
Magnesium metal	$MgCl_2$ (melt)	$Mg^{2+} + 2e^- \rightarrow Mg$	(cathode)
Chlorine gas	$MgCl_2$ (melt)	$2Cl^- \rightarrow 2e^- + Cl_2$	(anode)
		$Mg^{2+} + 2Cl^- \rightarrow Mg + Cl_2$	(overall)
Aluminum metal	Al_2O_3 (melt)	$Al^{3+} + 3e^- \rightarrow Al$	(cathode)
		$2O^{2-} \rightarrow O_2 + 4e^-$	(anode)
		$4Al^{3+} + 6O^{2-} \rightarrow 4Al + 3O_2$	(overall)
Hydrogen gas	$KF \cdot HF$ (melt)	$2H^+ + 2e^- \rightarrow H_2$	(cathode)
Fluorine gas	$KF \cdot HF$ (melt)	$2F^- \rightarrow 2e^- + F_2$	(anode)
		$2H^+ + 2F^- \rightarrow H_2 + F_2$	(overall)

Figure 12–3.
Electroplating with Silver

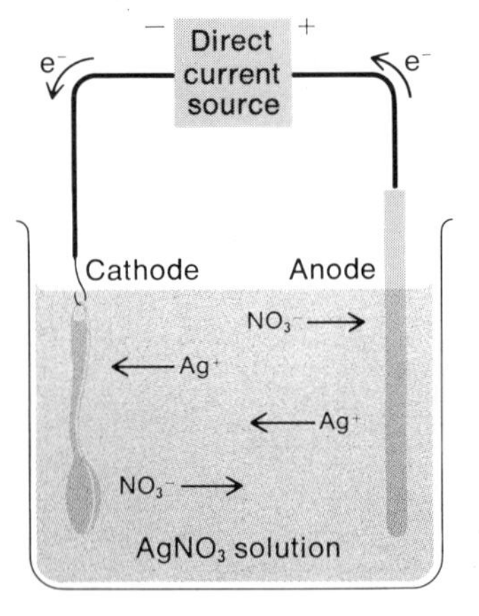

Cathode reaction $Ag^+ + 1e^- \rightarrow Ag$ Anode reaction $Ag \rightarrow Ag^+ + 1e^-$

The electroplating industry is also based on the electrolysis of melts or, usually, solutions. In this type of work, the cathode is an object on which a coating of metal is desired. The anode is generally made of the metal being coated on the object. The cathode reactions are similar to the ones in Table 12–1, and the anode reaction involves the oxidation of the metal. A typical electroplating system is shown in Figure 12–3.

Other electroplating processes are represented by the reactions given in Table 12–2.

Table 12–2. Metal Plating by Electrolysis

Plating material	*Plating solution*	*Reactions*	
Silver metal (decorative silverplate)	$AgCN + KCN$	$Ag^+ + 1e^- \rightarrow Ag$	(cathode)
		$Ag \rightarrow Ag^+ + 1e^-$	(anode)
Gold metal (jewelry)	$KAu(CN)_2 + KCN$	$Au(CN)_2^- + 1e^- \rightarrow Au + 2CN^-$	(cathode)
		$Au \rightarrow Au^+ + 1e^-$	(anode)
Platinum metal (jewelry)	$H_2PtCl_6 \cdot 6H_2O$	$Pt^{4+} + 4e^- \rightarrow Pt$	(cathode)
Palladium metal (jewelry)	$Na_2Pd(NO_2)_4$	$Pd(NO_2)_4^{2-} + 2e^- \rightarrow Pd + 4NO_2^-$	(cathode)
		$Pd \rightarrow Pd^{2+} + 2e^-$	(anode)
Chromium metal (decorative trim)	$Cr_2O_3 + H_2SO_4$	Not well known	
Copper metal (electrical copper)	$CuSO_4$	$Cu^{2+} + 2e^- \rightarrow Cu$	(cathode)
		$Cu \rightarrow Cu^{2+} + 2e^-$	(anode)
Tin metal (tin cans)	$SnSO_4$	$Sn^{2+} + 2e^- \rightarrow Sn$	(cathode)
		$Sn \rightarrow Sn^{2+} + 2e^-$	(anode)

The Chemical Production of Electricity

The second general application of electrochemistry, the chemical production of usable electric power, is important and widely used. This process is based on the idea of allowing substances with a tendency to react to do so under appropriate conditions. For example, if a piece of zinc metal is placed in a solution containing copper ions (Cu^{2+}), a coating of copper metal forms on the zinc, and at the same time some of the zinc dissolves to form zinc ions (Zn^{2+}). The reaction is

$$Zn + Cu^{2+} \rightarrow Zn^{2+} + Cu \qquad 12\text{–}7$$

Obviously, each reacting atom of zinc gives up two electrons to a reacting copper ion. This process occurs spontaneously and must be energetically favorable (energy is given up), since the entropy of the system will be about the same whether zinc or copper ions are in solution. When the reaction is carried out as given above, the energy is liberated as heat.

By arranging the reactants differently, it is possible to draw the energy off in the form of electricity. Imagine the reacting species are arranged as illustrated in Figure 12–4. The tendency or driving force for the zinc metal to give electrons to the copper ions still exists, but now the electrons must travel through an external wire to get from the zinc metal to the copper ions. If we insert an electrical motor into the external circuit and make the electrons flow through it on their way, we can derive useful work from the process. This approach is used in the construction of batteries and fuel cells. The reactants are physically separated but electrically connected, so that the reactions can proceed, but only by sending the electrons through an external

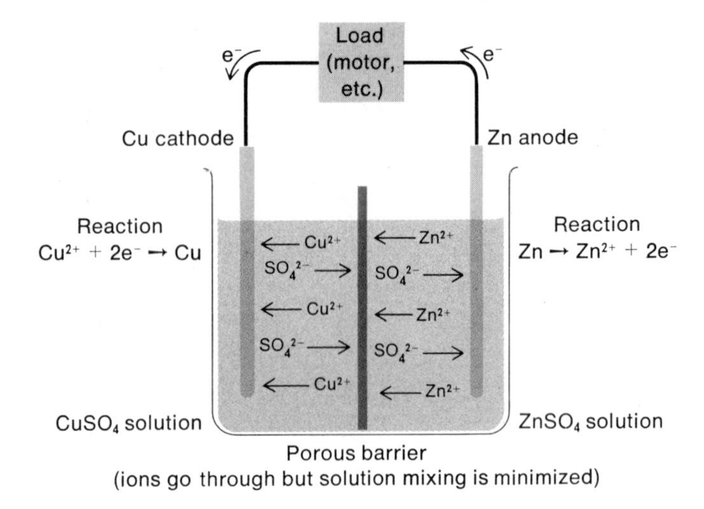

Figure 12–4.
Chemical Production of Electricity

circuit. When a single unit is used, as illustrated in Figure 12–4, it is called a **cell.** A group of electrically connected cells is correctly called a **battery.** However, in modern usage, the term *battery* is often used to describe single cells.

Useful Cells and Batteries

Cells or batteries used to produce useful electric energy are classified as **primary,** which cannot conveniently be recharged, and **storage,** which can be recharged. Figure 12–5 shows two common primary cells, together with the reactions occurring at each electrode. The le Clanche dry cell is the familiar flashlight battery. Ruben-Mallory mercury cells, combining small size with long life, are used in portable radios, hearing aids, and other electronic equipment.

In principle, the chemical reactions of primary cells such as those of Figure 12–5 are reversible. In practice, however, the cells have proved difficult to recharge. The reasons for the difficulty are apparently related to the kinds of products formed during discharge and the distribution of these products in the discharged cell.

A discharged cell appears to become recharged when it is heated. This is caused by an increased product diffusion rate at the higher temperature. During discharge, products collect around the electrodes and prevent or slow down further reactions. The discharge reactions can again take place only after these products diffuse away. The apparent effectiveness of primary-cell rechargers is due in a large part to this heating effect.

The lead cell is the most common storage cell used in the modern world. Millions of these durable energy sources provide starting power for automobiles—our most popular form of transportation. A single lead cell is represented in Figure 12–6. Each cell generates a potential of 2 volts. In

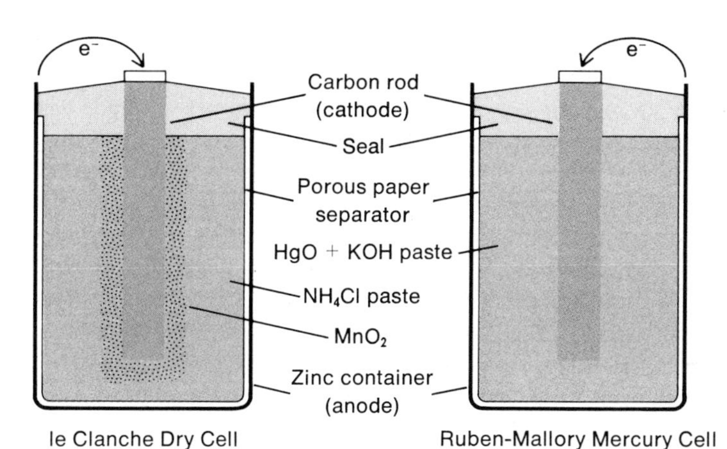

Figure 12–5.
Primary Cells

le Clanche Dry Cell	Ruben-Mallory Mercury Cell
Cathode reaction:	Cathode reaction:
$2MnO_2 + 2NH_4^+ + 2e^- \rightarrow$ $Mn_2O_3 + H_2O + 2NH_3$	$HgO + H_2O + 2e^- \rightarrow$ $Hg + 2OH$
Anode reaction:	Anode reaction:
$Zn \rightarrow Zn^{2+} + 2e^-$	$Zn + 2OH^- \rightarrow Zn(OH_2) + 2e^-$
Voltage: 1.5	Voltage: 1.35

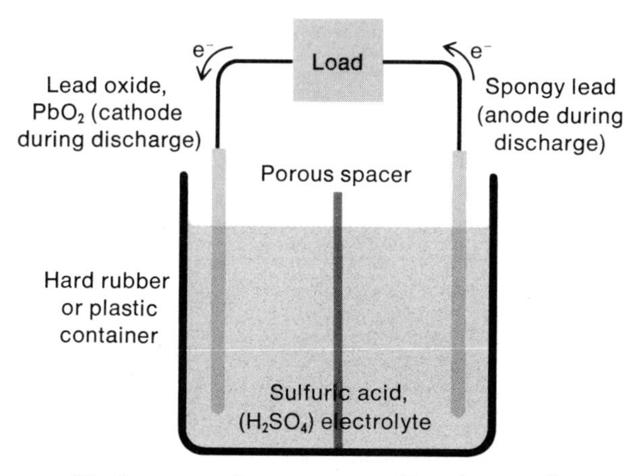

Figure 12–6.
The Lead Storage Cell

Discharge reactions	Charging reactions
Anode:	Cathode:
$Pb + SO_4^{2-} \rightarrow PbSO_4 + 2e^-$	$PbSO_4 + 2e^- \rightarrow Pb + SO_4^{2-}$
Cathode:	Anode:
$PbO_2 + 4H^+ + SO_4^{2-} + 2e^- \rightarrow$ $PbSO_4 + 2H_2O$	$PbSO_4 + 2H_2O \rightarrow$ $PbO_2 + 4H^+ + SO_4^{2-} + 2e^-$

practice, a number of these cells are connected in parallel groups (cathode to cathode and anode to anode) to increase the amount of current obtainable. Three or six such groups of cells are connected in series (cathode to anode) to give the familiar 6-volt or 12-volt batteries.

A lead storage battery will give excellent service if the water level is maintained, if it is never allowed to discharge completely, if it is kept from freezing, and if it is not subjected to "quick-charge" procedures which break down the spongy electrodes by the rapid evolution of gaseous hydrogen and oxygen. The hydrogen and oxygen are produced by the electrolysis of water (Table 12–1), which occurs as a side reaction during charging processes. During slow charges the gases are produced slowly and can easily escape from the interior of the porous electrodes. During quick charges, however, the gases are generated rapidly and break up the electrodes while escaping.

Another useful source of stored electrical energy is the nickel-cadmium battery, which has a longer life and delivers a more constant voltage than the lead battery. The anode is made of cadmium metal and the cathode of solid nickel hydroxide. They are immersed in a potassium hydroxide solution. The reactions are:

Anode:

$$Cd + 2OH^- \underset{\text{charge}}{\overset{\text{discharge}}{\rightleftharpoons}} Cd(OH)_2 + 2e^- \qquad 12\text{–}8$$

Cathode:

$$2Ni(OH)_3 + 2e^- \underset{\text{charge}}{\overset{\text{discharge}}{\rightleftharpoons}} 2Ni(OH)_2 + 2OH^- \qquad 12\text{–}9$$

A similar battery, called the Edison cell in honor of the inventor, contains iron anodes and nickel oxide cathodes in an alkaline electrolyte. The electrode reactions are:

Anode:

$$Fe + 2OH^- \underset{\text{charge}}{\overset{\text{discharge}}{\rightleftharpoons}} Fe(OH)_2 + 2e^- \qquad 12\text{–}10$$

Cathode:

$$NiO_2 + 2H_2O + 2e^- \underset{\text{charge}}{\overset{\text{discharge}}{\rightleftharpoons}} Ni(OH)_2 + 2OH^- \qquad 12\text{–}11$$

Fuel Cells

In recent years a great deal of research has been devoted to the perfection of a fuel cell. In this device, reactants (fuel) are continuously added and products are continuously removed. The reactants for these cells are fuels that normally would be burned to release energy; in the fuel cell the energy is released as an electric current.

Fuel cells utilize the fuel much more efficiently than conventional combustion devices. Efficiencies of 60–70 percent have been obtained for fuel cells as compared to values of 25–50 percent for combustion processes. Fuel cells, unlike conventional storage batteries, need no recharging but operate very much like a fueled engine. The main disadvantages of fuel cells are that expensive and easily poisoned (deactivated) catalysts must be used and that the fuels themselves are expensive and must be purified to remove catalyst-poisoning impurities. The most famous use of fuel cells has been in the moon rockets. These fuel cells use hydrogen and oxygen as reactants and produce water as the product. The reactions for a cell of this type (but not the one used on the moon ships) are:

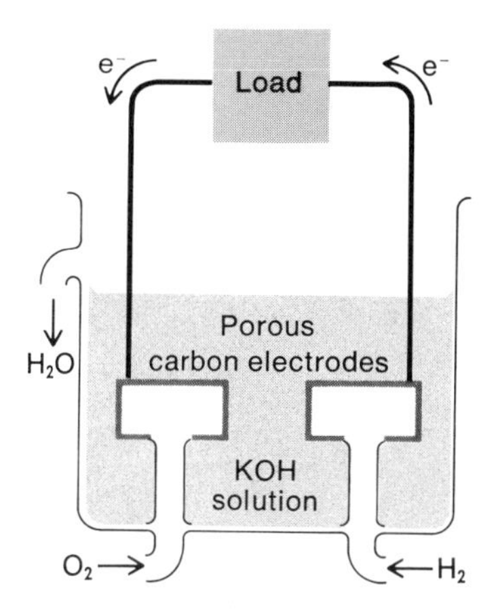

Figure 12–7.
A Fuel Cell

Anode:

$$2H_2 + 4OH^- \rightarrow 4H_2O + 4e^- \qquad 12\text{–}12$$

Cathode:

$$O_2 + 2H_2O + 4e^- \rightarrow 4OH^- \qquad 12\text{–}13$$

Net overall reaction:

$$2H_2 + O_2 \rightarrow 2H_2O \qquad 12\text{–}14$$

A fuel cell utilizing these reactions is diagrammed in Figure 12–7. A single cell of this type generates approximately 1 volt. Groups of these cells are connected in series to provide the desired voltages. This cell is operated at elevated temperatures; the water produced evaporates as it is formed and can be condensed and used if desired.

Review Questions

1. Compare the metallic and ionic conduction processes.
2. Lithium metal is in the same family of the periodic table as sodium, and it is also prepared by the electrolysis of molten salt. Write the cathode, anode, and overall reactions that would take place during the electrolysis of molten LiCl.
3. Write the cathode and anode half reactions that take place during a nickel plating process involving a solution of $NiCl_2$ as the electrolyte.
4. Discuss the significance of spontaneous chemical reactions in the construction of useful electricity-producing cells.
5. Contrast the terms *cell* and *battery*.
6. Differentiate between the designations *primary cell* and *storage cell*.
7. Why are modern le Clanche and Ruben-Mallory cells surrounded by a steel jacket? See Figure 12–5.
8. Draw a simple diagram to illustrate each of the following:
   a) Three lead storage cells are connected in parallel. What voltage does this arrangement produce?
   b) Three lead storage cells are connected in series. What voltage does this arrangement produce?
9. Suppose three identical flashlights were turned on at the same time. One was placed in a refrigerator, one was left in the room, and one was placed

in a warm oven. Which one would go out first? Which one would go out last? Explain your answers.

10. How do the following practices damage a lead storage battery?
    a) Hard tap water is used to maintain the water level.
    b) "Quick charges" are applied frequently.
11. A lead storage battery is found to be discharged. The electrolyte of such a battery is sulfuric acid. Can the discharged battery be rejuvenated by adding H_2SO_4? See Figure 12–6 and explain your answer.
12. The Edison storage cell (equations 12–10 and 12–11) utilizes an iron anode and a nickel oxide cathode. Draw a diagram of the cell and clearly indicate the following:
    a) The cathode
    b) The anode
    c) The direction of electron flow in the external circuit during discharge
    d) The flow of ions in the internal circuit during discharge
13. What advantage do fuel cells have compared to fuel combustion energy sources? What are the disadvantages of fuel cells at the present time? Give some useful applications for fuel cells.

Suggestions for Further Reading

Austin, L. G., "Fuel Cells," *Scientific American*, Oct. 1959.
Raisbeck, G., "The Solar Battery," *Scientific American*, Dec. 1955.
Yeager, E., "Fuel Cells," *Science*, Oct. 20, 1961.

13 Some Important Industrial Chemicals

Hundreds of thousands of different chemicals are known. Which are the most important? The answer depends upon the criteria used to define importance. One answer relates the importance of a chemical to the amount produced: those of greatest importance are the ones produced in largest amounts. We will use this approach here to discuss some of the important chemicals produced and used by man.

The Top Five Chemicals

Table 13–1 contains production figures for the five chemicals manufactured in largest amounts in the United States during 1971. Notice that two of the five are elements—O_2 and Cl_2. They are considered to be manufactured chemicals because they occur naturally in impure or combined forms and must be purified or produced industrially. All of the top five chemicals are classified as inorganic. This does not mean that inorganic chemicals are necessarily more important to man than organic chemicals; on the contrary, when the 50 chemicals produced in largest amounts are considered, 31 are found to be organic and only 19 inorganic.

Table 13–1. Production of the Top Five Chemicals

Rank	*Chemical name*	*Chemical formula*	*Annual production (billions of lb)*
1	Sulfuric acid	H_2SO_4	58.6
2	Ammonia	NH_3	27.4
3	Oxygen	O_2	25.9
4	Sodium hydroxide	$NaOH$	19.4
5	Chlorine	Cl_2	18.7

Only six elements—H, O, S, N, Na, and Cl—are contained in the top five chemicals. For this reason very few raw materials are required in the production of these important substances.

Sulfuric Acid

Sulfuric acid, a colorless, corrosive, oily liquid in the pure state, is the most widely used manufactured chemical in the United States. It is commonly used in the form of a concentrated water solution containing about 97 percent H_2SO_4. Produced in a quantity more than twice that of any other chemical, it rarely becomes a part of finished products but is used in a great number of industrial processes. Because of its wide usage and versatility, the remark is often made that the per capita use of sulfuric acid is a good index of the technical development of a country.

The chemical uses of H_2SO_4 depend primarily upon four of its properties: (1) it has a very low volatility—lower than that of most common acids, (2) it has a strong affinity for water and is therefore a good dehydrating agent, (3) it is a good oxidizing agent, and (4) it is a strong acid.

Uses of Sulfuric Acid

Table 13–2 lists eight uses of sulfuric acid. The first four represent the processes in which sulfuric acid is used in the largest quantities; the others illustrate the diversity of its use.

Table 13–2. Uses of Sulfuric Acid (1968)

Use	*Amount used annually (thousands of tons)*	*Percent of annual H_2SO_4 production*
Production of phosphate and superphosphate fertilizers	12,676	44.0
Production of ammonium sulfate fertilizer	1,815	6.3
Production of high-octane gasoline	1,332	4.6
Production of titanium dioxide	1,230	4.3
Iron and steel pickling	794	2.8
Production of aluminum sulfate	643	2.2
Uranium-ore processing	578	2.0
Storage batteries	128	0.4

The first two entries of Table 13–2 involve fertilizer production, and together they account for slightly more than 50 percent of all sulfuric acid usage. The fertilizer industry is concerned primarily with four elements—N, P, K, and S. The major elements supplied for use as plant nutrients are N, P, and K. In some soils sulfur fertilization is needed as well.

The prime source of the phosphorus used in fertilizers is phosphate rock, which contains phosphorus primarily in the form of $Ca_3(PO_4)_2$, which is highly insoluble and therefore useless to plants. The treatment of phosphate rock with H_2SO_4 results in the formation of soluble phosphorus compounds (superphosphate). Details and formulas involved in this process are given in Chapter 14.

Ammonium sulfate for fertilizer use is prepared by the acid-base reaction (a neutralization) between ammonia and sulfuric acid:

$$2NH_3 + H_2SO_4 \rightarrow (NH_4)_2SO_4 \qquad 13\text{–}1$$

The resulting $(NH_4)_2SO_4$ is relatively low in nitrogen (approximately 21 percent), but it contains 24 percent sulfur, which makes it desirable in areas needing regular sulfur fertilization. In addition, it is somewhat acidic, a property that makes it useful for alkaline soils.

Sulfuric acid is used in the petroleum industry during the production of certain components needed in high-octane fuels. These components are produced in a process that involves the combination of small petroleum molecules into larger molecules. Sulfuric acid acts as a catalyst for some of these reactions. More details concerning these processes are given in Chapter 19.

Titanium dioxide (TiO_2) is an important ingredient in many paints. It serves as an opaque, white pigment and has replaced the previously used toxic lead compounds in all inside and most outside paints.

The initial step in the production of TiO_2 from its ores involves a digestion in hot sulfuric acid. This treatment converts any titanium compounds into a soluble sulfate from which TiO_2 can be prepared.

The surfaces of iron and steel products must be completely free of oxides or greases before they can be galvanized or electroplated. This cleaning is accomplished by a **pickling** process in which the metal objects are dipped into hot sulfuric acid.

Aluminum sulfate, mentioned earlier (Chapter 10) as a chemical used in water treatment, is made by treating aluminum ore (bauxite-hydrated Al_2O_3) with H_2SO_4. The initial treatment of uranium ore also involves the use of H_2SO_4.

The lead storage battery represents one of the few products in which the consumer actually encounters sulfuric acid. We saw in Chapter 12 that the H_2SO_4 is produced during the charging reactions and serves as the battery electrolyte.

The strong affinity of H_2SO_4 for water makes it an efficient drying agent. Wet gases or liquids that do not react with H_2SO_4 can be dried by passage through it. Sulfuric acid is also used to absorb water formed in chemical reactions. This shifts the reaction equilibrium toward the products and increases the yield. Sulfuric acid is used this way in the production of nitroglycerin, a useful explosive.

Production of Sulfuric Acid

A number of the characteristics of H_2SO_4 are not unique to it; yet it is often used in favor of other available acids, chemicals, or different process technologies because it is the least expensive alternative. Its relatively low cost is related to the manufacturing methods used in its production.

In 1970, the **contact process** was used to manufacture 98–99 percent of all sulfuric acid produced in the United States. The raw materials required in this process are SO_2, air, and water. The chemical steps are: (1) SO_2 and O_2 from air are reacted together in the presence of a catalyst to give SO_3, (2) the SO_3 is reacted with water to give H_2SO_4. In the actual process, heat is evolved if step (2) is carried out. The increase in water temperature reduces the solubility of SO_3 and makes the step impractical, so an alternate technique is used. The SO_3 gas is dissolved in concentrated H_2SO_4 to produce fuming sulfuric acid or **oleum,** which is then diluted with water. The equations for these reactions are:

$$H_2SO_4 + SO_3 \rightarrow H_2S_2O_7 \text{ (oleum)} \qquad 13\text{–}2$$

$$H_2S_2O_7 + H_2O \rightarrow 2H_2SO_4 \qquad 13\text{–}3$$

The SO_2 required in the contact process is obtained by burning elemental sulfur in air. Low-cost elemental sulfur is obtained in the United States from underground deposits. These deposits are mined by the **Frasch process,** in which the sulfur is melted underground and forced to the surface in the liquid state. In the Frasch process three concentric pipes are imbedded in the sulfur deposit. Water, under pressure and at a temperature of about 160°C, is forced down the outer pipe; the inner pipe carries down compressed air. The sulfur melts (m.p. = 119°C) and is forced up the middle pipe by the compressed air. The molten sulfur is collected in a sump at the surface, separated from the water, and conveyed to vats, where it solidifies. The Frasch process is depicted in Figure 13–1.

Elemental sulfur from the Frasch process is burned in dry air to give a gaseous mixture containing 8–11 percent SO_2, which is used to prepare sulfuric acid. The production of sulfuric acid is summarized as follows:

$$\text{Sulfur} \xrightarrow{O_2} SO_2 \xrightarrow[\text{catalyst}]{O_2} SO_3 \xrightarrow{H_2SO_4} H_2S_2O_7 \xrightarrow{H_2O} H_2SO_4 \qquad 13\text{–}4$$

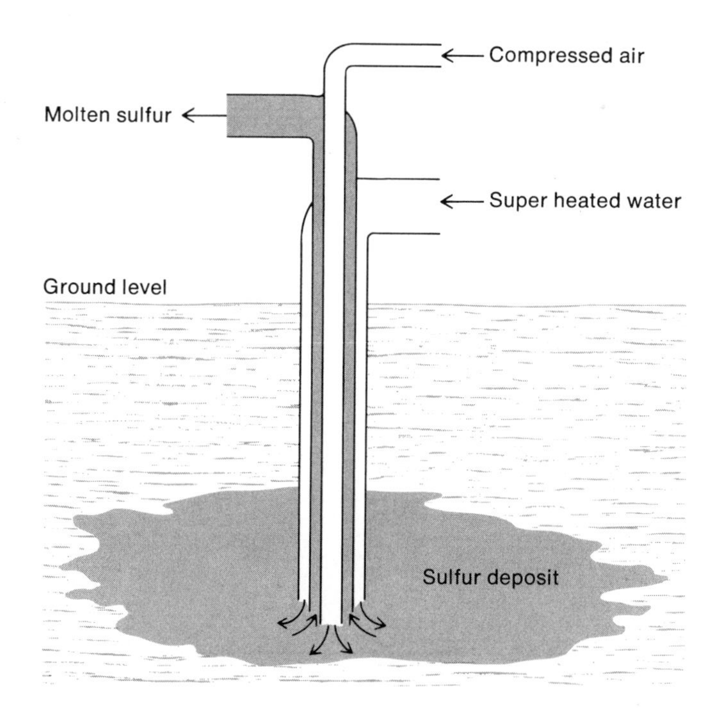

Figure 13–1.
Mining Sulfur by the Frasch Process

Ammonia

Ammonia is a colorless gas that has a strong penetrating odor. It is the basic raw material used in processes that require nitrogen in a "fixed" or chemically combined form. The major portion of naturally occurring nitrogen is in the elemental state and is found in the atmosphere. As already stated, the earth's atmosphere contains approximately 80 percent N_2 and 20 percent O_2 by volume. In the elemental form nitrogen is unreactive at normal temperatures, because the nitrogen atoms in N_2 molecules are bound together by a strong triple bond. Animals and most plants are unable to utilize elementary nitrogen; they must have it in a fixed form.

Some elemental nitrogen is fixed by natural processes. Legumes (a type of plant including beans, peas, alfalfa, and clover) harbor colonies of bacteria on their roots which are capable of converting elemental nitrogen into compounds. Lightning provides an important means for nitrogen fixation in the atmosphere. In the presence of an electrical discharge the two components of air, N_2 and O_2, react to produce fixed nitrogen. The equation for the reaction is:

$$N_2 + O_2 \xrightarrow{\text{elec.}} 2NO \qquad 13\text{–}5$$

The demand for fixed nitrogen exceeds the natural supplies, and man has been forced to develop nitrogen fixation processes to supply his additional needs. The nitrogen is fixed in the form of ammonia.

Uses of Ammonia

In the United States ammonia is used to provide the nitrogen contained in the products listed in Table 13–3. Once again, a single use dominates the consumption figures: most NH_3 goes into the production of fertilizers. We recall that this was also true for sulfuric acid. However, H_2SO_4 was used in the production of phosphorus-type fertilizers, while NH_3 is used to manufacture the nitrogen-containing types.

Table 13–3. Products Derived from Ammonia

Product	*Amount of NH_3 used annually (tons)*	*Percent of total NH_3 production used*
Fertilizers	7,790	72.0
Explosives	900	8.3
Synthetic fibers and plastics	610	5.6
Livestock feed	275	2.5
Miscellaneous	1,243	11.6
Total	10,818	100.0

Table 13-4. Types and Percentages of Ammonia-Derived Fertilizers Used in the United States

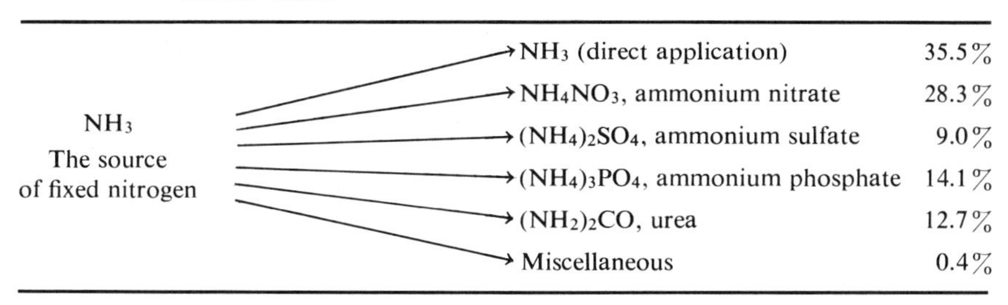

NH_3 The source of fixed nitrogen	NH_3 (direct application)	35.5%
	NH_4NO_3, ammonium nitrate	28.3%
	$(NH_4)_2SO_4$, ammonium sulfate	9.0%
	$(NH_4)_3PO_4$, ammonium phosphate	14.1%
	$(NH_2)_2CO$, urea	12.7%
	Miscellaneous	0.4%

Fertilizer production will continue to increase along with the world's population, since it is directly related to food production. As plants grow, they remove nutrients (mainly N, P, and K) from the soil. Plants undisturbed by humans die and decay, returning the used nutrients back to the soil for use by the next generation. When we cultivate and harvest crops on a large scale, the plants—and nutrients—are removed. These nutrients are restored through the use of fertilizers. Table 13–4 shows the nitrogen fertilizers commonly used.

Explosives manufacturing represents the second largest use of ammonia. An explosive is a substance which, under the stimulus of thermal or mechanical shock, can decompose almost instantaneously to liberate heat and large volumes of gases. When the explosive is detonated in a confined space, the gases expand because of the high temperature and create an extremely high pressure. The release of this pressure produces the well-known effects of an explosion. The explosive reaction of nitroglycerin is an example:

$$4C_3H_5(NO_3)_3 \rightarrow 12CO_2 + 10H_2O + 6N_2 + O_2 + 1{,}368 \text{ kcal} \quad 13\text{–}6$$

Even at room temperature the products of this reaction occupy a volume 1100 times that of the starting material, and the high temperature of the reaction causes the gaseous products to expand even more.

Many people tend to view explosives from a purely military standpoint. However, they are used in many nonmilitary applications including construction and mining.

Most common explosives contain nitrogen derived from nitric acid produced from ammonia (see equations 13–7 through 13–9). Table 13–5 contains examples of some common explosives.

Table 13–5. Some Common Nitrogen-Containing Explosives

Compound name	*Formula*	*Comments*
Nitroglycerin	$C_3H_5(NO_3)_2$	Dynamite consists of nitroglycerin absorbed into a cushioning agent to which an oxidizer is added. It usually contains 5 to 25 percent nitroglycerin.
Trinitrotoluene (TNT)	$C_7H_5(NO_2)_3$	The most important U.S. military explosive.
Picric acid	$C_6H_2(OH)(NO_2)_3$	Used extensively by the Japanese during World War II.
RDX	$C_3H_6O_6N_6$	One of the most powerful explosives known.

Ammonia is also the source of nitrogen for some compounds important in the manufacture of synthetic fibers and plastics. Two important compounds of this type are acrylonitrile (CH_2CHCN) and hexamethylenediamine ($NH_2(CH_2)_6NH_2$). Acrylonitrile is used in the formulation of Buna N, a synthetic rubber, and hexamethylenediamine is used to manufacture nylon (Chapter 20).

Certain animals, such as cattle, sheep, and goats, have a second stomach containing anaerobic bacteria that can convert nonprotein nitrogen materials into useful protein. Urea produced from ammonia is now being used extensively as a source of such nonprotein nitrogen. Its price is lower and more stable than that of natural protein sources such as corn, cottonseed, and soybean meals.

The miscellaneous uses of ammonia include two that are not often thought to have much in common—cleaning agents and refrigeration systems. Household ammonia, a popular cleaning agent, is a dilute solution of ammonia in water. Other agents, such as detergents, are often added to increase the cleaning effectiveness.

Ammonia is the refrigerant used in greatest tonnage because it is particularly well suited for large industrial installations. Its heat of vaporization is higher than that of any common liquid except water. Refrigeration operations are based on the fact that heat is absorbed from the surroundings when a liquid changes to a vapor, and a high heat of vaporization indicates a large capacity for absorbing heat.

In the operation of an industrial refrigeration system liquid NH_3 (b.p. = $-33°C$) is circulated through pipes that surround the area to be cooled. Heat is absorbed as the liquid vaporizes. The gaseous refrigerant is then circulated outside the refrigerator, where it is again liquefied by the combined effects of compression and cooling. Figure 13–2 is a diagram of a refrigeration system.

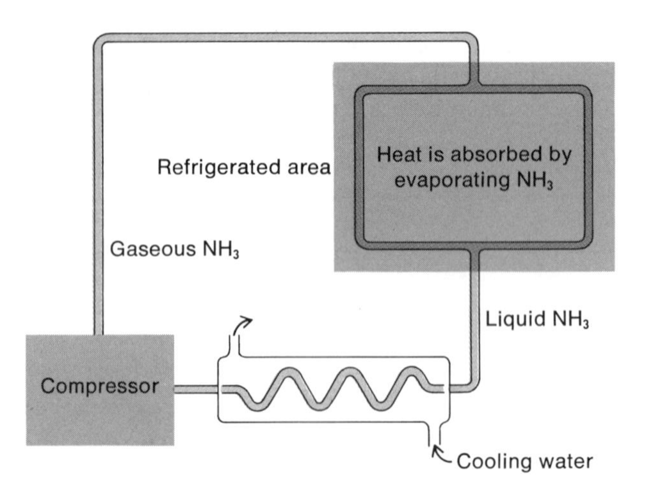

Figure 13–2.
A Refrigeration System

Ammonia is not used in household refrigerators because of its strong, choking odor and the possibility of leaks.

In many chemical processes, including some discussed previously, ammonia is converted into nitric acid, which is then used as the nitrogen source. The synthesis of nitric acid involves a stepwise oxidation of NH_3 through NO to NO_2, which is dissolved in water. The synthesis reactions are:

$$4NH_3 + 5O_2 \xrightarrow{\text{catalyst}} 4NO + 6H_2O \qquad 13\text{–}7$$

$$2NO + O_2 \rightarrow 2NO_2 \qquad 13\text{–}8$$

$$3NO_2 + H_2O \rightarrow 2HNO_3 + NO \text{ (recycled)} \qquad 13\text{–}9$$

Production of Ammonia

The most significant and widely used technique for the production of NH_3 is the **Haber process,** in which elementary nitrogen and hydrogen are reacted together:

$$N_2 + 3H_2 \rightarrow 2NH_3 \qquad 13\text{–}10$$

This reaction does not occur to an appreciable extent at room temperature and normal pressure. The reaction can be made to proceed at reasonable rates by the use of appropriate reaction conditions. Commercial processes use pressures of 200–600 atmospheres and temperatures of 450–600°C plus an iron oxide–potassium aluminate catalyst mixture.

Only two raw materials, N_2 and H_2, are needed for ammonia production. Air is directly or indirectly the source of N_2. In some ammonia manufacturing facilities N_2 from an air separation unit is used; in others air is the nitrogen source.

The source of hydrogen varies. The process from which hydrogen is most commonly obtained for industrial use is called catalytic steam-hydrocarbon reforming. In this process gaseous hydrocarbons (CH_4 and C_2H_6) or hydrocarbons easily vaporized at moderate temperatures (C_3H_8 and C_4H_{10}) are reacted with steam in the presence of a nickel catalyst at 650–1100°C. The products are carbon oxides and hydrogen. The hydrocarbons used in the process are obtained from natural gas or low-boiling fractions of petroleum (Chapter 19). At the temperatures used, all of the hydrocarbons decompose to methane (CH_4), and the net reforming reactions that take place can be represented as:

$$CH_4 + H_2O \rightarrow CO + 3H_2 \qquad 13\text{–}11$$

$$CO + H_2O \rightarrow CO_2 + H_2 \qquad 13\text{–}12$$

$$CH_4 + 2H_2O \rightarrow CO_2 + 4H_2 \qquad 13\text{–}13$$

The CO_2 is removed from the hydrogen in another step. When the H_2 is used for NH_3 production, air is added, and a secondary reforming step occurs in the presence of the same nickel catalyst. The oxygen of the air is consumed, and the nitrogen reacts to form ammonia. The air is introduced in an amount sufficient to produce a "raw" synthesis gas with a hydrogen:nitrogen ratio of 3:1, which is required for ammonia production (equation 13–10).

Oxygen

Oxygen is a colorless, odorless, tasteless gas which comprises about 20 percent of the atmosphere by volume. It is an active element that combines directly or indirectly with all other elements except the lower-molecular-weight rare gases. Compounds composed of elements that will combine with oxygen also react with oxygen, yielding the oxides of the constituent elements. The reaction of oxygen with fuels (combustion) provides much of the energy used by man. This is possible because most reactions involving oxygen are highly exothermic.

The oxygen molecule itself is generally not reactive at or near room temperature. Most materials have to be provided with activation energy through heating before they will react with oxygen at an appreciable rate. Catalysts, however, can effectively lower the necessary activation energy. The action of catalysts (enzymes—Chapter 24) allows oxygen to react and release energy in living cells during metabolic processes that take place at body temperature.

Oxygen for industrial chemical reactions may be supplied in the form of air, enriched air (containing added O_2), or pure oxygen. Air rather than pure oxygen has traditionally been used for most processes because nitrogen, the other major component of air, is quite unreactive in most instances and the reaction products are usually the same. The cost of pure oxygen also prevented its use in the past. During the last decade the use of pure oxygen in chemical processes has increased dramatically, as shown in Figure 13–3.

The use of pure oxygen offers a number of advantages. (1) The use of pure O_2 rather than air amounts to a five-fold increase in O_2 concentration, which in turn results in a large increase in production. (2) The rates of the reactions are increased and production times decreased, since reaction rates usually increase with increasing reactant concentrations. (3) Smaller production facilities can be used, since the volume of gaseous reactant is reduced to one-fifth that required with air. (4) Any possibility for interfering reactions with nitrogen is completely eliminated.

Most of the increased use of pure oxygen is the result of substituting pure O_2 for air in existing processes rather than the discovery of new processes. It has been found that the advantages compensate for the increased cost of pure O_2. Ironically, as more and more pure O_2 is used, the cost goes down, because the necessarily larger O_2 production facilities operate more efficiently.

Uses of Pure Oxygen

The important areas of high-purity oxygen use include the steelmaking industry, the welding and cutting of metals, the large-scale manufacture of chemicals, aerospace applications, and life-support systems.

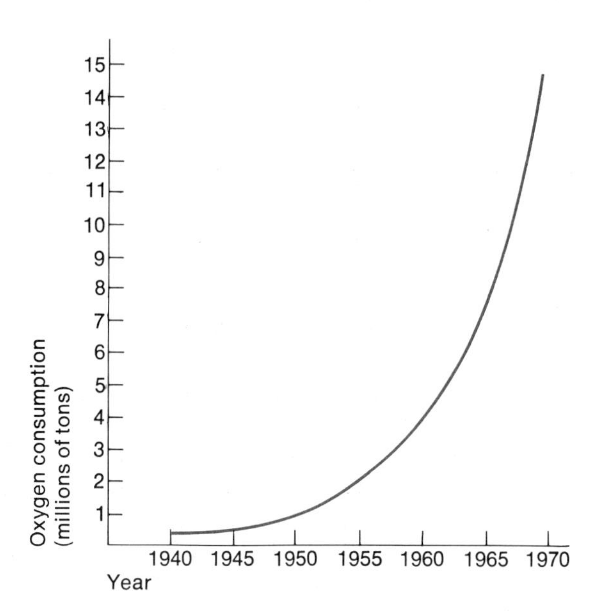

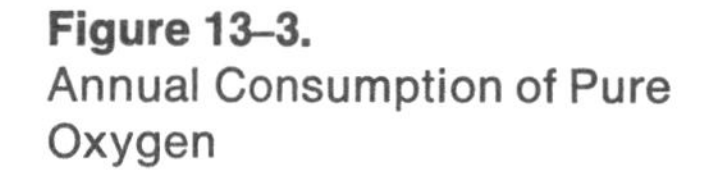
Figure 13–3.
Annual Consumption of Pure Oxygen

Just as fertilizer production consumes most of the manufactured H_2SO_4 and NH_3, the steelmaking industry is the dominant consumer of pure O_2. More than 70 percent of all pure oxygen is used to produce steel from molten iron—and the percentage is increasing. Steelmaking is fundamentally a process in which the impurities Si, Mn, C, S, and P are oxidized and removed from molten iron. Oxygen blown directly onto the melt speeds up the process and a complete "heat" (30 to 300 tons of steel) can be produced every hour. By contrast, older furnaces using air require 10–12 hours per heat. Steel production is discussed further in Chapter 15.

The cutting and welding of metals require temperatures much higher than those obtainable from combustion reactions involving air. Fuel-oxygen torches are often used—the fuel is usually acetylene or hydrogen. In the process known as oxygen cutting, the metal is heated to its combustion temperature by the fuel-oxygen torch. The fuel gas is then shut off and a stream of pure O_2 is directed at the hot metal. This causes the metal to burn and produces a narrow cut. The torch has literally burned through the metal. The necessary flame temperature of approximately 3000°C can be obtained by the combustion of acetylene with pure oxygen:

$$2C_2H_2 + 5O_2 \rightarrow 4CO_2 + 2H_2O \qquad 13\text{–}14$$

Acetylene (C_2H_2) and ethylene oxide (C_2H_4O) are important chemicals produced in large quantities by processes involving pure oxygen. Acetylene is made by the partial oxidation of methane (CH_4) and other similar compounds according to the following reactions:

$$2CH_4 + O_2 \rightarrow 2CO + 4H_2 \quad 13\text{–}15$$

$$2CH_4 \rightarrow C_2H_2 + 3H_2 \quad 13\text{–}16$$

The first reaction (equation 13–15) produces heat that causes the second reaction (equation 13–16) to take place.

Ethylene oxide is made by reacting ethylene (C_2H_4) with oxygen in the presence of a silver metal catalyst:

$$2C_2H_4 + O_2 \xrightarrow{\text{Ag cat.}} 2C_2H_4O \quad 13\text{–}17$$

Further reaction of C_2H_4O with an acid solution yields ethylene glycol, $C_2H_4(OH)_2$, which is used as an antifreeze for automobile cooling systems.

Liquid rocket propellant systems have been used extensively in space exploration vehicles. Systems of this type consist of an oxidizing agent (oxidant) and a fuel, which are kept in separate tanks until propulsion is required. At that time streams of the two liquids are brought together and the mixture is ignited; thrust is produced by the high-speed expulsion of large amounts of the gaseous combustion products. The pipes, pumps, and valves in the system must be designed to allow the entire combustion process to occur within a few seconds after the process is initiated. Figure 13–4 shows the engine components of a typical liquid propellant rocket.

Liquid oxygen has often been used as the oxidant in liquid propellant systems. It is readily prepared at a nominal cost and, unlike other oxidants such as fluorine or nitrogen oxides, its use does not create the possibility for hazardous spills that require decontamination. A disadvantage of oxygen is that it must be handled and stored as a liquid at very low temperatures. Gaseous oxygen cannot be used because of the large volume it would occupy. The firing of the first stage of a Saturn V rocket, for example, requires the equivalent of 38 million cubic feet of gaseous oxygen measured at normal atmospheric pressure and 0°C. Table 13–6 lists some common liquid-propellant pairs.

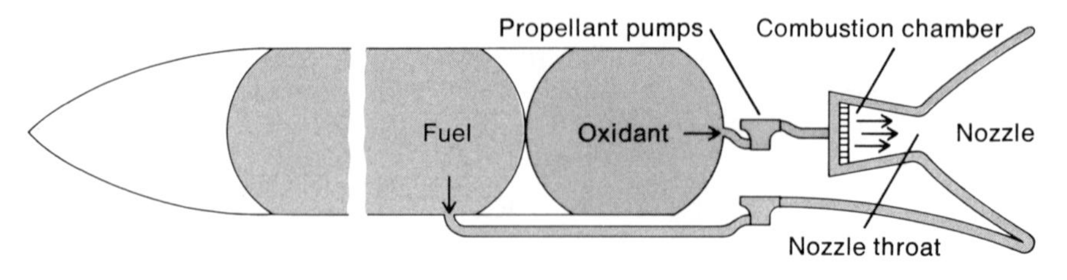

Figure 13–4.
Components of a Typical Liquid Propellant Rocket
Fuel and oxidant are pumped from their storage tanks to the combustion chamber. Circulation of the fuel through the walls of the combustion chamber and nozzle cools the surfaces of these structures.

From *Space Resources—Chemistry*, National Aeronautics and Space Administration, 1971, Publication NASA EP-87, p. 58.

Table 13–6. Common Pairs of Liquid Propellants

Oxidant (O)	*Liquid fuel (F)*	*Best weight ratio (O/F)*
Oxygen: O_2, −183°C	Alcohol: C_2H_5OH, 16°C	2.00
Nitrogen tetroxide: N_2O_4, 16°C	Hydrazine: N_2H_4, 16°C	1.30
Oxygen: O_2, −183°C	RP–1: kerosene, 16°C	2.60
Oxygen: O_2, −183°C	Hydrogen: H_2, −253°C	4.00

From *Space Resources—Chemistry*, National Aeronautics and Space Administration, 1971, Publication NASA EP–87, p. 58.

Oxygen is well-known for its use in life-support systems for space or underwater exploration. Man is accustomed to breathing a mixture of oxygen and nitrogen in which the pressure of oxygen is about one-fifth of an atmosphere or about 159 torr. Man can adapt to oxygen pressures in the range of 65–425 torr; he cannot survive if forced to breathe oxygen outside these pressure limits. In order to sustain a person in environments different from normal, we must give careful attention to the oxygen content of life-support gases that are used.

The total atmospheric pressure and consequently the oxygen pressure drops with increasing altitude, as shown in Figure 13–5. Adjustments must therefore

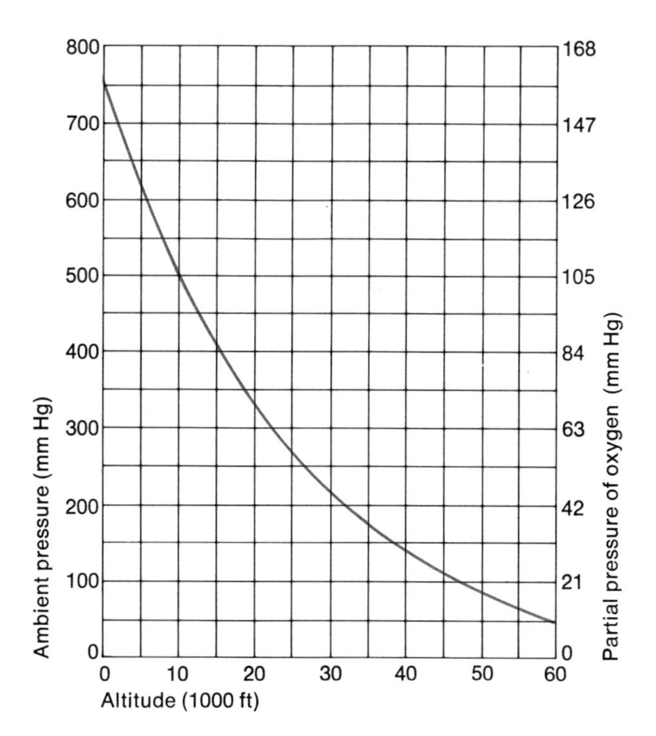

Figure 13–5.
The Change in Atmospheric and Oxygen Pressure with Altitude

From *Space Resources—Chemistry*, National Aeronautics and Space Administration, 1971, Publication NASA EP–87, p. 17.

be made if a person is to survive at high altitudes. In high-altitude commercial aircraft the oxygen in the cabin is usually maintained at a pressure equivalent to an 8000-foot elevation (118 torr). Supplemental oxygen masks are available for use in emergencies that result in decompression.

American astronauts who journey into space breathe pure oxygen at a pressure of 254 torr ($\frac{1}{3}$ atm). This greater-than-minimum pressure is used to allow time for emergency action in the event of a decompression. Pure oxygen is used to avoid the added weight of nitrogen gas and the equipment needed to maintain a constant ratio of the two gases.

Life-support systems supplied under pressure, as in deep-sea diving, require even more control. If air is used, it must be provided at a pressure sufficient to balance the water pressure, which can be 100 lb per square inch or higher. At such pressures, the oxygen pressure in the lungs might reach toxic levels. Even if this does not occur, dissolved nitrogen in the blood might lead to the bends (Chapter 11).

Production of Oxygen

Commercial production of pure oxygen involves its extraction from air. Because air is a mixture and not a chemical compound, it can be separated into its components by physical processes. Chemical reactions are not needed because no chemical bonds are broken. Separation is accomplished by the use of cryogenic (low-temperature) techniques. Any gas or gaseous mixture will liquefy if cooled to a low enough temperature. The commercial process for liquefying air begins by compressing dry, CO_2-free air until its pressure reaches about 20 atmospheres. The compression process heats the air, which is then cooled by passage through pipes submerged in cold water. The cool compressed air is pumped into a second-stage compressor where the pressure is raised to about 200 atmospheres. The air again becomes very warm, and the excess heat is once more removed by cooling in cold water. The cool, highly compressed air now enters a liquefying chamber where it is permitted to expand through a valve into a large pipe. During this process the pressure is suddenly reduced from 200 atmospheres to 20 atmospheres and the air cools. The expanded cold gas is drawn back to the second compressor through a large pipe which jackets the incoming high-pressure pipe. The incoming high-pressure air is continually cooled by the recycling cold air. After a few cycles the air in the high-pressure pipe is cooled to a low enough temperature that a portion of it liquefies upon expansion. The remainder is recycled and the process continues. The apparatus used to liquefy air is represented in Figure 13–6.

Liquid air, a mixture of components, has a boiling point of about $-200°C$. We can fractionate the liquid by taking advantage of the different boiling points of the components. When heat is applied to the liquid, nitrogen (b.p. = $-196°C$) and argon (b.p. = $-186°C$) boil off first, since they have lower boiling points than oxygen (b.p. = $-183°C$). The relatively pure oxygen (about 99.5 percent) obtained from the fractionation is stored, usually as a gas, under high pressure in steel cylinders.

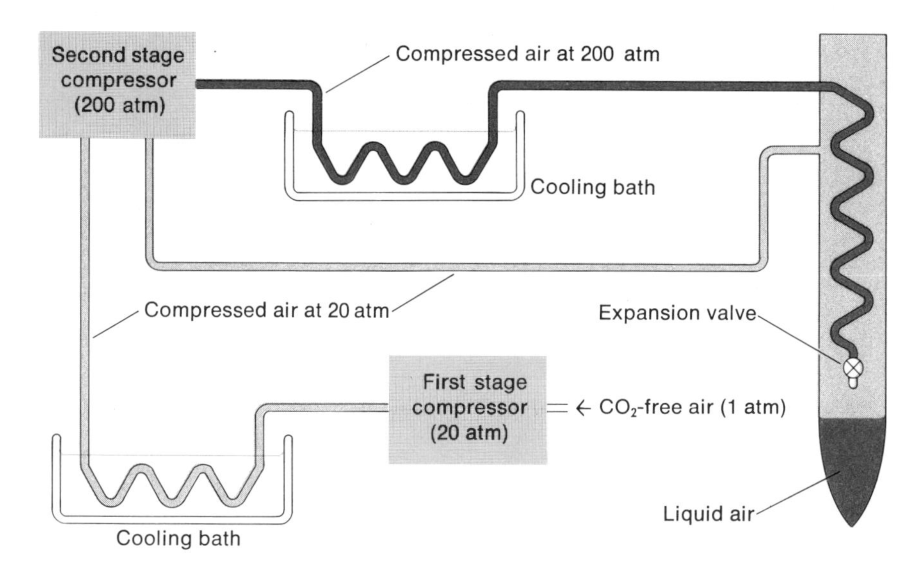

Figure 13–6.
The Production of Liquid Air

Sodium Hydroxide

Sodium hydroxide is a white, crystalline ionic substance composed of the Na^+ and OH^- ions. It readily absorbs moisture from its surroundings and will, upon exposure to moist air, rapidly absorb enough water to dissolve itself and form a liquid solution. It also absorbs carbon dioxide from air and forms Na_2CO_3. Sodium hydroxide is corrosive to the skin, a property that resulted in the widely used industrial name **caustic soda.** It dissolves readily in water to form a colorless solution that is strongly alkaline. Its strong basic properties are the result of its complete dissociation in water solutions. Commercial caustic soda is frequently shipped and used in the form of a 50 percent solution. Impure NaOH is sometimes called **lye** or **soda lye.**

Sodium hydroxide reacts with acidic substances and forms the corresponding salts and water (equation 13–18). It also dissolves a number of metals by forming soluble hydroxide complexes in water solutions (equation 13–19). Many of the uses of NaOH depend upon these two types of reactions:

$$2NaOH + H_2SO_4 \rightarrow Na_2SO_4 + 2H_2O \qquad 13\text{–}18$$

$$2NaOH + Zn + 2H_2O \rightarrow H_2 + 2Na^+ + Zn(OH)_4^{2-} \qquad 13\text{–}19$$

Uses of Sodium Hydroxide

Five industries use nearly three-fourths of the annual NaOH production in the United States, as indicated by the 1969 estimated data given in Table 13–7.

Table 13–7. Products Requiring NaOH in Their Preparation

Product	*Percent of total NaOH used*
Chemical intermediates	42.5
Pulp and paper	13.6
Aluminum	7.4
Rayon	5.4
Soaps and detergents	4.8
Miscellaneous	26.3
	100.0

Phenol (C_6H_6O) and glycerin (C_3H_8O) are examples of chemical intermediates that require NaOH in their production. Phenol is used mainly in the production of phenolic resins (plastics). These resins are important components of such items as appliance knobs, handles, and housings; washing machine agitators; and electrical devices (see Chapter 20). Glycerin is used in plastic production; it is also a component of some medicines, cosmetics, and tobacco products. It is used to retain moisture in medicinal creams and ointments. It serves as a moisture retainer, solvent, and lubricant in a variety of cosmetic creams and lotions, and as a humectant (moisture retainer) in tobacco products.

Sodium hydroxide is used in large quantities by the pulp and paper industry because of its ability to react with cellulose. The woody structure of plants consists of fibers of the natural polymer cellulose (see Chapter 23) which are held together by a natural "glue" called lignin. Sodium hydroxide has the ability to break down lignin.

The production of paper from wood requires a number of steps. The wood is shredded and then digested in a solution containing NaOH. This treatment breaks down the lignin and allows the cellulose fibers to separate and form pulp. The pulp is then bleached, and filler, sizing, and color are added. The filler (often a clay) occupies spaces between the cellulose fibers in the final product; the sizing imparts a resistance to penetration by liquids.

Bauxite, an impure hydrated oxide of aluminum ($Al_2O_3 \cdot 3H_2O$), is the principal ore from which aluminum is obtained. The ore usually contains 40–60 percent Al_2O_3. The remainder consists of iron oxide (Fe_2O_3) and silicate impurities that must be removed. This is done by treating the ore with NaOH. The aluminum oxide dissolves according to the following equation:

$$Al_2O_3 \cdot 3H_2O + 2NaOH \rightarrow \underset{\substack{\text{sodium}\\\text{aluminate}\\\text{(soluble)}}}{2NaAlO_2} + 4H_2O \qquad 13\text{–}20$$

The iron oxide and silicate impurities are insoluble in NaOH and can be removed. The purified aluminum oxide is then electrolytically reduced to aluminum metal (see Chapter 15).

The production of rayon from cellulose, and of soaps from fats and oils, also involve the use of NaOH. These processes are both discussed in Chapter 23.

Production of Sodium Hydroxide

Sodium hydroxide and chlorine gas (discussed next in this chapter) are produced in large amounts by the chlor-alkali industry. A small amount of hydrogen is also generated in the single process used to produce the two main products.

The basic raw materials are common salt (NaCl) and water. A large amount of relatively inexpensive electrical energy is also required. The process consists primarily of the electrolysis of concentrated salt water solutions, a reaction previously described in Chapter 12. During the electrolysis, chlorine gas is liberated at the anode. The cathode reaction and immediate products depend upon the design of the electrolysis cell, but the ultimate products are hydrogen gas and hydroxide ions. The net overall process thus produces gaseous hydrogen, gaseous chlorine, and a solution of sodium hydroxide:

$$\underset{\text{(from salt)}}{2Na^+} + \underset{\text{(from salt)}}{2Cl^-} + 2H_2O \xrightarrow{\text{elect.}} 2Na^+ + 2OH^- + H_2 + Cl_2 \qquad 13\text{–}21$$

Chlorine gas reacts readily with NaOH or H_2. In order to prevent such a reaction the chlor-alkali cells are designed to keep the cathode and anode products separated. The anode half reaction, regardless of cell design, is

$$2Cl^- \rightarrow Cl_2\ (gas) + 2e^- \qquad 13\text{–}22$$

In some cells a graphite cathode is used, and the cathode half reaction is a reduction of water:

$$2H_2O + 2e^- \rightarrow 2OH^- + H_2\ (gas) \qquad 13\text{–}23$$

The H_2 gas is given off, leaving the Na^+ and the newly formed OH^- ions in solution. Another type of cell has mercury cathodes. In these cells sodium is liberated at the cathode instead of hydrogen; the sodium dissolves in the mercury and forms an amalgam (a mercury alloy):

$$Na^+ + e^- \rightarrow Na\ (amalgam) \qquad 13\text{–}24$$

The amalgam is drained into a reaction vessel, where the contained sodium is reacted with water to produce H_2 gas and NaOH:

$$2Na(Hg) + 2H_2O \rightarrow 2NaOH + H_2\ (gas) + 2Hg \qquad 13\text{–}25$$

Figure 13–7 contains a diagram of a typical cell used in the chlor-alkali industry.

Nearly 100 percent of all NaOH and 97 percent of all Cl_2 are produced by the chlor-alkali process. However, the amount of H_2 produced as a by-product

Figure 13–7.
A Chlor-Alkali Electrolysis Cell

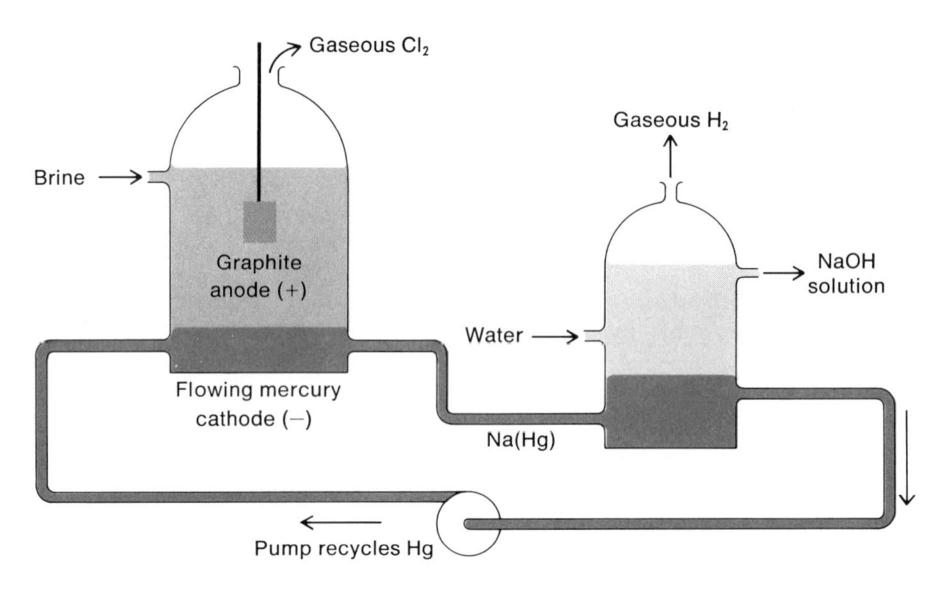

is only a small part of that used annually. Approximately 0.5 billion lb of hydrogen are produced, and the ammonia industry alone consumes 4.5 billion lb.

Chlorine

At normal temperature and pressure chlorine is a greenish-yellow gas characterized by a pungent, unpleasant odor. It has a density nearly 2.5 times that of air; and when released into the atmosphere it forms a layer which stays near the ground and is hard to dissipate. Chlorine is nonexplosive and noncombustible in oxygen, and it is a nonconductor of electricity. It will react with all metals, but in a number of cases only at elevated temperatures and pressures. It also reacts with many nonmetals to give covalent compounds.

Uses of Chlorine

Compounds known as chlorinated hydrocarbons provide the principal route by which chlorine reaches the open market and the consumer. The production of these compounds accounts for nearly 60 percent of all chlorine use, as shown by Table 13–8.

Hydrocarbons are organic chemicals that contain only carbon and hydrogen (see Chapter 17). Most of these compounds can be made to react with

Table 13–8. Chlorine Consumption (1969)

Product or use	*Percent of total chlorine used*
Chlorinated hydrocarbons	59
Paper and pulp	18
Other chemicals	11
Water treatment	3.6
Miscellaneous	8.4

chlorine, yielding many useful products or chemical intermediates. During many types of chlorination reactions a bond between carbon and hydrogen is broken, chlorine replaces the hydrogen atom in the hydrocarbon molecule, and HCl is produced as a by-product. Chlorination reactions can take place more than once on some hydrocarbons, thus increasing the number of different products that are possible. Chlorinated hydrocarbons are used, among other things, in the production of solvents, plastics, and pesticides. Two important chlorine compounds used as solvents are carbon tetrachloride (CCl_4) and chloroform ($CHCl_3$). Vinyl chloride (C_2H_3Cl) is the starting material for polyvinyl chloride, a plastic used in flooring, phonograph records, and similar products. Some chlorinated hydrocarbons make up a group of pesticides called organochlorine compounds; DDT is a member of this group.

The by-product HCl can be used for further chlorinating reactions, or it can be dissolved in water to form the important solution called hydrochloric acid; this is the main source of hydrochloric acid (over 80 percent). In addition to being a chlorinating agent, hydrochloric acid is used in metal pickling, oil-well acidizing (to increase the permeability of oil-bearing formations), and in some food processing industries.

The second largest consumer of chlorine is the pulp and paper industry. As mentioned previously, cellulose fibers are bleached before they are made into paper. The bleaching is done with a "liquor" made from chlorine and lime water, $Ca(OH)_2$:

$$2Cl_2 + 2Ca(OH)_2 \rightarrow Ca(OCl)_2 + CaCl_2 + 2H_2O \qquad 13\text{–}26$$

The resulting calcium hypochlorite, $Ca(OCl)_2$, is the active bleaching agent.

Chlorine is a commonly used disinfectant in water treatment. It effectively destroys or attenuates microorganisms of sanitary significance. Liquid chlorine is the least expensive form when large amounts are required, but in small water-treatment facilities and for emergency or specialized uses some chlorine-containing compounds are used (hypochlorites or chloramines, for example).

When chlorine is added to water, a mixture of hypochlorous acid (HOCl) and hydrochloric acid (HCl) is formed:

$$Cl_2 + H_2O \rightleftharpoons HOCl + H^+ + Cl^- \qquad 13\text{–}27$$

At ordinary water temperatures the reaction is completed in a matter of seconds. In dilute solutions and at a *p*H near 7, the equilibrium of equation 13–27 lies toward the right, and very little Cl_2 is found in solution. The HOCl that is formed dissociates into hydrogen and hypochlorite ions:

$$HOCl \rightleftharpoons H^+ + OCl^- \qquad 13\text{–}28$$

The extent of dissociation depends on the *p*H and the water temperature. The *p*H of chlorinated water is normally within a range that allows both HOCl and OCl^- to exist; the disinfecting power is due to the presence of the OCl^- ion, which is a strong oxidizing agent. The familiar odor frequently associated with chlorinated water is not caused by Cl_2, HOCl, or OCl^-; it is the result of chloramines (NH_2Cl, $NHCl_2$) that form when the OCl^- ion reacts with nitrogen-containing impurities.

Hypochlorite solutions are marketed as bleaches. Solutions containing 5–5½ percent sodium hypochlorite are sold under familiar trade names such as Purex and Clorox. More concentrated commercial solutions contain 12–15 percent NaOCl and are marketed for use in swimming pools (as a disinfectant), commercial laundries, and textile plants.

Production of Chlorine

The main source of Cl_2 is the chlor-alkali electrolysis process discussed earlier. Another 3 percent of the total is produced during the electrolysis of molten sodium chloride or magnesium chloride. The amount of Cl_2 available from these sources is governed by the market for the metals sodium and magnesium that are also generated in the process.

Review Questions

1. What four properties make sulfuric acid a very widely used chemical?
2. What is the advantage to a sulfuric acid manufacturer of shipping the product in the form of oleum, $H_2S_2O_7$, rather than as H_2SO_4?
3. Describe the Frasch process for mining sulfur.
4. By what fraction did pure oxygen consumption increase during each of the following five year periods? See Figure 13–3.
   a) 1945–1950 b) 1950–1955 c) 1955–1960 d) 1960–1965
5. What are the advantages of using pure O_2 rather than air in industrial processes?
6. List five areas in which high purity oxygen is used. Illustrate each type of use by a specific example and, where possible, with chemical equations.

7. What is the oxygen pressure at the following altitudes? How does this pressure compare to the minimum required by man? See Figure 13–5.
   a) 5,000 ft—Denver, Colorado
   b) 15,000 ft—altitude of some airliners
   c) 29,000 ft—top of Mount Everest
8. Briefly outline the steps involved in the production of pure oxygen from atmospheric air.
9. Calculate the percentage of nitrogen in each of the fertilizers given in Table 13–4.
10. What is meant by the term *fixed nitrogen?* Give two examples of substances that represent fixed nitrogen.
11. What are the necessary characteristics or properties of a substance that is to be used as an explosive?
12. Many useful nitrogen-containing products require nitric acid in their preparation. Show by means of equations how atmospheric nitrogen is converted into nitric acid.
13. List at least four properties of sodium hydroxide that make it a useful industrial chemical.
14. A cook is using a pickle recipe in which the cucumbers are soaked in a solution of lye. She puts the ingredients into an aluminum pan and finds to her amazement that the mixture "works"—it foams and gives off bubbles of gas. Explain what is probably happening. See equation 13–19.
15. How is sodium hydroxide prepared commercially? What raw materials and other resources would be needed to make the preparation economically feasible?
16. Chlorine was the first "war gas" used in combat. What properties made this use possible?
17. List five important chlorine-containing compounds. Indicate a use for each compound.
18. Why would it be a dangerous practice to mix vinegar (an acid) with laundry bleach (NaOCl)? See equations 13–27 and 13–28.
19. How is chlorine prepared commercially?
20. What environmental pollutant might be produced during the commercial preparation of NaOH and Cl_2?

Suggestions for Further Reading

Bowman, W. H., and R. M. Lawrence, "Life Support Systems for Manned Space Flights," *Journal of Chemical Education*, April 1971.

Bowman, W. H., and R. M. Lawrence, "The Cabin Atmosphere in Manned Space Vehicles," *Journal of Chemical Education*, March 1971.

Cloud, P., and A. Gibor, "The Oxygen Cycle," *Scientific American*, Sept. 1970.

Delwiche, C. C., "The Nitrogen Cycle," *Scientific American*, Sept. 1970.
Farber, E., "Oxygen—The Element with Two Faces," *Chemistry*, May 1966.
Manufacture of Sulfuric Acid (American Chemical Society Monograph #144), ed. by W. W. Duecker and J. R. West, Reinhold Publishing Corp., New York, 1959.
Pratt, C. J., "Sulfur," *Scientific American*, May 1970.
Stone, J. K., "Oxygen in Steelmaking," *Scientific American*, April 1968.
White, L., Jr., "Medieval Uses of Air," *Scientific American*, Aug. 1970.

14 Some Interesting and Useful Nonmetals

In this chapter we shall investigate three nonmetallic elements: phosphorus, carbon, and silicon.

Phosphorus

The element phosphorus was discovered in 1674 by Hennig Brand, a merchant of Hamburg. Brand was searching for the **philosophers' stone**—a mysterious substance which would turn base metals, such as lead, into gold. Many alchemists felt this substance would be found in connection with living material, especially the more repulsive sorts of living material. Brand eventually got around to working with urine and discovered a process for isolating phosphorus.

A number of pails of urine were heated and evaporated until the urine was reduced to a thick syrup. This syrup was collected and allowed to putrefy for a number of days in a warm cellar. The putrefied material was mixed with sand and heated in a way that allowed the vapors to be collected under water, where the elemental phosphorus condensed as a white, waxy-looking solid.

The element was immediately interesting to its discoverer because of two properties. It glows in the dark (hence the name), and it has a low ignition temperature, so that in a warm room a sample will smoke and finally burst into flame unless stored under water.

Phosphorus has been found to be an essential constituent of all animal and vegetable matter. The bones of most animals contain about 60 percent calcium phosphate, $Ca_3(PO_4)_2$.

The Modern Preparation of Phosphorus

The element does not occur free in nature because of its extreme reactivity toward oxygen. The main sources in nature are the minerals **apatite,** $CaF \cdot 3Ca_3(PO_4)_2$, and impure **rock phosphate,** $Ca_3(PO_4)_2$.

The preparation of the element from these sources involves a reaction with sand, SiO_2, to give phosphorus oxide, and then reduction with carbon to yield the free element. The reactions are:

$$2Ca_3(PO_4)_2 + 6SiO_2 \rightarrow 6CaSiO_3 + P_4O_{10} \qquad 14\text{–}1$$

$$P_4O_{10} + 10C \rightarrow P_4 + 10CO \qquad 14\text{–}2$$

The elemental phosphorus produced is a gas which is condensed under water to give the solid white form. This is one of several forms in which the element may exist. If the white form, which is very toxic, is heated to a temperature of 200°C while under water or otherwise protected from oxygen, it is converted into the nontoxic red phosphorus. Red phosphorus is still pure but has a dark red color and lacks the crystallinity of the white form.

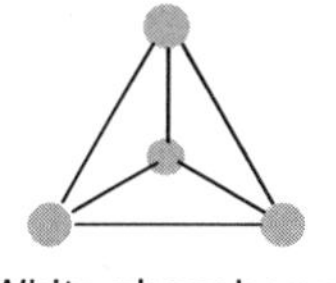

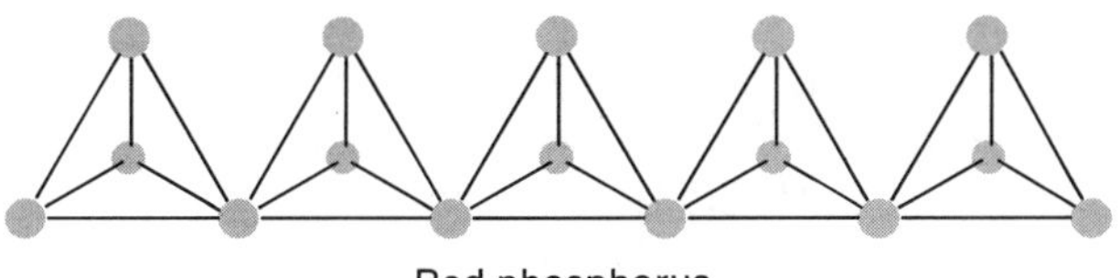

Figure 14–1.
Molecular Structure of Phosphorus Allotropes

The red form also lacks the low ignition temperature of the white and may be safely stored outside of water. The existence of an element in more than one elemental form is a property called **allotropy,** and the different forms are referred to as allotropic forms. Some other elements we shall study also show this property.

The difference in the forms given above is one of molecular structure. The white form is made up of individual tetrahedral-shaped P_4 molecules, while the red form consists of long chains of linked-together P_4 molecules, as shown in Figure 14–1. It is quite interesting to note the great differences in properties—especially the toxicity—caused by such a molecular difference.

The Use of Phosphorus in Matches

The most obvious chemical property of phosphorus is its reactivity with oxygen, leading to the spontaneous combustion of the white form in a warm room (35–45°C). Because of this property, phosphorus has been used extensively in the manufacture of matches. The very earliest matches contained no phosphorus; the first that did were prepared in about 1833. White phosphorus was used together with other chemicals, but the matches proved to be unpredictable and often ignited spontaneously. The use of white phosphorus also proved hazardous to those working in the match factories; many of the young girls employed there contracted a disease called *phossy jaw*, which was a rotting or necrosis of the bones, primarily of the lower jaw. Because of these hazards, the use of white phosphorus in matches was discontinued about 1848. Red phosphorus was substituted, and many of the difficulties in the match industry were overcome. The replacement match was the forerunner of the modern safety match. The modern strike-anywhere match and safety match are diagrammed in Figure 14–2.

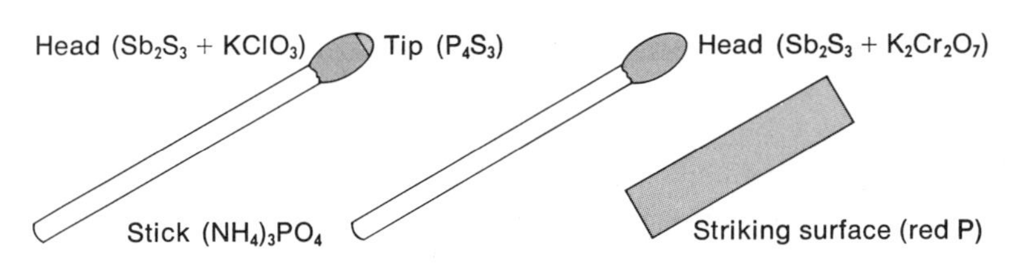

Figure 14–2.
Modern Matches

The tip material of strike-anywhere matches is easily ignited by friction. This in turn ignites the head, which then ignites the stick. The ammonium phosphate, $(NH_4)_3PO_4$, impregnated in the paper or wood of the stick keeps it from burning too fast and prevents it from glowing after the flame is extinguished.

The safety match contains no phosphorus in the head. The combustible materials of the head are ignited by rubbing the head over the striking surface, which does contain red phosphorus with its low kindling temperature. The heat of friction ignites the red phosphorus of the striking surface, which in turn ignites the match head.

Phosphorus in Fertilizers

The presence of phosphorus in all living material implies that it might be useful as a fertilizer. Almost all commercial fertilizers do contain some phosphorus compounds. Rock phosphate, $Ca_3(PO_4)_2$, is of little value as a fertilizer because of its low solubility in water. This raw material can be converted to soluble forms by proper chemical treatment. Two such treatments are given below:

$$Ca_3(PO_4)_2 + 2H_2SO_4 \rightarrow \underbrace{Ca(H_2PO_4)_2 + 2CaSO_4}_{\text{mixture is called superphosphate fertilizer}} \qquad 14\text{–}3$$

$$Ca_3(PO_4)_2 + 4H_3PO_4 \rightarrow \underbrace{3Ca(H_2PO_4)_2}_{\text{triple superphosphate}} \qquad 14\text{–}4$$

Carbon

Carbon is the second most abundant element in the human body—the most abundant being oxygen. Carbon occurs in all plant and animal tissue, mostly in the form of hydrogen- and oxygen-containing compounds. These compounds are classified as organic and are discussed in Chapters 17 through 20. Carbon occurs in nature in nonorganic (inorganic) compounds, such as carbonates and CO_2, and also in the free elemental state.

Allotropes of Carbon

Pure elemental carbon exists in two allotropic crystalline forms called graphite and diamond. They differ only in the way the carbon atoms are bound together to give the crystal.

Graphite is a soft, grayish-black solid which is often used as a component of black pencil leads and black paint, as a lubricant, and as a component of some electrical equipment. The crystal structure of the material is layered, with each carbon atom in a layer bonded to three neighbors. Within a layer, the atoms are tightly bound to each other, but the layers are only weakly bound and are able to slide over one another easily. This accounts for the softness of the allotrope. This structure is also characterized by the presence within a layer of loosely held electrons that are quite mobile. These mobile electrons give graphite the ability to conduct electricity, so that it can be used to make brushes in electric motors.

Diamond, on the other hand, is a hard, sometimes colorless, crystalline material. It is, in fact, the hardest naturally occurring substance known. Diamond crystals are used as gem stones and also as abrasives in the form of grinding wheels, rock drill bits, and other similar devices. The crystal structure of diamond is tetrahedral, with each atom bonded to four neighboring atoms which are located at the corners of a tetrahedron. This results in a very rigid, strong, three-dimensional crystal lattice. Figure 14–3 shows the structures of graphite and diamond.

Figure 14–3.
Allotropic Forms of Carbon

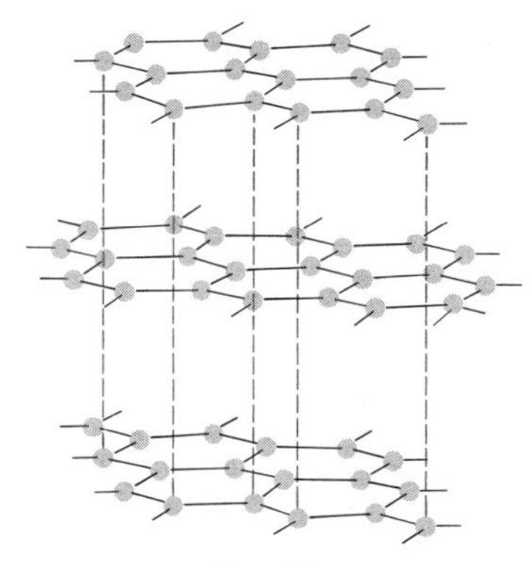

Graphite

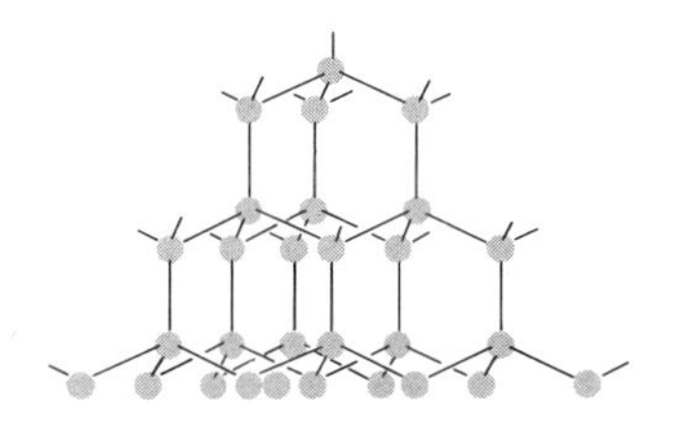

Diamond

Synthetic Diamonds

The fact that graphite and diamond are both forms of carbon led to numerous attempts to synthesize diamond from graphite. The first successful synthesis took place in 1954. Graphite was dissolved in iron sulfide, FeS; heated to a temperature of 1650°C; and subjected to a pressure of 95,000 atmospheres—about 1.3 million lb per square inch. The system, still under pressure, was then cooled to room temperature in about five minutes; and then the pressure was lowered to room pressure in about 18 minutes. The resulting diamonds had flaws and were dark, small ($\frac{1}{10}$ to 1 carat), and unsuitable for use as gems. But they were quite acceptable for industrial use. Gem-quality diamonds can now be produced, but at present they are more expensive than the naturally occurring variety.

The largest natural diamond ever discovered was the Cullinan, which weighed 3106 carats (one carat = 200 mg), or about $1\frac{1}{3}$ lb. This stone was cut into a number of smaller gems, including the world's largest cut gem, the Great Star of Africa, which weighs 530 carats.

Compounds of Carbon

The most common inorganic compounds of carbon are CO_2, CO (discussed in Chapters 21 and 25), and various carbonates. Carbon dioxide—a colorless, odorless gas with a density 1.5 times that of air—is a product of the combustion of carbon-containing fuels and of the respiration processes of plants and animals. The atmosphere contains about 0.035 percent or 35 parts in 10,000 of CO_2, but the amount is not constant because processes may locally remove or add the gas to the atmosphere. The amount can be decreased by plant absorption during photosynthesis, by the formation of water solutions—CO_2 is very water soluble—and by fixation in the form of carbonate

Figure 14–4.
The Acid-Soda Fire Extinguisher

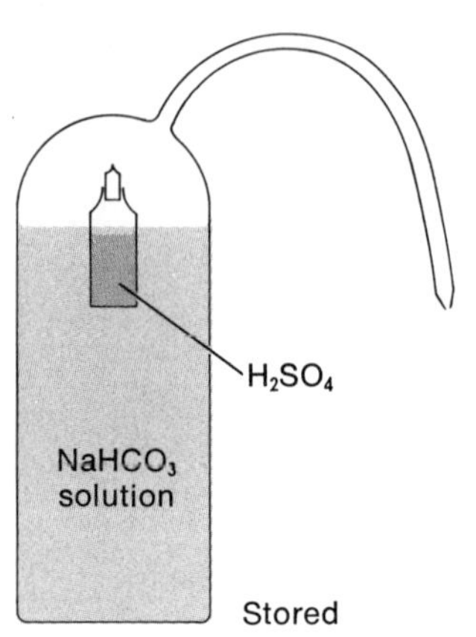

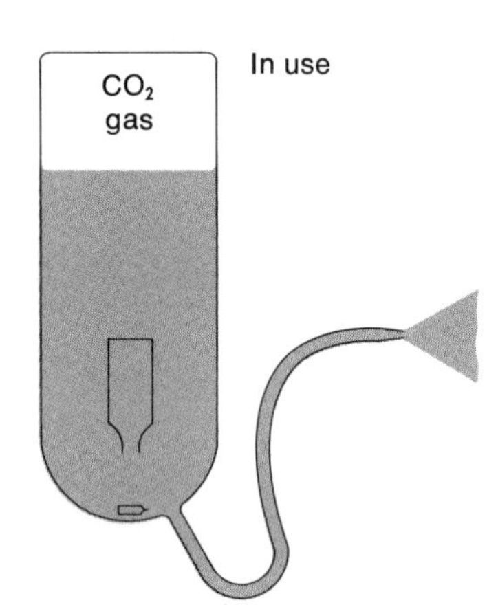

minerals such as $CaCO_3$ and $MgCO_3$. The amount is increased when carbon-containing compounds are burned or decayed, when fermentation or respiration takes place, and when volcanic gases are discharged into the atmosphere.

Carbon dioxide neither burns nor supports other combustion processes requiring oxygen. It is, in fact, used as a fire extinguisher. If the normal 21 percent oxygen content of the air is reduced to about 17 percent or less, most materials will cease to burn; carbon dioxide is used to accomplish this decrease in concentration.

Two types of fire extinguishers involve the use of CO_2. In the **acid-soda** type (Figure 14–4), the CO_2 acts as a propellant to push water out of the tank onto the fire. Sulfuric acid from a container mixes with a solution of sodium bicarbonate when the fire extinguisher is inverted. CO_2 is produced by the reaction:

$$H_2SO_4 + NaHCO_3 \rightarrow H_2O + CO_2 + NaHSO_4 \qquad 14\text{–}5$$

The CO_2 creates a pressure which forces the water solution out of a nozzle. This type of extinguisher acts on a fire in two ways: the liquid cools the burning materials below their kindling temperature, and dissolved CO_2 escapes from the warm liquid, blankets the fire, and keeps oxygen away.

The second type of CO_2 fire extinguisher consists of a steel container partially filled with liquid CO_2 under pressure (Figure 14–5). The pressure inside one of these containers is usually about 60 atm, or 850 lb per square inch. When the valve to the tank is opened, a stream of liquid CO_2 emerges and quickly cools itself by evaporation, because it cannot remain in the liquid state at a pressure of less than 5.3 atm. This cooling causes some of the liquid to freeze into a solid "snow," which cools the burning material and blankets it with CO_2 gas. This is the only type of extinguisher that should be used on electrical fires (the acid-soda type produces an electrically conductive liquid) and on oil fires (the oil will float and spread on water).

This same type of cooling by evaporation is used in the preparation of the popular "dry ice" refrigerant. The solid snow is collected and pressed into blocks. It is dry because it sublimes directly from the solid to gaseous state at one atmosphere of pressure. The temperature of dry ice is $-79°C$; this means that the vapor pressure of the solid reaches one atmosphere at $-79°C$, which is lower than the melting temperature.

Carbonate compounds are quite abundant in nature. In fact, an impure form of $CaCO_3$ known as **limestone** is the most common nonsilicate mineral. Most of the metallic carbonates are insoluble in water, the exceptions being the group IA metals.

Calcium carbonate is essentially insoluble in water but is appreciably soluble in water containing dissolved CO_2. Since the atmosphere contains about 0.04 percent CO_2, which is very soluble in water, essentially all ground waters contain dissolved CO_2. As these ground waters trickle through limestone, some $CaCO_3$ is dissolved according to the following equation:

$$CaCO_3\ (\text{solid}) + CO_2 + H_2O \rightarrow Ca^{2+} + 2HCO_3^- \qquad 14\text{–}6$$

The effects of this action are the creation of large caves in limestone-rich regions and the presence of Ca^{2+} ions (a hard-water ion) in most ground waters.

Many of the resulting caves have weird formations consisting of stalagmites (up from the floor) and stalactites (down from the ceiling). These formations result when ground water seeps through a fissure in the roof of the cave and hangs as a drop from the ceiling, or drops to the floor. As water and CO_2 evaporate from the drop, the reaction given in equation 14–6 is reversed and $CaCO_3$ is deposited. This process, repeated many times over long periods, builds up the deposits.

Figure 14–5.
Liquid CO_2 Fire Extinguisher

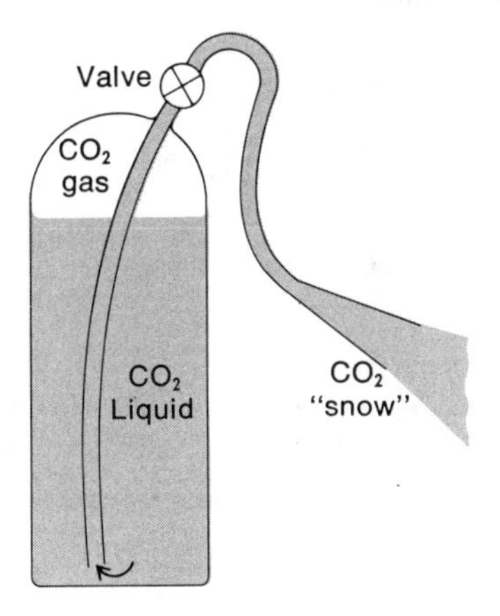

Silicon

Silicon is the second most abundant element in the earth's crust (26 percent). It has a high affinity for oxygen and is always found in nature combined with oxygen. The most common compound in nature is silicon dioxide or silica, SiO_2. Sand and sandstone are impure forms of SiO_2; quartz is a pure crystalline form.

Silicate minerals make up the bulk of most common rocks. The only exceptions are limestone and dolomite, which are carbonates ($CaCO_3$ and $MgCO_3 \cdot CaCO_3$, respectively).

The Structure of Silicates

The basic structure in silicate minerals is the silicate ion, SiO_4^{4-}. The ion is tetrahedral in structure, with the four oxygen atoms arranged at the corners and the silicon atom in the center. This basic unit is combined in various ways to give the variety of negative ions which are known to exist in various minerals. Figure 14–6 shows some of these ions. Each is negatively charged and is held together in the mineral by electrostatic attractions to positive metal ions, which must also be present. The chainlike anions give rise to fibrous minerals such as asbestos, and the layer types are involved in the sheetlike minerals such as mica.

In some cases the SiO_4^{4-} groups are arranged in a three-dimensional network, with each oxygen atom shared by two tetrahedra to give a silicon-to-oxygen ratio of 1:2. This rigid, strong arrangement is found in silica, SiO_2, itself (Figure 14–7). If some of the Si ions are replaced by aluminum ions, the neutral charge situation is upset, because Al^{3+} replaces Si^{4+}, and additional cations, usually Na^+, are needed to balance the charges. The resulting minerals, called zeolites, were mentioned earlier in connection with water softening.

Figure 14–6.
Structures of Some Silicate Anions

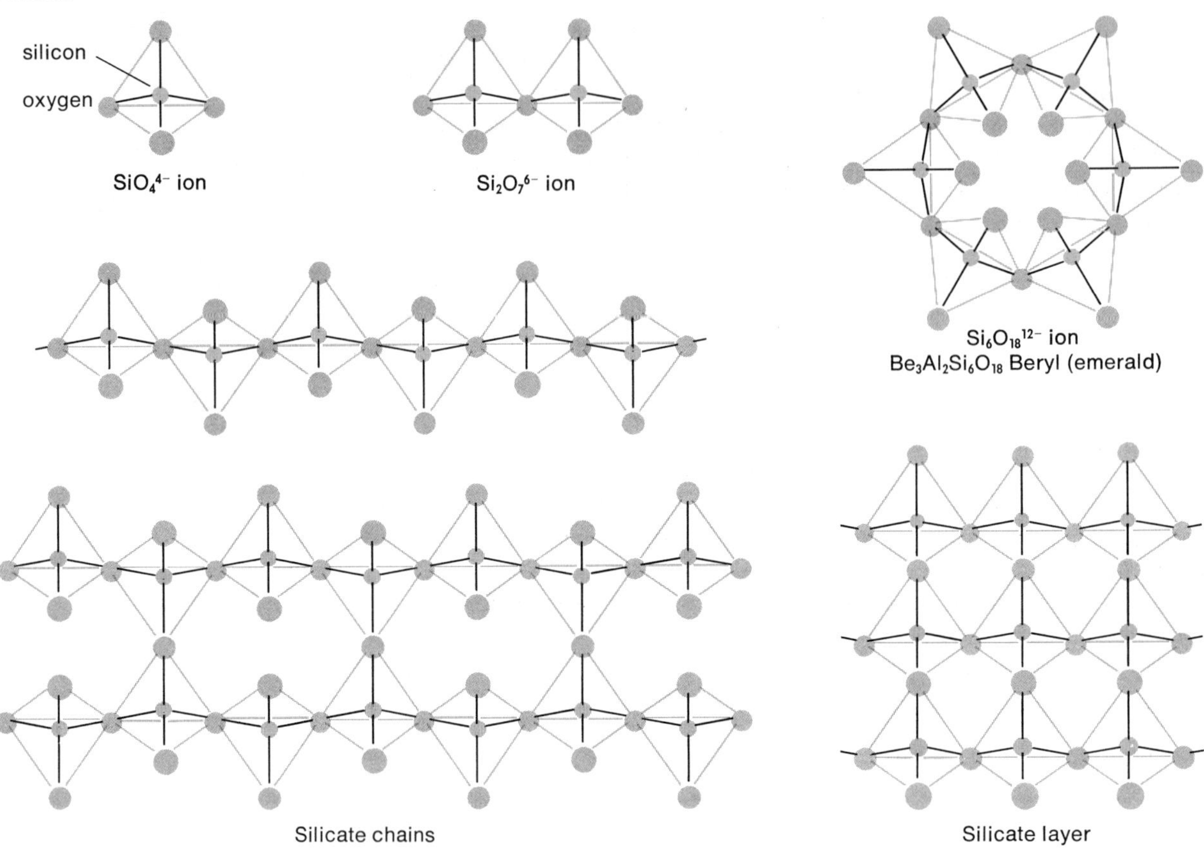

Silicon and Glass

The most widely used product of silicon is glass. This material, in one form or another, has been known for thousands of years. Glass is believed to date back to at least 1500 B.C.

Figure 14–7.
The Structure of Silica, SiO_2

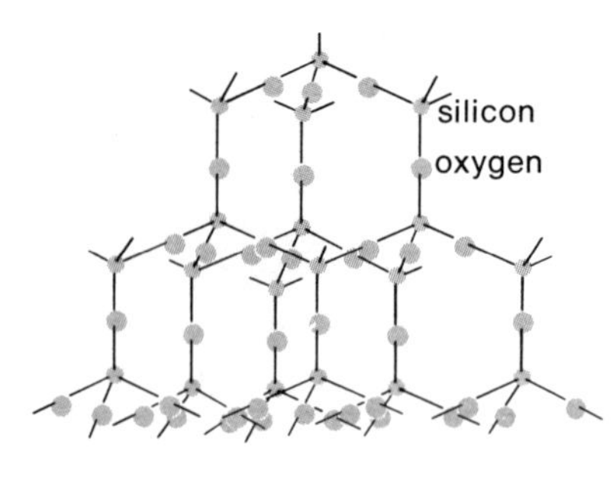

Glass is produced by melting together, in various proportions, limestone, sodium carbonate, and silica in the form of sand. The result is a clear, homogeneous mixture containing the silicates of sodium and calcium, which becomes rigid when it cools. Other compounds, such as Na_2O and CaO, can be substituted to provide the sodium and calcium. The reactions which take place are:

$$Na_2CO_3 + SiO_2 \rightarrow Na_2SiO_3 + CO_2 \quad 14\text{–}7$$

$$CaCO_3 + SiO_2 \rightarrow CaSiO_3 + CO_2 \quad 14\text{–}8$$

Table 14–1 lists the compositions, properties, and uses of some common glasses.

The molecular structure of glasses is very random, having little of the crystallinity associated with true solids. For this reason glasses are sometimes thought of as **supercooled liquids.**

Beyond its usual familiar uses, glass has many special applications. A product called safety glass is manufactured by cementing a thin sheet of pliable, transparent plastic between two sheets of ordinary glass. The plastic keeps the glass from scattering upon breaking. This material is required in the windshields of automobiles.

It has been possible to increase the amount of lead in glasses to rather high values and still obtain transparent materials. These high-lead glasses are used as viewing windows into radioactive areas; they provide good radiation protection and a reasonable look at the interior.

The etching or frosting of glass to produce scientific equipment and decorative items has been practiced for many years. Today, much of this work is done by machines, but some is still done by a chemical method. The glass surface is coated with wax, and the marks or design are scratched through the wax with a sharp tool; the surface is then exposed to a solution of hydrogen fluoride dissolved in water, called **hydrofluoric acid.** This solution attacks and dissolves the exposed glass, producing the desired effect. The reaction is:

$$\underset{\text{glass}}{CaSiO_3} + \underset{\substack{\text{hydrofluoric}\\\text{acid}}}{6HF} \rightarrow \underset{\substack{\text{soluble}\\\text{in } H_2O}}{CaF_2} + \underset{\text{a gas}}{SiF_4} + 3H_2O \qquad 14\text{–}9$$

Table 14–1. Some Common Glasses

Name	*Composition*	*Properties*	*Typical uses*
Soda lime glass	70% SiO_2, 17% Na_2O, 11% CaO	Expands upon heating; slight green color caused by impurities	Windows
Borosilicate glass (Pyrex)	80% SiO_2, 5% Na_2O, 13% B_2O_3	Expands only slightly upon heating	Cooking utensils, coffee makers, scientific glassware, glass sculptures
Flint glass	55% SiO_2, 11% K_2O, 33% PbO	Soft—easily scratched, cut, or polished	Lenses, cut glass articles

Review Questions

1. How does the modern preparation of elemental phosphorus differ from the original method used by H. Brand? In what ways are the processes similar?
2. What does the term *allotropy* mean? How do the two allotropes of phosphorus differ from one another?

3. Compare the composition of the modern strike-anywhere match with that of the early phosphorus-containing matches (called lucifers).
4. Why was the use of elemental white phosphorus in matches discontinued?
5. Calculate the percentage of phosphorus in superphosphate and triple superphosphate fertilizers.
6. Compare the properties of the two common allotropic forms of carbon.
7. Calculate the weight of the diamond known as the Great Star of Africa (530 carats) in grams and pounds. See Appendix III for weight relationships.
8. Why can graphite be used as a brush (a rubbing electrical contact) in electrical motors but diamond is totally unsuitable?
9. Three common carbon-containing compounds are CO_2, $NaHCO_3$, and $CaCO_3$. Give a practical use for each of these substances.
10. Explain from a molecular point of view why CO_2 and SiO_2 are so different from one another.
11. Diamond and SiO_2 are both hard crystalline solids at room temperature. Explain this fact in terms of the crystal structures of the two (Figures 14–3 and 14–7).
12. Why are glasses sometimes called supercooled liquids?
13. Why is borosilicate rather than soda lime glass used to manufacture cooking utensils?

Suggestions for Further Reading

"Asbestos, Asbestos Everywhere," *Chemistry*, May 1972.

Brill, R. H., "Ancient Glass," *Scientific American*, Nov. 1963.

Brunauer, S., and L. E. Copeland, "The Chemistry of Concrete," *Scientific American*, April 1964.

Charles, R. J., "The Nature of Glasses," *Scientific American*, Sept. 1967.

Daugherty, K. E., and J. Skalny, "The Slowest Polymerization Reaction," *Chemistry*, Jan. 1972.

Encke, F. L., "The Chemistry and Manufacturing of the Lead Pencil," *Journal of Chemical Education*, Aug. 1970.

Gilman, J. J., "The Nature of Ceramics," *Scientific American*, Sept. 1967.

Hall, H. T., "The Synthesis of Diamond," *Journal of Chemical Education*, Oct. 1961.

Olcott, J. S., "Chemical Strengthening of Glass," *Science*, June 14, 1963.

Pratt, C. J., "Chemical Fertilizers," *Scientific American*, June 1965.

"Structural Graphite Tailored for More and More Uses," *Chemistry*, April 1972.

Switzer, G. S., "Questing for Gems," *National Geographic*, Dec. 1971.

Weeks, M. E., and H. M. Leicester, *Discovery of the Elements*, 7th ed., Chapters 2, 3, and 12, Journal of Chemical Education Publication, Easton, Pa., 1967.

15 Some Interesting and Useful Metals

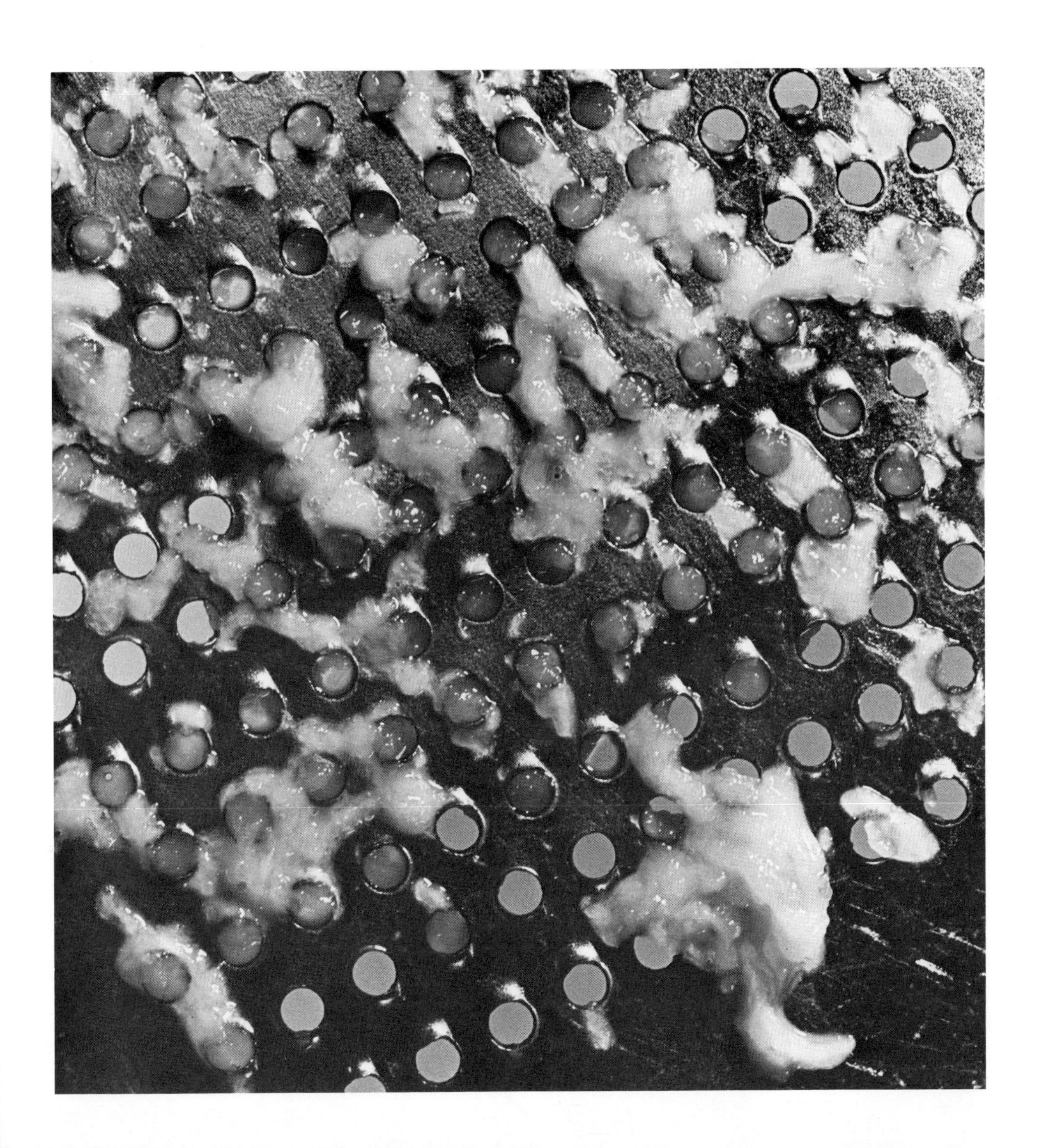

We saw in Chapter 3 that most of the known elements are classified as metals. Some were known to ancient man; others have only recently been discovered or synthesized (see Chapter 16). In this chapter we shall discuss a few metals that are especially interesting and useful.

Iron

The element iron is the second most abundant metal in the earth's crust and the most important industrial metal. More iron is used industrially than all other metals combined—fourteen times as much—for several reasons: the orcs arc abundant and widely distributed, the metal is easily and cheaply obtainable from the ores, and the properties of the metal can be varied over a wide range by the addition of relatively small quantities of other substances and by different treatment methods.

The important iron-bearing ores are **hematite,** Fe_2O_3, which gets its name from its red color, and **magnetite,** Fe_3O_4, which gets its name from the fact that some varieties such as lodestone are magnetic. About 90 percent of the iron ore mined in the United States comes from the Mesabi range in the Lake Superior region of Minnesota. Small amounts of iron have been found in the free elemental state, but most of this is believed to be of meteoric origin.

The Production of Iron

The metallurgy of iron—the process by which it is made into a useful form—involves three essential steps: The ore must first be converted into an impure iron called pig iron. The pig iron is converted into a desired form by the removal and addition of various substances. The desired form is then subjected to tempering or annealing processes that give it the desired physical properties.

Figure 15–1.
Blast Furnace Production of Pig Iron

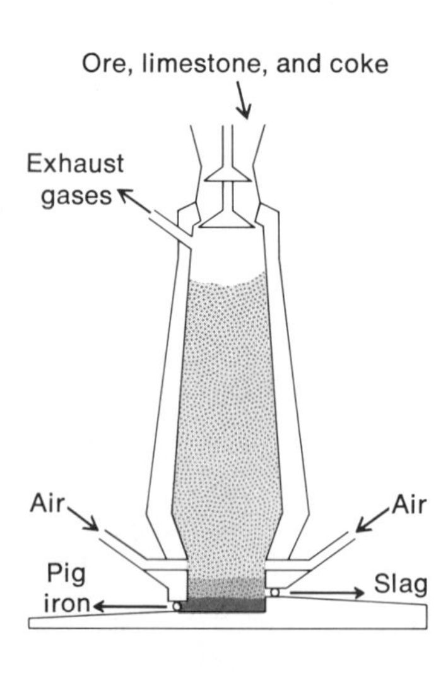

The ore is converted into pig iron in a large blast furnace, such as the one diagrammed in Figure 15–1. A charge of ore, coke (a form of carbon), and limestone is added through the top of the furnace. Hot air is blown into the bottom and passes up through the charge. At the high temperatures obtained, a number of reactions occur which produce pig iron and slag. The **slag,** a glassy molten material, contains and removes impurities such as sand, clay, and silicate minerals. The slag floats on top of the molten iron at the bottom of the furnace and is drawn off before the liquid iron is removed.

The reactions which take place to produce iron from the ore are:

$$\text{C (coke)} + O_2 \rightarrow CO_2 + \text{heat} \qquad 15\text{–}1$$

$$CO_2 + C \rightarrow 2CO \qquad 15\text{–}2$$

$$Fe_2O_3 + 3CO \rightarrow 2Fe + 3CO_2 \qquad 15\text{–}3$$

We see from these reactions that carbon monoxide, CO, is the actual reducing agent that produces elemental iron from the ore.

After the Fe_2O_3 has been converted into iron, another important reaction takes place inside the blast furnace. A small amount of the iron reacts with some of the carbon present to produce a compound called **cementite,** Fe_3C:

$$3Fe + C \rightarrow Fe_3C \qquad 15\text{–}4$$

The cementite dissolves in the molten iron—and this solution, either liquid or solid, is called **pig iron.**

The removal of mineral impurties in the form of slag is an important part of the blast-furnace process. The limestone in the original furnace charge is the slag-forming material. The reaction is:

$$CaCO_3 + SiO_2 \rightarrow CaSiO_3 + CO_2 \qquad 15\text{–}5$$

Notice that this is the same reaction given earlier (Chapter 14) for glass-making.

Cast Iron and Steel Production

Pig iron from the blast furnace contains 92–94 percent iron, about 4 percent carbon (partly in the form of dissolved cementite), 2 percent silicon, traces of sulfur and phosphorus, and between 1 and 2 percent manganese. These impurities cause the pig iron to be very brittle.

Pig iron is usually converted into either cast iron or one of a variety of steels. **Cast iron** is made by melting together scrap iron and pig iron. The melt is cast into molds and allowed to solidify into the final product. The resulting cast iron is still quite brittle and is used in items not normally subjected to sudden shock. Typical uses include iron drain pipe, steam radiators, and various parts for stoves and other similar objects. The cast iron can be produced in two forms with somewhat different properties, depending on the rate at which the liquid is allowed to cool. When slow cooling takes place, the dissolved carbon separates out and is deposited throughout the iron in the form of tiny graphite flakes. This gives the iron a gray color and causes it to be somewhat soft and tough. When rapid cooling occurs, the carbon does not separate out but remains dissolved in the form of cementite. The resulting material is whiter, harder, and more brittle than the gray cast iron.

The production of **steel,** the most used form of iron, from pig iron involves the removal of some materials and the addition of others to produce a steel with the desired properties. Most steel today is produced by the **basic oxygen process,** which was introduced in 1954. A charge of molten pig iron, solid iron, and steel scrap is placed in a refractory (very heat-resistant) lined, barrel-shaped furnace. A stream of pure oxygen is blown onto the surface at high velocity through a water-cooled lance. Lime, CaO, is added as a slag-making material. The equipment used is depicted in Figure 15–2.

Figure 15–2.
The Basic Oxygen Process

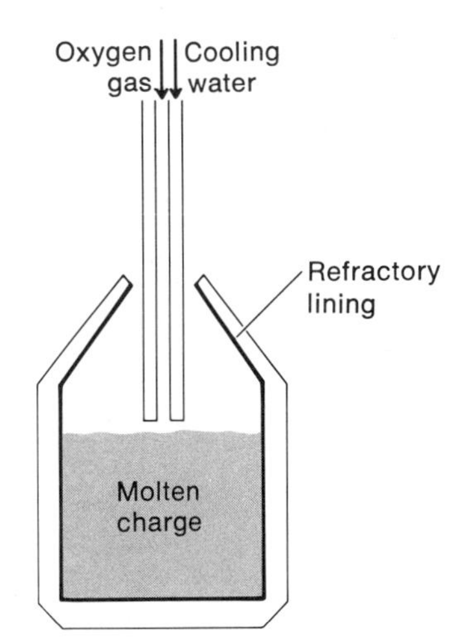

The high-velocity oxygen penetrates into, and vigorously mixes, the molten charge and slag. The reactions responsible for removing the impurities are:

$$2C + O_2 \rightarrow 2CO \quad \text{(gas)} \qquad 15\text{–}6$$

$$2Mn + O_2 \rightarrow 2MnO \quad \text{(dissolves in slag)} \qquad 15\text{–}7$$

$$Si + O_2 \rightarrow SiO_2 \qquad \text{(dissolves in slag)} \qquad 15\text{–}8$$

$$4P + 5O_2 \rightarrow P_4O_{10} \qquad \text{(dissolves in slag)} \qquad 15\text{–}9$$

$$2S + 2CaO \rightarrow O_2 + 2CaS \quad \text{(dissolves in slag)} \qquad 15\text{–}10$$

The reactions take place very rapidly under the prevailing conditions; the entire batch can be processed in less than one hour. Since the process occurs so rapidly, a computer is used to calculate in advance the amounts of the various raw materials needed to produce a product with the final desired composition. The computer is also fed analysis information obtained during the process, and on the basis of this determines the amount of alloying and other materials needed to give the required composition. The use of computers and the ready availability of oxygen have made this the most rapidly growing method for producing steel.

Types of Steel

The properties of steel depend upon the carbon content. Steels are often classified as low, medium, or high carbon steel. **Low carbon steel** contains less than 0.3 percent C and is used in products such as horseshoes, wire, and nails. **Medium carbon steel,** 0.3–0.8 percent C, is used in manufacturing such products as railroad rails and machine axles. **High carbon steel,** 0.8–2 percent C, is the type used in tools, springs, and files. Two properties very dependent on the carbon content and important in dictating the uses of the steel are **tensile strength,** measured by the force required to pull a rod into two pieces, and **ductility,** related to the ease with which the material may be formed into wire.

Figure 15–3 shows how these two properties vary with carbon content and how an increase in one results in a decrease in the other. Low carbon steels have low tensile strength but high ductility; this explains their use in wire and nails, where strength is not too important but ductility is. The medium carbon steels, having moderate to high strength and moderate to low ductility, are used where both strength and the ability to be "worked" or formed by tools are important. The high carbon steels, with low ductility and high tensile strength, are useful in applications such as tools and springs where strength is essential.

In some applications, such as axles with bearing surfaces which must be hard to resist wear, a process called **case hardening** is employed. A low or medium carbon steel is used; the parts to be case hardened are put into a closed container and covered with powdered carbon. The container is heated to a high temperature, at which the carbon reacts with the outer surface of the steel to form a thin coating of high carbon steel. In this way an axle or other part can be produced from a somewhat flexible type of steel, low to medium carbon type, but still be given very hard—case hardened—bearing surfaces where needed.

Another process used on steel is called **tempering.** This involves the altering

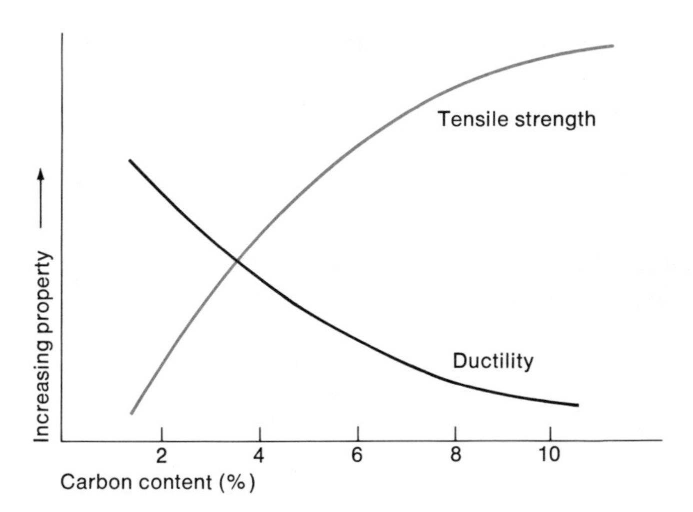

Figure 15–3.
The Dependence of Tensile Strength and Ductility on the Carbon Content of Steel

of the hardness and brittleness of steel by heat treatment. A specific type of tempering is **annealing,** in which a hard steel is softened. The steel is heated to a high temperature and then cooled very slowly, producing (if done properly) a steel that is soft and easily worked. The basis for this process is the fact that the solubility of solid cementite in solid iron is very temperature dependent. At the high temperature the cementite is soluble, and a homogeneous solution results. As the temperature comes down, the solubility of cementite decreases; if the cooing takes place slowly enough, the cementite crystallizes out and collects into fairly large crystals. Between these crystals of pure cementite are large areas of pure iron, which is soft. The steel sample then becomes soft because of the large areas of pure iron. On the other hand, rapid cooling prevents the cementite from forming large localized crystals, and it is distributed evenly throughout the steel. Since cementite is quite hard, the resulting steel is hard.

Corrosion of Iron and Steel

One of the biggest problems related to the use of iron and steel is **corrosion,** a general term applied to processes in which uncombined metals are changed into compounds. The most important corrosion problem of iron and steel products is **rusting,** where the iron is changed into a hydrated oxide. Rusting is a very serious problem economically. It has been estimated that as much as one-seventh of the annual production of iron goes simply to replace that lost by rusting.

The rusting process requires the presence of both moisture and oxygen. Iron will not rust in dry air or in water from which all dissolved oxygen has been removed. Because of these facts it is thought that the rusting process consists of two steps. In the first step, iron reacts with water to form iron hydroxide, $Fe(OH)_2$. The second step involves the reaction of $Fe(OH)_2$ with oxygen to form the hydrated Fe_2O_3 or rust. The generally accepted formula for rust is $Fe_2O_3 \cdot xH_2O$, since the water content varies. The overall reaction for rust formation is:

$$4Fe + 3O_2 + 2xH_2O \rightarrow 2(Fe_2O_3 \cdot xH_2O) \qquad 15\text{–}11$$

Unfortunately, the rust does not adhere to the surface of the metal and protect it from further reaction, but peels off, exposing a fresh surface of iron which will rust very rapidly.

A number of methods are used to protect iron and steel from rusting. A common approach is to cover the surface with paint, lacquer, or, if appearance is not very important, grease or asphalt. These types of coatings keep the air and moisture from the surface. Another approach produces an adherent oxide coating of Fe_3O_4 by exposing the iron or steel to superheated steam. The reaction is

$$3Fe + 4H_2O \rightarrow Fe_3O_4 + 4H_2 \qquad 15\text{–}12$$

The coating serves to keep the air and moisture from reaching the metal surface. A third method, which is used a great deal, puts a coating of some other metal on the iron or steel. The metal most commonly used is zinc; the specific process is called **galvanizing.** Almost 40 percent of the annual production of zinc metal is used to galvanize iron or steel products. The zinc coating is more protective than other materials used to cover the surface. Even if a break occurs in the zinc and exposes the iron underneath, no rusting will take place as long as the zinc is in contact with the iron. The zinc is more easily oxidized than the iron and will, through an electrolytic process, corrode while the iron remains unchanged.

A fourth technique used to prevent corrosion is the manufacturing of corrosion-resistant steels. When 14–18 percent chromium (Cr) and 7–9 percent nickel (Ni) are added to the iron, the result is a **stainless steel** that is very hard and corrosion resistant; only strong acids will attack it. This form of steel is quite expensive, so that its use is somewhat limited.

The Coinage Metals: Copper, Gold, and Silver

The three metals copper, gold, and silver are referred to as the coinage metals because of their extensive use—at least in the past—in coins. All three metals were known and used anciently, primarily because they occur free in nature or are very easily obtained from their ores. In fact, it is likely that the first production of silver and copper from ores was accidentally accomplished when rocks containing the minerals were used in a fireplace on the ground; silver and copper could have been produced by the heat and the reducing properties of the charcoal that was present.

Copper

Copper occurs naturally in both the free elemental state and various compounds. The most important commercial ores are the sulfides, Cu_2S and

$CuFeS_2$, from which more than 75 percent of the annual world copper production is obtained. Most of these ores are classified as low-grade because they contain less than 10 percent Cu.

Copper Production

The production of copper from low-grade ore involves a number of steps. The ore is first pulverized into a very fine state. Next, this finely divided material is concentrated by a process known as **froth flotation,** in which the pulverized ore is mixed with water to which a carefully selected oil has been added. A froth or foam is produced by blowing air through the mixture while it is vigorously stirred. The sulfide ore particles are attracted and cling to the oil droplets, while water is attracted to and coats the waste rock and clay particles. The oil-containing foam overflows the flotation cell and carries the ore particles to a filter where they are recovered. The resulting **concentrate** contains 20–50 percent copper. The waste rock and clay fall to the bottom of the flotation cell and are periodically drawn off.

Next, the concentrated ore is **roasted** in a furnace at a temperature below the melting point of the ore. This process drives off moisture and removes part of the sulfur in the form of sulfur dioxide. The resulting product, called **calcine,** contains, depending upon the composition of the original ore, all or some of the compounds Cu_2S, FeS, FeO, and SiO_2. The calcine is mixed with sand and limestone and heated to a temperature above the melting point of the mixture. This process is referred to as **smelting,** and the primary result is the removal of impurities in the form of a slag. The reactions are:

$$CaCO_3 + SiO_2 \rightarrow CaSiO_3 \text{ (slag)} + CO_2 \qquad 15\text{–}13$$

$$FeO + SiO_2 \rightarrow FeSiO_3 \text{ (slag)} + CO_2 \qquad 15\text{–}14$$

The molten slag is drawn off, leaving a material called **copper matte,** which consists of Cu_2S and small amounts of FeS. The copper matte is mixed with a little sand and transferred to a **converter,** where it is reacted with hot air. This process removes most of the remaining impurities as a slag and converts the Cu_2S into 99 percent pure copper. The reactions are:

$$2FeS + 3O_2 \rightarrow 2FeO + 2SO_2 \text{ (gas)} \qquad 15\text{–}15$$

$$FeO + SiO_2 \rightarrow FeSiO_3 \text{ (slag)} \qquad 15\text{–}16$$

$$2Cu_2S + 3O_2 \rightarrow 2Cu_2O + 2SO_2 \text{ (gas)} \qquad 15\text{–}17$$

$$2Cu_2O + Cu_2S \rightarrow 6Cu + SO_2 \text{ (gas)} \qquad 15\text{–}18$$

The gas produced in these reactions bubbles off the molten copper and leaves blisters on the surface of the solid product. The product is called **blister copper** because of the appearance of its surface.

Blister copper is cast into large anode plates and electrolytically refined to a purity of between 99.96 and 99.99 percent. This refining process is done in large electrolytic cells such as the one sketched in Figure 15–4. As current passes through the cell, pure copper is reduced and plated onto the cathode,

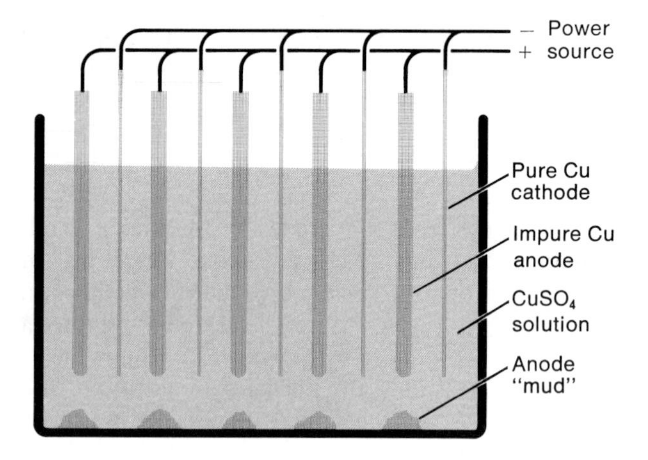

Figure 15–4.
Electrolytic Refining of Copper

and the anode is oxidized and slowly dissolves. The reactions are:

$$\text{Cathode:} \quad Cu^{2+} + 2e^- \rightarrow Cu \text{ (pure)} \qquad 15\text{–}19$$

$$\text{Anode:} \quad Cu \text{ (impure)} \rightarrow 2e^- + Cu^{2+} \qquad 15\text{–}20$$

During this operation the voltage is adjusted so that only copper will be oxidized and dissolve from the anode. Any impurities, such as gold, silver, and molybdenum, fall from the anode and collect under it as **anode mud.** The pure cathodes are ultimately removed, melted, and cast into desired forms such as wire ingots.

Uses of Copper

Copper is used in the pure form and as alloys. About 38 percent of the copper produced is made into electrical wire, and 56 percent is alloyed with other metals. Two of the most common alloys are copper plus up to 40 percent zinc, which is referred to as **brass,** and copper plus up to 18 percent tin, which is called **bronze.** These terms, *brass* and *bronze,* do not refer to alloys of a specific composition, but rather to alloys of copper with zinc and tin, respectively, in varying percentages.

Copper and its alloys resist corrosion because of the formation of a protective coating. In moist air copper first turns a dull brown, owing to the formation of a thin adherent film of oxide, CuO, or sulfide, CuS. Prolonged weathering results in the formation of a green film composed of basic copper carbonate. This is the familiar green coating found on many statues and buildings. The reaction is:

$$2Cu + O_2 + CO_2 + H_2O \rightarrow \underset{\substack{\text{green coating} \\ \text{(patina)}}}{Cu(OH)_2 \cdot CuCO_3} \qquad 15\text{–}21$$

The deliberate production of coatings such as those mentioned above, called **antiqueing,** is recurrently in vogue as an interior decorating technique. The motive may be that it is difficult to prevent copper surfaces from undergoing such reactions. Often the surface is thinly coated with plastic to protect it. A favorite place for copper articles, however, is near fireplaces, where the plastic might be damaged by heat—or it might be removed by a zealous housekeeper during vigorous cleaning procedures. The surface then tends to react and antique itself. If an interior decorator or architect can convince a home owner that artificial antiqueing is really very attractive and modern (plus charge a little extra for it), many problems are solved and everyone is made happy.

Gold

The element gold is always found free in nature; undoubtedly it was one of the first metals discovered and used by man. It is valued today as a medium of exchange or a basis for monetary systems. In its earliest applications, it was valued more for its beauty—color and luster—and corrosion resistance than for its monetary uses.

Today gold is used primarily as a monetary standard, in jewelry and other decorative objects, and as a noncorroding coating in certain electronic and scientific applications.

Gold is usually alloyed with other metals to increase its hardness and durability. The amount of gold in such alloys is indicated by a karat value (not the unit of weight for gems). Gold that is 100 percent pure is defined as 24 karat, and other percentages are compared to this value. Thus, 18-karat gold is $18/24 \times 100$ percent or 75 percent gold, and 12-karat is 50 percent gold.

The metals used to form gold alloys vary, depending upon the color and other properties that are desired. Table 15–1 lists the composition of a few typical 14-karat gold alloys.

Table 15–1. Colors of 14-Karat Gold Alloys

Alloy ingredients (%)	*Alloy color*			
	White	*Yellow*	*Red*	*Green*
Gold	58.33	58.33	58.33	58.33
Copper	17.00	25.58	27.96	6.50
Silver		13.33	6.54	35.00
Zinc	7.67	2.76	7.17	0.17
Nickel	17.00			

From *The Encyclopedia of the Chemical Elements*, ed. by Clifford A. Hampel, p. 249. © 1968 by Litton Educational Publishing, Inc. Reprinted by permission of Van Nostrand Reinhold Company.

Silver

The third coinage metal, silver, is found in both the free and combined states, but the free state is quite rare. It is quite easily obtained from its compounds by heating. A typical reaction is:

$$2Ag_2O \xrightarrow{\text{heat}} 4Ag + O_2 \qquad 15\text{–}22$$

It is often found as a minor component in other metal ores or in the metals themselves. The silver can be separated out by heating and oxidizing the other metals (lead, for example) in appropriate containers; the metal oxides are absorbed into the container walls, leaving the free silver. This is called a **cupellation** process.

Uses of Silver

Silver is used in ornaments, coins (a rapidly disappearing use), jewelry, flatware, electrical circuitry (especially where cost is not as important as reliability—as in a space vehicle), and the manufacture of photographic materials. Most of these uses are based on its beautiful white color and lustrous appearance. Electrical applications reflect the fact that silver is the best metallic conductor of electricity. Photographic applications take advantage of the characteristic reaction of silver salts upon exposure to light.

Silver is usually alloyed with other metals, often copper, to increase its hardness and wear resistance. Sterling silver, for example, is 92.5 percent silver and 7.5 percent copper. This silver is said to be 925 fine (925 parts Ag per 1000 parts alloy, or 92.5 percent). American coinage silver is 900 fine, since it contains 90 percent Ag and 10 percent Cu.

Silver is not attacked by oxygen in the air, but it does tarnish quickly in the presence of hydrogen sulfide gas, H_2S, or upon contact with sulfur-containing materials such as eggs, mustard, and some air pollutants (Chapter 21). The tarnish is a thin layer of silver sulfide Ag_2S. The equation for the reaction which takes place is:

$$4Ag + 2H_2S + O_2 \rightarrow 2Ag_2S + 2H_2O \qquad 15\text{–}23$$

Many people remove silver tarnish by the use of a scouring powder or abrasive cream and a lot of physical effort. This practice is inadvisable because the removal of the tarnish constitutes the removal of a small amount of silver. In time, the silver of a silver-plated object can be worn away and the underlying base metal exposed. A much better method is to change the tarnish back into silver. This may be accomplished by placing the silver objects in an aluminum pan of warm water in which a small amount of table salt or baking soda has been dissolved. After a few minutes the silver objects should be removed, rinsed in clean water, and dried. If an aluminum pan is not available, some crumpled aluminum foil or old aluminum or tin cans may be substituted. Best results are obtained when the aluminum and silver are in contact.

In this process the Ag_2S is reduced back to Ag, and some of the aluminum is oxidized to the sulfide as given in equation 15–24.

$$3Ag_2S + 2Al \rightarrow 6Ag + Al_2S_3 \quad 15\text{–}24$$

Silver and Photography

Approximately 20 percent of the annual silver production goes into the manufacture of silver halides (Cl, Br, I) which are used to prepare photographic films and papers. This application is based on the fact that a silver halide, most commonly AgBr, will slowly decompose to Ag and free halogen (Br_2) upon exposure to light. This reaction is used and carefully controlled in modern photographic processes.

A photographic film consists of a thin sheet of plastic coated with a gelatin suspension of small silver halide crystals. The size of the silver halide particles and the relative amounts of bromide and iodide used determine the sensitivity or "speed" of the film and its "grain" or ability to show detail when the final picture is enlarged.

Upon exposure, the light reflected from an object is focused on the film surface. Light-colored objects reflect more light than darker objects. The chemical processes involved are not well understood, but apparently when light strikes a tiny silver halide crystal, a few atoms of metallic silver are produced. These atoms activate the rest of the silver compounds in the crystal and make them easily reducible to free silver during the developing process. The activation reaction is represented as:

$$AgX + \text{light} \rightarrow AgX^*, \quad 15\text{–}25$$

where X represents a halogen and the asterisk refers to an activated state. The film is developed by being washed in a solution which reduces the AgX^* to metallic silver:

$$AgX^* + \text{developing solution} \rightarrow Ag + X^- \quad 15\text{–}26$$

The activated AgX is changed into metallic silver, which appears black on the developed film. The previously unactivated AgX must now be removed, or it would gradually change to black Ag upon exposure to light. This "fixing" process is carried out when the film is washed in a solution containing a material which will dissolve only the remaining AgX. Sodium thiosulfate or "hypo" has been widely used for this purpose. The reaction is:

$$AgX + 2S_2O_3^{2-} \rightarrow [Ag(S_2O_3)_2]^{3-} + X^- \quad 15\text{–}27$$

Both species on the right side of equation 15–27 are ions which are soluble in the water of the washing solution and are removed from the film. The developed and fixed film or negative contains black metallic silver wherever light from the photographed object struck it.

The negative is placed over a piece of paper on which a gelatin-AgX coating has been placed similar to that on the original film. Light is passed through the negative onto the paper. The dark areas of the negative, corresponding to light areas in the photographed object, block the light and prevent activation of the AgX on the paper. The exposed paper is developed and fixed in the same way as the film. The resulting dark and light areas are opposite to those on the negative and match the original photographed object. These steps are diagrammed in Figure 15–5.

Figure 15–5.
The Black and White Photographic Process

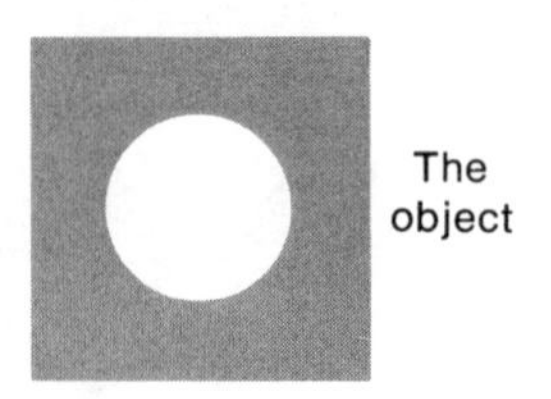

The object

AgBr

Unexposed film

AgBr* (latent image)

AgBr

Film after exposure

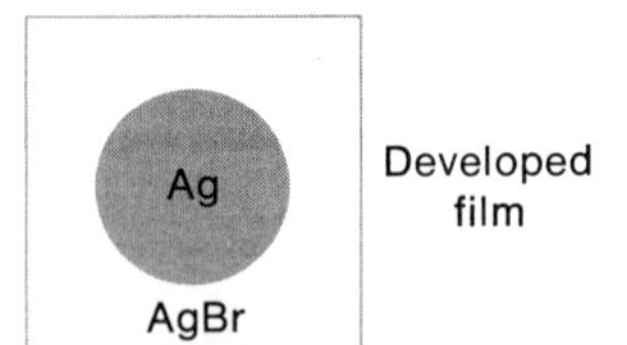

Developed film

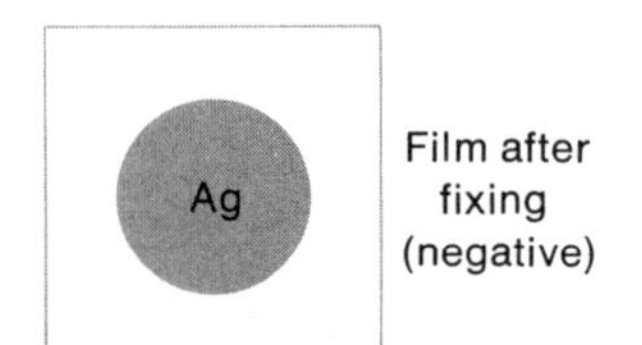

Film after fixing (negative)

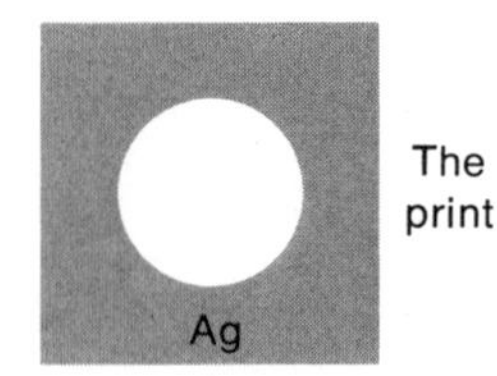

The print

Aluminum

Aluminum, a silvery white metal of low density, is the most abundant metal and third most abundant element in the earth's crust (see Figure 2–1). It was produced only in small quantities and had few practical uses until after 1886, when Charles M. Hall of the United States and P. L. T. Héroult of France simultaneously developed the commercial techniques needed to produce this useful metal in large quantities.

Occurrence of Aluminum

Aluminum occurs naturally only in the form of compounds. A large number of these compounds are known; those in greatest abundance are complex aluminum silicates. At present, no method is available for the economical extraction of aluminum from silicate minerals, and the source of nearly all of the metal is the mineral **bauxite,** a hydrated aluminum oxide, $Al_2O_3 \cdot xH_2O$. **Corundum** and **emery** are fine-grained varieties of Al_2O_3 that are almost as hard as diamonds; they are used as abrasives in grinding and polishing.

A number of aluminum-containing minerals are used as gems. Emerald is composed of the mineral **beryl,** which is a complex silicate with the formula $Be_3Al_2Si_6O_{18}$ (see Figure 14–6). Sapphire and ruby are both forms of Al_2O_3. The presence of small amounts of other metal oxides—Co or Ti in sapphire and Cr in ruby—produces the characteristic sapphire-blue and ruby-red colors.

Commercial Preparation of Aluminum

The process used to produce aluminum today is practically the same as that introduced in 1886 by Héroult and Hall. The bauxite is purified to remove most of the silicon dioxide and iron oxide normally found associated with it. The purified bauxite is then mixed with a mineral known as **cryolite,** Na_2AlF_6, and the mixture is melted in a graphite-lined steel container. An electric current is passed through the melt via carbon rod anodes, and the aluminum ions are reduced to aluminum metal on the graphite lining, which serves as a cathode (Figure 15–6).

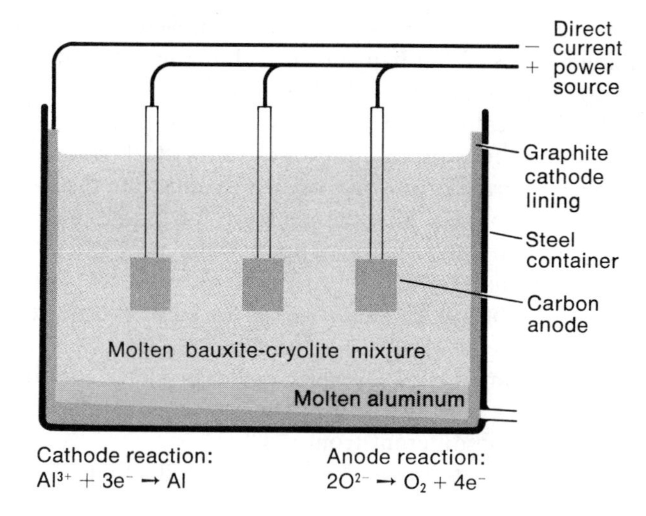

Figure 15–6.
The Electrolytic Production of Aluminum

Uses of Aluminum

Aluminum has especially useful properties: its density (2.70 g/ml) is less than half that of iron (7.86 g/ml), and, although it is a reactive metal, it is not noticeably corroded in air or water. The resistance to corrosion is provided by a very thin film of adherent oxide which forms on the surface and inhibits further attack.

Aluminum is used in both the pure and alloyed forms. A strong, light alloy of aluminum and magnesium makes up much of the metallic structure in modern aircraft. Aluminum is used in door and window frames and in decorative panels in many modern buildings. The metal is being used increasingly in automobiles—for engine parts as well as decorative trim. Large electrical and thermal conductivities make it useful in applications involving electrical wiring and heat-exchange systems.

Review Questions

1. Why is more iron used industrially than all other metals combined?
2. Use equations 15–1, 15–2, and 15–3 to calculate the weight of carbon needed to produce one ton (9.0×10^5 grams) of iron. Also, calculate the weight of CO_2 produced.
3. Name a useful consumer good that is manufactured from each of the following iron products:
   a) Cast iron
   b) Low carbon steel
   c) Medium carbon steel
   d) High carbon steel
   e) Stainless steel
   f) Galvanized iron or steel
4. Compare the following from the point of view of composition and properties: pig iron, cast iron, steel, stainless steel.

5. The following impurities of pig iron are removed in the basic oxygen steelmaking process. What happens to each impurity during the process?
   a) C b) Mn c) Si d) P e) S
6. What is the chemical meaning of the word *corrosion?*
7. Describe four methods used to prevent iron or steel from corroding.
8. Draw a block diagram or in some other way illustrate the steps involved in the production of very pure copper from low grade ore.
9. Discuss some common uses for copper. What properties of the metal account for these uses?
10. A common alloy of gold is described as 14 karat. What percentage of gold is found in the alloy?
11. Discuss three common uses for silver. What property accounts for each use?
12. How is silver tarnish different from rust?
13. Briefly describe the steps involved in the production of a black and white photograph.
14. What properties make aluminum a very useful metal?
15. During the production of most metals from their natural sources, one step is required in which the metal is obtained by a reduction process. Write an equation to represent this step for each of the following metals. How do the steps differ from each other?
    a) Iron c) Silver
    b) Copper d) Aluminum
16. Suppose you had to completely give up using any form of iron, copper, silver, gold, or aluminum. Indicate which one would be easiest and which one hardest to do without. Give reasons for your choices.

Suggestions for Further Reading

Fine, M. M., "The Beneficiation of Iron Ores," *Scientific American*, Jan. 1968.

Frieden, E., "The Biochemistry of Copper," *Scientific American*, May 1968.

Keller, E., "Photography, Part I: Images in Silver," *Chemistry*, Sept. 1970.

Keller, E., "Photography, Part II: Images in Color," *Chemistry*, Nov. 1970.

Leidheiser, H., "Corrosion," *Chemical and Engineering News*, April 5, 1965.

Parker, E. H., and V. F. Zackay, "Strong and Ductile Steels," *Scientific American*, Nov. 1968.

Putnum, G. L., "The Gold Content of the Sea," *Journal of Chemical Education*, Nov. 1953.

Stone, J. K., "Oxygen in Steelmaking," *Scientific American*, April 1968.

Weeks, M. E., and H. M. Leicester, *Discovery of the Elements*, 7th ed., Chapters 1 and 12, Journal of Chemical Education Publication, Easton, Pa., 1967.

Weissberger, A., "A Chemist's View of Color Photography," *American Scientist*, Nov./Dec. 1970.

Zackay, V. F., "The Strength of Steel," *Scientific American*, Aug. 1963.

16 Nuclear Processes

In 1896 Henri Becquerel, a French physicist, discovered that uranium compounds emitted radiation that could expose a photographic plate previously wrapped in lightproof paper. Forty-nine years later (July 16, 1945) the desert near Alamagordo, New Mexico, was seared by the most powerful explosion ever set off by man up to that time. Both processes were the result of reactions in the nuclei of uranium atoms.

Because the chemical behavior of atoms depends on the number and arrangement of extranuclear electrons, we have said little about the atomic nucleus. This chapter is devoted to a brief treatment of the topic.

Radioactive Nuclei

We found in Chapter 6 that in many spontaneous processes energy is given up. This corresponds to a change in which a system goes from a less stable (higher-energy) state to a more stable (lower-energy) state. When the nuclei of atoms undergo such changes, the liberated energy often appears as emitted radiation. The atoms containing such unstable nuclei are said to be **radioactive.**

Henri Becquerel discovered natural radioactivity as a result of some chance observations. Through research he found that the observed radiation was emitted by any compound of uranium. He found further that the intensity of the radiation was unaffected by the factors that normally influence the rate of chemical reactions: the temperature, pressure, and type of uranium compound used.

Subsequent studies by other investigators showed that the radiation emitted by uranium, or other radioactive elements discovered later, could be separated into three types by an electrical or magnetic field, as shown in Figure 16–1. The three types were characterized by their different electrical charges—positive, negative, and uncharged. Before they had been characterized further, the types were given the names that are still used today: alpha rays, beta rays, and gamma rays.

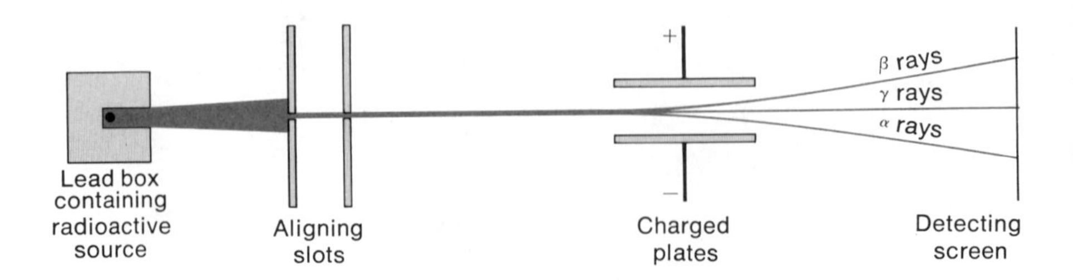

Figure 16–1. The Separation of Emitted Radiation into Three Types

It is now known that **alpha rays** consist of a stream of positively charged alpha particles. Each particle weighs 4 amu and carries a charge of +2. This means that each particle contains two protons and two neutrons and is identical to the nucleus of a helium atom. The notation used to represent alpha particles is ${}^{4}_{2}\alpha$.

Beta rays also consist of a stream of particles. The beta particles are negatively charged, and all their properties are identical to those of electrons. However, beta particles are not extranuclear electrons; they are electrons that have been produced inside the nucleus and then ejected. Beta particles have the characteristic electronic charge of -1 and a weight very close to zero on the amu scale. The symbol used to represent them is ${}^{0}_{-1}\beta$.

Gamma rays do not consist of particles but are composed of very high-energy radiation somewhat like light or X-rays. When radioactive elements emit alpha or beta particles, gamma rays are often given off as well. The symbol for gamma radiation is ${}^{0}_{0}\gamma$.

Equations for Nuclear Processes

Nuclear reactions can readily be represented by equations similar to those used to represent chemical reactions. The main difference is that the symbols used in nuclear equations always include the atomic number (or particle charge) and the mass number. A nuclear equation is balanced when the sum of the atomic numbers on the left side is equal to the sum on the right side, and the sum of the mass numbers on the left equals the sum on the right.

During the ejection of an alpha particle from a nucleus there is a decrease of two units in the atomic number and a decrease of four units in the mass number. The resulting product nucleus (daughter) must therefore have an atomic number of $Z - 2$ and a mass number of $A - 4$, where Z and A are the atomic and mass numbers of the decaying nucleus. This is represented in the general equation for alpha decay (equation 16–1) and the specific equation for the decay of ${}^{238}_{92}\mathrm{U}$ (equation 16–2).

$$ {}^{A}_{Z}\mathrm{X} \rightarrow {}^{4}_{2}\alpha + {}^{A-4}_{Z-2}\mathrm{Y} \qquad 16\text{–}1 $$

$$ {}^{238}_{92}\mathrm{U} \rightarrow {}^{4}_{2}\alpha + {}^{234}_{90}\mathrm{Th} \qquad 16\text{–}2 $$

Note that equation 16–2 is balanced, since $238 = 4 + 234$ and $92 = 2 + 90$.

The ejection of a beta particle leaves the mass number of a radioactive nucleus unchanged but increases the atomic number by one. The generalized equation and a specific example of beta decay are given in equations 16–3 and 16–4.

$$^{A}_{Z}X \rightarrow {}^{0}_{-1}\beta + {}^{A}_{Z+1}Y \qquad 16\text{–}3$$

$$^{234}_{90}Th \rightarrow {}^{0}_{-1}\beta + {}^{234}_{91}Pa \qquad 16\text{–}4$$

The emission of gamma rays ($^{0}_{0}\gamma$) changes neither the atomic number nor the mass number. For this reason, gamma emission is frequently neglected when nuclear equations are written. However, it should not be concluded that gamma emissions are unimportant. On the contrary, gamma rays are the most important of the three types in terms of radiation effects on biological systems.

Radioactive Disintegration Series

The product of the decay of a radioactive isotope is a new isotope of another element, as we have just seen. The new isotope may or may not be stable. If it is not, further radioactive decay takes place until a stable isotope is formed. The series of reactions that take place during the conversion of radioactive $^{238}_{92}U$ into stable $^{206}_{82}Pb$ is shown below. Such a series is known as a **radioactive disintegration series:**

$$^{238}_{92}U \xrightarrow[4.5 \times 10^{9}\ y]{\alpha} {}^{234}_{90}Th \xrightarrow[24\ d]{\beta,\ \gamma} {}^{234}_{91}Pa \xrightarrow[1\ m]{\beta,\ \gamma} {}^{234}_{92}U \xrightarrow[2.5 \times 10^{5}\ y]{\alpha} {}^{230}_{90}Th \xrightarrow[80\ y]{\alpha,\ \gamma} {}^{226}_{88}Ra$$

$$^{226}_{88}Ra \xrightarrow[1.6 \times 10^{3}\ y]{\alpha,\ \gamma} {}^{222}_{86}Rn \xrightarrow[4d]{\alpha} {}^{218}_{84}Po \xrightarrow[3\ m]{\alpha} {}^{214}_{82}Pb \xrightarrow[27\ m]{\beta,\ \gamma} {}^{214}_{83}Bi$$

$$^{214}_{83}Bi \xrightarrow[20\ m]{\alpha} {}^{210}_{81}Tl \xrightarrow[1.3\ m]{\beta} {}^{210}_{82}Pb$$

$$^{214}_{83}Bi \xrightarrow[20\ m]{\beta,\ \gamma} {}^{214}_{84}Po \xrightarrow[1.6 \times 10^{-4}\ s]{\alpha} {}^{210}_{82}Pb$$

$$^{210}_{82}Pb \xrightarrow[21\ y]{\beta,\ \gamma} {}^{210}_{83}Bi \xrightarrow[5\ d]{\beta} {}^{210}_{84}Po \xrightarrow[138\ d]{\alpha} {}^{206}_{82}Pb$$

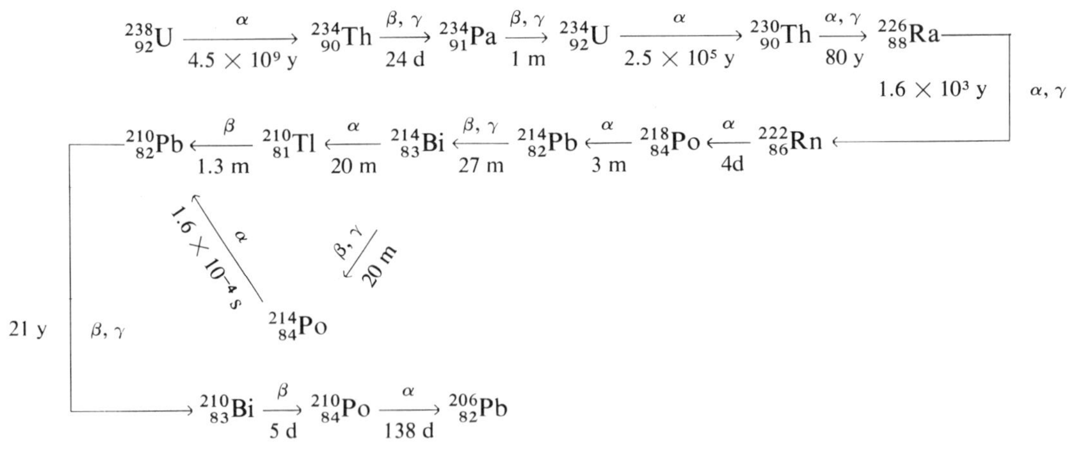

The symbols above the arrows indicate the radiation emitted, and the times under the arrows are the half-lives—a topic discussed later in the chapter (y = years, d = days, m = minutes, and s = seconds).

Induced Radioactivity

Before 1934 the study of radioactivity was limited to reactions of the relatively few radioisotopes that were found in nature. In that year Irene and Frederic Joliot-Curie found that radioactivity could be induced by bombarding nonradioactive nuclei with small particles. The first artificially produced radioisotope, $^{13}_{7}N$, was the result of bombarding $^{10}_{5}B$ with alpha particles obtained from a naturally radioactive isotope:

$$^{10}_{5}B + ^{4}_{2}\alpha \rightarrow ^{13}_{7}N + ^{1}_{0}n \qquad 16\text{–}5$$

Notice that a neutron, $^{1}_{0}n$, is produced in addition to $^{13}_{7}N$. Fifteen years earlier Ernest Rutherford had successfully used a similar reaction to produce nonradioactive $^{17}_{8}O$ from $^{14}_{7}N$. A small particle, a proton, was produced in addition to $^{17}_{8}O$:

$$^{14}_{7}N + ^{4}_{2}\alpha \rightarrow ^{17}_{8}O + ^{1}_{1}p \qquad 16\text{–}6$$

Today, hundreds of new radioactive isotopes have been produced in the laboratory by bombardment reactions. At least one such isotope has been prepared for every naturally occurring element. In addition, isotopes of at least 17 previously unknown elements have been formed in small quantities. We shall discuss these new elements later in this chapter.

Characteristics of Radioactive Nuclei

At one time it was thought that radioactive atoms contained nuclei too large and complex to be stable. It is true that no stable isotope is known with an atomic number greater than 83. However, with the discovery of very small and simple radioisotopes came the realization that nuclear complexity is only one of a number of factors that influence nuclear stability.

Some of the other factors become apparent when the number of neutrons is plotted versus the number of protons for each of the known stable isotopes. A rather definite pattern results (Figure 16–2). The following facts are obtained from Figure 16–2: (1) The stable nuclei fall into a rather narrow "zone of stability," indicating that the neutron-to-proton (n/p) ratio required for stability is restricted within narrow limits. (2) The n/p ratio increases with the size of the nucleus. (3) The n/p ratio is close to 1.0 (the line in the plot) for the lighter stable nuclei and increases to about 1.5 for the heavier elements. Thus, the greater the number of protons in a nucleus, the greater the number of neutrons required per proton to form a stable nucleus. (4) The upper limit to the number of protons found in a stable nucleus regardless of the number of neutrons present is 83, since the largest stable nucleus is that of $^{209}_{83}Bi$.

The existence of stable atomic nuclei is somewhat surprising when we remember that like charges repel each other and that most nuclei contain many protons—with identical positive charges—squeezed together into a

Figure 16–2.
The Stable Nuclei

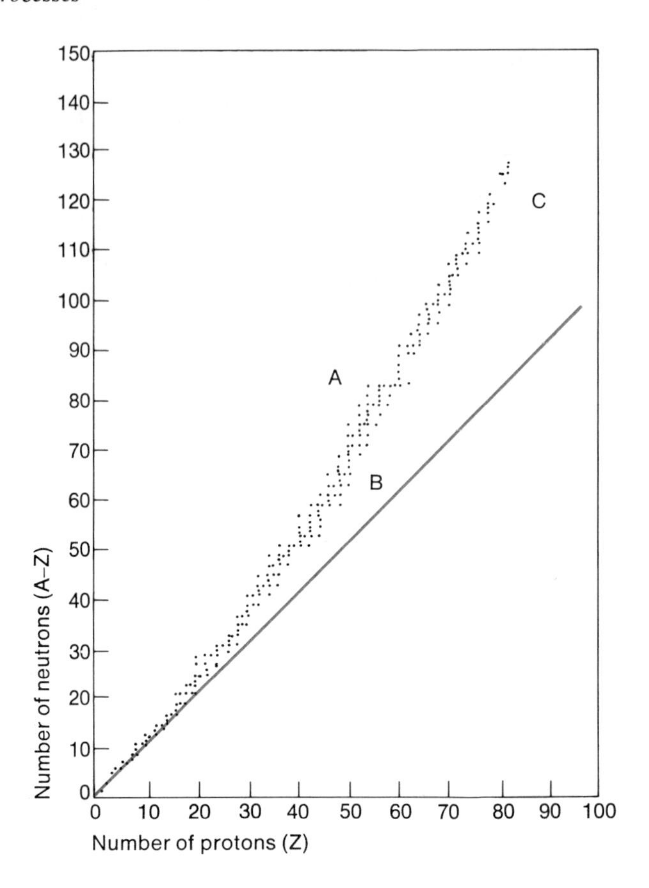

very small volume. The neutrons must play an important role in binding the nuclear particles together, since, as indicated by Figure 16–2, instability results when there are too many protons per neutron or too many neutrons per proton.

The Decay of Radioactive Elements

Three types of unstable nuclei are indicated by the information of Figure 16–2:

Those in which the n/p ratio is too high. The nucleus would fall in an area above the region of stability—point *A* of Figure 16–2.
Those in which the n/p ratio is too low. The nucleus is found below the region of stability—point *B* of Figure 16–2.
Those nuclei with a Z greater than 83—point *C* of Figure 16–2.

Consider each type of instability in terms of the changes which must take place to make the nucleus stable. In the case where the *n/p* ratio is too high, either n must be decreased, p increased, or both. This situation is corrected by the conversion of a neutron into a proton plus an electron, and the emission of the electron as a beta particle—a process called **beta emission.** The following reaction can be considered to take place in the nucleus:

$$^{1}_{0}\text{n} \rightarrow {}^{1}_{1}\text{p} + {}^{0}_{-1}\beta. \qquad 16\text{–}7$$

Notice the conservation-of-mass numbers (upper) and atomic numbers (lower) in the reaction. Beta emission is the type of decay usually found for radioisotopes with mass numbers larger than the elemental atomic weights. Two examples of beta emission are represented by the following reactions; note that the atomic weights of C and Mn are 12.0 and 55.0, respectively:

$$^{14}_{6}\text{C} \rightarrow {}^{14}_{7}\text{N} + {}^{0}_{-1}\beta \qquad 16\text{–}8$$

$$^{56}_{25}\text{Mn} \rightarrow {}^{56}_{26}\text{Fe} + {}^{0}_{-1}\beta \qquad 16\text{–}9$$

In a beta-emission process the mass number of the product nucleus is the same but the atomic number is increased by one. Therefore, p increased while n decreased and the n/p ratio has been decreased.

In the second situation, where the n/p ratio is too low, an increase in n, a decrease in p, or both will correct the situation. This change is accomplished in two ways. In one process a nuclear proton is converted into a neutron plus a positron, and the positron is emitted. This mode of decay is prevalent in light isotopes of elements with low atomic numbers. The reaction assumed to take place in the nucleus is:

$$^{1}_{1}\text{p} \rightarrow {}^{1}_{0}\text{n} + {}^{0}_{1}\beta \qquad 16\text{–}10$$

The emitted positron has a charge opposite that of an electron but the same weight. Two examples are represented by the following reactions:

$$^{15}_{8}\text{O} \rightarrow {}^{15}_{7}\text{N} + {}^{0}_{1}\beta \qquad 16\text{–}11$$

$$^{122}_{53}\text{I} \rightarrow {}^{122}_{52}\text{Te} + {}^{0}_{1}\beta \qquad 16\text{–}12$$

The process is called **positron emission;** the net effect is a decrease in atomic number while the mass number remains constant. In this way n is increased, p is decreased, and the n/p ratio is increased.

A second process, **electron capture,** is common in isotopes of heavy elements with mass numbers lower than the elemental atomic weights. Once again the result is an increased n/p ratio. In the process, an extranuclear electron of the atom is captured by the nucleus. The electron reacts with a proton and converts it into a neutron.

$$^{1}_{1}\text{p} + {}^{0}_{-1}\text{e} \rightarrow {}^{1}_{0}\text{n} \qquad 16\text{–}13$$

In this way n increases and p decreases, resulting in an increase in the n/p ratio. The following reactions are examples of electron-capture processes:

$$^{55}_{26}\text{Fe} + {}^{0}_{-1}\text{e} \rightarrow {}^{55}_{25}\text{Mn} \qquad 16\text{–}14$$

$$^{197}_{80}\text{Hg} + {}^{0}_{-1}\text{e} \rightarrow {}^{197}_{79}\text{Au} \qquad 16\text{–}15$$

The third situation in which the number of protons is too great for stability can be corrected by decreasing that number. This is often accomplished by **alpha-particle emission,** as represented by the following reaction:

$$^{212}_{84}Po \rightarrow ^{208}_{82}Pb + ^{4}_{2}\alpha \qquad 16\text{–}16$$

The rate at which a nucleus disintegrates radioactively is an indication of its stability. No two radioactive isotopes disintegrate at the same rate. The less stable nuclei disintegrate more rapidly than the more stable ones. The rate of disintegration or decay is commonly expressed in terms of the **half-life,** which is the time required for one-half of a given number of atoms in a sample to disintegrate. Some typical half-lives are given in Table 16–1.

Table 16–1. Half-Lives

Nucleus	*Half-life* ($t_{\frac{1}{2}}$)
$^{5}_{3}Li$	10^{-21} seconds
$^{15}_{8}O$	118 seconds
$^{13}_{7}N$	10.1 minutes
$^{24}_{11}Na$	15.0 hours
$^{238}_{92}U$	4.5×10^{9} years

Thus, at the end of 10.1 minutes, the number of radioactive nitrogen-13 atoms in a sample will be one-half the number present at the beginning. After two half-lives, one-half of the one-half, or one-fourth, of the original number remain. And after three half-lives, $\frac{1}{2} \times \frac{1}{2} \times \frac{1}{2} = \frac{1}{8}$ of the original atoms are left undecayed.

Because they are radioactive and are continually being disintegrated, any radioactive nuclei found in nature belong to one of three categories. They may have very long half-lives and so disintegrate very slowly ($^{238}_{92}U$), they may be radioactive nuclei produced from disintegrations of other nuclei ($^{222}_{86}Rn$), or they may be nuclei that are produced by natural processes such as the bombardment of stable nuclei by cosmic rays ($^{14}_{6}C$).

Effects of Radiation on Living Organisms

As radiation passes through matter, energy is transferred from the rays to the molecules, atoms, or ions that make up the matter. The rays are often collectively called **ionizing radiation,** because one common result of their passage is the conversion of neutral atoms or molecules into ions. This results when one or more electrons are knocked off atoms or molecules. The resulting ions are often very reactive, and they can seriously disrupt the normal functions of any living cell in which they are formed.

The three types of radiation are not equally dangerous to living organisms because of variations in penetrating ability. Alpha particles, the least penetrating, can be stopped by a thin sheet of paper, a thin film of water, or the outer layer of the skin. They can be much more damaging when they originate from sources that have been inhaled or ingested. Beta rays penetrate deeper and can cause severe skin burns if the source remains in contact with the skin for an appreciable time. Gamma rays are the most dangerous because of their extreme ability to penetrate matter. They have the ability to penetrate through organs, tissue, and bone. Normal clothing provides protection from both alpha and beta radiation but not from gamma radiation.

Two methods, physical and biological, are used to describe quantities of radiation. Physical measurements are related to the number of particles emitted per second by a source. A **curie,** the unit usually used, is equal to 3.7×10^{10} nuclear disintegrations (and particle emissions) per second. Biologically, radiation is measured in terms of the amount of ionization produced. A common unit used in conjunction with gamma rays—or X-rays—is the **roentgen,** which generates about 2×10^{9} ion pairs per cubic centimeter of air. An ion pair consists of the positive ion and the electron produced when radiation interacts with an atom or a molecule. When radiation passes through tissue, the ionization effect is greater; one roentgen generates about 1.8×10^{12} ion pairs per gram of tissue. Different units are used for types of radiation other than gamma rays or X-rays. A commonly used unit, the **rem** (roentgen equivalent in man), is the quantity of any type of radiation that produces the same effect on tissue as one roentgen of gamma rays or X-rays. Table 16–2 describes the health effects of radiation.

Table 16–2. Effects of Radiation on Man from Whole-Body Short-Term Exposures

Dose (rems)	*Effects*
0–25	No detectable clinical effects.
25–100	Slight short-term reduction in number of some blood cells, disabling sickness not common.
100–200	Nausea and fatigue, vomiting if dose is greater than 125 rems, longer-term reduction in number of some blood cells.
200–300	Nausea and vomiting first day of exposure, up to a two-week latent period followed by appetite loss, general malaise, sore throat, pallor, diarrhea and moderate emaciation. Recovery in about three months unless complicated by infection or injury.
300–600	Nausea, vomiting, and diarrhea in first few hours. Up to a one-week latent period followed by loss of appetite, fever and general malaise in the second week, followed by hemorrhage, inflammation of mouth and throat, diarrhea, and emaciation. Some deaths in two to six weeks. Eventual death for 50 percent if exposure is above 450 rems; others recover in about six months.
600 or more	Nausea, vomiting, and diarrhea in first few hours. Rapid emaciation and death as early as second week. Eventual death of nearly 100 percent.

From *Medical Aspects of Radiation Accidents*, E. L. Saenger, ed. (Washington, D.C.: United States Atomic Energy Commission, 1963), p. 9.

As a contrast, Table 16–3 contains the radiation dose equivalent received annually by the general population from various sources.

Table 16–3. Annual Radiation Dose Equivalents Received by the General Population

Dose (rems)	*Source*
0.100–0.140	Natural background radiation
0.020–0.050	Diagnostic X-rays
0.003–0.005	Therapeutic X-rays
0.0002	Medical radioisotopes
0.0015	Fallout (from atmospheric testing, 1954–1962)
0.00085	Radioactive pollutants from nuclear power plants (1970)
0.002	Smoking of one pack of cigarettes per day
0.170	Maximum annual limit proposed (by the International Commission on Radiological Protection, 1970) from all sources exclusive of medical and background

From *Elements of General and Biological Chemistry* by John R. Holum (New York: John Wiley and Sons, Inc., 1972), p. 535.

Induced Nuclear Reactions

Scientists have been able to artificially induce nuclear reactions by bombarding either stable or naturally radioactive nuclei with various high-energy particles. The results of these artificially induced reactions vary. The bombarded nucleus may be changed into another stable or radioactive nucleus. This is the **transmutation** of the elements that the ancient alchemists unsuccessfully attempted to perform—one element is changed into another. A second result of bombardment is the initiation of **nuclear fission,** a process involving the breakup of heavy, large nuclei into smaller nuclei. A third result of bombardment is **nuclear fusion.** In this process small nuclei are condensed together to form larger, heavier nuclei. The product or process resulting from bombardment experiments depends upon the kind of nucleus bombarded and the nature and energy of the bombarding particle.

Two types of particles, charged and uncharged (neutrons), are available for use in bombarding nuclei. It is believed that a bombarding particle will be drawn into the nucleus of an atom by strong nuclear forces if it can get to within 10^{-12} cm of the nucleus. The capture of neutrons occurs readily if their speeds are reduced enough to allow the nuclear forces to become effective. For this reason, substances such as graphite are used to slow down neutrons, and are called **moderators.** Charged particles present a different problem. If they are negatively charged ($_{-1}^{0}\beta$), they are repelled by the electron cloud around the nucleus and so must have sufficient energy to penetrate this cloud. Positive particles ($_{1}^{1}p$ and $_{2}^{4}\alpha$), on the other hand, are repelled by the positive charge of the nucleus itself and must have sufficient energy to overcome this repulsion; otherwise no reaction can take place.

Neutrons for bombardment reactions are obtained from radioactive materials or nuclear reactors and, together with proper moderators, are used as they come from the sources. Charged particles are obtained from radioactive sources or from ionization reactions—alpha particles are produced by removing two electrons from helium atoms. These particles are generally accelerated to high speeds—and energies—before bombarding the intended nuclei. Two types of particle accelerators, cyclic and linear, are often used.

The cyclotron, an example of the cyclic type, is represented in Figure 16–3. The particles to be accelerated enter the evacuated chamber at the center and, because of the magnetic field, move in a circular path toward the gap between the dees. Just as the particles reach the gap, the charge on the dees is adjusted so the particles are repelled by the dee they are leaving and attracted to the other one. The particles then coast around the dee until they again reach a gap, at which point the polarities are again adjusted to cause acceleration. This process continues and the particles are accelerated each time they pass through a gap. As the speed and energy of the particles increase, so does the radius of their circular path until they finally leave the dees and strike the target.

In a linear accelerator, the particles are accelerated through a series of charged tubes contained in an evacuated chamber (Figure 16–4). Each time a particle passes from one tube to another, it is accelerated in the same way

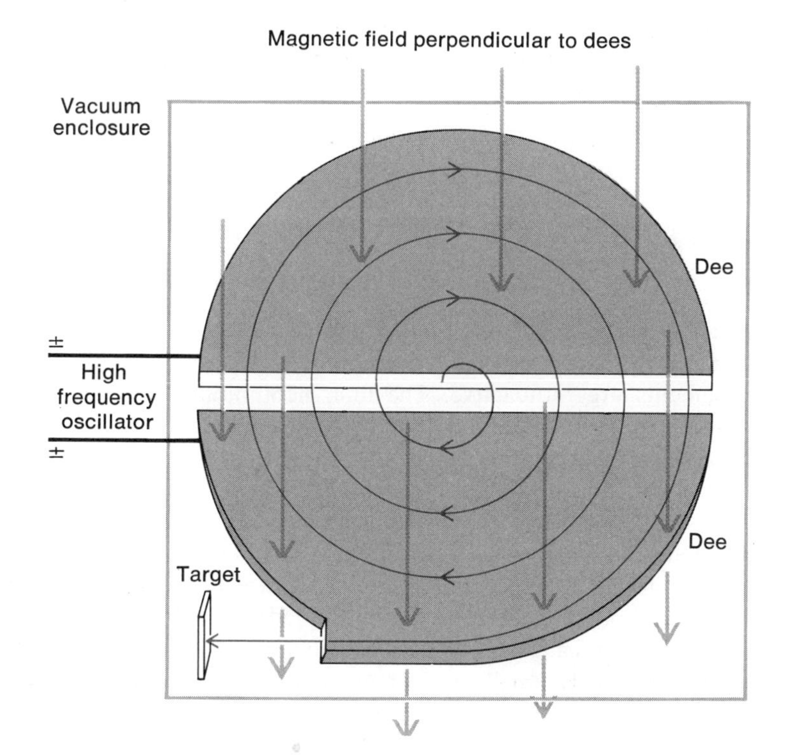

Figure 16–3. The Cyclotron

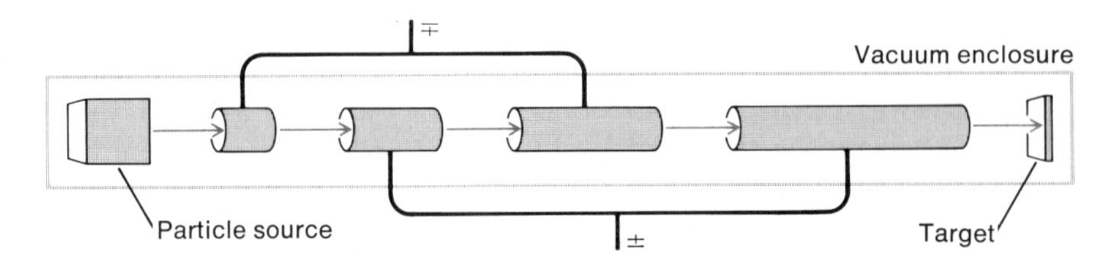

Figure 16–4.
A Linear Particle Accelerator

as particles passing between the dees of a cyclotron. The charges on the tubes are changed back and forth at proper time intervals to accomplish this acceleration. The tubes get successively longer to allow the particles the same residence or coasting time as they move toward the end at ever-increasing speeds. After acceleration, the particles exit and strike the target.

New Elements

Perhaps the most interesting transmutations are those resulting in the creation of completely new elements. Four of these elements, produced between 1937 and 1941, filled gaps in the periodic table for which no naturally occurring element had been found. These four are technicium (Tc, number 43), promethium (Pm, number 61), astatine (At, number 85), and francium (Fr, number 87). The reactions for their production are given below.

$$^{96}_{42}\text{Mo} + ^{2}_{1}\text{H} \rightarrow ^{1}_{0}\text{n} + ^{97}_{43}\text{Tc (half-life: } 2.6 \times 10^{6} \text{ years)} \qquad 16\text{–}17$$

$$^{142}_{60}\text{Nd} + ^{1}_{0}\text{n} \rightarrow ^{0}_{-1}\beta + ^{143}_{61}\text{Pm (half-life: 265 days)} \qquad 16\text{–}18$$

$$^{209}_{83}\text{Bi} + ^{4}_{2}\alpha \rightarrow 3\ ^{1}_{0}\text{n} + ^{210}_{85}\text{At (half-life: 8.3 hours)} \qquad 16\text{–}19$$

$$^{230}_{90}\text{Th} + ^{1}_{1}\text{p} \rightarrow 2\ ^{4}_{2}\alpha + ^{223}_{87}\text{Fr (half-life: 22 minutes)} \qquad 16\text{–}20$$

Man-made elements have also been produced which are heavier than the heaviest, naturally occurring element uranium. All isotopes of these **trans-uranium** elements are radioactive. The first, neptunium (Np, number 93), was discovered in 1940; since then twelve additional elements (numbers 94 through 105) have been synthesized (Table 16–4).

Significant quantities of only a few of these elements have been produced. Plutonium is synthesized in quantities large enough to make it useful in atomic weapons and reactors. In 1968 the entire world's supply of californium (10^{-4} g) was collected together and made into a target for bombardment by heavy, accelerated particles. These experiments led to the discovery of several elements heavier than californium. Usually the amounts of the new elements produced are extremely small—only hundreds of atoms in some cases—

Table 16–4. The Transuranium Elements

Name	*Symbol*	*Atomic number*	*Atomic weight of most stable isotope*	*Half-life of most stable isotope*	*Date of discovery*
Neptunium	Np	93	237	2.2×10^6 years	1940
Plutonium	Pu	94	242	3.8×10^5 years	1940
Americium	Am	95	243	8.0×10^3 years	1944
Curium	Cm	96	247	4.0×10^7 years	1944
Berkelium	Bk	97	247	104 years	1950
Californium	Cf	98	251	800 years	1950
Einsteinium	Es	99	254	480 days	1952
Fermium	Fm	100	253	4.5 days	1953
Mendelevium	Md	101	256	1.5 hours	1955
Nobelium	No	102	253	10 minutes	1958
Lawrencium	Lr	103	257	8 seconds	1961
Rutherfordium	Rf	104	257	4.5 seconds	1969
Hahnium	Ha	105	262	40 seconds	1970

because of the nature of the experiment. When the half-life is short, this small amount quickly disappears. Such tiny amounts of often short-lived elements cannot be detected and identified on the basis of chemical properties; identification is carried out by the use of instruments to analyze the characteristic radiation emitted by each new element.

Uses of Radioisotopes

Radioactive materials are widely used in science, industry, medicine, and related areas. A few examples will illustrate these applications.

In many applications radioisotopes are used as **tracers** to allow the progress of a reaction or process to be followed. Radioactive tracers are used by chemists to gain valuable information about chemical compounds and reactions. For example, it is well known to chemists that when sulfur is boiled in a solution containing a sulfite (Na_2SO_3), the thiosulfate ion or salt is produced. The reaction is:

$$S + SO_3^{2-} \rightarrow S_2O_3^{2-} \qquad 16\text{–}21$$

At one time there was a question concerning the position of the added sulfur atom in the $S_2O_3^{2-}$ ion. Are the two sulfur atoms equivalent? It was found experimentally that they are not. Radioactive sulfur was boiled with sulfite ion and the resulting thiosulfate was then decomposed to SO_2, H_2O, and free sulfur by the addition of acid. All of the radioactive isotope was found in the free sulfur resulting from decomposition and none in the SO_2. The reactions are given below, where the radioisotope is indicated by an asterisk:

$$S^* + SO_3^{2-} \rightarrow S_2^*O_3^{2-} \qquad 16\text{–}22$$

$$S_2^*O_3^{2-} + 2H^+ \rightarrow S^* + SO_2 + H_2O \qquad 16\text{–}23$$

This, along with other evidence, leads to the conclusion that the structure of the thiosulfate ion is

$$\left[\begin{array}{ccc} & S & \\ & | & \\ O- & S & -O \\ & | & \\ & O & \end{array}\right]^{2-}$$

It is apparent that the two sulfur atoms are not equivalent.

Interesting uses for radioactive tracers are found in industry. The petroleum industry, for example, uses radioisotopes in pipelines to indicate the boundary between different products moving through the lines [Figure 16–5(a)]. A tracer added to fluids moving in pipes makes it easy to detect leaks [Figure 16–5(b)]. The effectiveness of lubricants has been studied by the use of radioactive metal isotopes as components of metal parts. As the metal part wears, the isotope shows up in the lubricant, and the amount of wear can be determined [Figure 16–5(c)].

Radioactive tracers are used in a variety of ways in medicine. The circulation of blood in the body can be followed by using radioactive sodium-23 as a tracer. A small amount of this isotope is injected into the bloodstream in the form of a sodium chloride solution. The movement of the isotope through the circulation system can easily be followed with radiation detection equipment.

The functioning of the thyroid gland is evaluated by administering iodine-131 to a patient. The iodine is preferentially absorbed by the thyroid gland

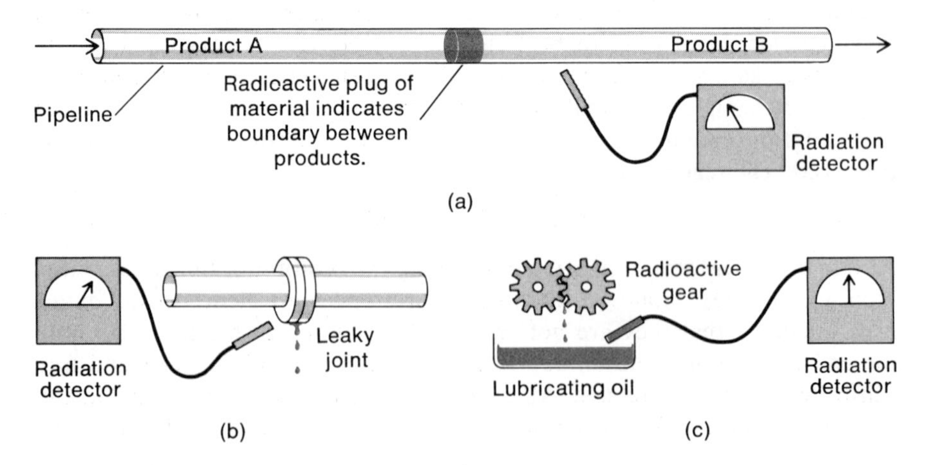

Figure 16–5.
Radioactive Tracers in Industry

at a rate related to the activity of the gland. The rate of absorption is determined by measuring the increase in radioactive emissions from the thyroid as a function of time.

Radioisotopes are also useful in forms other than tracers. The thickness of manufactured materials such as metal foils or sheets can be determined continuously by the use of a radioisotope. The equipment used is represented in Figure 16–6. The amount of radiation getting through the material to the detector is related to the thickness of the material.

Objects such as clocks and watches have luminous dials, some of which are produced by mixing zinc sulfide with a very tiny amount of radioactive material. When the radioactive particles or rays strike the ZnS, the substance gives off visible light.

The use of radioisotopes in cancer therapy is well known. For instance, cancer of the thyroid gland has been treated by radioactive iodine therapy. The same radioactive isotope of iodine mentioned before as a diagnostic aid is given to the patient and collects in the thyroid. The localized radioiodine can then destroy the cancerous thyroid cells, while doing little damage elsewhere in the body.

A very interesting application of radioisotopes involves the use of naturally occurring radioactive materials to determine the ages of rocks and artifacts. Perhaps the most widely known example is the use of carbon-14. It is produced in the atmosphere when cosmic-ray neutrons strike nitrogen nuclei. The reaction is:

$$^{14}_{7}N + ^{1}_{0}n \rightarrow ^{14}_{6}C + ^{1}_{1}H \qquad 16\text{–}24$$

The $^{14}_{6}C$ is converted to $^{14}_{6}CO_2$. In this form the radioactive carbon is absorbed by plants and converted into cellulose or other carbohydrates during the process of photosynthesis. As long as the plant lives, it will take in $^{14}_{6}CO_2$ at what is assumed to be a constant rate. The carbon-14 is constantly decaying and decreasing in concentration within the plant. As long as the plant is alive, a constant amount of carbon-14 will be present per unit amount of carbon in the plant. This is caused by an equilibrium between the rate of carbon-14 uptake in the plant and the rate of disappearance through disintegration.

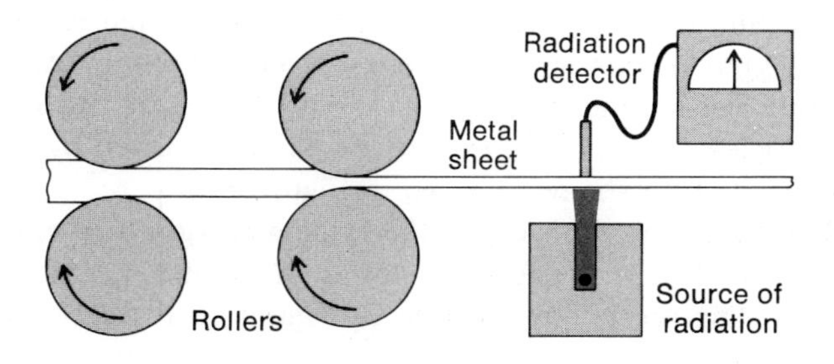

Figure 16–6.
Radioisotopes Used as a Thickness Gauge

Another way to state this is that the fraction of carbon-14 to total carbon will remain constant as long as the plant lives. Upon the death of the plant the ${}^{14}_{6}CO_2$ intake stops and the fractional amount of carbon-14 begins to decrease. From a measurement of this amount in a wooden object, and a knowledge of the half-life (5600 years), it is possible to calculate the age of the object—since the tree died when it was cut down to produce the object. An object containing only about one-eighth as much carbon-14 as a fresh wood sample from a living tree would be about three half-lives or 16,800 years old.

This method is limited to objects less than about 50,000 years old because of difficulties in measuring the small amount of carbon-14 present. For this reason other methods have been developed. One method involves potassium, ${}^{40}_{19}K$, and argon, ${}^{40}_{18}Ar$. The half-life of potassium-40, which absorbs one of its own inner electrons to produce argon-40, is 1.3×10^9 years. By determining the amount of argon-40 in a potassium-containing mineral, it is possible to estimate the age of the mineral. The reaction is:

$$ {}^{40}_{19}K + {}^{0}_{-1}e \rightarrow {}^{40}_{18}Ar \qquad 16\text{–}25 $$

Nuclear Bombs and Energy Sources

Theories developed in the 1930s predicted that it might be possible to cause some heavy nuclei to split into smaller nuclei by bombarding them with neutrons. As the nuclei split into smaller particles, energy would be released. The total weight of the product nuclei and particles would be slightly less than the total weight of reacting nuclei and particles. This weight difference shows up as an equivalent amount of energy, given by Einstein's relationship $E = \Delta wc^2$, where E is the energy, Δw the weight difference, and c the speed of light. According to this equation huge amounts of energy would be generated by the disappearance—conversion—of relatively small amounts of matter. One gram of matter, upon complete conversion, would generate an amount of energy equivalent to that produced by burning 2500 tons of coal.

This idea can be clarified a little by considering the concept of binding energy of the nucleus. Consider a number of individual protons and neutrons separated in space. Suppose the total weight of the particles is w_1. Now, bring all these particles together and condense them into a small nucleus. The fact that the nucleus remains intact indicates it is of lower energy than the separated particles. Suppose the weight of the nucleus is determined to be w_2. We find that w_1 is greater than w_2. Suppose $w_1 - w_2 = \Delta w$; then Δw represents the weight equivalent to the energy given up in forming the nucleus from the individual particles. The value of the energy would be given by the Einstein relationship. This process is represented in Figure 16–7.

The total energy given up is called the **binding energy;** the value obtained by dividing the total binding energy by the total number of particles bound is called the **binding energy per nucleon.** In Figure 16–8 the binding energy

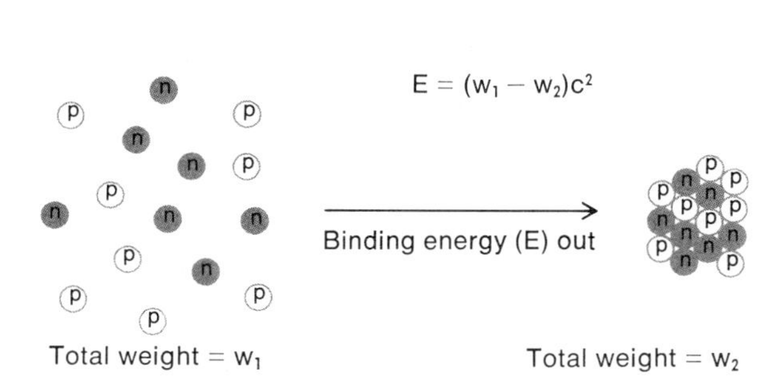

Figure 16–7.
Binding Energy of the Nucleus

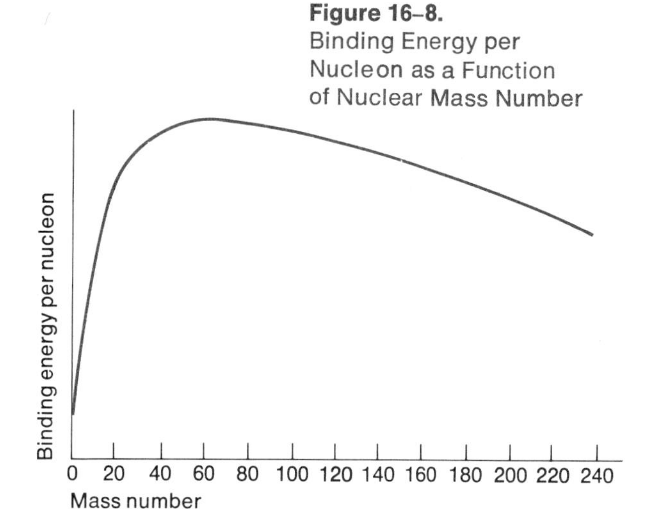

Figure 16–8.
Binding Energy per Nucleon as a Function of Nuclear Mass Number

per nucleon is plotted against the mass number of the nuclei. The maximum in the plot occurs at a mass number of about 56, indicating that this nucleus should be the most stable—the most energy per nucleon is holding it together. The binding energy decreases for higher mass numbers. If an atom of high mass number absorbs enough energy in the form of a bombarding neutron, it will break into smaller particles. Since the smaller particles have higher binding energies, the difference in binding energies will show up as energy given off during the reaction. This is the process involved in **nuclear fission.**

Nuclear Fission

It was found that an isotope of uranium, $^{235}_{92}U$, would undergo fission by reactions such as the following:

$$^{235}_{92}U + ^{1}_{0}n \rightarrow ^{135}_{53}I + ^{97}_{39}Y + 4\,^{1}_{0}n \qquad 16\text{–}26$$

$$^{235}_{92}U + ^{1}_{0}n \rightarrow ^{139}_{56}Ba + ^{94}_{36}Kr + 3\,^{1}_{0}n \qquad 16\text{–}27$$

$$^{235}_{92}U + ^{1}_{0}n \rightarrow ^{103}_{42}Mo + ^{131}_{50}Sn + 2\,^{1}_{0}n \qquad 16\text{–}28$$

$$^{235}_{92}U + ^{1}_{0}n \rightarrow ^{139}_{54}Xe + ^{95}_{38}Sr + 2\,^{1}_{0}n \qquad 16\text{–}29$$

For every uranium-235 atom disintegrated, more than one neutron is generated, which opens up the possibility for a chain reaction to take place. If only one of the neutrons produced each time reacted with another uranium-235 atom, the process would continue at a constant rate. When this circumstance

Figure 16–9.
Nuclear Chain Reactions

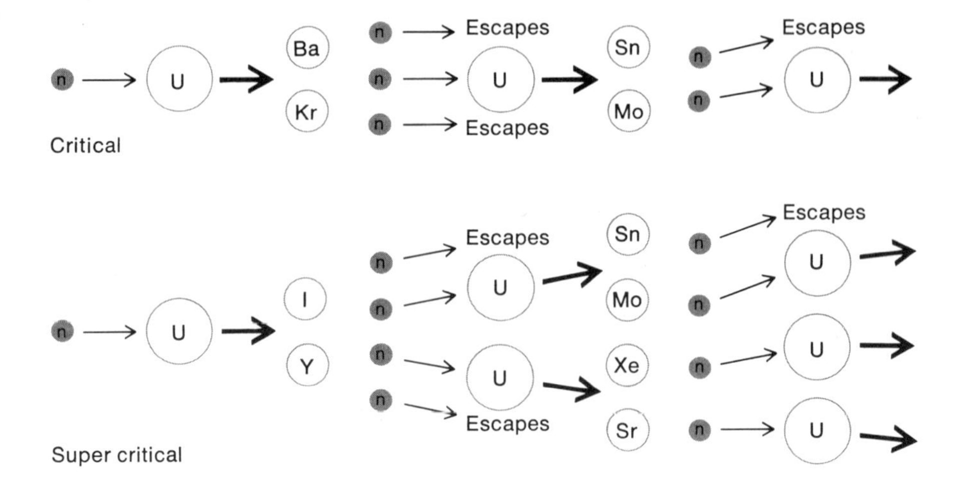

occurs, the process is said to be self-propagating or **critical.** If more than one of the neutrons generated per disintegration produces another disintegration, the chain reaction of the critical situation becomes an expanding or branching chain reaction, creating a **supercritical** situation which then leads to an explosion. Critical and supercritical reactions are represented in Figure 16–9.

The first critical chain reaction was carried out in an atomic pile, which was literally a stack or pile of rods containing uranium-235 fuel and slabs of graphite moderator. The pile was constructed in a squash court at the University of Chicago, and the experiments were carried out by a team of scientists led by Enrico Fermi. In December 1942 the team observed that nuclear reactions in the pile had become self-sustaining or critical. The essential components of an atomic pile are given in Figure 16–10.

Figure 16–10.
An Atomic Pile—The World's First Nuclear Reactor

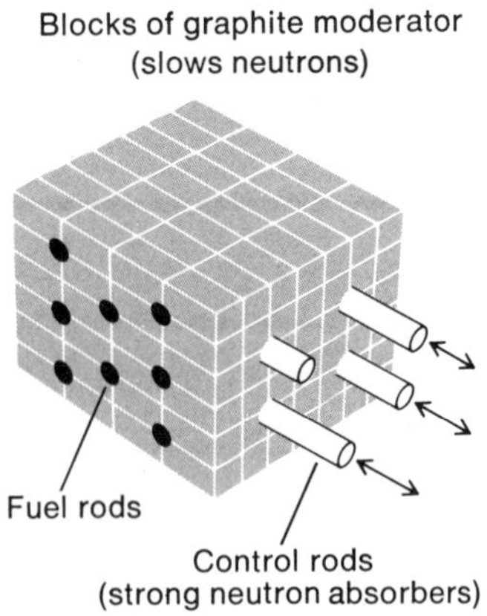

Blocks of graphite moderator (slows neutrons)
Fuel rods
Control rods (strong neutron absorbers)

Once it had been shown experimentally that uranium-235 would undergo self-sustaining fission, the construction of a bomb was fairly simple. All that was needed was a supercritical amount or mass of fissionable materials. A **subcritical** amount is small enough so that, on the average, less than one of the neutrons produced per disintegration collides and reacts with another uranium-235 nucleus. As the amount is increased, the chances of neutron escape grow smaller, and finally critical and supercritical masses are reached.

The big difficulty encountered in building a bomb was the size of device needed to bring together the subcritical masses. The configurations of the first two devices finally used are shown in Figure 16–11. The spherical configuration resulted in a short, wide bomb just small enough around the middle to be dropped through the bomb bay of a B-29 bomber, the largest available at that time. The long tubular configuration gave rise to a long narrow bomb just barely short enough to be delivered by a B-29.

On August 6, 1945, the United States became the first country to use a nuclear weapon in combat. Approximately 70,000 inhabitants of Hiroshima,

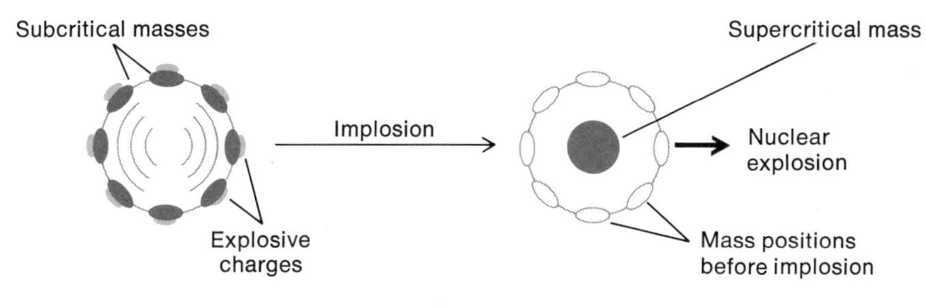

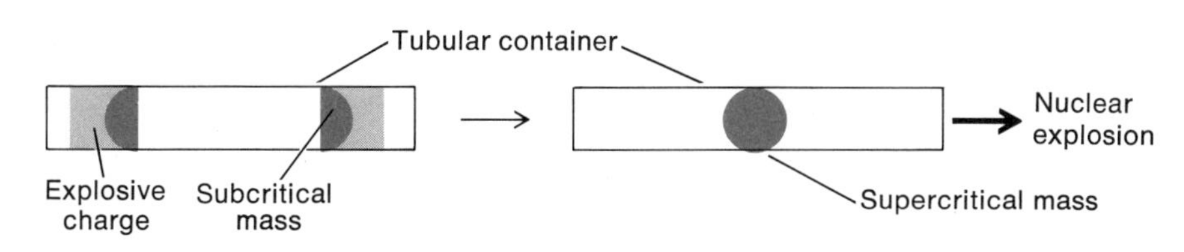

Figure 16–11.
Configurations Used in Fission Bombs

Japan, were killed in the explosion; an equal number were seriously injured. A short time later a second bomb was dropped on the city of Nagasaki with similar results. One of the bombs used was made of plutonium, an artificial fissionable material produced in breeder reactors.

The fission reaction can be used to provide energy in forms other than bombs. Energy in the form of heat can be extracted from the interior of a critical pile or reactor by means of a circulating heat-exchange fluid—liquid metals such as sodium have been used. The hot fluid produces steam in a boiler, which can then be used to drive electrical generators. This type of facility is diagrammed in Figure 16–12.

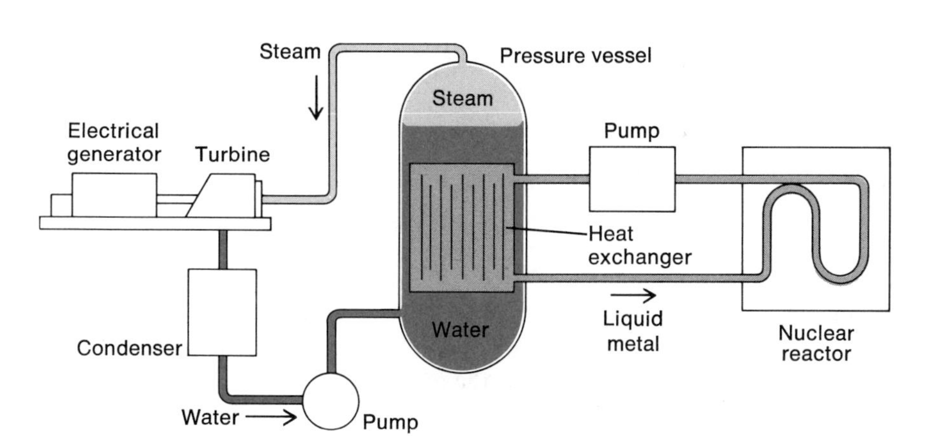

Figure 16–12.
Electrical Energy from a Nuclear Reactor

Under proper conditions an atomic reactor can be made to produce fissionable fuel along with energy. In these **breeder reactors** some neutrons produced by the fission reaction do not propagate the chain but are absorbed by nonfissionable nuclei, which are converted into fissionable isotopes. For example, uranium-238 is the most abundant naturally occurring isotope of uranium, but it is useless as a direct energy source because it will not undergo fission. However, when it is placed in a neutron-rich pile, the following reactions take place:

$$^{238}_{92}\text{U} + ^{1}_{0}\text{n} \rightarrow ^{239}_{92}\text{U} \qquad 16\text{–}30$$

$$^{239}_{92}\text{U} \rightarrow ^{239}_{93}\text{Np} + ^{0}_{-1}\beta \qquad 16\text{–}31$$

$$^{239}_{93}\text{Np} \rightarrow ^{239}_{94}\text{Pu} + ^{0}_{-1}\beta \qquad 16\text{–}32$$

The resulting plutonium-239 is capable of fission and was used in one of the first atomic bombs. Thus a fissionable energy source has been produced from a nonfissionable and somewhat abundant material.

Nuclear Fusion

A higher-yield nuclear energy source is available. This involves a **nuclear fusion** reaction in which small nuclei fuse together into larger nuclei. This behavior is predicted by the data of Figure 16–8. As small nuclei fuse together and form larger nuclei, the binding energy per nucleon increases and the products of the reaction are stabilized. The energy output of fusion reactions is higher than that of fission reactions. Only about 0.1 percent of the weight involved in fission reactions is converted into energy, but in fusion reactions the percentage is between 0.4 and 0.7. Thus, on a weight basis, fusion reactions produce four to seven times as much energy as fission reactions.

A fusion-type process, represented by the following reactions, is thought to be responsible for the energy output of the sun.

$$2\,^{1}_{1}\text{H} \rightarrow ^{2}_{1}\text{H} + ^{0}_{1}\beta \qquad 16\text{–}33$$

$$^{1}_{1}\text{H} + ^{2}_{1}\text{H} \rightarrow ^{3}_{2}\text{He} \qquad 16\text{–}34$$

Nuclear fusion reactions require high temperature to get started. In the so-called hydrogen bomb, a fission device is used to start the fusion reactions. The fusion reactions of a hydrogen bomb are:

$$^{3}_{1}\text{H} + ^{2}_{1}\text{H} \rightarrow ^{4}_{2}\text{He} + ^{1}_{0}\text{n} \qquad 16\text{–}35$$

$$^{6}_{3}\text{Li} + ^{1}_{0}\text{n} \rightarrow ^{4}_{2}\text{He} + ^{3}_{1}\text{H} \qquad 16\text{–}36$$

$$^{2}_{1}\text{H} + ^{3}_{1}\text{H} \rightarrow ^{4}_{2}\text{He} + ^{1}_{0}\text{n} \qquad 16\text{–}37$$

Figure 16–13 is a schematic diagram of a hydrogen bomb assembly.

In principle, fusion reactions can be used as controlled energy sources the same as fission reactions. However, the temperatures created by fusion reactions are so high that no known substance can be used as a container. Work

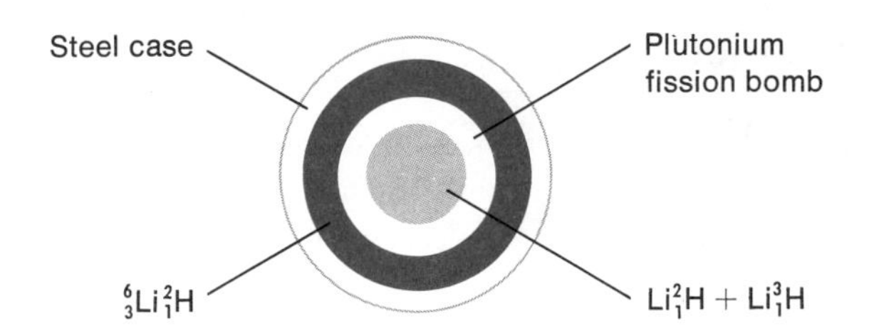

Figure 16–13. Components of a Hydrogen Bomb

is being done toward developing magnetic containers in which the high-temperature reactants are held away from the walls by strong magnetic fields. When such processes are perfected, man will have almost literally isolated a bit of the sun to use as he wishes.

A Comparison of Nuclear and Chemical Reactions

The nuclear reactions studied in this chapter differ from ordinary chemical reactions in a number of significant ways, which are listed below.

CHEMICAL REACTION	NUCLEAR REACTION
Different isotopes of an element have practically identical chemical properties.	Different isotopes of an element have different properties in nuclear processes.
The chemical reactivity of an element depends on the element's state of combination (free element, compound, etc.).	The nuclear reactivity of an element is independent of the state of chemical combination.
Elements retain their identity in chemical reactions.	Elements may be changed into other elements during nuclear reactions.
Energy changes that accompany chemical reactions are relatively small.	Nuclear reactions involve energy changes that are several orders of magnitude larger than those of chemical reactions.

Review Questions

1. Characterize by charge, weight, and energy (relative) the three common types of radiation known as alpha, beta, and gamma rays.
2. How do the atomic number and mass number of daughter nuclei differ from the original nuclei after:
   a) An alpha particle is emitted
   b) A beta particle is emitted
   c) A gamma ray is emitted
3. Discuss the three types of unstable nuclei indicated by A, B, and C in Figure 16–2. What type of radiation is characteristic of each type?
4. What effect does each of the following processes have on the n/p ratio of a nucleus?
   a) Beta particle emission
   b) Positron emission
   c) Electron capture
   d) Gamma ray emission
5. Complete the following reactions which represent radioactive decay processes. Use appropriate notations and formulas.
   a) ${}^{10}_{4}\text{Be} \rightarrow \_\_\_\_ + {}^{10}_{5}\text{B}$
   b) ${}^{210}_{83}\text{Bi} \rightarrow {}^{4}_{2}\alpha + \_\_\_\_$
   c) ${}^{41}_{20}\text{Ca} + {}^{0}_{-1}\text{e} \rightarrow \_\_\_\_$
   d) $\_\_\_\_ \rightarrow {}^{0}_{1}\beta + {}^{15}_{7}\text{N}$
   e) ${}^{44}_{22}\text{Ti} + \_\_\_\_ \rightarrow {}^{44}_{21}\text{Sc}$
   f) ${}^{53}_{26}\text{Fe} \rightarrow {}^{0}_{1}\beta + \_\_\_\_$
   g) $\_\_\_\_ \rightarrow {}^{4}_{2}\alpha + {}^{222}_{86}\text{Rn}$
   h) ${}^{236}_{93}\text{Np} \rightarrow \_\_\_\_ + {}^{236}_{94}\text{Pu}$
6. Show how a radioactive element could be converted into an isotope of itself by successively emitting three particles. Note that this process occurs in the radioactive decay series given in this chapter.
7. What is meant by the term *half-life* when it is applied to a radioisotope?
8. A sample of a radioisotope produces a nonradioactive daughter. The sample is found to lose 87.5 percent of its original radioactivity in 90 minutes. What is the half-life of the radioisotope?
9. What fraction of a sample of ${}^{24}_{11}\text{Na}$ would remain undecayed after 60 hours? See Table 16–1 for the half-life.
10. What three categories of radioactive nuclei are found in nature?
11. Numerous units are used to measure quantities of radiation. Compare the three known as the curie, the roentgen, and the rem.
12. Suppose you received a whole body dose of radiation equivalent to 230 rems. What symptoms and effects on your health would you expect?
13. What are three possible results of the bombardment of target nuclei by high energy particles?
14. In what ways are cyclic and linear particle accelerators similar? In what ways are they different?

15. Complete the following reactions. Each one represents a transmutation caused by bombarding a target with a neutron or accelerated particle.
    a) $^{24}_{12}Mg + \_\_\_\_\_ \rightarrow ^{27}_{14}Si + ^{1}_{0}n$
    b) $^{27}_{13}Al + ^{2}_{1}H \rightarrow \_\_\_\_\_ + ^{4}_{2}\alpha$
    c) $\_\_\_\_\_ + ^{1}_{1}p \rightarrow 2\ ^{4}_{2}He$
    d) $^{14}_{7}N + ^{4}_{2}\alpha \rightarrow \_\_\_\_\_ + ^{1}_{1}p$
    e) $\_\_\_\_\_ + ^{11}_{5}B \rightarrow ^{257}_{103}Lr + 4\ ^{1}_{0}n$
    f) $^{9}_{4}Be + \_\_\_\_\_ \rightarrow ^{12}_{6}C + ^{1}_{0}n$
16. Describe at least one use for radioisotopes in each of the following categories:
    a) Molecular structure determinations
    b) Industrial applications
    c) Diagnostic medicine
    d) Therapeutic medicine (treatment)
    e) Archeology
    f) Studies of the moon
17. What is meant by the term *binding energy?*
18. Discuss the relationship between binding energy and nuclear stability as represented in Figure 16–8.
19. Discuss the process of nuclear fission. Relate the process to Figure 16–8. Include the ideas of critical and supercritical mass.
20. Discuss the process of nuclear fusion. Relate the process to Figure 16–8. Does the critical mass concept enter into a fusion process?
21. Discuss the uses of fission and fusion processes to generate useful electrical energy. What problems are associated with these uses?

Suggestions for Further Reading

Bump, T. R., "A Third Generation of Breeder Reactors," *Scientific American*, May 1967.

Chappin, G. R., "Nuclear Fission," *Chemistry*, July 1967.

Chen, F. F., "The Leakage Problem in Fusion Reactors," *Scientific American*, July 1967.

"Element 105," *Chemistry*, June 1970.

Evans, D. M., and A. Bradford, "Under the Rug," *Environment*, Oct. 1969.

"Fossil-Track Dating," *Chemistry*, July 1970.

Fowler, T. K., and R. F. Post, "Progress Toward Fusion Power," *Scientific American*, Dec. 1966.

Gofman, J. W., and A. R. Tamplin, "Radiation: The Invisible Casualties," *Environment*, April 1970.

Gough, W. C., and B. J. Eastlund, "The Prospects of Fusion Power," *Scientific American*, Feb. 1971.

Johnsen, R. H., "Radiation Chemistry," *Chemistry*, July 1967.
Kamen, M. D., "Tracers," *Scientific American*, Feb. 1949.
Leachman, R. B., "Nuclear Fission," *Scientific American*, Aug. 1965.
Morgan, K. Z., "Never Do Harm," *Environment*, Jan./Feb. 1971.
"Nuclear Power and By-Product Plutonium," *Chemistry*, March 1967.
"Radiocarbon Dating," *Chemistry*, July 1970.
Seaborg, G. T., and J. L. Bloom, "Fast Breeder Reactors," *Scientific American*, Nov. 1970.
Seaborg, G. T., and J. L. Bloom, "The Synthetic Elements," *Scientific American*, April 1969.
Weinberg, A. M., "Breeder Reactors," *Scientific American*, Jan. 1960.
Wilson, R. R., "Particle Accelerators," *Scientific American*, March 1958.
Wolfgang, R., "Chemistry at High Velocities," *Scientific American*, Jan. 1966.

17 Hydrocarbons

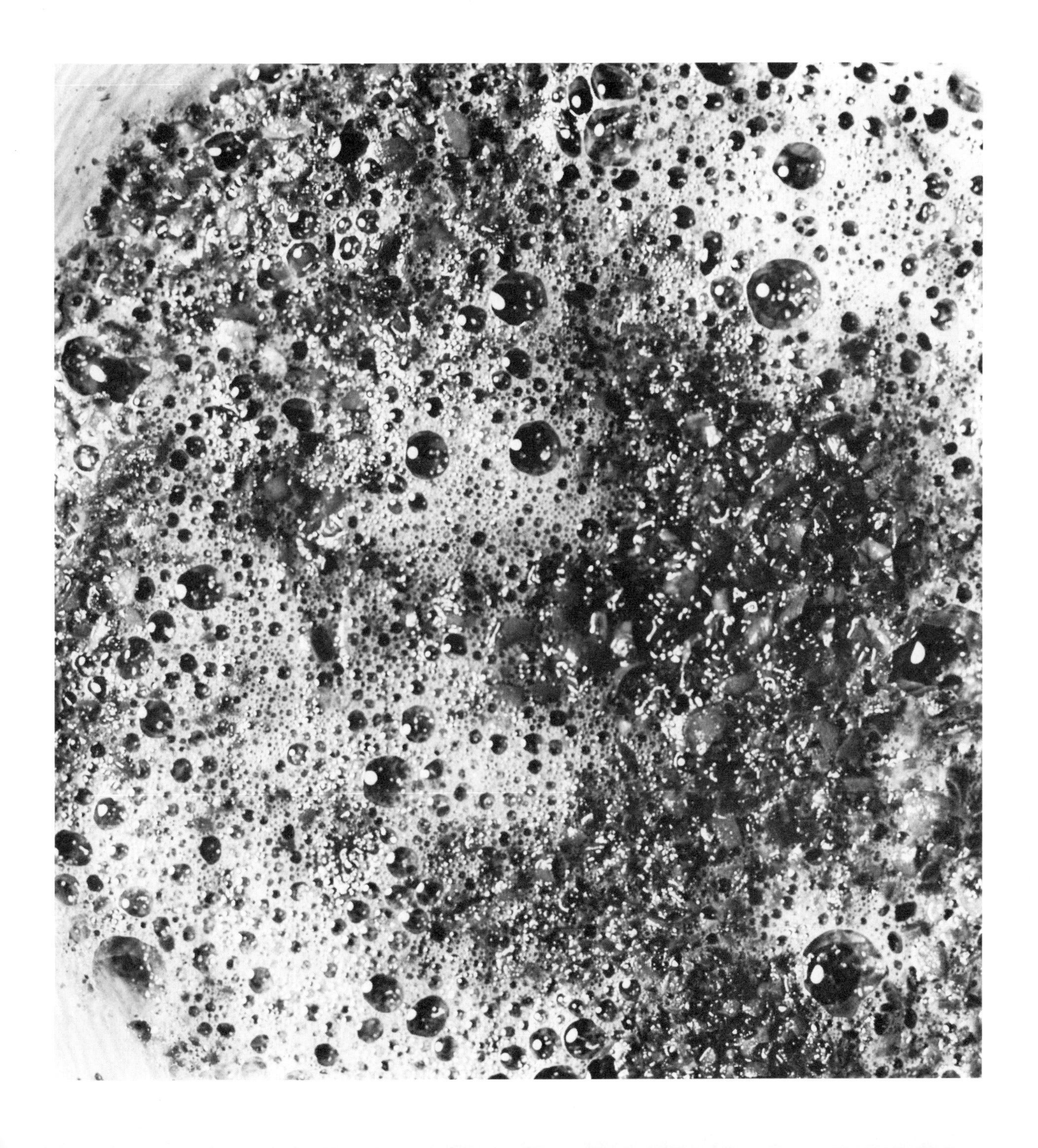

Carbon and some of its compounds were discussed in Chapter 14, but only in terms of the inorganic chemistry of carbon. Another entire area, called **organic chemistry,** is devoted to the study of compounds in which carbon generally demonstrates a property known as **catenation.** This is the ability of carbon atoms to form bonds with other carbon atoms, producing large networks of connected atoms. This property, unique to carbon, is shown in all organic compounds, except those consisting of simple one-carbon molecules.

One result of this property is an almost unlimited variety of carbon-atom arrangements. Carbon is also able to combine with nearly every other element at various sites within a given compound, thus creating the possibility for an extremely large number of different carbon-containing molecules. It is estimated, at present, that between three and four million such compounds are known; about 75,000 new ones are being produced every year. Both the large number and the many useful and interesting properties of these compounds have made them a separate area of study.

At one time, chemists thought that only living organisms—plants or animals—could produce the type of compounds now called organic. All organic compounds were extracted from living or at least once-living material until the early nineteenth century. In 1828 a chemist named Friedrich Wöhler demonstrated that the organic compound urea could be prepared by heating ammonium cyanate obtained from mineral sources. After this time scientists gradually gave up the idea that organic compounds had to be obtained from organic sources, and they began to synthesize them from inorganic raw materials. Today, the compounds studied in organic chemistry include all those in which a carbon atom is bonded to other carbon atoms. Many of these compounds do not occur naturally and were unknown until they were synthesized from inorganic materials, or from previously known organic materials, or from combinations of both.

Figure 17–1.
Structure of Methane, the Simplest Hydrocarbon

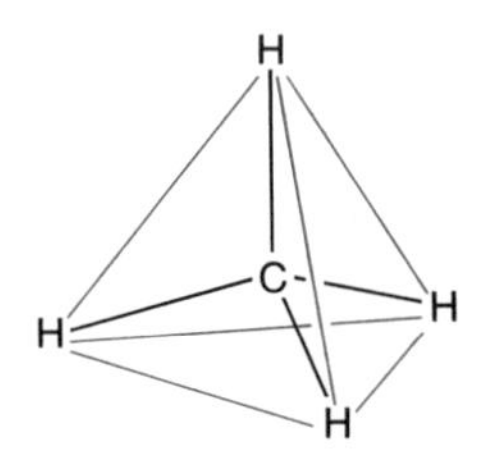

(a) Tetrahedral structure

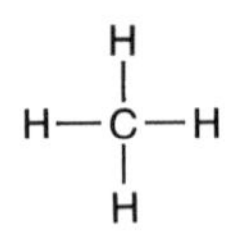

(b) Planar representation

The Alkanes

The simplest class of organic compounds is the **alkane hydrocarbons.** These compounds, which contain only carbon and hydrogen, are like other organic compounds in that each carbon atom shares four pairs of electrons to form four covalent bonds. The bonds in all hydrocarbons are either between two carbon atoms or between a carbon atom and a hydrogen atom. In alkanes, the four bonds of each carbon atom are single bonds that are directed toward the corners of a regular tetrahedron.

The first member of this class of compounds has the molecular formula CH_4 and the structural formula given in Figure 17–1, where each bar represents a covalent bond between a hydrogen atom and a carbon atom. Even though the usual representation (b) is drawn in two dimensions, we must remember that the hydrogen atoms are actually located at the corners of a regular tetrahedron with the carbon atom at the center (a). This arrangement makes each hydrogen atom geometrically equivalent to the others in the molecule. This compound is called methane.

The next member of the series is ethane with the molecular formula of C_2H_6 and the structural formula

$$\begin{array}{ccccccc} & & H & & H & & \\ & & | & & | & & \\ H & — & C & — & C & — & H \\ & & | & & | & & \\ & & H & & H & & \end{array}$$

This molecule can be thought of as a methane molecule with one hydrogen removed and a $\begin{array}{ccc} & H & \\ & | & \\ — & C & —H \\ & | & \\ & H & \end{array}$ put in its place.

The third member of the alkanes is propane, C_3H_8, with the structural formula

$$\begin{array}{ccccccccc} & & H & & H & & H & & \\ & & | & & | & & | & & \\ H & — & C & — & C & — & C & — & H \\ & & | & & | & & | & & \\ & & H & & H & & H & & \end{array}$$

Again, this formula can be produced by removing a hydrogen atom from the previous member and substituting a $\begin{array}{ccc} & H & \\ & | & \\ — & C & —H \\ & | & \\ & H & \end{array}$ in its place. All six hydrogen atoms of ethane are geometrically equivalent, so the one replaced makes no difference in the resulting propane.

The fourth member of the series, called butane, and having the formula C_4H_{10}, can be produced in the same way as before, but all the hydrogens of propane are not geometrically equivalent, and more than one position is available at which the substitution can be made. If the substitution is made on one of the end carbons of propane, the result is the structural formula

$$\begin{array}{ccccccccccc} & & H & & H & & H & & H & & \\ & & | & & | & & | & & | & & \\ H & — & C & — & C & — & C & — & C & — & H \\ & & | & & | & & | & & | & & \\ & & H & & H & & H & & H & & \end{array}$$

or, more compactly, $CH_3—CH_2—CH_2—CH_3$. However, substitution on the center carbon atom of propane gives the structure

$$\begin{array}{ccccccccc} & & H & & H & & H & & \\ & & | & & | & & | & & \\ H & — & C & — & C & — & C & — & H \\ & & | & & | & & | & & \\ & & H & & | & & H & & \\ & & H & — & C & — & H & & \\ & & & & | & & & & \\ & & & & H & & & & \end{array}$$

or $\begin{array}{ccccc} CH_3 & - & CH & - & CH_3 \\ & & | & & \\ & & CH_3 & & \end{array}$. Notice that both the linear and branched structures have the molecular formula C_4H_{10}. Whenever it is possible to write different structural formulas for the same molecular formula, the compounds involved are called **structural isomers.** We see, then, that two butanes are possible which are isomers of each other. The butane represented by the straight chain is called a **normal alkane,** the other a **branched alkane.**

The number of possible isomers increases rapidly with the number of carbon atoms in an alkane. Three are possible for pentane, C_5H_{12}, the next member of the series:

$$CH_3-CH_2-CH_2-CH_2-CH_3 \qquad \begin{array}{ccccccc} & & & & CH_3 & & \\ & & & & | & & \\ CH_3 & - & CH_2 & - & CH & - & CH_3 \end{array} \qquad \begin{array}{ccccc} & & CH_3 & & \\ & & | & & \\ CH_3 & - & C & - & CH_3 \\ & & | & & \\ & & CH_3 & & \end{array}$$

Notice that

$$\begin{array}{ccccccc} & & & & CH_3 & & \\ & & & & | & & \\ CH_3 & - & CH_2 & - & CH & - & CH_3, \end{array} \quad \begin{array}{ccccccc} & & CH_3 & & & & \\ & & | & & & & \\ CH_3 & - & CH & - & CH_2 & - & CH_3, \end{array} \quad \begin{array}{ccccccc} CH_3 & - & CH & - & CH_2 & - & CH_3, \\ & & | & & & & \\ & & CH_3 & & & & \end{array}$$

and

$$\begin{array}{ccccccc} CH_3 & - & CH_2 & - & CH & - & CH_3 \\ & & & & | & & \\ & & & & CH_3 & & \end{array}$$

all represent the same structural formula merely flipped over or turned around on the page. Five isomers are possible for C_6H_{14}, 75 for $C_{10}H_{22}$, 366,319 for $C_{20}H_{42}$, and 4,111,846,763 for $C_{30}H_{62}$. Obviously, all of these isomeric compounds, though theoretically possible, have not been prepared, since only between three and four million organic compounds are known.

Compounds of the alkane series of hydrocarbons or the paraffin series, as they are sometimes called, can be represented by a general formula C_nH_{2n+2}. All members of this series are called **saturated compounds.** This term is used to describe any organic molecule that contains only single bonds between carbon atoms.

Nomenclature of the Alkanes

When only a relatively few compounds were known, it was the practice of chemists to give them what are today called trivial names. The names methane, ethane, propane, and butane are examples. However, as the number of compounds increased, it became more difficult to think up names and much more difficult to commit them to memory; obviously, a systematic method was needed. Such a method has been developed by the International Union of Pure and Applied Chemistry (IUPAC). Sometimes the application of the method leads to unwieldy results, but the names derived for each compound are unique and tell everything necessary to draw a structural formula. Rules for naming the alkanes based on the IUPAC system are listed below.

1. The longest continuous carbon-atom chain is chosen as the basis for the name. This chain is named according to the stem name plus the ending *-ane*, as given in the following series. The first four members of the series reflect a retention of the trivial name, while all subsequent stems are obtained from the Greek numerical prefixes. The stem is in bold type.

CH_4	**meth**ane	C_5H_{12}	**pent**ane	C_9H_{20}	**non**ane
C_2H_6	**eth**ane	C_6H_{14}	**hex**ane	$C_{10}H_{22}$	**dec**ane
C_3H_8	**prop**ane	C_7H_{16}	**hept**ane	$C_{11}H_{24}$	**undec**ane
C_4H_{10}	**but**ane	C_8H_{18}	**oct**ane	$C_{12}H_{26}$	**dodec**ane

2. The carbon atoms in the longest chain are numbered consecutively from the end of the chain nearest a branch.

$$\overset{4}{C}H_3-\overset{3}{C}H_2-\overset{2}{C}H(CH_3)-\overset{1}{C}H_3 \quad \text{and not} \quad \overset{1}{C}H_3-\overset{2}{C}H_2-\overset{3}{C}H(CH_3)-\overset{4}{C}H_3$$

3. Each branch is located by the number of the carbon atom to which it is attached on the chain. The branches are called **alkyl groups** and are named by using the stem of the parent alkane plus the ending *-yl*. The position and name of alkyl branches appear first in the final name.

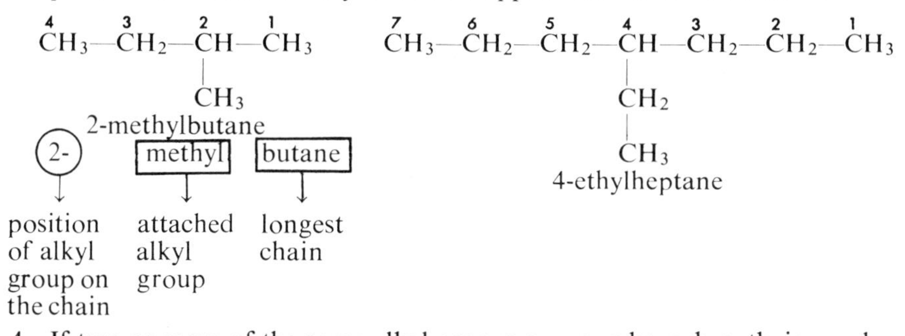

$$\overset{7}{C}H_3-\overset{6}{C}H_2-\overset{5}{C}H_2-\overset{4}{C}H(CH_2CH_3)-\overset{3}{C}H_2-\overset{2}{C}H_2-\overset{1}{C}H_3$$

4-ethylheptane

4. If two or more of the same alkyl groups occur as branches, their number is indicated by the prefixes di-, tri-, tetra-, penta-, etc., and the location of each is indicated as before by a number:

$$CH_3-CH(CH_3)-CH_2-CH(CH_3)-CH_3$$

2,4-dimethylpentane

$$CH_3-CH_2-CH(CH_2CH_3)-CH(CH_2CH_3)-CH_2-CH_2-CH_2-CH_3$$

3,4-diethyloctane

5. Numbers designating the positions of alkyl groups are placed immediately before the names of the groups, and hyphens are placed before and after the numbers when necessary. If two or more numbers occur together, commas are placed between them.
6. If two or more different alkyl groups are present, their names are placed in alphabetical order and prefixed onto the name of the basic alkane as before. The entire name is written as a single word.

Although the names obtained by this method are sometimes cumbersome, they are unique and specific for each compound. Let's apply the rules to the structural formula given below.

```
                         CH3
                          |
      CH3                CH2                CH3
       |                  |                  |
CH3—CH—CH—CH2—CH—CH—CH2—CH—CH2
            |                 |                  |
           CH3               CH3                CH2
                                                 |
                                                CH3
```

Step 1. The longest chain contains 11 carbon atoms, so the compound is an undecane.

Step 2. The numbers assigned to the carbon atoms of the longest chain are given below.

```
                         CH3
                          |
      CH3                CH2                CH3
 1    2|     3      4    5|     6     7     8|    9
CH3—CH—CH—CH2—CH—CH—CH2—CH—CH2
            |                 |                10|
           CH3               CH3                CH2
                                               11|
                                                CH3
```

Notice that bending the chain on the end does not eliminate the number 10 and 11 carbon atoms from being part of the longest chain. The chain is numbered from left to right, since the branch nearest an end is on the left.

Step 3. There is an ethyl group on carbon 5; it will appear in the entire name as 5-ethyl.

Step 4. There are four methyl groups on carbons 2, 3, 6, and 8; these will be designated 2,3,6,8-tetramethyl.

Step 5. Assembly of all components of the name gives

5-ethyl-2,3,6,8-tetramethylundecane

One must use these rules carefully to avoid obtaining incorrect, nonunique names. Especially be careful to choose the longest carbon chain and number it correctly. For example:

$$\begin{array}{c} \qquad\qquad\qquad\qquad\qquad\quad CH_3 \\ \qquad\qquad\qquad\qquad\qquad\quad | \\ CH_3-CH_2-CH_2-C-CH_3 \\ \qquad\qquad\qquad\qquad\qquad\quad | \\ \qquad\qquad\qquad\qquad\qquad\quad CH_3 \end{array}$$

is called 2,2-dimethylpentane (numbered right to left) and not 4,4-dimethylpentane (numbered left to right). Also,

$$\begin{array}{l} \qquad\qquad\qquad\quad CH_3 \\ \qquad\qquad\qquad\quad | \\ CH_3-CH-CH-CH_2-CH_2-CH_3 \\ \qquad\quad\; | \\ \qquad\quad\; CH_2 \\ \qquad\quad\; | \\ \qquad\quad\; CH_3 \end{array}$$

is 3,4-dimethylheptane and not 2-ethyl-3-methylhexane.

A structural formula can easily be constructed from a correct name, such as 3,3-diethyl-5-methyl-5-propyldecane.

Step 1. Draw a decane carbon skeleton and number it.

$$\overset{1}{C}-\overset{2}{C}-\overset{3}{C}-\overset{4}{C}-\overset{5}{C}-\overset{6}{C}-\overset{7}{C}-\overset{8}{C}-\overset{9}{C}-\overset{10}{C}$$

Step 2. Place two ethyl groups on carbon number 3.

$$C-C-\underset{\displaystyle CH_2CH_3}{\overset{\displaystyle CH_2CH_3}{\overset{|}{\underset{|}{C}}}}-C-C-C-C-C-C-C$$

Step 3. Place a methyl group on carbon number 5.

$$C-C-\underset{\displaystyle CH_2CH_3}{\overset{\displaystyle CH_2CH_3}{\overset{|}{\underset{|}{C}}}}-C-\overset{\displaystyle CH_3}{\overset{|}{C}}-C-C-C-C-C$$

Step 4. Place a propyl group on carbon number 5.

$$C-C-\underset{\displaystyle CH_2CH_3}{\overset{\displaystyle CH_2CH_3}{\overset{|}{\underset{|}{C}}}}-C-\underset{\displaystyle CH_2CH_2CH_3}{\overset{\displaystyle CH_3}{\overset{|}{\underset{|}{C}}}}-C-C-C-C-C$$

Step 5. Fill in with necessary hydrogen atoms so that each carbon atom has four bonds.

$$CH_3-CH_2-\underset{\displaystyle CH_2CH_3}{\overset{\displaystyle CH_2CH_3}{\overset{|}{\underset{|}{C}}}}-CH_2-\underset{\displaystyle CH_2CH_2CH_3}{\overset{\displaystyle CH_3}{\overset{|}{\underset{|}{C}}}}-CH_2-CH_2-CH_2-CH_2-CH_3$$

In discussions to follow, common names, names based on the IUPAC system, and names based on other systems will be used. When the IUPAC name is used, it will be underlined.

Physical Properties of Alkanes

The normal or straight-chain alkanes form a set of compounds called a **homologous series,** a name applied to any series in which each member differs from the previous member by only a CH_2 unit. The most striking physical properties of such a series are the melting and boiling points. Table 17–1 lists these properties for the normal alkanes.

Table 17–1. Melting and Boiling Points for Normal Alkanes

Name	*Melting point (°C)*	*Boiling point (°C)*
Methane	−182	−162
Ethane	−183	−89
Propane	−190	−45
Butane	−135	−1
Pentane	−130	36
Hexane	−95	69
Heptane	−91	98
Octane	−57	126
Nonane	−54	151
Decane	−30	171

Room temperature is about 20°C, so we see that all of these compounds are either liquids (pentane through decane) or gases (methane through butane) at room temperature.

Melting and boiling points are functions of molecular weight, size and shape, and polarity of molecules. These molecules are all nonpolar, and it is expected that the melting and boiling points should depend on weight, size, and shape. The data bear this out: both boiling and melting points increase with molecular weight. Some higher-molecular-weight members of the series would thus correctly be expected to be solids at room temperature.

Uses of Alkanes

Alkanes are used in the largest amounts in the form of mixtures, containing compounds of various molecular weights. Most of these mixtures are used as solvents or fuels. The primary source of alkanes is petroleum, a topic of Chapter 19, where a more detailed discussion of their uses is given.

The Alkenes and Alkynes

Two series of hydrocarbons contain fewer hydrogen atoms per carbon atom than the alkanes: the **alkenes** and **alkynes**. In these, each carbon atom must still possess four covalent bonds, and the lower H-to-C ratio makes multiple bonding between some carbon atoms a necessity. The term **unsaturated** is used to indicate compounds that contain such multiple bonds. In the alkene series at least two carbon atoms in a molecule share a double bond; in the alkyne series at least one triple bond is found per molecule. The structural formulas and names of the first, second, and third unbranched members of the alkene and alkyne series are:

$CH_2{=}CH_2$	$CH_2{=}CH{-}CH_3$	$CH_2{=}CH{-}CH_2{-}CH_3$
ethylene or ethene	propylene or 1-propene	butylene or 1-butene
$HC{\equiv}CH$	$CH_3{-}C{\equiv}CH$	$CH{\equiv}C{-}CH_2{-}CH_3$
acetylene or ethyne	1-propyne	1-butyne

Nomenclature of Alkenes and Alkynes

IUPAC names for these compounds are especially useful for branched and higher-molecular-weight members of these series. The rules are similar to those used for naming alkanes. The characteristic name endings are *-ene* for alkenes and *-yne* for alkynes. The root name is derived from the longest chain of carbon atoms containing the multiple bond. Chains are numbered so the carbon atoms connected by the multiple bond have numbers as low as possible. The multiple-bond position is indicated by the number of the lowest-numbered carbon atom bound by the multiple bond; this number is placed before the root name. Attached alkyl groups are handled as they were with alkanes. The few examples below illustrate the method.

$CH_3{-}CH{=}CH{-}CH_3$

2-butene

$CH_3{-}CH_2{-}C(CH_3){=}CH_2$

2-methyl-1-butene

$CH_3{-}CH_2{-}C(CH_2{-}CH_3){=}C(CH_2{-}CH_3){-}CH_2{-}CH_3$

3,4-diethyl-3-hexene

ALKENES

$CH_3{-}C{\equiv}C{-}CH_3$

2-butyne

$CH{\equiv}C{-}CH_2{-}C(CH_3)_2{-}CH_2{-}CH_3$

4,4-dimethyl-1-hexyne

$CH_3{-}C{\equiv}C{-}CH(CH_3){-}CH_3$

4-methyl-2-pentyne

ALKYNES

Uses of Alkenes and Alkynes

Alkenes and alkynes are used primarily as fuels and as starting materials for the preparation of other organic compounds. Acetylene or ethyne, for example, is used to produce a hot flame for welding. Ethylene or ethene is a starting material for many products, including plastics such as polyethylene. These and other uses are described further in later chapters.

Before leaving the topic of unsaturated hydrocarbons, we should note that some compounds contain more than one double bond per molecule. Molecules of this type are important components of natural and synthetic rubber and other useful materials.

The nomenclature of these compounds is the same as for the alkenes with one double bond, except that the ending *-diene*, *-triene*, etc. is used to denote the number of double bonds. Also, the locations of the bonds must be indicated, as in the examples below.

$$CH_2{=}\overset{\overset{\large CH_3}{|}}{C}{-}CH{=}CH_2 \quad CH_2{=}CH{-}CH{=}CH_2 \quad H_2C{=}\underset{\underset{\large CH_3}{|}}{C}{-}\overset{\overset{\large CH_3}{|}}{C}{=}CH{-}CH{=}CH_2$$

2-methyl-1,3-butadiene or Isoprene — 1,3-butadiene — 2,3-dimethyl-1,3,5-hexatriene

Cyclic Hydrocarbons

According to what has thus far been said about hydrocarbons, the formula C_3H_6 could not represent a saturated alkane. It appears that at least one multiple bond is needed to satisfy the requirement of four bonds per carbon atom. For example, the structural formula $CH_3{-}CH{=}CH_2$ would be satisfactory. No multiple bonds are necessary, however, if we allow the carbon atoms to form a ringlike or cyclic structure rather than the open-chain structure:

C—C—C (open chain) — C—C with both bonded to a third C below (cyclic)

The resulting saturated cyclic compound, called cyclopropane, has the structural formula

CH_2 at the apex of a triangle, bonded to H_2C and CH_2 at the base, which are bonded to each other: $H_2C{-}CH_2$

Both saturated and unsaturated cyclic hydrocarbons are known. They are named the same as open-chain hydrocarbons, except that *cyclo-* precedes the name of the corresponding open-chain hydrocarbon with the same number of carbon atoms as the ring.

For convenience, carbon rings are often represented by simple geometric figures: a triangle for a three-carbon ring, a square for a four-carbon ring, and so on. It must be remembered that each carbon atom still possesses four bonds. Hydrogen is assumed to be bonded to the carbon atoms unless something else is indicated. The examples below illustrate the nomenclature and geometrical representation of cyclic hydrocarbons.

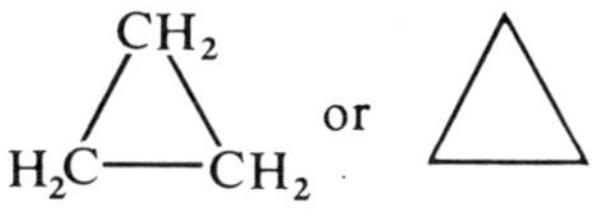

cyclopropane

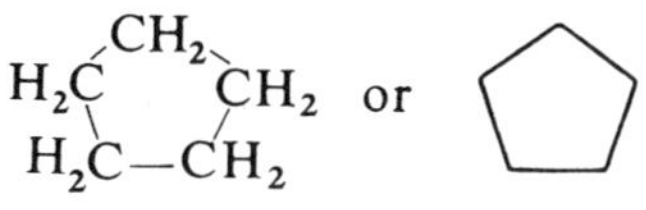

cyclopentane

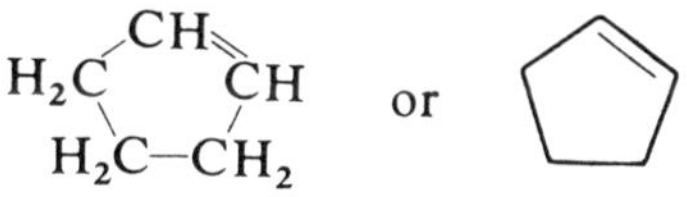

cyclopentene

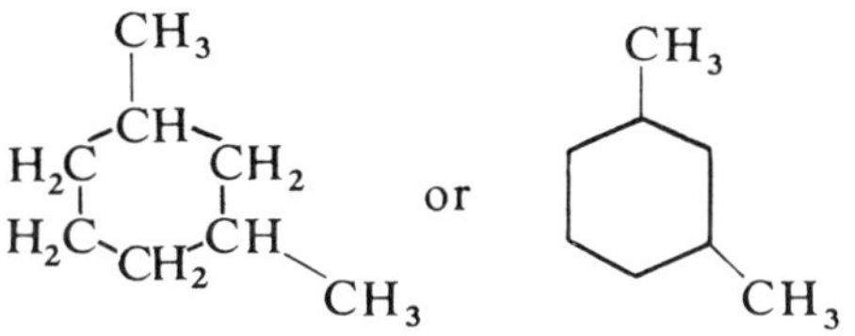

1,3-dimethylcyclohexane

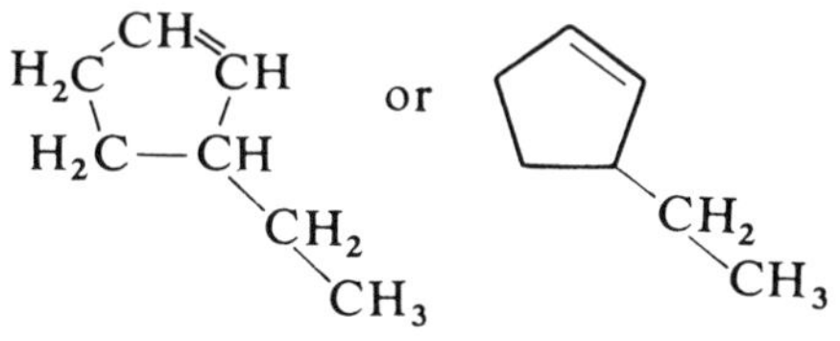

3-ethylcyclopentene

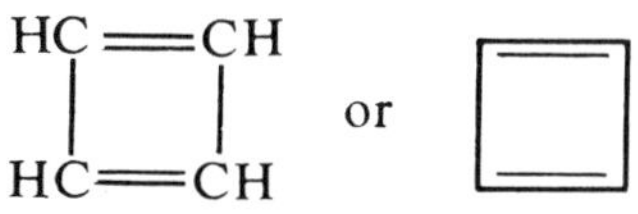

1,3-cyclobutadiene

Aromatic Hydrocarbons

The term **aromatic** was once used to describe a group of compounds possessing a characteristic fragrance. This original meaning has no significance when applied to aromatic compounds today. Many of these compounds are in fact vile smelling. Aromatic compounds include benzene and compounds with benzenelike structures that give them chemical properties similar to those of benzene.

Benzene has a formula, C_6H_6, obtained from chemical analysis and molecular weight determinations. The low hydrogen-to-carbon ratio suggests that the compound must contain some multiple bonds. The chemical properties, however, are quite different from those of compounds containing double or triple bonds. This problem plagued chemists for many years, and not until 1865 was the problem resolved. In that year F. A. Kekulé suggested that the benzene compound might be represented by a ring arrangement of carbon atoms. The idea came to him as he slept in front of his fireplace:

I was sitting writing at my textbook, but the work did not progress; my thoughts were elsewhere. I turned my chair to the fire, and dozed. Again the atoms were gamboling before my eyes. This time the smaller groups kept modestly in the background. My mental eye, rendered more acute by repeated visions of this kind, could now distinguish larger structures of manifold conformations; long rows, sometimes more closely fitted together; all twisting and turning in snake-like motion. But look! What was that? One of the snakes had seized hold of its own tail, and the form whirled mockingly before my eyes. As if by a flash of lightning I woke; . . . I spent the rest of the night working out the consequences of the hypothesis. Let us learn to dream, gentlemen, and then perhaps we shall learn the truth.

Kekulé proposed the following molecular structure for benzene:

```
          H
          |
    H    C    H
     \ /   \\ /
      C      C
      ||     |
      C      C
     / \    // \
    H    C     H
          |
          H
```

He later suggested that the double bonds alternate in their positions between carbon atoms to give two equivalent structures:

⇌

Figure 17–2.
Volumes for Free (Delocalized) Electron Movement in the Benzene Molecule

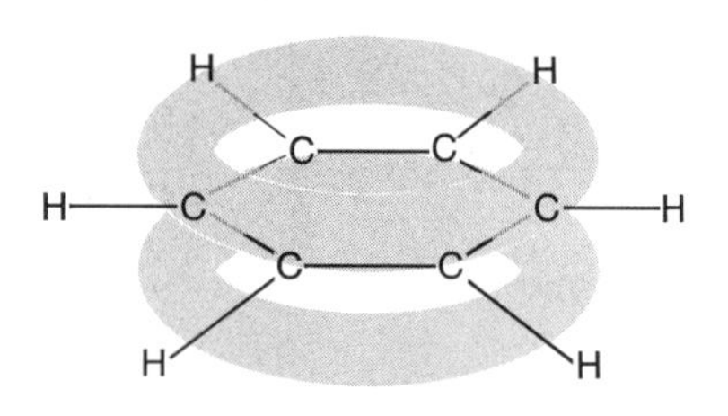

Remember that in these representations a carbon atom is at each point of the hexagon, and hydrogen atoms are bonded to carbon atoms to complete the necessary four carbon-atom bonds.

A modern interpretation of the benzene structure leads to the conclusion that six of the bonding electrons, rather than forming double bonds, are distributed equally throughout the molecule. They are quite free to move in a space or volume which can be represented as a pair of donuts, one below and one above the plane of the carbon atoms (Figure 17–2).

Because of the distribution of these electrons over the entire ring, the benzene structure is sometimes represented as

where the circle represents the evenly distributed, delocalized electrons. This interpretation of the structure enables chemists today to understand and explain quite well the chemical properties of benzene and other aromatic compounds.

Nomenclature of Aromatic Hydrocarbons

Many of the common compounds in the aromatic class have been given trivial names, some of which persist today. The general class name of the aromatic compounds is **arene,** and the removal of a hydrogen atom from an arene compound produces a fragment called an **aryl** group. In the specific case of benzene, the benzene ring with one hydrogen removed is called a **phenyl** group.

In those compounds in which a benzene ring is attached to a large chain, the term **phenyl** is used in the same way as methyl was used to represent a CH_3 group attached to a chain. Thus the compound

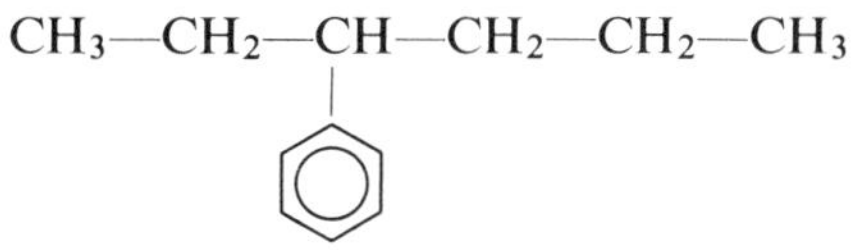

$$CH_3—CH_2—CH(C_6H_5)—CH_2—CH_2—CH_3$$

would be called 3-phenylhexane.

Molecules such as 3-phenylhexane may also be thought of as substituted benzene molecules, where a hydrogen atom of a ring carbon has been replaced by an alkyl or other group. These compounds (illustrated below) can be named as substituted benzenes, although some are well known by common names, some of which are approved by the IUPAC.

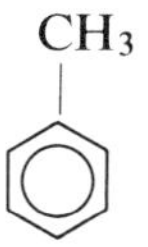

$C_6H_5—CH_3$: methylbenzene, or toluene

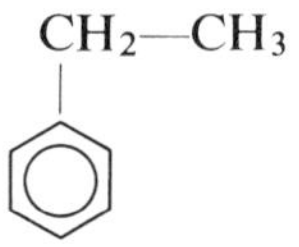

$C_6H_5—CH_2—CH_3$: ethylbenzene

$C_6H_5—CH_2—CH_2—CH_3$: propylbenzene

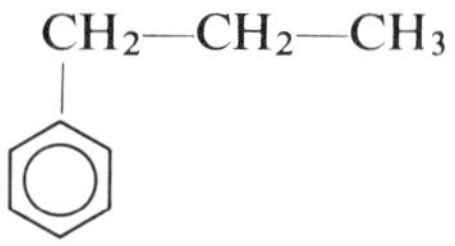

$C_6H_5—CH(CH_3)—CH_3$: isopropylbenzene, or cumene

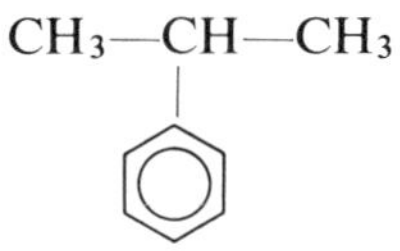

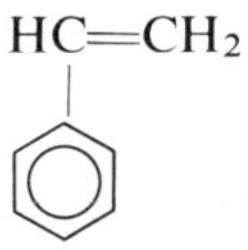

$C_6H_5—HC{=}CH_2$: vinylbenzene, or styrene

When two hydrogen atoms of ring carbons are replaced, the replacing groups can be arranged in several ways, yielding isomers. Three arrangements are possible, as shown below, where methyl and ethyl groups have been added. These compounds are named as disubstituted benzenes, with the group positions on the ring indicated by numbers or by the names **ortho** (groups on adjacent ring carbon atoms), **meta** (one ring carbon atom between groups), and **para** (two ring carbon atoms between groups).

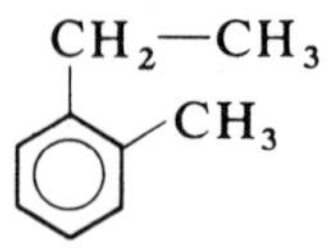

$CH_2{-}CH_3$, CH_3 (ortho)

2-methylethylbenzene, o-methylethylbenzene, or o-ethyltoluene

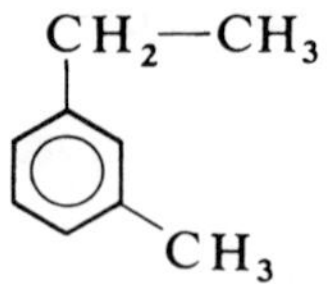

$CH_2{-}CH_3$, CH_3 (meta)

3-methylethylbenzene, m-methylethylbenzene, or m-ethyltoluene

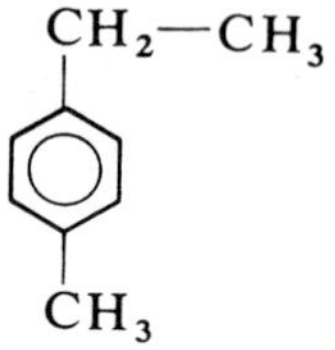

$CH_2{-}CH_3$, CH_3 (para)

4-methylethylbenzene, p-methylethylbenzene, or p-ethyltoluene

When three or more groups are attached to a benzene ring, the carbon atoms in the ring are given numbers, which are used to indicate the positions of the groups. The starting point for numbering is the carbon to which the largest group is attached, unless the compound is to be called by a common name, in which case the group upon which the common name is based will be carbon atom number 1. These examples will help clarify this idea:

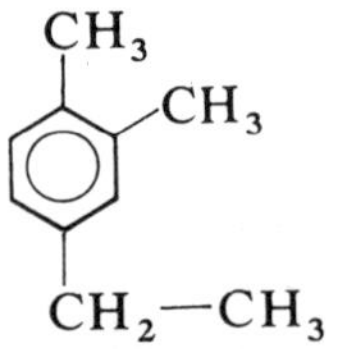

CH_3, CH_3, $CH_2{-}CH_3$

1-ethyl-3,4-dimethylbenzene, 3,4-dimethylethylbenzene, or 4-ethyl-2-methyltoluene

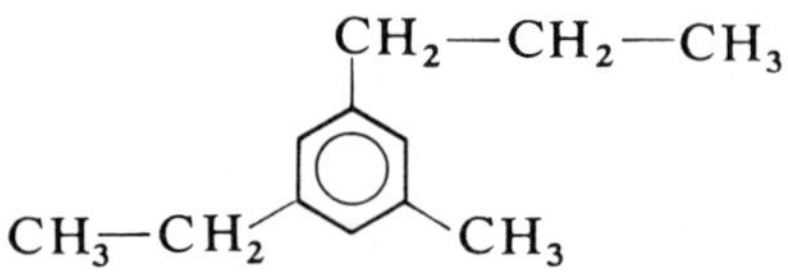

$CH_2{-}CH_2{-}CH_3$, $CH_3{-}CH_2$, CH_3

5-ethyl-3-methyl-1-propylbenzene, 5-ethyl-3-methylpropylbenzene, or 3-ethyl-5-propyltoluene

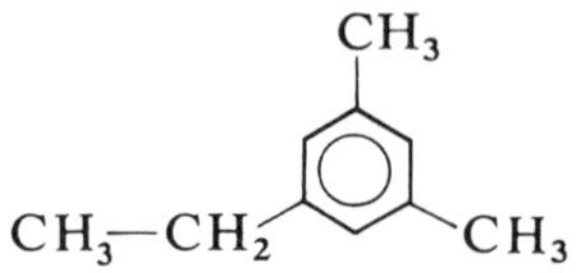

CH_3, $CH_3{-}CH_2$, CH_3

1-ethyl-3,5-dimethylbenzene, 3,5-dimethylethylbenzene, or 3-ethyl-5-methyltoluene

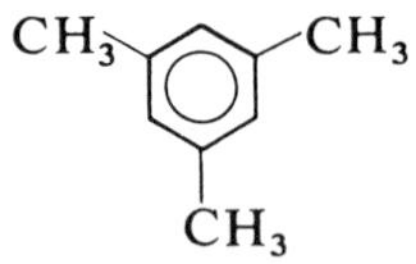

CH_3, CH_3, CH_3

1,3,5-trimethylbenzene, 3,5-dimethyltoluene, or mesitylene

Fused-Ring Compounds

A special group of aromatic compounds contain two or more benzene rings which share one or more sides in common. They are called **fused-ring, condensed-ring,** or **polynuclear** aromatic compounds. Examples are given in Table 17–2. As with all aromatics, substituted compounds are possible.

Table 17–2. **Fused-Ring Aromatic Hydrocarbons**

Structural formula	*Name*	*Some uses and characteristics*
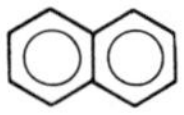	Naphthalene	Used in some mothballs.
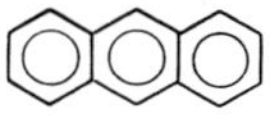	Anthracene	Fluoresces blue in ultraviolet light. Used in dyes.
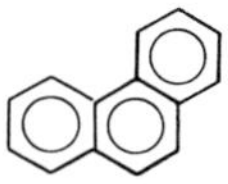	Phenanthrene	Used in dyes, explosives, and synthesis of drugs.
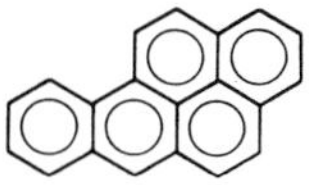	3,4-Benzpyrene	An active carcinogen (cancer-producing agent).
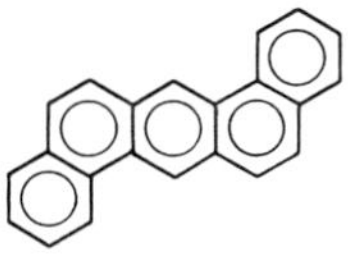	1,2,5,6-Dibenzanthracene	An active carcinogen.

Uses of Aromatic Hydrocarbons

Aromatic hydrocarbons are used as fuels, solvents, and as starting materials for the synthesis of a multitude of products, ranging from low-lead gasoline to broad-spectrum antibiotics. A number of these compounds and uses are discussed in later chapters.

Heterocyclic Compounds

Heterocyclics are not hydrocarbons, but we mention them here because of the close relationship between their ring structures and those of hydrocarbons.

A characteristic of **heterocyclic** compounds is a cyclic structure that includes in the ring some atom other than carbon. Some of the ring structures are aromatic; others are not. The most common noncarbon elements included in the ring are nitrogen, oxygen, and sulfur.

Heterocyclic structures are commonly found in compounds important to living organisms. A number of these compounds are discussed in the biochemistry part of this text (Chapters 23–26). Table 17–3 gives a sampling of heterocyclic compounds to illustrate the diversity of structures and characteristics involved.

Table 17-3. **Some Heterocyclic Compounds**

Structural formula	*Name*	*Characteristics*
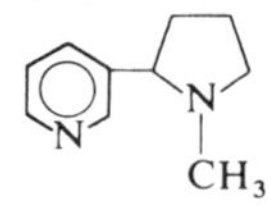	Nicotine	A highly toxic material found in tobacco.
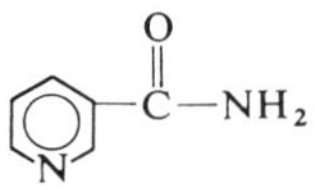	Nicotinamide	One of the B vitamins (niacin).
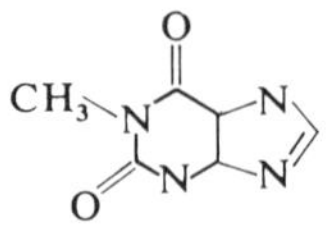	Caffeine	A stimulant found in coffee, tea, and cola drinks.
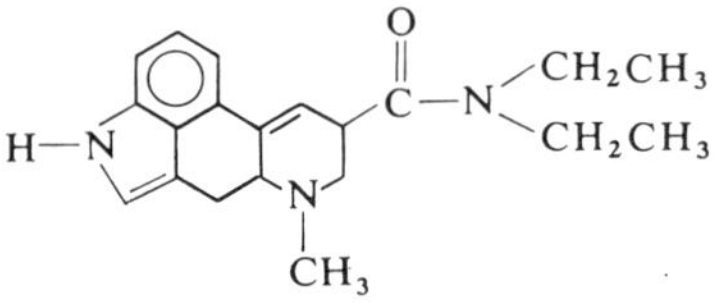	Lysergic acid diethylamide (LSD)	A hallucinogenic drug effective in very small doses. Side effects are not completely known.
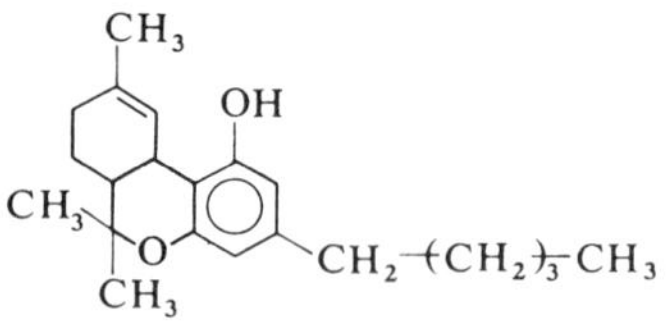	Tetrahydrocannabinol	Believed to be the major active constituent of marijuana.

Review Questions

1. Draw structural formulas and write the name of each of the following:
   a) A three carbon alkane
   b) A three carbon alkene
   c) A three carbon alkyne
2. The general formula for an alkane is given in this chapter as C_nH_{2n+2}. Write a similar formula for
   a) A mono-alkene (one double bond)
   b) A di-alkene (two double bonds)
   c) A mono-alkyne (one triple bond)
   d) A di-alkyne (two triple bonds)
   e) A cyclic alkane (no multiple bonds)
   f) A cyclic mono-alkene (one double bond)

3. What is an alkyl group?
4. Why is the name 2-ethyl-2,5-dimethylhexane not suitable for the compound

```
        CH3                  CH3
        |                    |
CH3—C—CH2—CH2—CH—CH3
        |
        CH2—CH3
```

5. Draw structural formulas for all isomers of the alkane C_6H_{14}. Name each isomer.
6. A compound has the formula C_4H_6. Draw structural formulas for compounds with the given formula that would be classified as
   a) An alkyne
   b) A di-alkene
   c) A cyclic alkene
7. Draw structural formulas for all the different molecules that could be produced by attaching one methyl and two ethyl groups to a benzene ring. Name each compound.
8. Draw structural formulas for all possible isomers of the mono-alkene (one double bond) C_4H_8. Name each isomer.
9. Name the following hydrocarbons:

a)
```
              CH3
              |
CH3—CH2—CH—CH3
```

b)
```
       CH2—CH3
       |
CH3—CH—CH—CH3
            |
            CH3
```

c)
```
        CH3
        |
CH2—CH
|       |
CH2—CH
        |
        CH3
```

d)
```
   CH2
  /  \
CH2—CH—CH2—CH2—CH3
```

e)
```
CH3—CH=CH—CH—CH3
              |
              CH3
```

f)
```
                CH3
                |
CH2=CH—C—CH2—CH3
                |
                CH3
```

g)
```
CH2=CH—CH=C—CH3
              |
              CH3
```

h)
```
   CH
  //   \
CH      C—CH2—CH3
  \     ||
   CH2—CH2
```

i)
```
      CH3             CH3
      |               |
CH3—CH—C≡C—CH—CH3
```

j)
```
              CH3
              |
CH3—CH2—CH—C≡C—CH—CH3
                    |
                    CH3
```

10. Name the following aromatic hydrocarbons. If more than one name is used, underline those acceptable in the IUPAC system.

a) 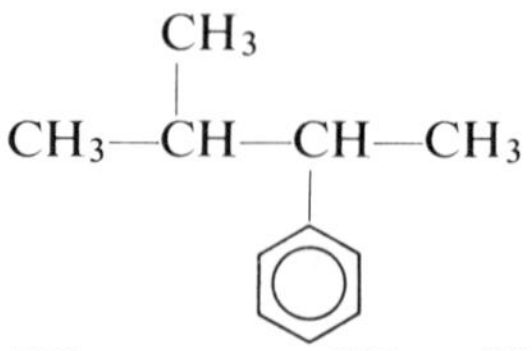 CH_3 | $CH_3—CH—CH—CH_3$ (with a benzene ring on the second CH)

b) 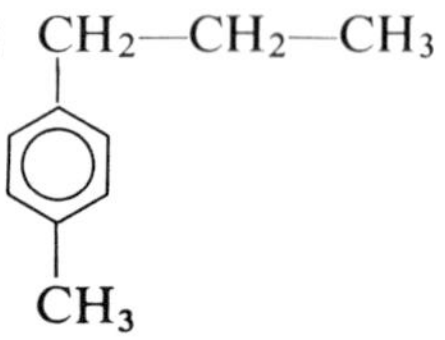 $CH_2—CH_2—CH_3$ on a benzene ring, with CH_3 para

c) 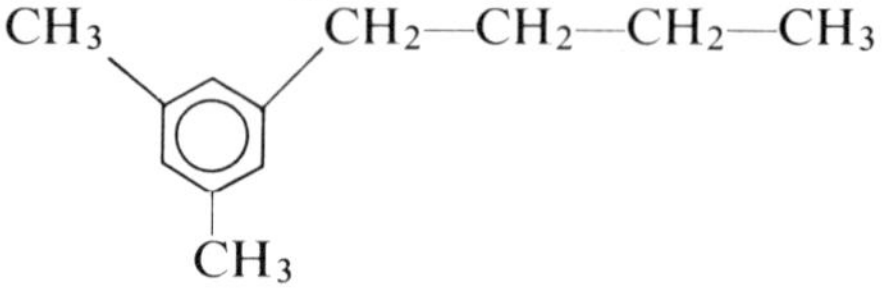 CH_3, $CH_2—CH_2—CH_2—CH_3$ and CH_3 on a benzene ring

d) 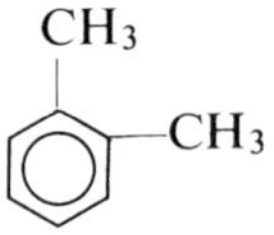 CH_3 and CH_3 on adjacent positions of a benzene ring

e) 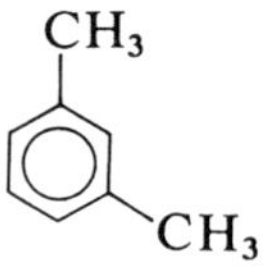 CH_3 and CH_3 on a benzene ring (meta)

f) 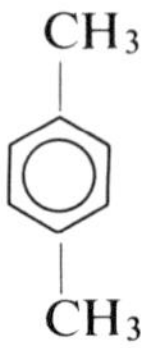 CH_3 and CH_3 on a benzene ring (para)

11. Draw structural formulas for the following compounds:
    a) 2,4-dimethylnonane
    b) 4-ethyl-3-heptene
    c) 4-ethyl-3,3-dimethyl-5-propyldodecane
    d) 4,4-diethyl-1-hexyne
    e) 4-methyl-3-phenylheptane
    f) ethylcyclohexane
    g) 3-ethyl-1,4-pentadiene
    h) 3-methylcyclopentene
    i) 1,3-dimethylcyclobutane
    j) 1-methyl-3-propylbenzene

Suggestions for Further Reading

Lambert, J. R., "The Shapes of Organic Molecules," *Scientific American*, Jan. 1970.

Mills, G. A., "Ubiquitous Hydrocarbons," *Chemistry*, Feb. 1971.

Von Tamelen, E. E., "Benzene, The Story of Its Formula, 1865–1965," *Chemistry*, Jan. 1965.

18 Functional Groups and Organic Reactions

In light of the very large number of compounds involved, the study of organic reactions might seem an almost hopeless activity. Each compound can presumably undergo a variety of reactions to produce numerous products: how can anyone hope to remember all of the possible reactions and products? This apparent difficulty is much reduced by the introduction of the concept of functional groups.

Functional Groups

With the exception of saturated hydrocarbons, each organic molecule contains some atom or group of atoms that defines the class to which it belongs and, at the same time, determines its chemistry (reactions). The unsaturated hydrocarbons, studied in the preceding chapter, have such functional groups. The C═C and C≡C multiple bonds are responsible for the characteristic reactions of alkenes and alkynes, and we also remember that the presence of these bonds led to the classification of each compound as an alkene or alkyne.

When an organic molecule undergoes a reaction, only a small portion, the functional group, is usually changed. Therefore, by remembering the behavior of one functional group in a reaction, we may predict the products of many reactions. For example, we observe that a mixture of ethyl alcohol and hydrogen bromide reacts to give water and ethyl bromide:

$$CH_3{-}CH_2{-}\underset{\text{functional group}}{\boxed{OH}} + HBr \rightarrow H_2O + CH_3{-}CH_2{-}\underset{\text{new functional group}}{\boxed{Br}}$$

The functional groups involved are discussed later in this chapter. Let us note an important fact however: in this reaction only the functional groups changed. The alkyl part of each molecule remains the same.

From this reaction and the functional-group concept we can write a general reaction:

$$R{-}OH + HBr \rightarrow H_2O + R{-}Br \qquad 18\text{–}2$$

and assume the reaction will occur as long as R is a straight-chain alkyl group. If propyl alcohol is used, propyl bromide is the product.

A large part of organic chemistry is concerned with the transformation of one functional group into another, just as we transformed an alcohol into a bromide. In this chapter we shall study a few of these transformations. We shall generally limit our discussion to those transformations needed to understand the material contained in the chapters that follow.

Functional groups, other than those of the hydrocarbons, contain elements other than carbon and hydrogen; the most common are oxygen, nitrogen, and sulfur. The functional groups we shall include in this discussion are

Table 18–1. Functional Groups

Functional group name	*Characteristic group*	*General formula*	*Specific example*	*Example names*
Alcohol	—OH	R—OH	CH_3—OH	Methyl alcohol, methanol
Aldehyde	—C(=O)H	R—C(=O)H	CH_3—C(=O)H	Acetaldehyde, ethanal
Amide*	—C(=O)—NH_2	R—C(=O)—NH_2	CH_3—C(=O)—NH_2	Acetamide, ethanamide
Amine*	—NH_2	R—NH_2	CH_3—NH_2	Methyl amine, aminomethane
Carboxylic acid	—C(=O)—OH	R—C(=O)—OH	CH_3—C(=O)—OH	Acetic acid, ethanoic acid
Ester	—C(=O)—O—	R—C(=O)—O—R′	CH_3—C(=O)—O—CH_2—CH_3	Ethyl acetate, ethyl ethanoate
Ether	—O—	R—O—R′	CH_3—CH_2—O—CH_2—CH_3	Diethyl ether, ethoxyethane
Halide	—X (Cl, Br, I)	R—X	CH_3Br	Methyl bromide, bromoethane
Ketone	—C(=O)—	R—C(=O)—R′	CH_3—C(=O)—CH_3	Acetone, propanone
Sulfonic acid	—S(=O)(=O)—OH	R—S(=O)(=O)—OH	C_6H_5—S(=O)(=O)—OH	Benzenesulfonic acid

* An R group (alkyl or aryl) may replace one or more of the H's in amides and amines.

given in Table 18–1 along with general formulas, specific examples, and names. IUPAC names are in color as before. The abbreviations R and R′ correspond to either alkyl or aryl groups.

Alcohols

These compounds, containing the —OH functional group, vary in complexity from the simplest, which contains a single carbon atom, to some containing hundreds of carbon atoms or more. They can, in many cases, be thought of as hydrocarbons with one or more hydrogens replaced by the —OH group. The IUPAC name of alcohols containing a single —OH group is obtained by dropping the final *e* and adding the suffix *-ol* to the name of the hydrocarbon containing the same number of carbon atoms. The position of the —OH group is indicated by the number of the carbon atom to which it is attached. Table 18–2 gives examples of nomenclature; these names and common names will be used in this discussion.

The simplest alcohol is methanol, methyl alcohol, or wood alcohol with the formula CH_3OH. For many years the principal source of this compound was wood distillation, hence the name wood alcohol. Today, it is synthetically produced in large quantities through a reaction of hydrogen gas and carbon monoxide:

$$CO + 2H_2 \rightarrow CH_3OH \qquad 18\text{–}3$$

The reaction is carried out in the gas phase in the presence of a catalyst. The compound is very poisonous, causing blindness and death. It is used as a solvent for varnishes and shellacs and as a denaturant for ethyl alcohol.

The second member of the group, ethanol, ethyl alcohol, or grain alcohol, CH_3CH_2OH, is probably the most familiar because it is the essential ingredient in alcoholic beverages. It is also used in the pharmaceutical industry as a solvent (tinctures are ethanol solutions) and medicinal ingredient, and in industry as a solvent and antifreeze.

Ethyl alcohol has a low toxicity, but long-term, excessive use is known to cause undesirable effects such as cirrhosis of the liver. When ethyl alcohol is destined for beverage use, it is carefully controlled and heavily taxed. Any that is used industrially is essentially tax-free but is often denatured to prevent it from being ingested. **Denaturing** is accomplished by the addition of some agent disagreeable to the taste and possibly poisonous. Methyl alcohol is sometimes used as a denaturant.

Most ethyl alcohol is produced by yeast fermentation of liquids containing simple sugars or carbohydrates, such as starch or cellulose. A reaction for the fermentation of glucose, grape sugar, is

$$\underset{\text{glucose}}{C_6H_{12}O_6} \xrightarrow{\text{yeast}} \underset{\text{ethyl alcohol}}{2C_2H_5OH} + 2CO_2 \qquad 18\text{–}4$$

In addition to ethyl alcohol the fermentation produces some higher molecular weight alcohols in small quantities. These compounds, mostly isoamyl alcohol, $C_5H_{11}OH$, are collectively called **fusel oil.**

Ethyl alcohol destined for nonbeverage uses is distilled (see Chapter 19) from the fermentation mixture as a solution containing 96 percent ethanol and 4 percent water. For some applications it is further purified up to 100 percent.

Alcoholic beverages produced by fermentation are of two types, undistilled and distilled. The principal undistilled beverages are beer, champagne, and wine. Undistilled beverages have a much lower alcohol content than the distilled. Included among the distilled beverages are whiskey, gin, rum, and brandy.

Beer is manufactured by fermenting malted (germinated) grain. It usually contains 3 to 5 percent alcohol. The addition of hops, the dried cones of a special vine, imparts a bitter taste to the product.

Wine and champagne are fermented grape juice. The alcoholic content of natural wines, produced by the fermentation process alone, is usually no higher than 15 percent. Fermentation stops when the ethanol concentration reaches that level. This corresponds to a **proof** value of 30, the proof being double the alcoholic percentage by volume. Some wines contain added alcohol. These **fortified wines** may have alcoholic contents as high as 20 percent (40 proof). Champagne is wine bottled so that it retains dissolved carbon dioxide. Some low-cost champagnes are artificially carbonated, much like a soft drink.

The distilled beverages, including whiskey, brandy, and gin, are prepared from various fermented materials. Whiskey is obtained from grains, brandy from fruit juices, and rum from molasses or sugar-cane juice. Gin is prepared by distilling alcohol with juniper berries and adding other flavoring agents. These beverages have alcohol concentrations which are typically quite high—86 proof or 43 percent, for example.

Compounds containing more than one —OH group are also classed as alcohols. Among these are ethylene glycol, used as an antifreeze, and glycerine, used in cosmetics and as a raw material for some explosives. Notice the use of the *-diol* and *-triol* suffixes in their IUPAC names, as given in Table 18–2.

Table 18–2. Examples of Alcohols

Names	*Structural formula*	*Characteristics and typical uses*
Isopropyl alcohol, 2-propanol	$CH_3—CH—CH_3$ | OH	Rubbing alcohol, solvent
n-Butyl alcohol, butanol	$CH_3—CH_2—CH_2—CH_2—OH$	Solvent, hydraulic fluid
Tertiary amyl alcohol, 2-methyl-2-butanol	CH_3 | $CH_3—CH_2—C—CH_3$ | OH	Solvent, medicines, flotation agent
Phenol	OH | [OH on benzene ring]	Germicide, component of some plastics. Reactions are quite different from those of nonaromatic alcohols.
Ethylene glycol, 1,2-ethanediol	$CH_2—OH$ | $CH_2—OH$	Automobile antifreeze
Glycerin, 1,2,3-propanetriol	$CH_2—OH$ | $CH—OH$ | $CH_2—OH$	Cosmetics, pharmaceuticals, inks, candy

Amines

Amines, the important basic-behaving compounds of organic chemistry, can be thought of as alkyl derivatives of the inorganic base ammonia, NH_3. A variety of amines results when one or more of the hydrogen atoms of ammonia is replaced by an alkyl group.

$$\underset{\text{ammonia}}{H-\overset{H}{\overset{|}{N}}-H} \xrightarrow[\text{by R}]{\text{replace H}} \underset{\text{an amine}}{R-\overset{H}{\overset{|}{N}}-H} \xrightarrow[\text{by R}']{\text{replace H}} \underset{\text{an amine}}{R-\overset{R'}{\overset{|}{N}}-H} \xrightarrow[\text{by R}'']{\text{replace H}} \underset{\text{an amine}}{R-\overset{R'}{\overset{|}{N}}-R''} \quad 18\text{–}5$$

These relatives of ammonia are often associated with animal wastes or decay products, and in some cases the characteristic odor of such material is partly due to the presence of amines. Amines play important roles in the chemical processes of living organisms (see Chapters 24 and 25).

Names for amines may be obtained in two ways according to the IUPAC system. The alkyl group involved can be named and followed by *-amine*, or the amine group can be treated as a substituent on an alkane or other hydrocarbon. In the latter method, *amino-* is used as a prefix on the hydrocarbon name. Table 18–3 gives examples.

Table 18–3. Examples of Amines

Names	*Structural formula*	*Characteristics and typical uses*
Methylamine, aminomethane	CH_3-NH_2	Characteristic fish odor
Putrescine, 1,4-diaminobutane	$H_2N-CH_2-CH_2-CH_2-CH_2-NH_2$	Formed from decaying flesh, unpleasant odor
Cadaverine, 1,5-diaminopentane	$\underset{NH_2}{\underset{\vert}{CH_2}}-CH_2-CH_2-CH_2-\underset{NH_2}{\underset{\vert}{CH_2}}$	Formed from decaying flesh, unpleasant odor
Trimethylamine	$CH_3-\overset{CH_3}{\overset{\vert}{N}}-CH_3$	Present in fish—characteristic odor
Aniline, aminobenzene	$C_6H_5-NH_2$ (benzene ring with NH_2)	The basic starting material for many dyes

Halides

The compound that results when one or more hydrogen atoms of a hydrocarbon are replaced with halogen atoms is called an organic halide. These compounds are very rare in nature, but many have been synthesized by man. Halides are named by adding the prefix *chloro-*, *bromo-*, etc. to the name of the parent hydrocarbon. The number of added halide atoms is indicated by an appropriate prefix as before. Examples are given in Table 18–4.

Table 18–4. Examples of Organic Halides

Names	*Structural formula*	*Characteristics and typical uses*
Carbon tetrachloride, tetrachloromethane	CCl_4	Good solvent
Chloroform, trichloromethane	$CHCl_3$	Good solvent, general anesthetic
Freon-12, dichlorodifluoromethane	CF_2Cl_2	Refrigerant fluid, propellant for nonfood aerosol products
Halothane, 2-bromo-2-chloro-1,1,1-trifluoroethane	Br F HC—C—F Cl F	General inhalation anesthetic
Paradichlorobenzene, 1,4-dichlorobenzene	Cl (benzene ring) Cl	A common mothball ingredient
Lindane, hexachlorocyclohexane	Cl Cl Cl (cyclohexane ring) Cl Cl Cl	Agricultural insecticide

Carboxylic Acids

Carboxylic acids behave in many ways like inorganic acids, such as giving H_3O^+ ions in solution and reacting with bases to form salts. In these reactions they generally behave as weak acids. Like the inorganic acids, they have a sour taste and are often responsible for sour tastes in foods. The higher-molecular-weight acids have pungent, sometimes disagreeable odors and are the source of such odors in some foods.

In the IUPAC system, carboxylic acids are named by dropping the *e* and adding the suffix *-oic* to the name of the parent hydrocarbon containing the same total number of carbon atoms. This resulting name is then followed by *acid.* Table 18–5 gives examples.

Perhaps the most familiar form of organic acids or their products is **soap.** A true soap is a salt, usually sodium, of a long-chain acid. Stearic acid is commonly used, since it is quite available—it is obtained from animal fat. A reaction between stearic acid and sodium hydroxide will yield a soap:

$$\underset{\text{stearic acid}}{CH_3(CH_2)_{16}C(=O)OH} + NaOH \rightarrow \underset{\text{soap}}{CH_3(CH_2)_{16}C(=O)ONa} + H_2O \qquad 18\text{–}6$$

Table 18–5. Examples of Carboxylic Acids

Names	*Structural formula*	*Characteristics and typical uses*
Formic acid, methanoic acid	$HC(=O)OH$	Stinging agent of red ants, used in medicine and food preservation
Acetic acid, ethanoic acid	$CH_3{-}C(=O)OH$	Active ingredient in vinegar, used as a solvent and in photography
Propionic acid, propanoic acid	$CH_3CH_2C(=O)OH$	Salts used as mold inhibitors and emulsifiers
Butyric acid, butanoic acid	$CH_3CH_2CH_2C(=O)OH$	Odor-causing agent in rancid butter; flavoring agent
Caproic acid, hexanoic acid	$CH_3(CH_2)_4C(=O)OH$	Characteristic odor of limburger cheese; used as a flavoring agent
Capric acid, decanoic acid	$CH_3(CH_2)_8C(=O)OH$	Characteristic odor of goats
Oxalic acid, ethanedioic acid	$HO(O=)C{-}C(=O)OH$	Poisonous material in leaves of some plants such as rhubarb; used as a cleaning agent
Citric acid, 2-hydroxy-1,2,3-propanetrioic acid	$CH_2{-}C(=O)OH$ / $HO{-}C{-}C(=O)OH$ / $CH_2{-}C(=O)OH$	Present in citrus fruits; used as flavoring agent in foods
Benzoic acid	$C_6H_5{-}C(=O)OH$	Acid or salts used as a food preservative

The soap will ionize in water, giving a stearate anion and a sodium cation,

$$CH_3(CH_2)_{16}C(=O)ONa \rightarrow CH_3(CH_2)_{16}C(=O)O^- + Na^+ \qquad 18\text{–}7$$

The anion is responsible for the cleaning action of soap, which was discussed earlier (Chapter 10).

Sulfonic Acids

Sulfonic acids are also important organic acids. Unlike carboxylic acids, these compounds behave as strong acids in ionization, neutralization, and other typical acid reactions. They are usually encountered as aromatic compounds. Names are derived by adding the suffix *-sulfonic acid* to the name of the compound to which the $-SO_3H$ group is attached as in the examples below.

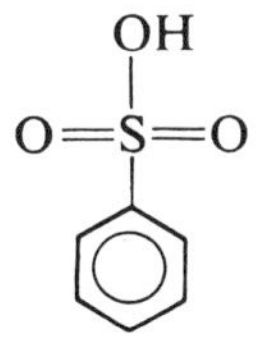

Benzenesulfonic acid

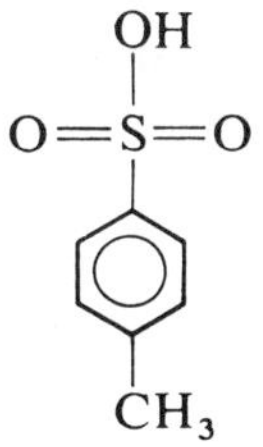

p-Methylbenzenesulfonic acid

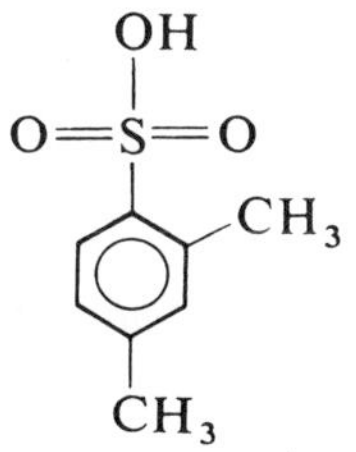

2,4-Dimethylbenzenesulfonic acid

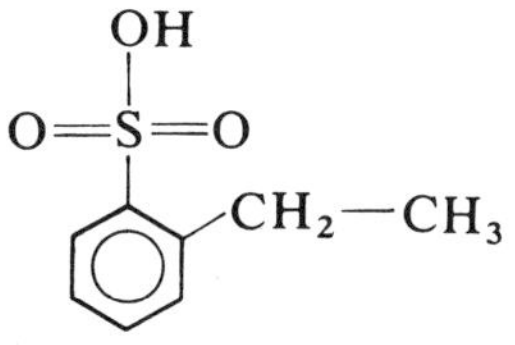

o-Ethylbenzenesulfonic acid

Sulfonic acids are the parent compounds for two widely used products: sulfa drugs and laundry detergents. Sulfa drugs are discussed in Chapter 26. Detergents are salts of long-chain sulfonic acids; a typical formula is

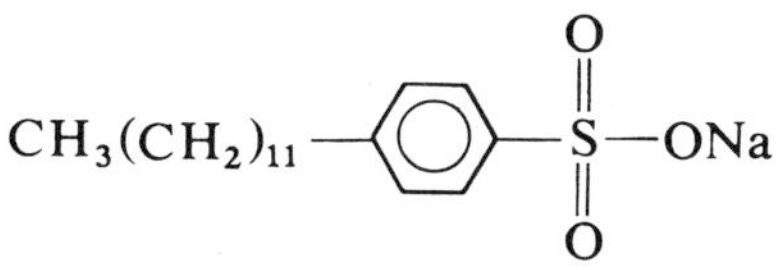

$$CH_3(CH_2)_{11}-C_6H_4-S(=O)_2-ONa$$

Sodium dodecylbenzenesulfonate

Detergents work very well as cleaning agents. Unfortunately, the bacteria in nature which had previously metabolized and disposed of soaps placed in natural waters refused to do the same with the early detergents. As a result, the concentration of these "hard" detergents began to increase in rivers and other natural waters. The housewives in cities on the lower end of rivers could almost do their laundry in water as it came from the tap. A glass of water in some areas began to be served with a detergent foam head. This problem was solved with the introduction of "soft" (biodegradable) detergents. The only difference between the original hard detergents and the soft is that the chain attached to the benzene ring is straight or normal in the soft and branched in the hard detergents. The formula given above is that of a soft detergent.

Aldehydes and Ketones

Both aldehydes and ketones possess an oxygen atom bonded to a carbon atom by a double bond. This structure is located at the end of the carbon chain in aldehydes but may occur on any carbon except end ones in ketones. Most chemical reactions characteristic of these compounds involve this oxygen-carbon bond.

Aldehydes are named in the IUPAC system by dropping the *e* and adding the suffix *-al* to the name of the parent hydrocarbon. The characteristic ending for ketone names in this system is *-one*, and the position of the oxygen is indicated by the number of the carbon atom to which it is bonded. Examples of these compounds, their names, and uses are given in Table 18–6.

Esters

Esters are important derivatives of carboxylic or other organic acids, and alcohols. The reaction for their formation is given later. Esters are widely distributed in nature and are especially abundant in fats, oils, and animal and vegetable waxes. Unlike the acids involved in their formation, simple esters often have pleasant odors, and some are responsible for the tastes and odors of fruits.

A general ester formula is

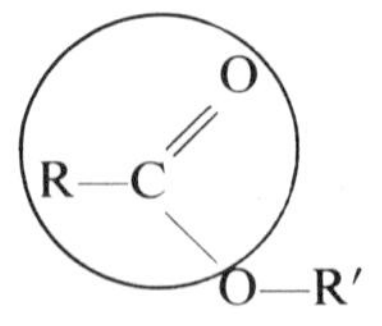

where the circled portion is considered the acid part of the molecule and the —O—R′ is the alkyl or aryl part which came from the alcohol.

In the naming of an ester, the alkyl or aryl name is given first and followed by the name of the acid part. In the IUPAC system, the *-ic* ending of the acid is replaced by *-ate*. Table 18–7 gives some examples.

Table 18–6. Examples of Aldehydes and Ketones

Names	*Structural formula*	*Characteristics and typical uses*
Formaldehyde, methanal	$\overset{O}{\overset{\|}{CH_2}}$	Plastics, germicide, preservative, embalming fluid
Acetaldehyde, ethanal	$CH_3\overset{O}{\overset{\|}{C}}H$	Starting material for other compounds including drugs and perfumes
Isobutyraldehyde, 2-methylpropanal	$CH_3{-}\overset{CH_3}{\overset{\vert}{C}}H{-}\overset{O}{\overset{\|}{C}}H$	Used in synthesis of other organic compounds
Cinnamaldehyde	$C_6H_5{-}CH{=}CH{-}\overset{O}{\overset{\|}{C}}H$	Causes odor and flavor of cinnamon bark
Citral	$(H_3C)_2C{=}CH{-}CH_2{-}CH_2{-}C(CH_3){=}CH{-}\overset{O}{\overset{\|}{C}}H$	Found in rinds of citrus fruits, characteristic odor of lemons
Acetone, 2-propanone	$CH_3{-}\overset{O}{\overset{\|}{C}}{-}CH_3$	An important solvent, also used to prepare other organic compounds
Methylethyl ketone, 2-butanone	$CH_3{-}\overset{O}{\overset{\|}{C}}{-}CH_2{-}CH_3$	A solvent, also used in manufacture of smokeless powder, dyes, lacquer removers
Menthone	$(H_3C)_2CH{-}C_6H_8O{-}CH_3$ (cyclohexanone ring)	A ketone found in mint leaves

Amides

Amides, like esters, are derivatives of organic acids. They result when an acid reacts with ammonia or an amine under proper conditions. In the reaction with ammonia, the —OH of the acid is replaced by an $—NH_2$ (amino) group. The resulting bond between carbon and nitrogen is called an **amide linkage.** This bond is very stable and is found in proteins as well as nylon and other synthetic polymers (Chapters 20 and 24).

Amides are named as acid derivatives. The *-ic* ending of the common name or *-oic* ending of the IUPAC name is dropped and replaced by the suffix *-amide*. Groups other than hydrogen attached to the nitrogen are indicated as *N-alkyl* or *aryl*. Table 18–8 gives examples.

Table 18–7. Some Simple Esters

Names	*Structural formula*	*Characteristics and typical uses*
Ethyl formate, ethylmethanoate	$HC(=O)-O-CH_2-CH_3$	Rum flavor and odor
Isobutyl formate, 2-methyl propylmethanoate	$HC(=O)-O-CH_2-CH(CH_3)-CH_3$	Raspberry flavor and odor
Ethylacetate, ethylethanoate	$CH_3-C(=O)-O-CH_2-CH_3$	Solvent, fingernail polish remover
n-Pentyl acetate, *n*-pentylethanoate	$CH_3-C(=O)-O-(CH_2)_4-CH_3$	Banana flavor and odor
n-Pentyl propionate, *n*-pentylpropanoate	$CH_3-CH_2-C(=O)-O-(CH_2)_4-CH_3$	Apricot flavor and odor
Ethyl butyrate, ethylbutanoate	$CH_3-(CH_2)_2-C(=O)-O-CH_2-CH_3$	Used in artificial peach, pineapple, and apricot flavors

Ethers

Ethers can be thought of as derivatives of water, in which both hydrogen atoms have been replaced by an alkyl or aryl group. Alternatively, they may be considered to be alcohols with the hydroxyl hydrogen replaced by an alkyl or aryl group.

These compounds are commonly named by naming the two groups attached to the oxygen and following with *ether*—methyl ethyl ether, for example. IUPAC names are obtained by naming ethers as alkoxy derivatives of a parent molecule—methoxyethane, for example. Table 18–9 gives examples.

Table 18–8. Examples of Amides

Names	*Structural Formula*	*Characteristics and typical uses*
Formamide, methanamide	$H-C(=O)-NH_2$	Good solvent
Acetamide, ethanamide	$CH_3-C(=O)-NH_2$	Soldering flux, wetting agent, solvent
Acetanilide, *n*-phenyl ethanamide	$CH_3-C(=O)-N(H)-C_6H_5$	A drug
Xylocaine	$(CH_3-CH_2)_2N-CH_2-C(=O)-N(H)-C_6H_3(CH_3)_2$	Local anesthetic
Lysergic acid diethylamide, LSD	$H-N$ (ring system) $N-CH_3$; $-C(=O)-N(CH_2CH_3)_2$	Hallucinogenic drug

Table 18–9. Ether Examples

Names	*Structural formula*	*Characteristics and typical uses*
Dimethyl ether, methoxymethane	CH_3-O-CH_3	Solvent, refrigeration medium; flammable
Ethyl ether, diethyl ether, ethoxyethane	$CH_3-CH_2-O-CH_2-CH_3$	General inhalation anesthetic, solvent; very flammable
Methyl-*n*-propyl ether, methoxypropane	$CH_3-O-CH_2-CH_2-CH_3$	Solvent
Anisole, methyl phenyl ether, methoxybenzene	$CH_3-O-C_6H_5$	Solvent, used in perfumery, vermicide
Diphenyl ether, phenoxybenzene	$C_6H_5-O-C_6H_5$	Used in perfumery, heat-transfer medium

Organic Reactions

Molecules containing functional groups undergo numerous reactions. However, we will be concerned only with a few that will be useful in subsequent discussions.

In a number of the reactions that follow, two functional groups react with each other, and the reactions are considered to be characteristic of both groups. It should also be noted that more than one set of products can be produced from a given set of reactants. The products formed often depend upon the reaction conditions, such as temperature, pressure, or the presence of catalysts. These conditions, when important, will be indicated.

Reactions of Alkanes

Only one reaction is given here for alkanes. They undergo combustion to give carbon oxides and water.

$$2CH_3{-}CH_2{-}CH_2{-}CH_3 + 13O_2 \xrightarrow[O_2]{\text{excess}} 8CO_2 + 10H_2O \qquad 18\text{–}8$$

If excess oxygen is not available, a mixture of carbon monoxide and carbon dioxide is usually formed.

Reactions of Alkenes and Alkynes

These compounds, like the alkanes, undergo combustion. Also, they can be hydrogenated under appropriate conditions to give alkenes and alkanes.

$$\underset{\text{alkyne}}{R{-}C{\equiv}CH} \xrightarrow[\text{Ni cat.}]{H_2} \underset{\text{alkene}}{R{-}\overset{H}{\overset{|}{C}}{=}\overset{H}{\overset{|}{C}}H} \xrightarrow[\text{Ni cat.}]{H_2} \underset{\text{alkane}}{R{-}\underset{H}{\underset{|}{\overset{H}{\overset{|}{C}}}}{-}\underset{H}{\underset{|}{\overset{H}{\overset{|}{C}}}}H} \qquad 18\text{–}9$$

Reactions of Carboxylic Acids

Carboxylic acids react with alcohols, amines (or ammonia), and bases to give esters, amides, and salts, respectively.

$$\underset{\text{carboxylic acid}}{R{-}C{\overset{\nearrow O}{\underset{\searrow OH}{}}}} + \underset{\text{alcohol}}{R'{-}OH} \xrightarrow{H^+} \underset{\text{ester}}{R{-}\overset{O}{\overset{\|}{C}}{-}O{-}R'} + H_2O \qquad 18\text{–}10$$

$$\underset{\text{carboxylic acid}}{R\text{—}C(=O)\text{—}OH} + \underset{\text{amine or ammonia}}{H\text{—}N(H)\text{—}R'} \xrightarrow{100°C} \underset{\text{amide}}{R\text{—}C(=O)\text{—}N(H)\text{—}R'} + H_2O \qquad 18\text{–}11$$

In practice, amides are usually formed by reacting acid chlorides with amines.

$$\underset{\text{acid chloride}}{R\text{—}C(=O)\text{—}Cl} + \underset{\text{amine or ammonia}}{H\text{—}N(H)\text{—}R'} \rightarrow \underset{\text{amide}}{R\text{—}C(=O)\text{—}N(H)\text{—}R'} + HCl \qquad 18\text{–}12$$

$$\underset{\text{carboxylic acid}}{R\text{—}C(=O)\text{—}OH} + \underset{\text{base}}{NaOH} \rightarrow \underset{\text{acid salt}}{R\text{—}C(=O)\text{—}O^-Na^+} + H_2O \qquad 18\text{–}13$$

Salts formed by this reaction readily ionize in water.

$$\underset{\text{acid salt}}{R\text{—}C(=O)\text{—}ONa} \xrightarrow{H_2O} \underset{\text{acid anion}}{R\text{—}C(=O)\text{—}O^-} + Na^+ \qquad 18\text{–}14$$

Reactions of Alcohols

In addition to the reaction with organic acids given above (equation 18–10), alcohols react to give halogen compounds and ethers.

$$\underset{\text{alcohol}}{3\ R\text{—}OH} + \underset{\text{phosphorus trihalide}}{PX_3} \rightarrow \underset{\text{alkyl halide}}{3\ R\text{—}X} + \underset{\text{phosphorus acid}}{H_3PO_3} \qquad 18\text{–}15$$

$$\underset{\text{alcohol}}{R\text{—}OH + R\text{—}OH} \xrightarrow[135°C]{H_2SO_4} \underset{\text{ether}}{R\text{—}O\text{—}R} + H_2O \qquad 18\text{–}16$$

Reactions of Amines

Amines will react with organic acids to form amides (equation 18–11); also, they form salts with inorganic acids.

$$\underset{\text{amine}}{R—NH_2} + \underset{\text{acid}}{HCl} \rightarrow \underset{\text{salt}}{R—NH_3^+Cl^-} \qquad 18\text{–}17$$

Reactions of Amides

Amides, formed in reactions such as those represented by equations 18–11 and 18–12, can be reacted with water to produce the original acid and amine. Notice that this corresponds to a reversal of equation 18–11. This reaction, called **hydrolysis,** is very important in the digestion of food by the body. It is catalyzed by acids, bases, or enzymes (biochemical catalysts).

$$\underset{\text{amide}}{R—\overset{\overset{O}{\|}}{C}—\overset{\overset{H}{|}}{N}—R'} + H_2O \xrightarrow{\text{catalyst}} \underset{\text{carboxylic acid}}{R—\overset{\overset{O}{\|}}{C}—OH} + \underset{\text{amine}}{H—\overset{\overset{H}{|}}{N}—R'} \qquad 18\text{–}18$$

Reactions of Esters

Esters may also undergo a hydrolysis reaction which corresponds to a reversal of equation 18–10. In the presence of an acid catalyst, the original organic acid and alcohol are formed. When a strong base is present, the organic salt is formed along with the alcohol.

$$\underset{\text{ester}}{R—\overset{\overset{O}{\|}}{C}—O—R'} + H_2O \xrightarrow{H^+} \underset{\text{carboxylic acid}}{R—\overset{\overset{O}{\|}}{C}—OH} + \underset{\text{alcohol}}{R'—OH} \qquad 18\text{–}19$$

$$\underset{\text{ester}}{R—\overset{\overset{O}{\|}}{C}—O—R'} + \underset{\text{base}}{NaOH} \longrightarrow \underset{\text{carboxylic acid salt}}{R—\overset{\overset{O}{\|}}{C}—O^-Na^+} + \underset{\text{alcohol}}{R'—OH} \qquad 18\text{–}20$$

Review Questions

1. Explain how the functional group concept aids in the study of organic chemical reactions.
2. Represent each functional group of Table 18–1 by a compound formula that contains a total of six carbon atoms. Name each of the resulting compounds.

3. Write an equation to represent a reaction between each of the following pairs of compounds. Name the products in each case.
   a) Butanol and propanoic acid
   b) Propylamine and ethanoic acid
   c) 2-methyl-2-butanol and PCl_3
   d) Propanol and propanol
   e) *n*-Phenyl ethanamide and water
   f) Ethylbutanoate and water
4. Identify the functional group or groups present in each of the following compounds:
   a) Norepinephrine (involved in the transport of messages along nerves):

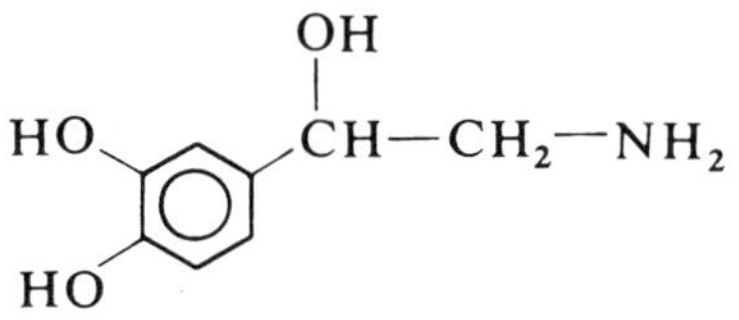

   $$(HO)_2C_6H_3{-}CH(OH){-}CH_2{-}NH_2$$

   b) Oil of wintergreen (a topical pain killer and a flavoring agent):

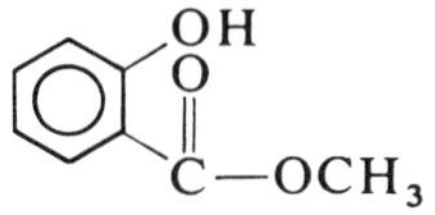

   $$C_6H_4(OH){-}\overset{O}{\overset{\|}{C}}{-}OCH_3$$

   c) Chloral hydrate (a sedative): $Cl{-}CCl_2{-}CH(OH)_2$

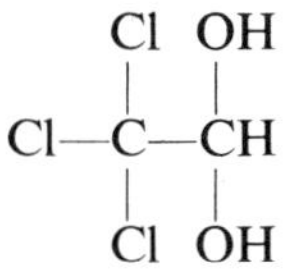

   d) Alanine (a constituent of protein): $CH_3{-}CH(NH_2){-}C(=O){-}OH$

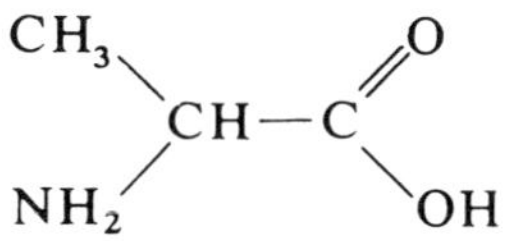

   e) Vanillin (vanilla flavoring agent):

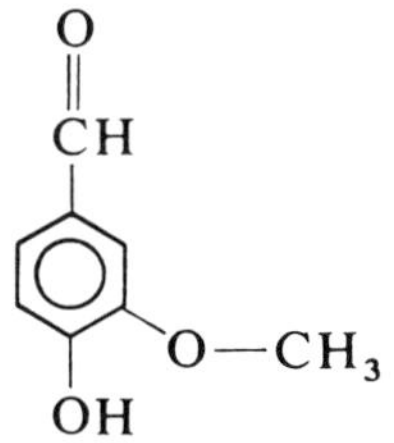

   $$C_6H_3(\overset{O}{\overset{\|}{C}}H)(OH){-}O{-}CH_3$$

f) Fructose (a sugar):

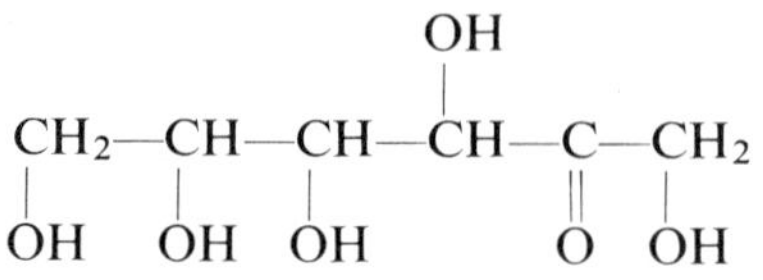

$$\begin{array}{cccccc} & & & OH & & \\ & & & | & & \\ CH_2- & CH- & CH- & CH- & C- & CH_2 \\ | & | & | & & \| & | \\ OH & OH & OH & & O & OH \end{array}$$

g) Urea (a body waste found in urine):

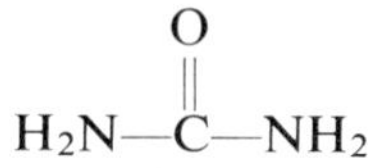

$$\begin{array}{c} O \\ \| \\ H_2N-C-NH_2 \end{array}$$

h) LSD (a hallucinogen—see Table 18–8)

5. Draw a structural formula to represent each of the following compounds:
   a) Amphetamine, 2-amino phenylpropane (a stimulant)
   b) Cyclohexylamine (a carcinogen thought to be formed in the body from cyclamates)
   c) Lactic acid, 2-hydroxypropanoic acid (found in sour milk)
   d) 4-amino benzoic acid (an active ingredient in lotions used to prevent sunburn)
   e) 1,3-dihydroxy-2-propanone (a sugar)
   f) 2-methyl propylmethanoate (partially responsible for odor of raspberries)
   g) 2,2,2-trichloro-1,1-dimethylethanol (a sedative)
6. Draw a structural formula for each of the four isomers of pentanoic acid. Name each isomer.
7. Write equations for the preparation of the following compounds:
   a) *n*-Butane from 2-butene
   b) *n*-Pentylethanoate
   c) *n*-Methyl propanamide
   d) Sodium acetate (acetic acid salt)
   e) Ethoxyethane
   f) Sodium-2-ethylbenzenesulfonate (acid salt)
8. Name the following aromatic compounds:

a) 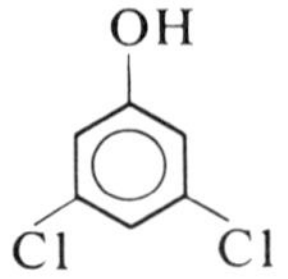 OH, Cl, Cl

b) 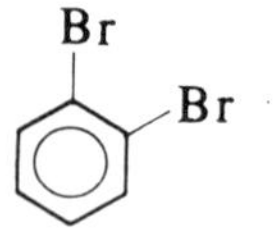 Br, Br

c) 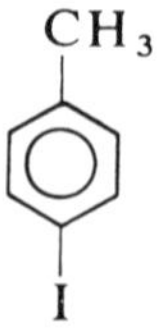 CH_3, I

d) 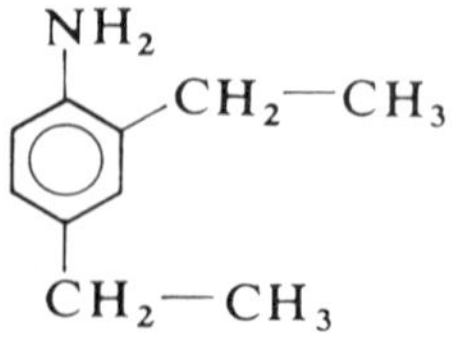 NH_2, CH_2-CH_3, CH_2-CH_3

e) 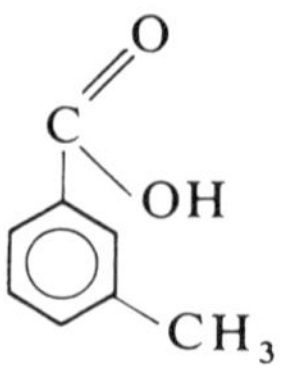 C, O, OH, CH_3

Suggestions for Further Reading

Amarine, M. A., "Wine," *Scientific American*, Aug. 1964.

Amoore, J. E., J. W. Johnston, and M. Rubin, "The Stereochemical Theory of Odor," *Scientific American*, Feb. 1964.

Eglinton, G., and M. Calvin, "Chemical Fossils," *Scientific American*, Jan. 1967.

Mason, B., "Organic Matter from Space," *Scientific American*, March 1963.

Roberts, J. D., "Organic Chemical Reactions," *Scientific American*, Nov. 1957.

Rose, A. H., "Beer," *Scientific American*, June 1959.

Song, S. K., and E. Rubin, "Ethanol Produces Muscle Damage in Human Volunteers," *Science*, Jan. 21, 1972.

19 Petroleum and Coal

Petroleum and coal, chemical gifts from ages long past, play vital roles in our modern society. In addition to serving as prime energy-producing fuels, these two substances are the sources of nearly all the valuable synthetic organic compounds—polymers, medicines, dyes, etc.—produced today. Vast industries, employing millions of people, have grown around the need to obtain from nature, refine, and distribute these products.

Petroleum

The dark-colored, viscous substance called petroleum was known and used by people of ancient civilizations. The Egyptians were using asphalt derived from petroleum as an embalming agent in 5000 B.C. Archeological findings indicate that asphalt was used in 4000 B.C. as a waterproofing agent and mortar by civilizations along the Euphrates River. Ships were caulked with asphalt as early as 3000 B.C. Indians of North America used petroleum for war paint and medication.

Until 1900, the main product obtained from petroleum was kerosene, which was used in lamps and stoves. Fuel oils and lubricants were prepared in smaller amounts, and gasoline was an undesirable side product that created disposal problems. The evolution of gasoline from an unwanted by-product in 1900 to an indispensable part of modern life was truly remarkable. The demand for this product has so increased the use of petroleum that some people estimate the known reserves will last only for another two or three decades. These reserves, as of December 31, 1971, are given in Table 19–1, and their geographical distribution is shown in Figure 19–1.

Table 19–1. World Petroleum Reserves, December 31, 1971

Location of reserves	*Millions of barrels in reserve (1 barrel = 42 gal.)*	*Percent of world total*
North America	47,225	14.3
Latin America	22,632	6.9
Western Europe	1,771	0.5
Middle East	216,766	65.6
Asia-Pacific	10,863	3.3
China	323	0.1
Russia	29,900	9.0
Other	911	0.3
World total	330,391	100.0

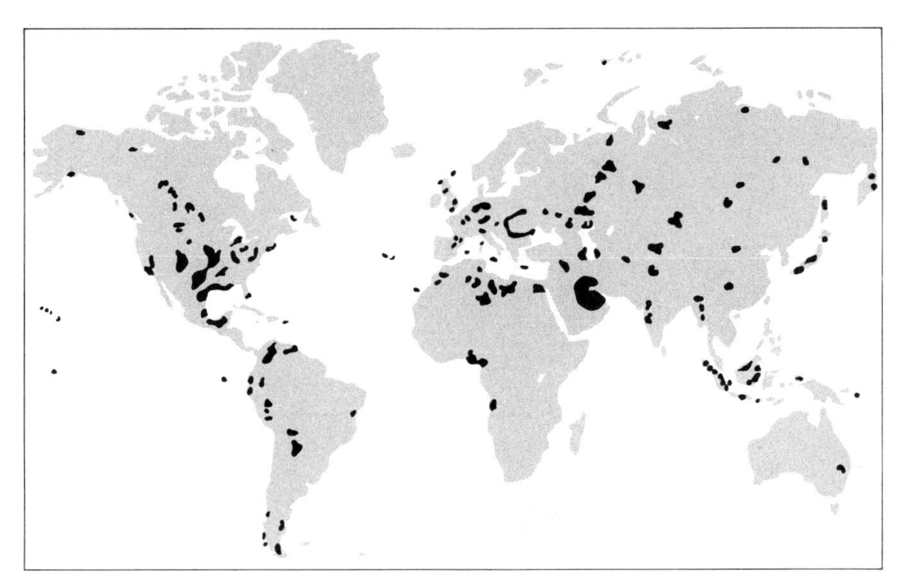

From *The Encyclopedia of Chemical Technology*, 2nd ed., vol. 14 (New York: John Wiley and Sons, Inc., 1967), p. 860.

Figure 19–1.
Geographical Distribution of Known Crude Oil Reserves

Petroleum Refining

Petroleum, composed mainly of hydrocarbons, is used today mostly as a fuel and lubricant for internal-combustion engines. In order to be used for these purposes, the various components of the crude petroleum must be separated from one another. This process, called **refining,** is the basis for a very important industry. The components of crude petroleum have different boiling points, and a rough separation can be accomplished by the process of **fractional distillation.** In this process the mixture is subjected to a number of successive vaporizations and condensations. The process can be illustrated by following the events that occur as a mixture of different components is boiled in an apparatus such as that in Figure 19–2.

As the mixture in the boiler is heated, it will boil, and the vapors will pass into the fractionating column, which is filled with some material of large surface area. The temperature inside the column will decrease with increasing distance from the boiler. When the vapor gets high enough in the column, the temperature will be low enough to condense the higher-boiling parts of the vapor back to a liquid. The lower-boiling parts will continue on up the column until they also reach an area where they will condense. Some of the lower-boiling components go right on through the column and condense back to a liquid in the condenser. As the lower-boiling materials are removed from the mixture, the temperature at each level of the fractionating column increases. This allows a higher-boiling component to reach the condenser and come off as a liquid. As the process continues, the components of the mixture are separated and removed in order of increasing boiling points.

Figure 19–2.
Apparatus for Fractional Distillation

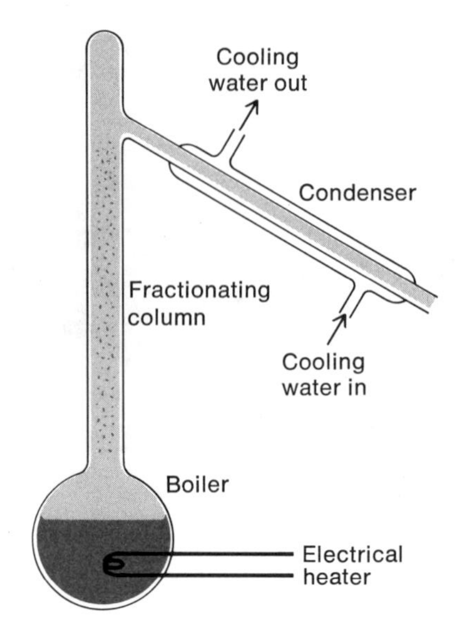

The same type of process is used with petroleum except that the liquids are allowed to condense at various levels in the column and are collected and drawn off to storage tanks. The arrangement used in petroleum refining is represented schematically in Figure 19–3. The various products or fractions obtained are tabulated in Table 19–2 according to use, molecular size, and boiling range.

Figure 19–3.
Fractional Distillation of Petroleum

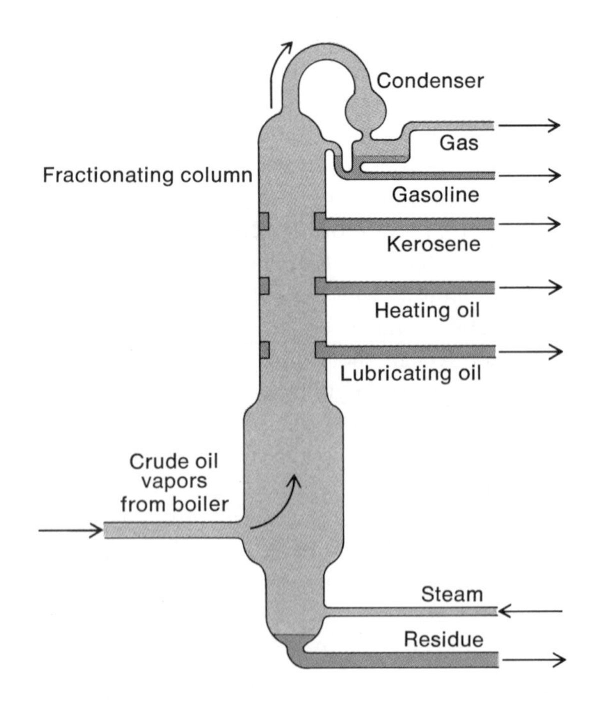

Table 19–2. Products of Petroleum Distillation

Fraction	*Molecular size range*	*Boiling-point range (°C)*	*Typical uses*
Gas	C_1–C_5	−164 to 30	Gaseous fuel
Petroleum ether	C_5–C_7	30 to 90	Solvent, dry cleaning
Straight-run gasoline	C_5–C_{12}	30 to 200	Motor fuel
Kerosene	C_{12}–C_{16}	175 to 275	Fuel for stoves, diesel and jet engines
Gas oil or fuel oil	C_{15}–C_{18}	Up to 375	Furnace oil
Lubricating oils	C_{16}–C_{20}	350 and up	Lubrication
Greases	C_{18}–up	Semisolid	Lubrication
Paraffin (wax)	C_{20}–up	Melts at 52–57	Candles
Pitch and tar	high	Residue in boiler	Roofing, paving

Gasoline Production

When crude petroleum is distilled, 25–45 percent shows up in the gasoline fraction. Today, the large amount of gasoline used would create huge surpluses of other petroleum products if fractional distillation alone were used to satisfy the demand. This situation began around 1912, and since that time the petroleum industry has been devising ways to convert a larger fraction of crude petroleum into gasoline.

Three processes widely used today are cracking, polymerization, and combination. In the cracking process, gasoline-sized molecules are produced by breaking up larger molecules. Both polymerization and combination involve reactions between smaller molecules to produce gasoline molecules.

In the **cracking** process, the heavier, higher-boiling fractions are passed over catalysts at temperatures in excess of 400°C and at pressures greater than two atmospheres. Under these conditions, gasoline-sized molecules are obtained. A typical cracking reaction is:

$$\underset{\text{a tetradecane}}{C_{14}H_{30}} \xrightarrow[\text{high press. \& temp.}]{\text{catalyst}} \underset{\text{an octane}}{C_8H_{18}} + \underset{\text{a hexane}}{C_6H_{12}} \qquad 19\text{–}1$$

The **polymerization** process involves a reaction between unsaturated molecules in the presence of a catalyst at moderate temperatures of 150°C or lower. An example of such a reaction is:

$$\underset{\text{2-methylpropene}}{CH_2{=}\overset{\overset{CH_3}{|}}{C}{-}CH_3} + \underset{\text{2-methylpropene}}{CH_2{=}\overset{\overset{CH_3}{|}}{C}{-}CH_3} \xrightarrow[150^\circ]{\text{cat.}} \underset{\text{2,4,4-trimethyl-1-pentene}}{CH_2{=}\overset{\overset{CH_3}{|}}{C}{-}CH_2{-}\overset{\overset{CH_3}{|}}{\underset{\underset{CH_3}{|}}{C}}{-}CH_3} \qquad 19\text{–}2$$

Combination processes make use of reactions between different small molecules to produce gasoline-sized molecules. The reacting molecules need not all be unsaturated. An example of such a reaction is:

$$\underset{\text{2-methylpropane}}{CH_3-\underset{\displaystyle CH_3}{\underset{|}{CH}}-CH_3} + \underset{\text{1-propene}}{CH_2=CH-CH_3} \rightarrow \underset{\text{2,2-dimethylpentane}}{CH_3-CH_2-CH_2-\overset{\displaystyle CH_3}{\overset{|}{\underset{\displaystyle CH_3}{\underset{|}{C}}}}-CH_3} \qquad 19\text{–}3$$

Modern Gasoline

Today, gasoline is a blend of hydrocarbons with various boiling points and numerous additives. The hydrocarbon blend includes some low-molecular-weight, low-boiling components for easy starting in cold weather, and some components with intermediate and higher molecular weights for good warmup and mileage characteristics. Most gasolines are specially blended to perform well under the conditions characteristic of the geographical areas in which they are used.

The reaction which takes place inside the internal-combustion engine is a burning or oxidation of fuel, usually initiated by a spark from a spark plug. The reaction for such a process is given below, where gasoline is represented by one of its components, heptane.

$$C_7H_{16} + 11O_2 \rightarrow 7CO_2 + 8H_2O + \text{heat} \qquad 19\text{–}4$$

This reaction is idealized and assumes sufficient oxygen is present for complete combustion. This is not generally true, and small amounts of CO are also produced.

During the operation of a typical internal-combustion automobile engine, each cylinder is periodically filled with a mixture of gasoline vapor and air. The mixture is ignited at the proper time—if the engine is well tuned. The heat liberated during combustion expands the gaseous combustion products, which push on a piston and finally, through various mechanical linkages, turn the wheels of the automobile. When straight-run (unblended) gasoline is used, the rate of burning in the cylinder is uniform until perhaps two-thirds to three-fourths of the fuel mixture has burned. At this point the rate increases rapidly and culminates in a detonation or explosion. When this happens, the engine is said to be *knocking* or *pinging*.

The tendency of a fuel to cause knocking during engine operation is described by an assigned **octane number.** High values for octane numbers indicate low knocking tendencies for a fuel. Octane numbers were originally based on the knocking tendencies of two pure compounds. An octane number of 0 was given to *n*-heptane, a poor fuel. The much better fuel 2,2,4-trimethylpentane or isooctane was given an octane number of 100. The octane rating of a gasoline was determined by comparing its knocking tendencies in a standard engine to those of mixtures of the two reference materials. The octane number of a gasoline was set as the percentage of isooctane in a reference mixture with the same knocking tendencies. Thus, a gasoline would be assigned an octane number of 85 if it knocked as much as a mixture containing 85 percent isooctane and 15 percent *n*-heptane. On this basis, straight-run gasoline has an octane number of 50–55. For some applications today, fuels are needed with octane numbers higher than 100. Numbers are assigned to these fuels by comparing their knocking tendencies to those of pure isooctane to which known quantities of antiknock additives—usually tetraethyl lead—have been added. These additives are discussed later in this chapter.

The knocking tendency of a gasoline depends upon several factors. Branched molecules have lower tendencies than comparable straight-chain compounds. Aromatic compounds have higher octane ratings than non-aromatics with similar molecular weights. Thus, the structural nature of the compounds in a gasoline determines to some degree the octane number. The knocking tendency is also influenced by the amount of squeezing a fuel-air mixture undergoes before ignition in the engine. This is measured by the **compression ratio** of the engine. A compression ratio of 8:1, or 8 means that the fuel-air mixture is reduced to one-eighth the original volume before being ignited. More power can be put into an engine by increasing the compression ratio, and the development of modern automobiles has followed a general trend toward more powerful engines, made possible largely by increased compression ratios. In order to avoid knocking in the higher-compression engines, higher octane gasolines must be used. The relationship between compression ratio, knocking tendencies, and octane number is given in Figure 19–4.

The high-octane gasolines on the market today are produced in several ways. The gasoline may be put through a sort of cracking process in which the degree of branching of the molecules is greatly increased. This **reforming process** involves passing the gasoline over an aluminum chloride catalyst at about 120°C and under high pressures of up to 50 atm. A typical reforming reaction is

$$CH_3-CH_2-CH_2-CH_2-CH_2-CH_3 \xrightarrow[120°C,\ 50\ atm]{AlCl_3} CH_3-\underset{CH_3}{\overset{CH_3}{\underset{|}{\overset{|}{C}}}}-CH_2-CH_3 \qquad 19\text{–}5$$

n-hexane 2,2-dimethylbutane

In another type of catalytic reforming, alkanes are converted to aromatic hydrocarbons. The alkane vapors are passed over a platinum catalyst at 500°C. An example of the reaction is the conversion of *n*-heptane to toluene:

$$CH_3-CH_2-CH_2-CH_2-CH_2-CH_2-CH_3 \xrightarrow[500°C]{Pt} C_6H_5-CH_3 + 4H_2 \qquad 19\text{–}6$$

n-heptane toluene

Figure 19–4.
Influence of Compression Ratio on Fuel Octane Requirements

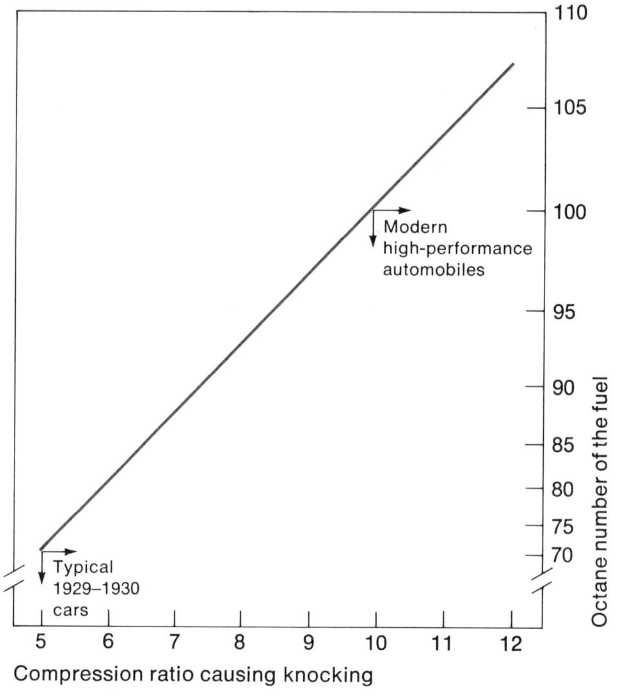

From *The Encyclopedia of Chemical Technology*, 2nd ed., vol. 10 (New York: John Wiley and Sons, Inc., 1967), p. 469.

or in steps,

$$\underset{n\text{-heptane}}{C_7H_{16}} \longrightarrow H_2 + \underset{\text{methylcyclohexane}}{C_6H_{11}CH_3} \longrightarrow 3H_2 + \underset{\text{methylbenzene or toluene}}{C_6H_5CH_3} \qquad \text{19–7}$$

A second way to increase octane numbers is to add antiknock agents to the gasoline. These agents are compounds with the ability to prevent detonation and knocking in the cylinders. The most widely used is **tetraethyl lead,** $Pb(C_2H_5)_4$, which will prevent knocking when present in gasoline at a concentration of 0.01 percent. The name *Ethyl* gasoline was derived from this additive.

Nonleaded gasoline contains no tetraethyl lead but its octane number is increased to acceptable levels by the blending in of larger proportions of branched and aromatic components. These components are prepared separately at some expense, hence the higher price of nonleaded gasoline.

At present, there is a great deal of concern over the role of the internal-combustion engine and its fuels in polluting the environment, a topic dealt with in some detail in Chapter 21.

Coal

It is known that coal was used to a limited extent in Greece, Italy, and China more than 2000 years ago. Marco Polo, upon returning from China, reported the use by the Chinese of "black rocks" for fuel. Coal mining in Germany began in about the tenth century. England was shipping coal from various deposits in the thirteenth century, and production increased rapidly in the sixteenth century because of wood shortages. Coal was first mined in the United States in the beginning of the eighteenth century. By 1860 world production reached 134 million metric tons and has been increasing since (see Table 19–3). Note that one metric ton is equal to 1000 Kg or 2204 lb.

The reserves of coal far exceed the known reserves of all other fossil fuels—natural gas, petroleum, oil shales, and tar sands. The estimated coal consumption in the next twenty years will reduce known coal reserves by less than 5 percent. It seems evident that, where feasible, a trend away from the use of petroleum and natural gas toward the use of coal is likely in the future. Estimates of the world coal reserves are given in Table 19–4.

Table 19–3. Annual World Coal Production

Date	*Production in millions of metric tons/year*
1860	134
1913	1,257
1954	1,631
1960	2,414
1964	2,600

Table 19–4. World Coal Reserves

Location of reserves	*Billions of metric tons in reserve*	*Percent of world total*
North America	798	34.4
Latin America	10	0.4
Europe	301	13.0
Asia-Pacific	41	1.8
China	506	21.8
Russia	600	25.8
Africa	35	1.5
Australia	29	1.3
World total	2320	100.0

From *The Encyclopedia of Chemical Technology*, 2nd ed., vol. 5 (New York: John Wiley and Sons, Inc., 1967), p. 613.

Composition of Coal

Coal is thought to have originated from plants which died, decayed, and were transformed by heat and pressure. Chemically, coal is a complicated mixture of organic compounds with small amounts of inorganic substances sometimes present.

Coal apparently develops from dead plants to the final anthracite or hard-coal state through a series of stages. Each stage represents an increase in carbonization (percent carbon). Substances typical of four stages are given in Table 19–5.

Table 19–5. Stages in the Formation of Coal

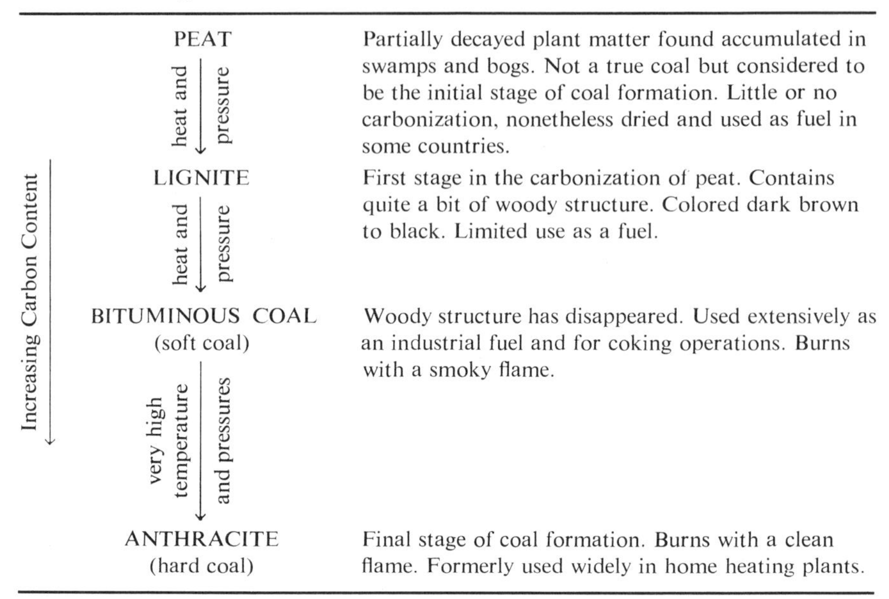

Stage (Increasing Carbon Content ↓)	Description
PEAT	Partially decayed plant matter found accumulated in swamps and bogs. Not a true coal but considered to be the initial stage of coal formation. Little or no carbonization, nonetheless dried and used as fuel in some countries.
↓ heat and pressure	
LIGNITE	First stage in the carbonization of peat. Contains quite a bit of woody structure. Colored dark brown to black. Limited use as a fuel.
↓ heat and pressure	
BITUMINOUS COAL (soft coal)	Woody structure has disappeared. Used extensively as an industrial fuel and for coking operations. Burns with a smoky flame.
↓ very high temperature and pressures	
ANTHRACITE (hard coal)	Final stage of coal formation. Burns with a clean flame. Formerly used widely in home heating plants.

Chemically, the main functional groups present in coal are aldehydes, and aromatic alcohols (phenols). Condensed aromatic and cyclic alkane ring structures are also present in large amounts. Nitrogen is present as a component of heterocyclic ring compounds. Small amounts of sulfur are also present in the form of organic compounds, but most sulfur occurs as inorganic mineral impurities.

Uses of Coal

About 70 percent of the coal produced in the world is burned to provide heat. This heat is used to generate electricity, supply steam for transportation and industry, and provide heat for comfort. Most of the remaining 30 percent is strongly heated in the absence of air and carbonized to produce **coke** for use in the iron and steel industry. A by-product of this coking operation is coal tar, discussed below.

The magnitude of world coal reserves compared to the dwindling petroleum and natural gas reserves makes the idea of using coal as a substitute for oil and natural gas very attractive. A direct substitution of coal is difficult because distribution systems and final consumers are geared to liquid or gaseous products. For this reason, vigorous research is in progress aimed at the economical conversion of coal into oil and fuel gas. Such conversions are possible. In 1914 a man named Bergius converted coal into oil by reacting hydrogen gas with finely divided coal at 450°C under a pressure of 200 atm. During World War II, Germany used this method to produce gasoline of good quality. The drawback today is cost: natural gas and petroleum are cheaper than converted coal products. Optimistic estimates are that by the early 1980's, converted coal products will become competitive, at least in some locations.

Coal Tar

Coal tar, a by-product of the coking process, is a valuable mixture of chemical compounds. During the coking process the complex compounds making up bituminous coal are changed into carbon (coke) and simpler compounds that distill off. Some of the material distilling off is condensed as **coal tar;** the remaining gaseous distillate is called **coal gas** and can be used as a fuel. About 70 percent of the coal is converted to coke in this process, 6–10 percent becomes coal tar, and the remainder is coal gas.

Coal tar is of value because it contains 65–90 percent aromatic compounds, including some heterocyclics. Plants and animals produce aromatic compounds, but only in very small quantities; consequently, coal tar is the richest natural source of aromatic compounds used to manufacture many useful products. A few of the products derived from coal tar are dyes and coloring agents, plastics, pesticides, explosives, textiles, preservatives, and drugs and medicines. Table 19–6 gives a few examples of the many aromatic compounds found in coal tar.

Table 19–6. Some Aromatic Compounds of Coal Tar

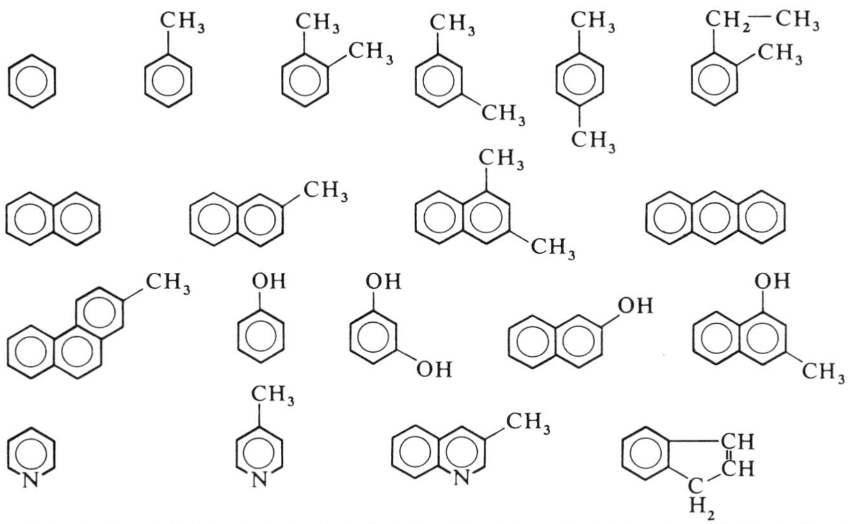

CH_3 CH_3 CH_3 CH_3 CH_3 CH_3 CH_3 CH_3 CH_2-CH_3 CH_3

CH_3 CH_3 CH_3

CH_3 OH OH OH OH OH OH CH_3

N CH_3 N CH_3 N CH CH $C H_2$

Review Questions

1. Discuss the importance to the petroleum industry (and petroleum consumers) of the processes known as cracking, polymerization, and combination.
2. A gasoline produces the same knocking tendency in a standard engine as a mixture containing 40 percent *n*-heptane and 60 percent isooctane. What is the octane number of the gasoline? Discuss methods by which this octane number could be increased.
3. How are octane numbers greater than 100 determined for gasolines?
4. Assume the petroleum products of Table 19–2 are all in the form of saturated alkanes with formulas of C_nH_{2n+2}. Calculate the molecular weight range for each product.
5. The compression ratio of internal combustion automobile engines was doubled between 1930 and the present. What effect did this have on the octane number of the fuel used? (See Figure 19–4.)
6. By what percentage did annual coal production increase from 1954 to 1960? What was the increase from 1960 to 1964? (See Table 19–3.)
7. Why are attempts being made to discover economical methods for converting coal into liquid or gaseous products?
8. A large number of useful compounds are found in coal tar. Name any eight of those given in Table 19–6.

Suggestions for Further Reading

Ayres, E., "The Fuel Situation," *Scientific American*, Oct. 1956.

Bengelsdorf, I., "Are We Running Out of Fuel?," *National Wildlife*, Feb./March 1971.

Gaucher, L. P., "Energy in Perspective," *Chemical Technology*, March 1971.

Gilluly, R. H., "The Competitive Comeback of Coal," *Science News*, Jan. 30, 1971.

Greek, B. F., "Gasoline," *Chemical and Engineering News*, Nov. 9, 1970.

Haensel, V., and R. L. Burwell, Jr., "Catalysis," *Scientific American*, Dec. 1971.

Lessing, L. P., "Coal," *Scientific American*, July 1955.

Mills, A., "Gas from Coal," *Environmental Science and Technology*, Dec. 1971.

Starr, C., "Energy and Power," *Scientific American*, Sept. 1971.

20 Polymers: Natural and Man-Made

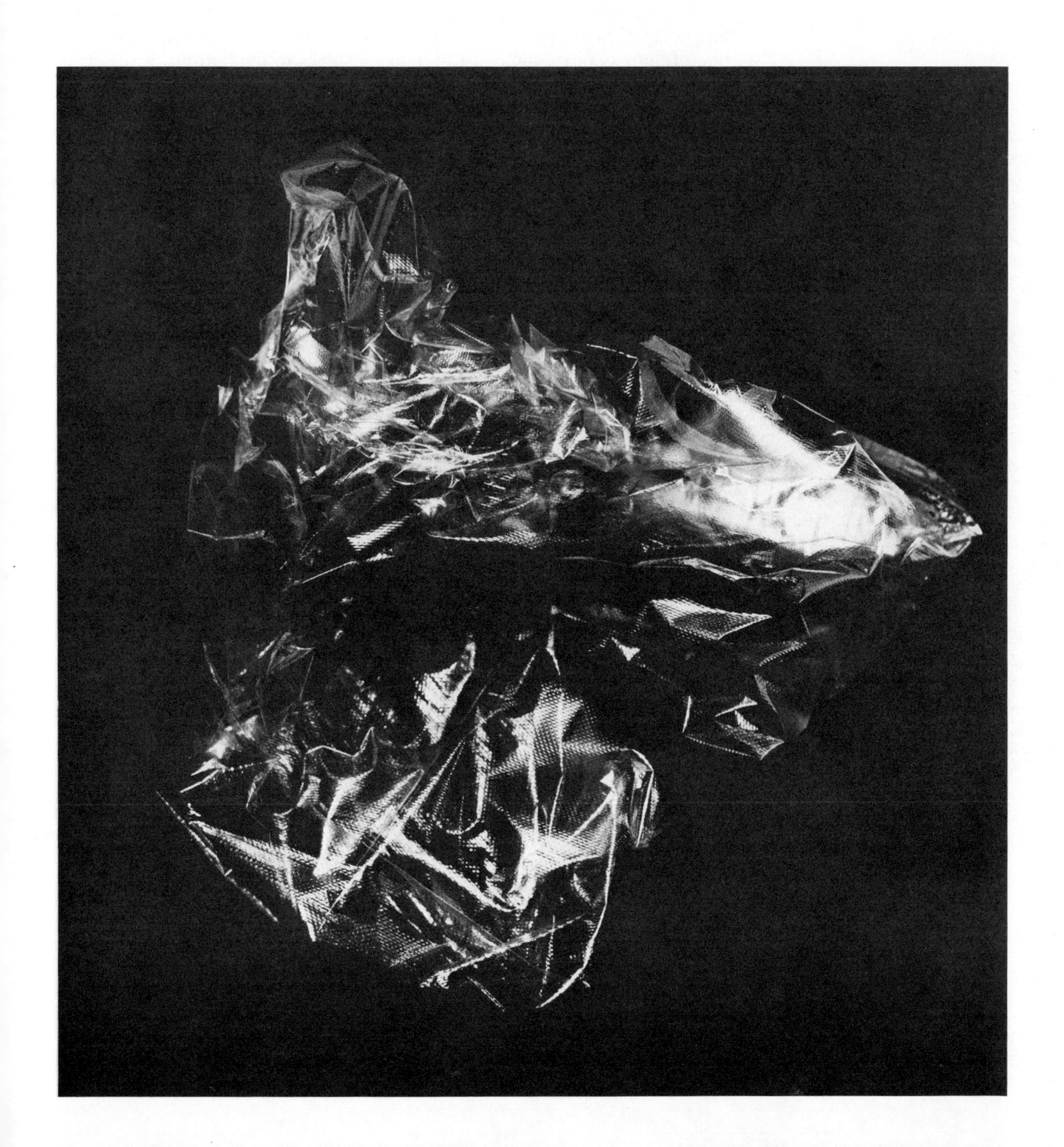

Chemists of the late nineteenth century were aware that some organic substances from natural sources behaved differently than "pure" organic compounds. One obvious difference became apparent when attempts were made to melt a substance such as rubber. Instead of melting sharply like a pure compound, rubber gradually softened as it was heated over a wide temperature range. Compounds with similar unsatisfactory characteristics were sometimes found as tarry, difficult-to-remove residues in the bottom of vessels in which organic reactions had been carried out.

Because of these and other difficulties, organic chemists of the early twentieth century busied themselves with the task of preparing and isolating "pure" compounds. It was not until the 1920's that much attention was given to these "other" compounds, many of which are now known to be polymers.

Formation of Polymers

The term **polymer** means *many parts*, and is used because it is now understood that large polymers consist of many smaller units called **monomers.** In order to form polymers, monomer molecules must have the capacity to form two or more bonds with each other. This allows long chains or sheets of connected monomer units to form. If monomers can form only one bond, the reaction stops after a dimer—a two-monomer-unit molecule—is formed:

$$\underset{\text{monomer}}{\mathrm{M—}} + \underset{\text{monomer}}{\mathrm{—M}} \rightarrow \underset{\text{dimer}}{\mathrm{M—M}}$$

Monomers with a capacity to form two bonds form chainlike polymers:

$$\underset{\text{monomer}}{\mathrm{—M—}} + \underset{\text{monomer}}{\mathrm{—M—}} \rightarrow \underset{\text{dimer}}{\mathrm{—M—M—}} \xrightarrow{\text{add —M—}} \underset{\text{trimer}}{\mathrm{—M—M—M—}}$$

$$\underset{\text{trimer}}{\mathrm{—M—M—M—}} \xrightarrow{\text{add —M—}} \underset{\text{tetramer}}{\mathrm{—M—M—M—M—}} \xrightarrow{\text{add many —M—}}$$

$$\underset{\text{polymer}}{\text{etc. —M—M—M—M— etc.}} \qquad \text{20–1}$$

When monomers form more than two bonds, two-dimensional, cross-linked polymers can result:

$$\underset{\text{monomer}}{\overset{\mathrm{—M—}}{|}} + \underset{\text{monomer}}{\overset{\mathrm{—M—}}{|}} \rightarrow \underset{\text{dimer}}{\overset{\mathrm{—M—M—}}{|\quad|}} \xrightarrow{\text{add }\underset{|}{\mathrm{—M—}}} \underset{\text{trimer}}{\overset{\mathrm{—M—M—M—}}{|\quad|\quad|}} \xrightarrow[\underset{|}{\mathrm{—M—}}]{\text{add many}} \text{polymer chain}$$

$$\underset{\text{polymer chains}}{\begin{array}{c}\text{etc. —M—M—M— etc.}\\ |\quad|\quad| \\ \\ |\quad|\quad| \\ \text{etc. —M—M—M— etc.}\end{array}} + \underset{|}{\mathrm{—M—}} \rightarrow \underset{\text{cross-linked polymer}}{\begin{array}{c}\text{etc. —M—M—M— etc.}\\ |\quad|\quad| \\ \qquad\quad \mathrm{M—} \\ |\quad|\quad| \\ \text{etc. —M—M—M— etc.}\end{array}} \qquad \text{20–2}$$

Addition Polymers

Polymers can be classified into two categories, addition and condensation, depending upon the type of reaction that takes place between the monomer molecules. **Addition polymers** are often formed by unsaturated monomers, and the polymer is the only product. The reaction involves the unpairing of the electrons in one or more of the multiple bonds in the monomer.

The simplest unsaturated hydrocarbon is ethylene (ethene), which will polymerize when placed under pressure at a temperature of 100–300°C. The reaction requires the presence of an **initiator** to get it started. Peroxides are often used as initiators. The reactions leading to the polymerization of ethylene are:

$$R-\ddot{\underset{..}{O}}-\ddot{\underset{..}{O}}-R \rightarrow 2R-\ddot{\underset{..}{O}}\cdot \qquad 20\text{–}3$$

peroxide — peroxide-free radicals with unpaired electrons

$$R-\ddot{\underset{..}{O}}\cdot + H_2C{::}CH_2 \rightarrow R-\ddot{\underset{..}{O}}-CH_2-CH_2\cdot \qquad 20\text{–}4$$

ethylene — peroxide-ethylene product

(labels: new bond; remaining electron pair forms single bond; unpaired electron)

$$R-\ddot{\underset{..}{O}}-CH_2-CH_2\cdot + H_2C{::}CH_2 \rightarrow$$

$$R-\ddot{\underset{..}{O}}-CH_2-CH_2-CH_2-CH_2\cdot \xrightarrow[\text{ethylene}]{\text{add}} R-\ddot{\underset{..}{O}}\left[-CH_2-CH_2-\right]_n \qquad 20\text{–}5$$

(labels: new bond; long-chain polymer)

In addition to chain formation, cross-links between chains can form when hydrogen atoms are extracted from the chain. For example:

$$R-\ddot{\underset{..}{O}}-CH_2-CH_2-CH_2-CH_2\cdot \rightarrow R-O-CH_2-\dot{C}H-CH_2-CH_2-H \qquad 20\text{–}6$$

(note the position change of the unpaired electron)

Two such chains or one chain and another ethylene molecule can react:

```
        H  H  H  H               H  H  H  H
   ..   |  |  |  |          ..   |  |  |  |
R—O—C—C—C—C—H        R—O—C—C—C—CH
   ..   |  (·) |  |          ..   |  |  |  |                 20–7
        H  |  H  H               H  |  H  H
           |           →            |←———————— new bond
        H  |  H  H               H  |  H  H
   ..   |  (·) |  |          ..   |  |  |  |
R—O—C—C—C—C—H        R—O—C—C—C—CH
   ..   |  |  |  |          ..   |  |  |  |
        H  H  H  H               H  H  H  H
                              cross-linked polymer
```

or

```
        H  H  H  H      H  H                  H  H  H  H
   ..   |  |  |  |      |  |             ..   |  |  |  |
R—O—C—C—C—C—H + C::C → R—O—C—C—C—CH
   ..   |  ·  |  |      |  |             ..   |  |  |  |        20–8
        H     H  H      H  H                  H  |  H  H
                                                 |↖
                                               HCH      new bond
                                                 |
                                               HCH
                                                 ·
                                          branched radical
```

The growth of addition polymers stops when the unpaired electrons of the growing molecule react with a radical and give a product which has no more free electrons—with an initiator radical, for example.

Polymers with different properties can be synthesized by starting with different monomers and controlling the final molecular weight and amount of cross-linking.

Several common addition polymers are described in Table 20–1. The polymer formulas are given in terms of the repeating unit in the structure. The *n* signifies a large number of repeating units.

Naturally occurring polymers were known and used before the development of synthetics. The most common example is natural rubber. This material is an addition polymer of 2-methyl-1,3-butadiene, commonly called isoprene. The polymer occurs naturally as a colloidal suspension of particles in the sap or latex of the rubber tree. The colloidal particles coagulate upon the addition of acid, but the resulting rubber is soft and sticky when warm, and hard when cool. This undesirable property is removed by a process called **vulcanization** in which the rubber is heated with sulfur.

The polymerization of isoprene can occur by several different mechanisms, each of which results in a structurally different polymer. We shall consider natural rubber to be produced by 1,4 addition. The reactions are given below.

Table 20–1. Some Common Addition Polymers

Monomer formula and name	*Polymer formula*	*Polymer names*	*Typical uses*
$H_2C{=}CH_2$ ethylene	$-[H_2C-CH_2]_n-$	Polyethylene	Bottles, toys, packaging
$H_2C{=}CH(CH_3)$ propylene	$-[H_2C-CH(CH_3)]_n-$	Polypropylene	Carpet fiber, pipes, valves, bottles
$H_2C{=}CH(Cl)$ vinyl chloride	$-[H_2C-CH(Cl)]_n-$	Polyvinylchloride, PVC, vinyl	Floor tile, pipes, packaging, phonograph records
$F_2C{=}CF_2$ tetrafluoroethylene	$-[F_2C-CF_2]_n-$	Polytetrafluoroethylene, or Teflon	Cooking utensil coatings, valves, gaskets, electrical insulation
$H_2C{=}CH(CN)$ acrylonitrile	$-[H_2C-CH(CN)]_n-$	Polyacrylonitrile, Orlon, or Acrilan	Textile fibers
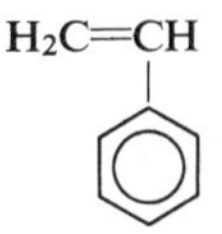 $H_2C{=}CH(C_6H_5)$ styrene	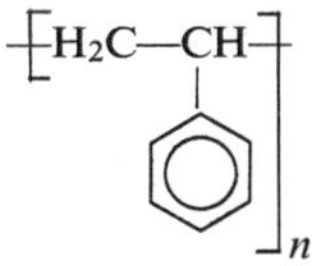 $-[H_2C-CH(C_6H_5)]_n-$	Polystyrene, or styrene	Toys, models, styrofoam, knobs
$H_2C{=}C(CH_3)-C(=O)-O-CH_3$ methyl methacrylate	$-[H_2C-C(CH_3)(C(=O)-O-CH_3)]_n-$	Polymethyl methacrylate, Acrylic, Lucite, or Plexiglas	Windows, costume jewelry, handles, decorative objects

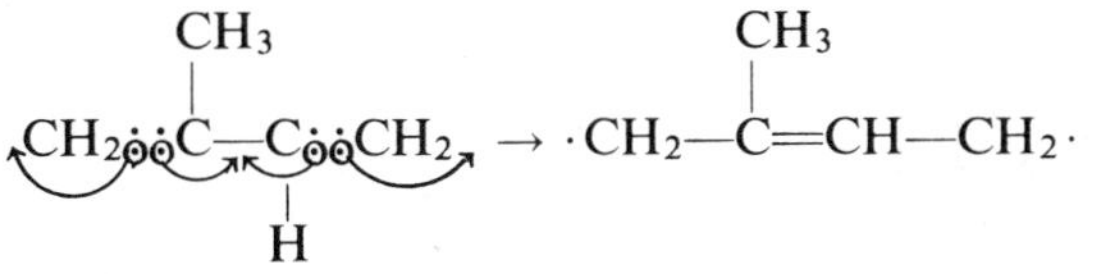

$$CH_2{:}{:}C(CH_3)-C(H){:}{:}CH_2 \rightarrow \cdot CH_2-C(CH_3){=}CH-CH_2\cdot \qquad 20\text{–}9$$

isoprene monomer → active radical with unpaired electrons

$$2\ \cdot CH_2\text{—}\overset{\overset{\displaystyle CH_3}{|}}{C}\text{=}CH\text{—}CH_2\cdot \rightarrow$$

$$\odot CH_2\text{—}\overset{\overset{\displaystyle CH_3}{|}}{C}\text{=}CH\text{—}CH_2\text{—}CH_2\text{—}\overset{\overset{\displaystyle CH_3}{|}}{C}\text{=}CH\text{—}CH_2\odot \qquad \text{20–10}$$

new bond; unpaired electron; unpaired electron

dimer

$$\text{Reaction with more isoprene units} \rightarrow \left[CH_2\text{—}\overset{\overset{\displaystyle CH_3}{|}}{C}\text{=}CH\text{—}CH_2\right]_n \qquad \text{20–11}$$

polymer

A single polymer molecule of rubber may contain 2000–7000 monomer units and have a molecular weight of 136,000–500,000 amu.

The vulcanization process produces cross-links between polymer chains; the links are made up of sulfur atoms. In Figure 20–1 natural, vulcanized, and stretched rubber are compared schematically. Chemists have been able to synthesize polymers that have elastomer properties similar to those of natural rubber. Some of these synthetics are better than natural rubber for particular applications—more oil resistant, etc.

Condensation Polymers

The second type of polymer, **condensation,** is produced most often when two different bifunctional monomer molecules react with one another. A small molecule, often water, is produced as well as the polymer. One of the best known of the condensation polymers is nylon. This name is applied to a polymer produced by the reaction of a monomer containing two amine groups

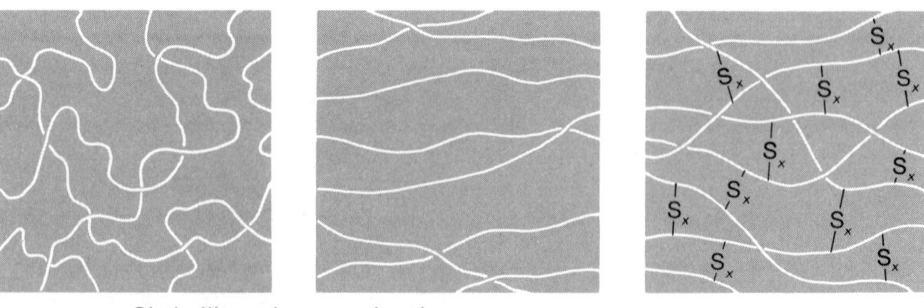

Figure 20–1.
Comparison of Natural and Vulcanized Rubber

with a monomer containing two carboxylic acid groups. The resultant bond or linkage is due to the formation of an amide, and the polymers are called **polyamides.** One type of commercial nylon is produced by reacting adipic acid, $HOOC{-}(CH_2)_4{-}COOH$, with hexamethylenediamine, $H_2N{-}(CH_2)_6{-}NH_2$. The reaction is

$$HO{-}\overset{\displaystyle O}{\overset{\|}{C}}{-}(CH_2)_4{-}\overset{\displaystyle O}{\overset{\|}{C}}{-}OH + H{-}\overset{\displaystyle H}{\overset{|}{N}}{-}(CH_2)_6{-}NH_2 \rightarrow$$

$$HO{-}\overset{\displaystyle O}{\overset{\|}{C}}{-}(CH_2)_4{-}\overset{\displaystyle O}{\overset{\|}{C}}{-}\overset{\displaystyle H}{\overset{|}{N}}{-}(CH_2)_6{-}NH_2 + H_2O \qquad \text{20–12}$$

new bond

The acid group on the left reacts with another amine group, and the amine group on the right with another acid group. A continuation of the process produces the polymer. The bonds holding monomeric units together are amide linkages. The resultant polymer, Nylon 66 (each monomer contains six C atoms), is represented by the formula

$$\left[\overset{\displaystyle O}{\overset{\|}{C}}{-}(CH_2)_4{-}\overset{\displaystyle O}{\overset{\|}{C}}{-}\overset{\displaystyle H}{\overset{|}{N}}{-}(CH_2)_6{-}\overset{\displaystyle H}{\overset{|}{N}}\right]_n$$

Another type of condensation polymer is a polyester produced by the reaction of a diacid with a dialcohol. Such a product, Dacron or Mylar, is produced from ethylene glycol and terephthalic acid. The reaction is

$$\underset{\text{ethylene glycol}}{HO{-}CH_2{-}CH_2{-}OH} + \underset{\text{terephthalic acid}}{HO{-}\overset{\displaystyle O}{\overset{\|}{C}}{-}C_6H_4{-}\overset{\displaystyle O}{\overset{\|}{C}}{-}OH} \rightarrow$$

$$HO{-}CH_2{-}CH_2{-}O{-}\overset{\displaystyle O}{\overset{\|}{C}}{-}C_6H_4{-}\overset{\displaystyle O}{\overset{\|}{C}}{-}OH + H_2O \qquad \text{20–13}$$

new bond

dimer

Each end of the dimer can react further with appropriate groups giving the polymer

$$\left[CH_2{-}CH_2{-}O{-}\overset{\displaystyle O}{\overset{\|}{C}}{-}C_6H_4{-}\overset{\displaystyle O}{\overset{\|}{C}}{-}O\right]_n$$

When the trihydroxy alcohol glycerol (glycerin) is used instead of ethylene glycol, a cross-linked polymer, Glyptal, is produced. The reactions involved are

$$\underset{\text{glycerin}}{HO{-}CH_2{-}\underset{\displaystyle \overset{|}{OH}}{CH}{-}CH_2{-}OH} + \underset{\text{terephthalic acid}}{HO{-}\overset{\displaystyle O}{\overset{\|}{C}}{-}C_6H_4{-}\overset{\displaystyle O}{\overset{\|}{C}}{-}OH}$$

$$\text{etc.}{-}O{-}\overset{O}{\overset{\|}{C}}{-}C_6H_4{-}\overset{O}{\overset{\|}{C}}{-}O{-}CH_2{-}\underset{\displaystyle \overset{|}{O}{-}\overset{}{C}({=}O){-}C_6H_4{-}C({=}O){-}OH}{CH}{-}CH_2{-}O{-}\overset{O}{\overset{\|}{C}}{-}C_6H_4{-}\overset{O}{\overset{\|}{C}}{-}O{-}\text{etc.}$$

$$\text{etc.}{-}O{-}\overset{O}{\overset{\|}{C}}{-}C_6H_4{-}\overset{O}{\overset{\|}{C}}{-}O{-}CH_2{-}\underset{\displaystyle \overset{|}{O}{-}C({=}O){-}C_6H_4{-}C({=}O){-}O{-}}{CH}{-}CH_2{-}O{-}\overset{O}{\overset{\|}{C}}{-}C_6H_4{-}\overset{O}{\overset{\|}{C}}{-}O{-}\text{etc.}$$

$$\text{etc.}{-}O{-}\overset{O}{\overset{\|}{C}}{-}C_6H_4{-}\overset{O}{\overset{\|}{C}}{-}O{-}CH_2{-}\overset{|}{CH}{-}CH_2{-}O{-}\overset{O}{\overset{\|}{C}}{-}C_6H_4{-}\overset{O}{\overset{\|}{C}}{-}O{-}CH_2{-}\underset{\displaystyle \overset{|}{O}{-}C({=}O){-}C_6H_4{-}C({=}O){-}O{-}\text{etc.}}{CH}{-}CH_2{-}O{-}\overset{O}{\overset{\|}{C}}{-}C_6H_4{-}\overset{O}{\overset{\|}{C}}{-}O{-}\text{etc.} \qquad \text{20–14}$$

A third example of condensation polymerization is represented by the phenol-formaldehyde polymer called Bakelite. It was one of the first condensation polymers produced and used commercially. The reaction between phenol and formaldehyde does not involve bifunctional molecules but is, nevertheless, a condensation process. The reactions are:

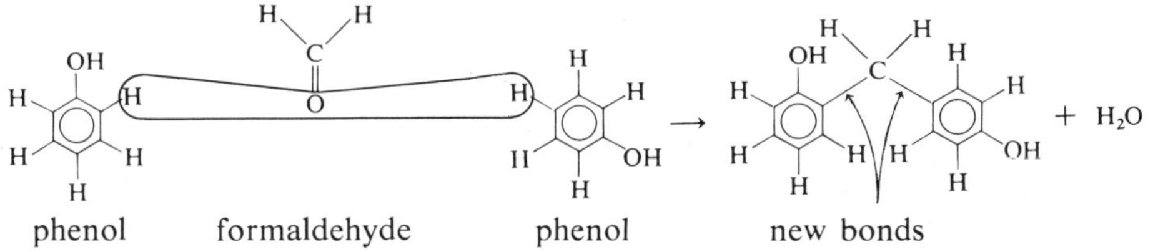

phenol + formaldehyde + phenol → new bonds + H_2O (20–15)

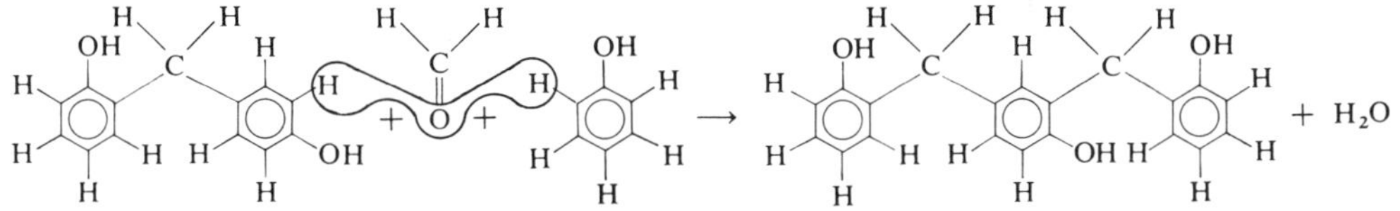

→ + H_2O (20–16)

This process continues, giving the polymer

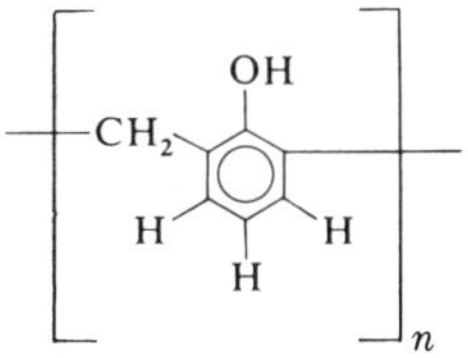

$$\left[-CH_2-C_6H_3(OH)- \right]_n$$

Cross-linking is also possible in phenol-formaldehyde polymers. The process takes place when formaldehyde molecules react with hydrogen atoms of two different polymer chains:

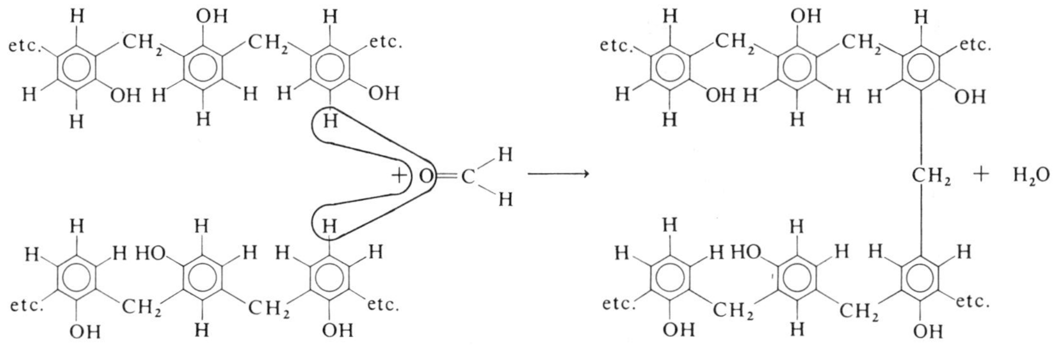

→ CH_2 + H_2O (20–17)

Table 20–2 lists some common synthetic condensation polymers. The position of the linkage between monomer units is indicated by the dotted line.

Table 20–2. Condensation Polymers

Monomer components	*Polymer*	*Polymer uses*
$HOOC-(CH_2)_4-COOH$ adipic acid; $H_2N-(CH_2)_6-NH_2$ hexamethylenediamine	$[-C(=O)-(CH_2)_4-C(=O)-N(H)-(CH_2)_6-N(H)-]_n$ Nylon 66	Textile fibers, cars, valves
$HO-CH_2-CH_2-OH$ ethylene glycol; $HOOC-C_6H_4-COOH$ terephthalic acid	$[-CH_2-CH_2-O-C(=O)-C_6H_4-C(=O)-O-]_n$ Dacron or Mylar	Textile fibers, packaging films, recording tape
$HO-CH_2-CH(OH)-CH_2-OH$ glycerol; $HOOC-C_6H_4-COOH$ terephthalic acid	$[-CH_2-CH(-)-CH_2-O-C(=O)-C_6H_4-C(=O)-O-]_n$ (CH: cross-linked) Glyptal	Baked enamels, molded objects
$H_2C{=}O$ formaldehyde; C_6H_5-OH phenol	$[-H_2C-C_6H_2(OH)-]_n$ (ring: cross-linked) Bakelite, Formica, Micarta	Handles, table tops, cabinets, wall panels
$H_2C{=}O$ formaldehyde; $(H_2N)_2C{=}O$ urea	$[-H_2C-N(-)-C(=O)-N(H)-]_n$ (N: cross-linked) urea formaldehyde resin	Adhesives, wall panels, bottle caps
$H_2C{=}O$ formaldehyde; $C_3N_3(NH_2)_3$ melamine	$[-H_2C-N(H)-C_3N_3(-NH-CH_2-)-N(H)-]_n$ (CH_2: cross-linked) Melmac	Dishes, buttons, panel boards

Thermoplastic and Thermosetting Polymers

Polymers have been classified as addition or condensation on the basis of formation reactions. Another useful classification is based on their behavior when heated. Polymers classified as **thermoplastic** become soft and remolded by heating. **Thermosetting** polymers harden or set when heated and remain hard and rigid even when heated again.

The essential difference is that thermoplastic polymers undergo little cross-linking, while thermosetting polymers, when set, are extensively cross-linked. The behavior of thermosetting polymers indicates that heating causes or accelerates the cross-linking process.

The uses made of polymers obviously depend a great deal on this behavior when heated. Many items are made by **injection molding,** a process in which melted polymer is injected under high pressure into a mold. Upon cooling, the polymer retains the shape of the mold. It is apparent that only thermoplastic polymers would be suitable for use in this process. In another application of polymers, Glyptal is mixed in liquid form with pigments to form enamels. After application to a surface, the mixture is heated (cured), the Glyptal cross-links, and a hard lustrous finish is produced.

Durable-press garments owe their unique property to cross-linking. Most durable-press textiles contain a mixture of man-made polyester (such as Dacron) and cotton (which is a form of cellulose). The fabric is folded and pleated into the desired shape, then treated with a chemical capable of forming cross-links in the cotton. After treatment, the formed cross-links lock the desired folds and pleats into the fabric.

Natural Polymers

Rubber, already mentioned, is a natural addition polymer. A number of condensation polymers occur naturally and are of great importance to man. These biologically important substances include cellulose (rayon is regenerated cellulose), starch, and proteins; all are discussed in detail in Chapters 23 and 24.

Review Questions

1. Illustrate the addition polymerization and cross-linking of vinyl chloride (Table 20–1) by means of equations similar to 20–3 through 20–8.
2. Show that all of the monomers of Table 20–1 can be considered to be derivatives of ethylene.
3. The first synthetic rubber, known as neoprene, was produced by the polymerization of 2-chloro-1,3-butadiene. Draw the structural formula of the monomer and illustrate the polymerization process by equations similar to 20–9 through 20–11.

4. The toy known as a super ball is composed of polybutadiene, $\text{+}CH_2\text{—}CH\text{=}CH\text{—}CH_2\text{+}_n$. The polymer forms by a 1,4 addition process. Draw the structural formula of the monomer and represent the polymerization by appropriate equations.
5. What is the vulcanization process?
6. By means of equations illustrate the condensation polymerization and cross-linking of formaldehyde and melamine to form Melmac (Table 20–2).
7. Silicon is found in the same family of the periodic table as carbon. Although silicon does not possess the property of catenation to the degree shown by carbon, a number of similar compounds of the two elements are known. One of these, $HO\text{—}Si(CH_3)_2\text{—}OH$, will undergo a condensation polymerization (water is eliminated) to form polymers—"silicons." Use equations to illustrate the reaction.
8. Differentiate between thermoplastic and thermosetting polymers. Illustrate a use for each type.

Suggestions for Further Reading

Anderson, E. V., "Rubber," *Chemical and Engineering News*, July 14, 1969.

Carpenter, F. L., "Rubber in the Seventies," *Chemical and Engineering News*, April 27, 1970.

Fisher, H. L., "Rubber," *Scientific American*, Nov. 1956.

Frazer, A. H., "High Temperature Plastics," *Scientific American*, July 1969.

Mark, H. F., "Giant Molecules," *Scientific American*, Sept. 1957.

Mark, H. F., "New Concept—New Polymers—New Applications," *American Scientist*, Sept. 1967.

Mark, H. F., "The Nature of Polymeric Materials," *Scientific American*, Sept. 1967.

Morton, M., "Big Molecules," *Chemistry*, Jan. 1964.

"Teflon—From Nonstick Frying Pans to Space Vehicles," *Chemistry*, June 1965.

21 Environmental Chemistry I: Air Pollution

What is meant by the term *air pollution?* A comparison of unpolluted and polluted air provides a general answer and a logical starting point for a study of air pollution.

A Comparison of Unpolluted and Polluted Air

Air is loosely defined to mean the mixture of gases that exists in a relatively thin layer around the earth. Its composition is not absolutely constant. Components that vary most in concentration are water vapor and carbon dioxide. The amount of water vapor varies from 6 percent down to a few tenths of 1 percent, depending on the weather and temperature. The concentration of CO_2 is always low—a value of about three hundredths of 1 percent (0.03 percent) is common. This concentration is increased somewhat (but still measured in hundredths of 1 percent) in the vicinity of CO_2 sources such as decaying vegetation or burning fuels. Relatively low concentrations are found over fields of growing vegetation or in air that has recently passed over the ocean. These low values are caused by the high solubility of CO_2 in water and the ability of plants to absorb CO_2 during photosynthesis. Because of the small amounts involved, the effect of either variation on the overall CO_2 concentration is minimal.

The composition of dry air, from which absolutely all water vapor has been removed, is practically constant over the entire world. The composition of such clean, dry air collected near sea level is given in Table 21–1. All components making up at least 0.0001 percent of the total volume are included. Two units are used to specify concentrations: **percent by volume** is the volume of the specific component contained in 100 volumes of air, and **parts per million (ppm)** represents the volume of a specific component in one million volumes of air. Both units are used, but parts per million is especially useful in the expression of the very low concentrations common in pollution work. Notice that concentrations in ppm are merely 10,000 times the volume percentage. In addition to the components listed in Table 21–1, trace amounts (less than 1 ppm) of other gases are also found in air.

Table 21–1. Composition of Clean, Dry Air

Gaseous component	*Formula*	*Volume percentage*	*Parts per million*
Nitrogen	N_2	78.08	780,800
Oxygen	O_2	20.95	209,500
Argon	Ar	0.934	9,340
Carbon dioxide	CO_2	0.0314	314
Neon	Ne	0.00182	18
Helium	He	0.000524	5
Methane	CH_4	0.0002	2
Krypton	Kr	0.000114	1

Air is never found completely clean in nature. Gases such as SO_2, H_2S, and CO are continually released into the air as by-products of natural occurrences such as volcanic activity, vegetation decay, and forest fires. In addition, tiny particles of solids or liquids are distributed throughout the air by winds, volcanic explosions, and other natural processes.

Man's activities add other substances to these "natural pollutants." It is estimated that nearly 300 million tons of man-made pollutants now enter the air annually in the United States. This amounts to about 1.5 tons per person each year, or between 8 and 9 pounds per person daily. It is interesting to contemplate what each of us would do with his daily share, if it were delivered on the doorstep along with the morning milk and paper.

The yearly total of 300 million tons seems like a massive amount of material, until we realize that 3×10^{15} tons of air are present over the continental United States at any one time. If all 300 million tons of pollutants were uniformly distributed in this huge air mass, the resulting concentration of pollutants would be about 0.1 ppm by weight.

Anyone experienced in the delights of inhaling polluted air knows that something is wrong with this reasoning. The major error, of course, is the assumption that all the air over the United States is available to dilute the pollutants. The fact is that very little atmospheric mixing of surface air occurs beyond 10,000–12,000 feet above ground level; many pollutants never rise beyond 2000 feet. Geologic and man-made barriers often limit lateral air movement and thereby greatly decrease mixing and diluting effects.

Thus, the reasons for the well-known buildup of pollutants beyond trace-level concentrations become apparent. Pollutant concentrations of 50–100 ppm have been observed but are not common (yet). However, concentrations in the 10–50 ppm range are not unusual.

Types and Sources of Air Pollutants

Five types of substances, known as primary pollutants, account for more than 90 percent of the nationwide air pollution. They are carbon monoxide (CO), nitrogen oxides (NO_x), hydrocarbons (HC), sulfur oxides (SO_x), and particulates (part.). We shall examine these more closely after discussing some general considerations.

Three questions frequently asked about air pollution are: (1) What source puts the largest amount of pollutant into the air? (2) What single air pollutant is present in the largest amount? (3) What are the effects of the various air pollutants on man and the surroundings? The answers to these important questions provide an overall perspective on the seriousness and extent of air pollution.

The major sources and amounts of each primary pollutant are summarized in Table 21–2. According to this information, transportation is the main source of air pollution—a conclusion reached by comparing the 144.4 million tons of total pollutants from transportation to the annual amounts from other sources. Carbon monoxide is the major individual pollutant, its tonnage being twice that of all other pollutants combined. For these reasons, CO and transportation are often the first items mentioned when air pollution is discussed.

Table 21–2. Primary Pollutant Sources and Amounts (millions of tons/year), 1969

Pollutant source	*Weight of pollutant produced*					*Total weight of pollutant produced by each source*
	CO	*NO_x*	*HC*	*SO_x*	*Part.*	
Transportation	111.5	11.2	19.8	1.1	0.8	144.4
Fuel combustion (stationary sources)	1.8	10.0	0.9	24.4	7.2	44.3
Industrial processes	12.0	0.2	5.5	7.5	14.4	39.6
Solid-waste disposal	7.9	0.4	2.0	0.2	1.4	11.9
Miscellaneous	18.2	2.0	9.2	0.2	11.4	41.0
Total weight of each pollutant produced	151.4	23.8	37.4	33.4	35.2	281.2

From *Environmental Quality*—2nd Annual Report of the Council on Environmental Quality (Washington, D.C.: U.S. Government Printing Office, 1971), p. 212.

Evaluation of pollutants and sources solely in terms of tonnage has one serious drawback: it overlooks the possibility that one pollutant might be much more harmful or dangerous than another. Some attempts have been made to account for this possibility by assigning each pollutant a weighting factor. The value of the weighting factor increases with increasing effects of a pollutant on the total environment. The weighting factors used in one approach, based on proposed air-quality standards for California, are summarized in Table 21–3. The weighting factors in this case are the relative toxicities of the pollutants.

Table 21–3. Pollutant Weighting Factors

Pollutant	*Tolerance levels*		*Relative toxicity (weighting factor)*
	ppm	*$\mu g/m^3$*	
CO	32.0	40,000	1.00
HC		19,300	2.07
SO_x	0.50	1,430	28.0
NO_x	0.25	514	77.8
part.		375	106.7

Adapted from *Journal of the Air Pollution Control Association*, 20:658 (1970).

Table 21–3 introduces a new concentration unit. This unit, **micrograms per cubic meter,** is the weight of pollutant in micrograms (10^{-6} g) per unit volume of air (one cubic meter).

The weighting factors of Table 21–3 are somewhat arbitrary, and conclusions they support should be considered qualitative rather than quantitative. Nevertheless, they do provide a more reasonable indication of the air-pollution problem than the oversimplified approach based entirely on total weights of pollutants.

Multiplication of the weight of each pollutant in Table 21–2 by an appropriate factor from Table 21–3 provides an adjusted weight for each pollutant. These adjusted weights can then be used to arrive at summation totals for each pollutant and source. Table 21–4 compares the two methods (weight basis and adjusted weight basis). All numbers are given as a percentage of the total.

Table 21–4. Total Emissions by Source—Weight vs. Adjusted-Weight Basis

	Percentage of total emissions	
Pollutant source	*Weight basis*	*Adjusted-weight basis*
Transportation	51.3	16.8
Fuel combustion (stationary sources)	15.8	33.0
Industrial processes	14.1	26.4
Solid-waste disposal	4.2	2.9
Miscellaneous	14.6	20.9

Conclusions based on the adjusted-weight calculations are somewhat different from those based only on total weights of pollutants. Transportation (mainly automobiles), the traditional villain of air pollution, becomes only the third most serious source when adjusted values are used, and stationary combustion sources take over the number-one position. Even though these results are qualitative, they demonstrate that the pollution problem can and should be looked at from more than one point of view.

Effects of Air Pollutants

The buildup of air pollutants to the 10–50 ppm concentration range is of concern because of the consequences. However, any determination of the damage directly attributed to air pollution is rudimentary at best. The evidence is seldom clear-cut, and research on many environmental effects is only now beginning. The determination of pollution effects is difficult because: (1) pollutants are numerous, varied, and in many cases difficult to detect, (2) in many cases it is nearly impossible to determine the precise degree of exposure of a given receptor (human, plant, etc.) to a specific pollutant, and (3) some pollutants do not in themselves cause problems but may be dangerous in combination with other pollutants. Despite these limitations, available information indicates that air pollutants cause problems in the following areas:

Human health: Eye irritation, respiratory-system irritation and damage, and oxygen deficiency in blood
Materials: Corrosion, loss of strength, loss of color, and weathering
Soiling: Clothes, homes, public buildings, and monuments
Esthetics: Low visibility, odor
Vegetation: Leaf spotting, leaf dropping, tissue death and decay, decreased rate of photosynthesis, and decreased crop yields
Animals: Central nervous system effects, respiratory problems

Table 21–5 shows the areas where effects have been identified for various air pollutants. Some selected pollutants present in trace amounts are listed in addition to the primary pollutants already discussed. Notice the wide variety of substances that are classified as air pollutants.

Carbon Monoxide Pollution

Table 21–2 shows CO to be the most prevalent of all air pollutants. The main chemical process resulting in CO formation is the combustion of carbon-containing materials (mainly fuels) in air. In simplified terms, the combustion of the carbon in these materials proceeds by a two-step process:

$$2C + O_2 \rightarrow 2CO \qquad 21\text{–}1$$

$$2CO + O_2 \rightarrow 2CO_2 \qquad 21\text{–}2$$

The final product of this combustion is CO_2; however, the first reaction (equation 21–1) goes about ten times faster than the second. Therefore, carbon monoxide is an intermediate in the overall reaction and can appear as an end product if insufficient O_2 is present to complete the second reaction. Carbon monoxide also appears in the final product, even when sufficient oxygen is present for complete combustion, if the fuel and air are poorly mixed, since poor mixing leads to localized areas of oxygen deficiency.

As a result of man's activities, enough CO is put into the air annually to double its concentration in the worldwide atmosphere every 4 or 5 years. However, the yearly increase in worldwide CO concentration has been found to be much less than expected on this basis, indicating the existence of a natural CO removal mechanism. Recent research has found that common soil microorganisms can rapidly remove CO from the air. The removal of CO from the air therefore involves a natural soil sink that is dependent upon particular soil microorganisms for its operation. It is estimated that the soil

Table 21-5. Air Pollutants and Their Effects

Air pollutants	*Receptors*					
	Health	*Materials*	*Soiling*	*Esthetics*	*Vegetation*	*Animal*
Particulates	■	■	■	■	■	■
Sulfur oxides	■	■		■	■	■
Oxidants	■	■		■	■	
Carbon monoxide	■					
Hydrocarbons	■			■	■	
Nitrogen oxides	■	■		■	■	
Fluorides	■				■	■
Lead	■				■	■
Polycyclic, org. matter	■					■
Odors (incl. H_2S)	■	■		■		
Asbestos	■					
Beryllium	■				■	■
Hydrogen chloride	■	■			■	
Chlorine	■	■			■	■
Arsenic	■				■	■
Cadmium	■					■
Vanadium	■					
Nickel	■					
Manganese	■					
Zinc	■					■
Copper	■					
Barium	■					
Boron	■					
Mercury	■	■			■	■
Selenium	■					■
Chromium	■	■				
Pesticides	■				■	■
Radioactive substances	■					■
Aeroallergens	■					

From *Environmental Quality*—2nd Annual Report of the Council on Environmental Quality (Washington, D.C.: U.S. Government Printing Office, 1971), p. 105.

sink in the United States alone has the capacity to remove more than 500 million tons of CO per year—over three times the annual U.S. discharge into the atmosphere.

Despite this large capability for removal, CO pollution at significant concentrations is still found in the atmosphere. The primary reason is the nonuniform distribution of atmospheric CO and the soil sink. In fact, the largest CO-producing areas often have the least amount of available soil sink.

Because transportation is the largest single source of CO pollution (77.2 percent), highly populated urban areas show the highest atmospheric concentrations. Within these communities the concentration of atmospheric CO follows a regular daily pattern clearly related to human activities and traffic volume. Daily concentration patterns show little variation with the day of the week except for weekends. Weekday concentrations are higher than those recorded on Saturdays, which in turn are higher than those on Sundays.

Atmospheric concentrations of CO are usually determined on the basis of 8-hour averages. The 8-hour period is used because the body absorbs and desorbs CO slowly, and at current atmospheric levels 4 to 12 hours are required for the establishment of an equilibrium between CO concentrations in the body and in the inhaled air (see Chapter 25 for the toxic effect in the body). Table 21–6 gives 8-hour average CO concentrations for various locations. These are averages of measurements obtained at many air-monitoring stations throughout the United States. The numbers in the first column indicate maximum concentrations that have occurred, since only the most highly polluted sites are included. The numbers in the second column provide a rough measurement of median concentrations.

Table 21–6. Eight-Hour Average CO Concentrations (ppm)

Location	*Eight-hour maximum value exceeded at 5% of sites*	*Eight-hour maximum value exceeded at 50% of sites*
Inside vehicles in downtown traffic	115	70
Inside vehicles on expressways	75	50
Commercial areas	50	17
Residential areas	23	16
Background in relatively clean, unpolluted air	0.025 to 1.0 ppm	

Data selected from *Air Quality Criteria for Carbon Monoxide*, U.S. Department of Health, Education, and Welfare, p. 6–22.

Much effort is being devoted to the development of controls for atmospheric CO pollution. Most of this effort is directed toward the automobile. The greatest improvements in pre-1971 automobile emission performance were accomplished by engine modifications. Further changes of this type, by themselves, will not satisfy future Federal emission standards.

In order to meet the standards set for 1975, attempts are being made to develop exhaust-system reactors. Two types are being considered. In the **thermal exhaust reactor,** CO is oxidized to CO_2 in a high-temperature chamber. As hot exhaust gases pass through the chamber, outside air is added to provide oxygen for the completion of the combustion process. The main problem with thermal reactors is the lack of an economical heat- and corrosion-resistant material to use in the chamber construction. The second type, the **catalytic reactor,** utilizes a bed of granular catalyst material. Exhaust gases are mixed with air and passed over the catalyst. In this process the oxidation of CO to CO_2 is completed at a much lower temperature than is possible in a thermal reactor. The greatest problem with these reactors is the lack of a durable catalytic material (one that will last 50,000 driven miles). The catalysts now in use are subject to poisoning (deactivation) by the

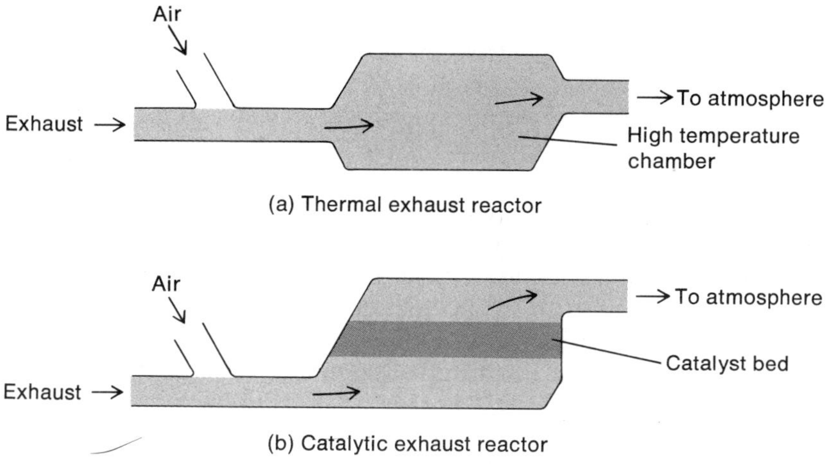

(a) Thermal exhaust reactor

(b) Catalytic exhaust reactor

Figure 21–1.
Exhaust Reactors

adsorption of materials on their surfaces. One of the most effective catalyst poisons is lead, offering another reason for the development of lead-free gasolines. Thermal and catalytic reactors are diagrammed in Figure 21–1.

Nitrogen Oxides as Pollutants

The formula NO_x is used to represent the gaseous atmospheric pollutants nitric oxide (NO) and nitrogen dioxide (NO_2). Although other nitrogen oxides exist, these are the ones primarily involved in air pollution. NO is emitted into the atmosphere in much larger quantities than NO_2.

The formation of NO_x involves a reaction between atmospheric nitrogen and oxygen to give NO. The NO then reacts further with O_2 to give NO_2. The equations for these steps are:

$$N_2 + O_2 \rightarrow 2NO \qquad 21\text{–}3$$

$$2NO + O_2 \rightarrow 2NO_2 \qquad 21\text{–}4$$

At room temperature, N_2 and O_2 have very little tendency to react with each other. At higher temperatures, such as those reached during combustion (1200–1800°C), they react (equation 21–3) to produce significant amounts—in terms of pollutant concentrations—of NO. This may be thought of as a side reaction that takes place during combustion. Table 21–2 shows the major sources of NO_x pollution to be transportation and stationary-source fuel combustion.

The reaction of NO with O_2 to give NO_2 does not account for much NO_2, even in the presence of excess air. This fact is related to the particular temperature and concentration dependence of the rate of the reaction—one of the few that slow down as the temperature increases. Therefore, at high combustion temperatures the rate of the reaction is very low. At the lower temperatures encountered outside the combustion zone the rate increases. This rate increase is counterbalanced, however, by the concentration decrease that takes place as the exhaust gases leave the combustion zone and become diluted with air. The dilution lowers the reactant concentrations and the reaction rate.

Continuous monitoring shows that peak NO concentrations of 1–2 ppm are common, but NO_2 concentrations are usually about 0.5 ppm. These concentrations are many times lower than those common for CO, but nitrogen oxides are more toxic than CO at equivalent concentrations. Both nitrogen oxides are potential health hazards.

On the basis of estimated global background levels and annual emission rates for NO_x, the average residence time of NO_2 in the atmosphere is estimated to be about three days and that of NO about four days. These residence times indicate that natural processes are active in removing the oxides from the atmosphere. The ultimate end product of NO_x pollution is nitric acid (HNO_3), which is precipitated as nitrate salts in either rainfall or dust. Field studies and laboratory investigations have successfully linked these reaction products (nitrates) to the corrosion and weakening of a variety of materials.

Many of the serious effects of NO_x pollution are caused not by the oxides themselves, but by products resulting from the involvement of NO_x in photochemical reactions. These products, called photochemical oxidants, are the more harmful components of smog and are produced when other pollutants take part in a group of naturally occurring atmospheric reactions involving NO and NO_2. These reactions, collectively known as the NO_2 photolytic cycle, are a direct consequence of an interaction between sunlight and NO_2. The steps in this cycle, illustrated in Figure 21–2, are:

Figure 21–2.
The Photolytic NO_2 Cycle

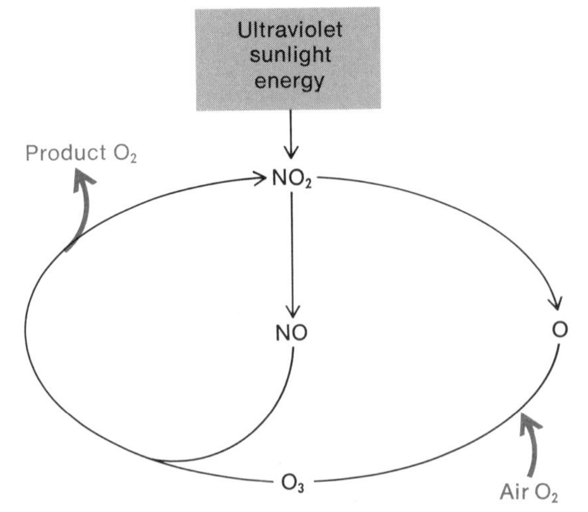

Redrawn from *Air Quality Criteria for Nitrogen Oxides*, U.S. Department of Health, Education, and Welfare, p. 2–3.

1. NO_2 absorbs energy from the sun in the form of ultraviolet light.
2. The absorbed energy causes the NO_2 molecules to break into NO molecules and oxygen atoms (O). The resulting atomic oxygen is extremely reactive.
3. The atomic oxygen reacts with atmospheric oxygen (O_2) to produce ozone (O_3), a secondary pollutant.
4. The ozone reacts with NO to give NO_2 and O_2, and the cycle is completed.

The net effect of this cycle is the rapid cycling of NO_2; if it were not for competing reactions in the atmosphere, the cycle would produce no net overall effect. The atmospheric NO and NO_2 concentrations would not change, because O_2 and NO would be formed and destroyed in equal quantities.

The competing reactions involve hydrocarbons, which are often emitted from the same sources as NO_x. The hydrocarbons interact in a way that causes the cycle to become unbalanced, so that NO is converted into NO_2 faster than NO_2 is dissociated into NO and O. In addition, the imbalance causes a buildup of ozone in the atmosphere. Photochemical pollutants are discussed in more detail in the next section.

The control of NO_x pollution has proved to be quite difficult, and no widely applicable methods have been developed. Since the primary source is automobile exhaust, catalytic reactors have been considered to be possible answers. In these reactors a reduction catalyst would convert NO and NO_2 into N_2.

Hydrocarbons and Photochemical Oxidants

Hydrocarbons and photochemical oxidants are two separate but related categories of pollutants. Hydrocarbons are **primary pollutants** introduced directly into the atmosphere. Photochemical oxidants are **secondary pollutants** originating in the atmosphere from reactions involving primary pollutants. Both are discussed together here, because the majority of photochemical oxidants result from reactions involving hydrocarbons either directly or indirectly. Any estimate of the effects of atmospheric hydrocarbon pollution must include the contributions made by photochemical oxidants. In fact, most of the principal effects attributed to hydrocarbon pollution are thought to be caused by compounds resulting from atmospheric reactions of hydrocarbons.

We noted in Chapter 17 that an almost unlimited number of hydrocarbon compounds exist. Those most important in air pollution are gases at room temperature or highly volatile (have high vapor pressures). Most have relatively simple structures and contain twelve or fewer carbon atoms per molecule.

Table 21–2 indicates that, once again, transportation is the main source of this category of pollutants. This is understandable when we remember that gasoline, the main fuel involved, is a complex mixture of hydrocarbons (see

Chapter 19). Hydrocarbon pollutants from automobiles reach the atmosphere through evaporation from fuel tanks, carburetors, and spills. In addition, incomplete fuel combustion produces exhaust gases containing unreacted hydrocarbons.

The term **photochemical oxidant** (photooxidant) is used to describe an atmospheric substance, produced by a photochemical process (a chemical process that requires light), which will oxidize materials not readily oxidized by gaseous oxygen. Important substances in this category are ozone (O_3) and the family of compounds known as peroxyacylnitrates—commonly called PAN compounds. The simplest and most important member of the PAN family, peroxyacetylnitrate, has the molecular structure

$$CH_3-\overset{\overset{\displaystyle O}{\|}}{C}-O-ONO_2$$

As mentioned earlier, photooxidant production results from an unbalancing of the NO_2 photolytic cycle. In this cycle several substances are generated which readily react with hydrocarbons to produce photooxidants. Hydrocarbons react very rapidly with atomic oxygen to produce a reactive intermediate called a hydrocarbon free radical (RO_2). Free radicals of this type then react with a number of different substances, including NO, NO_2, O_2, O_3, and other hydrocarbons. Many of the resulting products are photooxidants. The process of photooxidant production is shown in Figure 21–3.

Figure 21–3.
Hydrocarbon Disruption of the NO_2 Photolytic Cycle

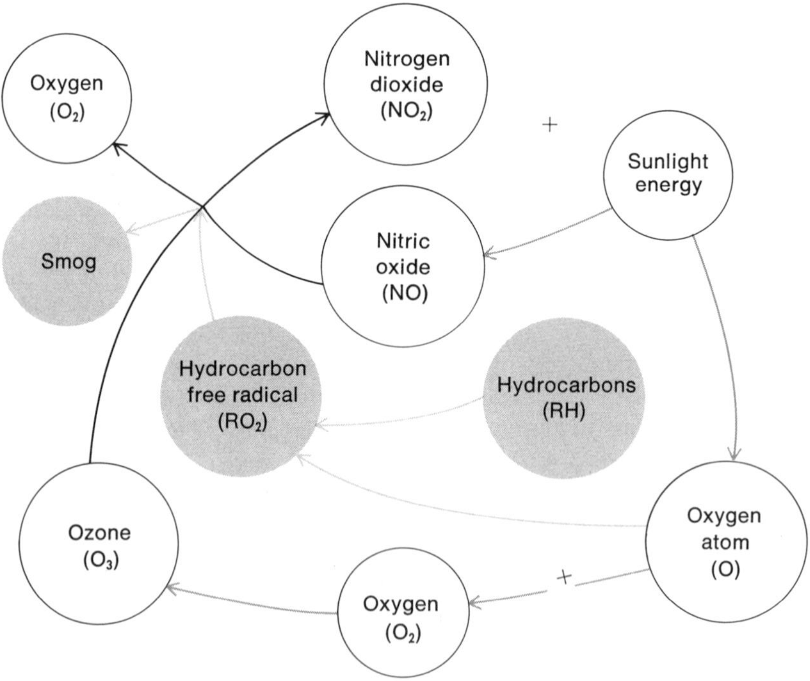

Redrawn, with modifications, from *Air Quality Criteria for Photochemical Oxidants*, U.S. Department of Health, Education, and Welfare, p. 3–1.

Compounds such as O_3 and PAN are known to cause injury to vegetation and are active irritants in smog. Many researchers feel that much of the degradation of materials now attributed to weathering is actually the result of photooxidant attack. Organic polymers, including rubber and natural textiles, are particularly susceptible to attack. Two different effects are produced by these reactions: carbon-chain breaking and carbon-chain cross-linking. In the first, the long carbon-atom chains making up the polymer are broken and the material loses strength. The second effect results in the formation of new cross-links between parallel carbon chains, making the material less elastic and more brittle.

Control of the secondary pollutants ozone and PAN depends ultimately upon control of their primary sources, hydrocarbons and nitrogen oxides. The control of NO_2 has been discussed. The control of hydrocarbon emissions from automobiles has taken the form of evaporation control and the conversion of unburned hydrocarbons in the exhaust into harmless products. Evaporation control (begun nationwide in 1971) essentially involves the collection of vapors from the fuel tank and carburetor. The vapors are returned to the fuel induction system and burned in the engine. Catalytic exhaust reactors are again being investigated as a possible solution to the exhaust problem. In these reactors the hydrocarbons are oxidized to H_2O and CO_2.

Sulfur Oxide Pollution

The two oxides of sulfur, SO_2 and SO_3, are designated collectively by the formula SO_x. The combustion of any sulfur-containing material produces both oxides. The relative amount of each does not depend to a large degree upon the amount of oxygen present; even in the presence of excess air, sulfur dioxide is formed in the largest amount. The quantity of sulfur trioxide formed varies between 1 and 10 percent of the total SO_x. A simplified mechanism for the formation of SO_x can be represented by a two-step process:

$$S + O_2 \rightarrow SO_2 \qquad 21\text{–}5$$

$$2SO_2 + O_2 \rightarrow 2SO_3 \qquad 21\text{–}6$$

The small amount of SO_3 generally produced is the result of two factors, both related to equation 21–6. The first is the rate at which the reaction proceeds; the second, the concentration of SO_3 in the resulting equilibrium mixture. The reaction rate is low at low temperatures and increases with temperature; thus, the rate factor favors SO_3 production at high temperatures. The amount of SO_3 in the equilibrium mixture decreases with increasing temperature, and the equilibrium factor therefore favors SO_3 production at lower temperatures. The small amount of SO_3 produced is the result of the mutual cancellation of these two effects during combustion (high temperature) and removal of exhaust gases (lower temperature).

Gaseous SO_3 can exist in the air only if the concentration of water vapor is very low. When sufficient water vapor is present (as normally it is), the SO_3 and water combine immediately to form droplets of sulfuric acid (H_2SO_4):

$$SO_3 + H_2O \rightarrow H_2SO_4 \qquad 21\text{–}7$$

Thus, H_2SO_4 rather than SO_3 is the compound normally found in an SO_x-polluted atmosphere. The amount of atmospheric H_2SO_4 is greater than can be justified on the basis of primary SO_3 emissions alone. This indicates that another mechanism exists for the production of atmospheric H_2SO_4. One proposed mechanism involves the catalytic oxidation of SO_2 to SO_3 in the presence of sunlight. The catalyst is NO_2, an effective catalyst and pollutant also found in the atmosphere.

Transportation in this case is *not* the main source of SO_x pollution (Table 21–2). Fuel combustion in stationary sources is the main culprit, accounting for 73.1 percent of all man-made SO_x pollution. The ultimate source is sulfur contamination of coal, the most commonly used fuel. Coal now being mined in the United States contains an average of about 2.7 percent sulfur.

Table 21–2 shows that industrial processes rank second to stationary sources in SO_x production. Many useful elements occur naturally in the form of sulfide ores, which, for example, are important sources of copper ($CuFeS_2$ and Cu_2S), zinc (ZnS), mercury (HgS), and lead (PbS). During the extraction of the metals, the sulfur is usually converted into oxides, some of which escape into the atmosphere.

SO_x concentrations in urban atmospheres occasionally reach 2 ppm, but maximum concentrations in the vicinity of 1 ppm are much more common. Injury to sensitive types of vegetation can occur at SO_x levels of 0.5 ppm, and plant injury is, therefore, an area of concern related to this pollutant. The levels of SO_x pollution needed to produce detectable effects in animals and healthy humans are much higher than current atmospheric concentrations. The main effect of SO_x pollution on man is irritation of the respiratory system.

In spite of the high levels of SO_2 needed to produce injurious effects in healthy adults, many health authorities consider SO_2 the most hazardous single air pollutant. The reason is not the effect of SO_2 on the general population, but its effect on the aged and those suffering from chronic diseases of the respiratory and cardiovascular systems. These individuals are highly susceptible to prolonged SO_2 exposure at levels characteristic of air-pollution episodes (0.1–0.4 ppm).

Much of the SO_x damage to materials is caused by the highly reactive sulfuric acid produced when SO_3 reacts with atmospheric water vapor. Corrosion rates of most metals are accelerated by SO_x-polluted environments. High concentrations of H_2SO_4 are capable of attacking a wide variety of building materials. Carbonate-containing materials such as marble, limestone, roofing slate, and mortar are especially susceptible. The carbonates in these materials are converted to sulfates, which are water-soluble:

$$CaCO_3 \text{ (limestone)} + H_2SO_4 \rightarrow CaSO_4 + CO_2 + H_2O \qquad 21\text{–}8$$

Much of the effort toward solving SO_x pollution problems has involved processes for removing sulfur from fuels before combustion. In one process under study, coal is converted into gaseous compounds. The sulfur-containing compounds and CO_2 are removed, and the remaining gases are used as fuel.

Particulates

Although the air pollutants discussed up to this point have all been gaseous, it should not be assumed that all air pollutants are found in that state. Small solid particles and liquid droplets, collectively called particulates, are also present in the atmosphere in great numbers and at times constitute a serious pollution problem.

So many different polluting substances are found in the form of particulates that a general discussion of their chemical properties is very difficult. A number of physical properties, however, are exhibited by particulates in general. The most important of these is size, which ranges from diameters of about 0.0002 μ (about the size of a small molecule) to a diameter of 500 μ. The micron (μ), a unit commonly used in pollution work, is equal to 10^{-6} meters. Figure 21–4 shows the range of sizes characteristic of some selected airborne particles. Notice the area in the lower right, which contains reference sizes and the size range visible to the unaided eye. The table also illustrates the variety of substances classified as particulates.

In the size ranges included in Figure 21–4, particles have a lifetime in the suspended state of between a few seconds and several months. This lifetime depends upon the settling rate, which in turn depends upon the size and density of the particles and the turbulence of the air.

A second general property of particulates is their ability to act as sites for sorption. This property is a function of surface area, which for most particulates is large. The sorption process takes place when an individual molecule impacts on the surface of a particle and sticks or sorbs instead of rebounding. Sorption takes place in three ways. When the impacting molecule is physically attracted and held to the particulate surface, **adsorption** has taken place. **Chemisorption** is the name given to sorption involving a chemical interaction between the impacting molecule and the particulate surface. If the impacting molecule dissolves in the particulate, **absorption** has occurred. These sorption processes help determine the effects of particulate pollution on health.

Optical behavior makes up the third general property of particulates. Those with diameters less than 0.1 μ are sufficiently small, compared to the wavelength of visible light, to affect light much as molecules do and, for example, refract it. Particles with diameters greater than 1 μ are so much larger than the wavelength of light that they obey the same optical laws as macroscopic objects and intercept or scatter light. This property is important in determining the effects of atmospheric particulates on solar radiation and visibility.

Figure 21–4. Sizes of Airborne Particulates

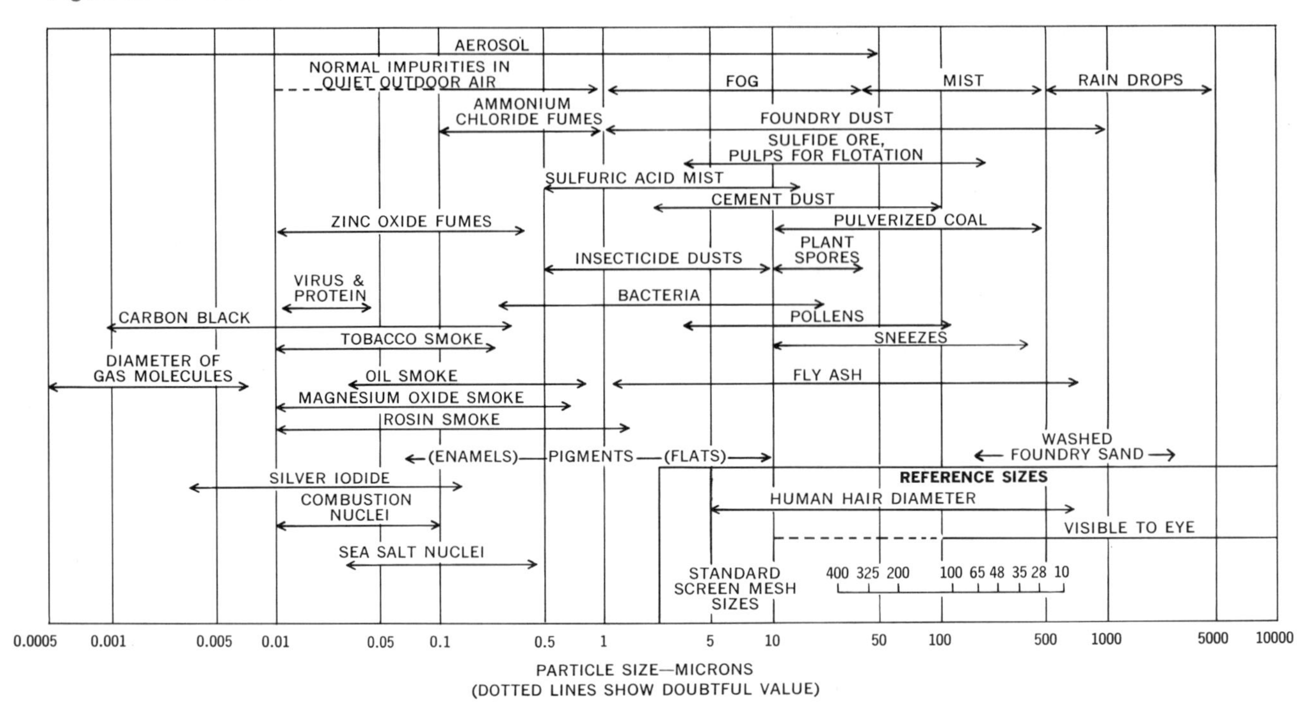

Redrawn, with permission, from L. Byers, "Controlling Atmospheric Particulates," *Technology Tutor* 1:43 (1971).

Table 21–2 shows that industrial processes are the main sources of particulate pollution, with fuel combustion in stationary sources also making a significant contribution. Relationships exist between the source and size of particulates. Particles larger than 10 μ in diameter are produced by mechanical processes such as wind erosion, grinding, spraying, and the pulverizing of materials by vehicles and pedestrians. Particles between 1 and 10 μ in diameter usually include soil, process dusts, and combustion products from local industries. Particles with diameters between 0.1 and 1 μ are primarily combustion products and photochemical aerosols.

Particulate pollutants enter the human body almost exclusively by way of the respiratory system. Particles that enter and remain in the lungs can exert toxic effects in three different ways: (1) The particles themselves may be intrinsically toxic because of chemical or physical characteristics (heavy metals, asbestos fibrils, and sulfuric acid aerosols). (2) The particles may be inert but, once in the respiratory tract, may interfere with the removal of other more harmful materials. (3) The particles may carry adsorbed or absorbed irritating gas molecules and thus enable these molecules to reach and remain in sensitive areas of the lungs (carbon in the form of soot is a good sorber). A great deal of concern has been expressed about particulates in category (1). Intrinsically toxic particulates are not commonly found in the atmosphere in high concentrations, but they do occur in trace amounts. The study of the effects of these trace amounts, especially of metallic particulates, is an area of particular research activity at present. Table 21–7 gives a list of trace metals that may be health hazards.

Table 21–7. Trace Metals That May Pose Health Hazards in the Environment

Element	*Sources*	*Health effects*
Nickel	Diesel oil, residual oil, coal, tobacco smoke, chemicals and catalysts, steel and nonferrous alloys	Lung cancer (as carbonyl)
Beryllium	Coal, industry (new uses proposed in nuclear power industry, as rocket fuel)	Acute and chronic system poison, cancer
Boron	Coal, cleaning agents, medicinals, glass making, other industrial	Nontoxic except as borane
Germanium	Coal	Little innate toxicity
Arsenic	Coal, petroleum, detergents, pesticides, mine tailings	Hazard disputed, may cause cancer
Selenium	Coal, sulfur	May cause dental caries, carcinogenic in rats, essential to mammals in low doses
Yttrium	Coal, petroleum	Carcinogenic in mice over long-term exposure
Mercury	Coal, electrical batteries, other industrial	Nerve damage and death
Vanadium	Petroleum (Venezuela, Iran), chemicals and catalysts, steel and nonferrous alloys	Probably no hazard at current levels
Cadmium	Coal, zinc mining, water mains and pipes, tobacco smoke	Cardiovascular disease and hypertension in humans suspected, interferes with zinc and copper metabolism
Antimony	Industry	Shortened lifespan in rats
Lead	Auto exhaust (from gasoline), paints (prior to about 1948)	Brain damage, convulsions, behavioral disorders, death

Redrawn, with permission, from table in *Chemical and Engineering News*, 49:30 (July 19, 1971).

Particulates in the atmosphere also exert a definite influence on the amount and type of sunlight that reaches the earth. A principal effect is decreased visibility, which may be dangerous (interferes with aircraft or automobile operation) or merely annoying (sightseeing).

The climate is affected in two ways by particulate pollution. It is believed that particulates, by acting as condensation sites, influence the formation of clouds, rain and snow. The reduction in the amount of sunlight reaching the earth's surface could possibly upset the earth's heat balance and cause the atmosphere to cool. An extreme result, according to some, would be a new ice age on the earth.

Techniques for controlling particulate emissions are based mainly on the idea of capturing the particulates before they enter the atmosphere. Several different processes and devices are used, depending on the size of particles involved. Among these processes and devices are gravity settling chambers, wet scrubbers, electrostatic precipitators, and cyclone collectors.

Temperature Inversions

Temperature inversions can cause serious pollution problems, not because they represent a source of pollution, but because they cause pollutants to accumulate in the lower atmosphere instead of dispersing. Many of the most serious air-pollution episodes (occurrences of extremely adverse health effects) have occurred in this country during temperature inversions.

The movement of air in the atmosphere can take place vertically or horizontally. Horizontal movement is governed mainly by prevailing winds. If these winds are active and of sufficient force, pollutants have little chance to build up before they are dispersed. Surrounding mountains, hills, or even buildings in a large city slow down and break up winds and lessen the horizontal movement of air. With limited horizontal movement, pollutant dispersion becomes dependent on the vertical movement of air.

The temperature of air above the earth normally decreases with altitude. Air closest to the earth's surface is warmed by the earth, expands, and becomes less dense than the cooler air above it. The warm, less dense air then rises through the cooler air, which flows in to replace it. This new lower air is warmed, expands, and in turn rises. Air currents are created this way, and pollutants are dispersed.

Occasionally, this normal pattern is disrupted when a cool layer of air, perhaps from the sea, flows in at a low altitude and displaces the warmer air up to a higher altitude. When this occurs, the air temperature will be found to decrease from the earth's surface to some altitude (for example, 1500 or 3000 feet). This normal behavior then gives way to an abnormal one in which the air temperature from 3000 to 5000 or 6000 feet increases with altitude. Beyond this layer, normal behavior again occurs as the air temperature decreases with altitude. The warm layer starting at 3000 feet and extending to 5000 or 6000 feet is the inversion layer, which effectively creates a lid for the air under it.

The presence of an inversion layer prevents vertical atmospheric circulation, because the cooler air cannot rise through the warm inversion layer. Pollutants put into the air are then trapped in the lower, noncirculating air, as illustrated in Figure 21–5. Situations of this type may continue for days until weather conditions change and the inversion layer breaks up. Inversion layers cause an added pollution problem in the form of increased photochemical activity. The inversion layer is usually warm, dry, and cloudless, and so transmits a maximum amount of sunlight, which interacts photochemically with

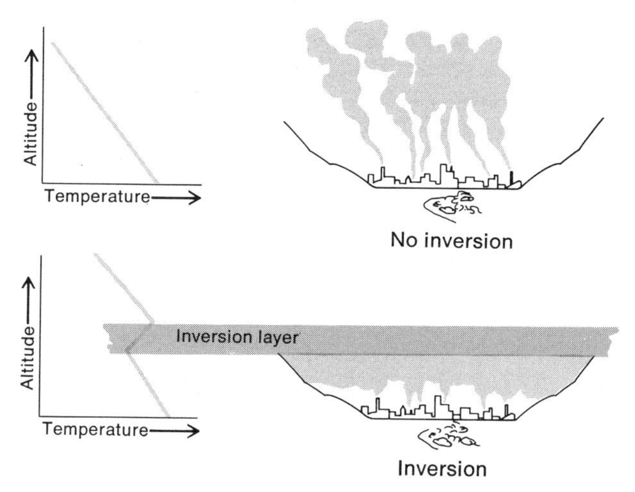

Adapted from *Population, Resources, Environment: Issues in Human Ecology*, 2nd ed., by Paul R. Ehrlich and Anne H. Ehrlich. W. H. Freeman and Company. Copyright © 1972.

Figure 21–5.
Temperature Inversion
A layer of warm air over a layer of cooler air holds air pollution close to the ground.

the trapped pollutants to form extreme amounts of smog. Thus, high levels of smog are usually associated with air-pollution episodes involving temperature inversions.

The Greenhouse Effect

Carbon dioxide is not normally considered to be an air pollutant, because it is a naturally occurring component of air. Carbon dioxide is continually being cycled into and out of the atmosphere in a process involving the activities of plants and animals. In this carbon cycle, plants, by means of photosynthesis, use light energy to react CO_2 from the air with water and produce carbohydrates and oxygen. The carbohydrates are stored in the plants, and the oxygen is released into the atmosphere. When plants are oxidized by natural decomposition, burning, or consumption by animals, oxygen is absorbed from the air and CO_2 is released back into the atmosphere. This carbon cycle of nature (described here in a simplified fashion) results in an essentially constant atmospheric CO_2 level if it is not upset by man.

Man has disrupted the carbon cycle by clearing land, which decreases the available plants; by burning fossil fuels; and by converting limestone into cement. The first activity decreases nature's ability to remove atmospheric CO_2, while the latter two increase the amount in the atmosphere. The net effect is an increasing atmospheric CO_2 level. The most significant activity of the three is the burning of fossil fuels.

The greenhouse effect results from an interaction between the increasing amount of atmospheric CO_2 and solar radiation. Although sunlight consists of many wavelengths, much of the radiation reaching the earth's surface is in the range of visible light. Ozone, which exists naturally in the upper atmosphere, filters out most ultraviolet light (wavelengths shorter than visible); atmospheric water vapor and CO_2 absorb much of the incoming infrared light (wavelengths longer than visible), which we detect on our skin as heat.

Approximately one-third of the light reaching the earth's surface is reflected back into space. Most of the remaining two-thirds is absorbed by such inanimate matter as rocks, concrete, and so forth. This absorbed light is reradiated, when the earth cools, in the form of long-wavelength infrared radiation or heat. Light with these longer wavelengths is absorbed by atmospheric CO_2, releasing heat that raises the temperature of the atmosphere. The CO_2 effectively behaves as a one-way filter, allowing visible light to pass through in one direction, but preventing light of a longer wavelength from passing in the opposite direction. This behavior is represented in Figure 21–6. (Although H_2O and ozone act as filters in the same way as CO_2, their concentrations are not being appreciably affected by man, and so their contribution to the atmospheric temperature remains constant.)

The one-way filtering action of CO_2 has led to some predictions that the temperature of the atmosphere and earth will increase. This is the origin of the term **greenhouse effect,** since the temperature in greenhouses is increased by the use of glass panels which act as one-way filters. The same result can be experienced inside an automobile on a sunny day.

Between 1885 and 1940 the worldwide average temperature did increase by 0.9°F. However, from 1940–1960 a decline of 0.2°F was noted, although Europe and North America experienced additional slight increases. The drop in the worldwide average is attributed to a cyclic variation in airborne particulates, which reflect sunlight. The average temperature has been found to drop following major volcanic eruptions such as those of 1953 in Alaska and 1956 in Russia.

Concern is expressed by some scientists that the greenhouse effect could cause ultimate melting of the polar ice caps. This would raise ocean levels and flood coastal areas of the continents. Other scientists seem to feel that an ice age, caused by particulate pollution, is more likely.

Figure 21–6.
The Greenhouse Effect

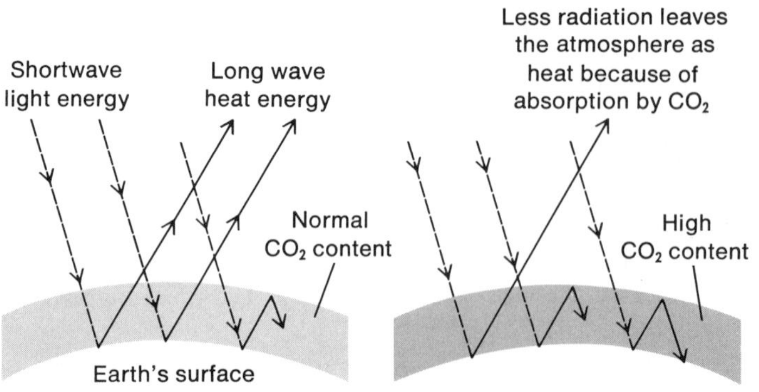

Redrawn, with permission, from Richard H. Wagner, *Environment and Man* (New York: W. W. Norton & Company, Inc., © 1971 by W. W. Norton & Company, Inc.), p. 190.

Review Questions

1. How does the composition of polluted air differ from that of clean, dry air?
2. The seriousness and extent of air pollution can often be estimated by answering three questions. What are the three questions?
3. What is the main source for each of the five primary air pollutants (Table 21–2)?
4. Discuss the significance of using adjusted weights as a basis for reporting pollution contributions by source.
5. In what six areas are air pollutants known to cause problems? Give an example of a typical problem in each area.
6. What is the source of the largest part of the total CO pollution? What chemical reaction is responsible for CO production by this source?
7. How is CO removed from the air naturally? What approach is being investigated to prevent CO pollution?
8. Discuss the reactions and conditions that lead to NO_x air pollution.
9. Explain the effects of the NO_2 photolytic cycle (Figure 21–2) on air pollution.
10. Under what conditions does the NO_2 photolytic cycle become a source of damaging air pollutants? Explain the process (Figure 21–3).
11. Distinguish between primary and secondary air pollutants.
12. Name two important constituents of photochemical smog. What single term is used to describe both of these substances? List some effects on the environment that are caused by these substances.
13. How can the production of photochemical smog be eliminated or at least minimized? Has anything been done toward accomplishing this goal?
14. Discuss the process and conditions that lead to the production of SO_x air pollutants.
15. In what compound form is SO_x pollution commonly found? How does this compound affect building materials?
16. How do the proposed control methods for SO_x pollution differ from those contemplated for CO and NO_x?
17. Particulate air pollutants are varied and cannot easily be classified chemically. What physical properties do they have in common?
18. Which of the physical properties asked for in question 17 is most involved in the effects of particulates on health? Why do you suppose this is the case?
19. What single commonly used substance given in Table 21–7 is the source of the widest variety of possibly hazardous trace metal pollutants?
20. What effects do particulate pollutants exert on the climate?
21. Describe the processes that lead to a temperature inversion. What are the pollution effects of temperature inversions?
22. Discuss the greenhouse effect and its possible influence on the climate.

Suggestions for Further Reading

Chisolm, J. J., Jr., "Lead Poisoning," *Scientific American*, Feb. 1971.

Haagen-Smit, A. J., "The Control of Air Pollution," *Scientific American*, Jan. 1964.

Hall, H. J., and W. Bartok, "NO_x Control from Stationary Sources," *Environmental Science and Technology*, April 1971.

Hall, S. K., "Sulfur Compounds in the Atmosphere," *Chemistry*, March 1972.

Lowry, W. P., "The Climate of Cities," *Scientific American*, Aug. 1967.

McDermott, W., "Air Pollution and Public Health," *Scientific American*, Oct. 1961.

Middleton, J. T., "Planning Against Air Pollution," *American Scientist*, March/April 1971.

Newell, R. E., "The Global Circulation of Atmospheric Pollutants," *Scientific American*, Jan. 1971.

"Rivers of Dust," *Chemistry*, November 1969.

Stoker, H. S., and S. L. Seager, *Environmental Chemistry: Air and Water Pollution*, Scott Foresman and Co., Glenview, Illinois, 1972.

22 Environmental Chemistry II: Water Pollution

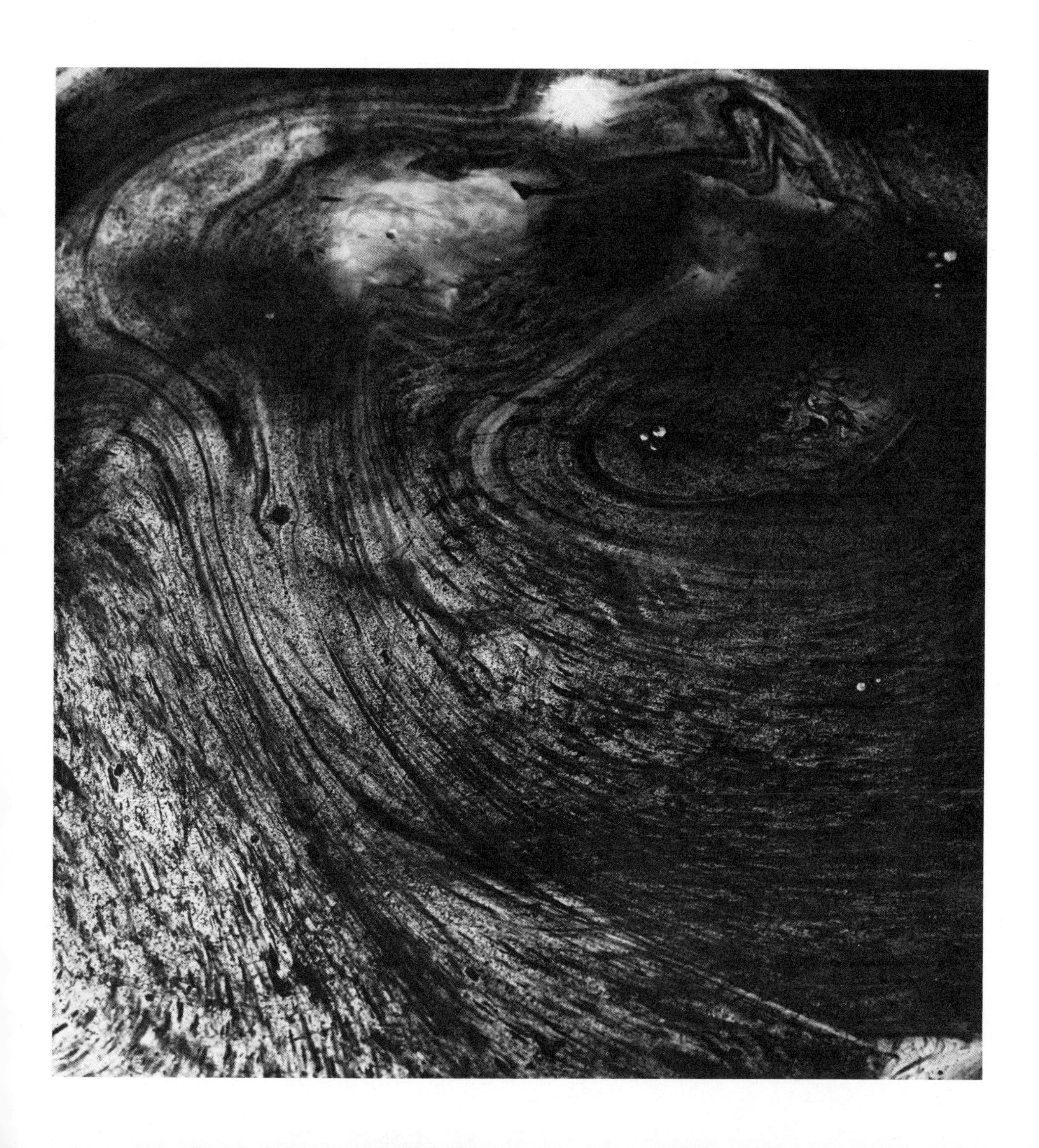

In this chapter the chemical aspects of water pollution will be discussed. Once again, a comparison of the unpolluted and polluted states serves as a useful starting point for the discussion.

A Comparison of Unpolluted and Polluted Water

Strictly speaking, pollution is any departure from purity. When environmental pollution is the topic, the term has come to mean a departure from a normal, rather than from a pure state. This is particularly true for water. This widely distributed substance is such a good solvent that it is never found naturally in a completely pure state.

Even in the most unpolluted geographical areas, rainwater contains dissolved CO_2, O_2, and N_2, and may also carry in suspension dust or other particulates picked up from the atmosphere. Surface and well waters usually contain dissolved compounds of metals such as Na, Mg, Ca, and Fe. The term **hard water** is used to describe water that contains appreciable amounts of such compounds (see Chapter 10). Even drinking water is not pure in a chemical sense. Suspended solids have been removed and harmful bacteria destroyed, but many substances still remain in solution. Indeed, absolutely pure water would not be pleasant to drink, for it is the impurities that give water the characteristic "taste" by which it is recognized.

In light of these facts the term **pure,** in a water-pollution context, will mean a state of water in which no substance is present in sufficient concentration to prevent the water from being used for purposes thought of as normal, including (1) recreation and aesthetics, (2) public water supply, (3) fish, other aquatic life, and wildlife, (4) agriculture, and (5) industry. Any substance that prevents the normal use of water must be considered a water pollutant. The water-pollution problem is complex, partly because the normal uses of water are so varied. Water that is suited for some uses, and therefore unpolluted, may have to be considered polluted when other uses are contemplated.

The Extent of Water Pollution

The United States Environmental Protection Agency estimates that almost one-third of the stream miles in the nation are characteristically polluted in the sense that Federal water-quality criteria are violated. In this estimate, pollution is defined as a demonstrable and recurrent breach of any of the physical or chemical criteria of purity that apply to water bodies, and not merely as a violation of regulations imposed upon waste dischargers. The estimated prevalence of U.S. water pollution, by region, for 1970 is given in Table 22–1.

Table 22–1. Regional Water Pollution in the United States

Region	Percent of stream miles polluted	*Percent of watersheds in pollution status*			
		Predominantly polluted[a]	Extensively polluted[b]	Locally polluted[c]	Slightly polluted[d]
Pacific Coast	33.9	14.8	59.3	22.2	3.7
Northern Plains	40.0	37.5	33.3	25.0	4.2
Southern Plains	38.8	27.3	51.5	18.2	6.1
Southeast	23.3	14.3	41.1	16.1	28.6
Central	36.6	23.2	51.8	21.4	3.6
Northeast	43.9	36.1	55.6	5.6	2.8
E. of Mississippi R.	31.6	23.0	48.7	15.5	12.8
W. of Mississippi R.	35.5	24.1	47.1	20.7	4.6
United States	32.6	23.7	48.5	17.7	9.9

[a] Predominantly polluted: ≥ 50 percent of stream miles polluted.
[b] Extensively polluted: 20–49.9 percent of stream miles polluted.
[c] Locally polluted: 10–19.9 percent of stream miles polluted.
[d] Slightly polluted: <10 percent of stream miles polluted.

From *Environmental Quality*—2nd Annual Report of the Council on Environmental Quality (Washington, D.C.: U.S. Government Printing Office, 1971), p. 220.

Column 2 gives the total percentage of stream miles polluted within each region. Columns 3–6 classify the drainage basins within a region according to the percentage of the basins suffering from given degrees of pollution. For example, the first value under column 3 shows that 14.8 percent of the basins in the Pacific Coast region have 50 percent or more of their stream miles polluted. The statistics of Table 22–1 show that water pollution is not a regional but a nationwide problem in the United States. The distribution of states within the regions is illustrated in Figure 22–1.

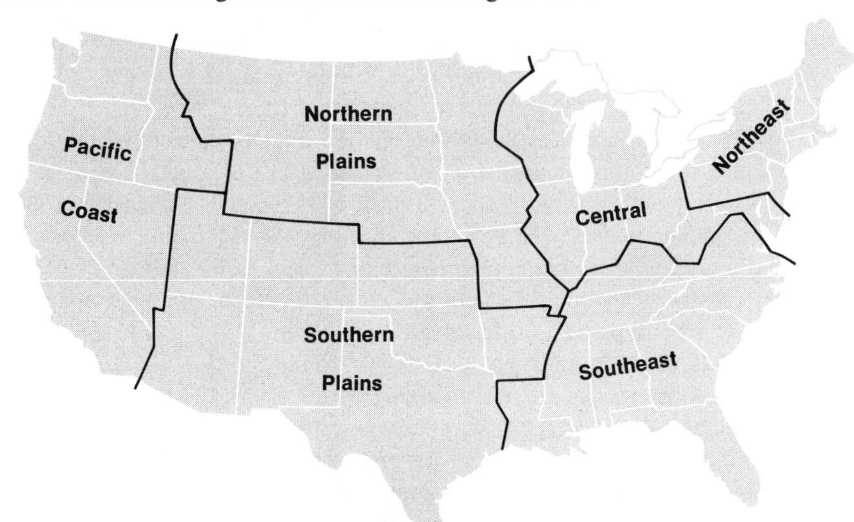

Figure 22–1.
Water Pollution Comparison Regions

From *Environmental Quality*—2nd Annual Report of the Council on Environmental Quality (Washington, D.C.: U.S. Government Printing Office, 1971), p. 212.

Sources of Water Pollution

Water pollution is caused by many different types of substances. An indication of their wide diversity is given by the varied outward signs of their effects, such as bad-tasting water, masses of aquatic weeds growing unchecked, emission of disgusting odors, floating scum, and the disappearance of game and commercial fish. It is unlikely that these varied effects result from a single cause. It seems much more reasonable to attribute them to numerous sources and causes.

The classification of water pollutants into categories allows us to discuss them systematically. Nine categories will serve our purposes in this book: (1) oxygen-demanding wastes, (2) disease-causing agents, (3) plant nutrients, (4) synthetic organic compounds, (5) oil, (6) inorganic chemicals and mineral substances, (7) sediments, (8) radioactive materials, (9) heat. Some overlap between categories is unavoidable, because some wastes contain more than one pollutant. Raw sewage, for example, is an oxygen-demanding waste which may contain disease causing agents and plant nutrients.

Oxygen-Demanding Wastes

Dissolved oxygen is a fundamental requirement for the plant and animal population in any given body of water. Their survival depends upon the water's ability to maintain certain minimal concentrations of this vital substance. Fish require the highest levels, invertebrates lower levels, and bacteria the least. For a diversified warm-water biota, including game fish, the dissolved-oxygen (DO) concentrations should be at least 5 mg/l (5 ppm). For a cold-water biota, DO concentrations at or near saturation values are desirable. The minimum level should be no lower than 6 ppm. The amount of DO at saturation varies with water temperature and altitude. At 20°C and 1 atmosphere of pressure, the value is 9 ppm. High mountain lakes may contain 20 to 40 percent less DO than similar lakes at sea level.

A body of water is classified as polluted when the DO concentration drops below the level necessary to sustain a normal biota for that water. The primary cause of water deoxygenation is the presence of substances collectively called **oxygen-demanding wastes.** These are substances easily broken down or decayed by bacterial activity in the presence of oxygen. The available dissolved oxygen is consumed by bacterial activity; thus, the presence of such materials quickly leads to a depletion of dissolved oxygen.

Although some inorganic substances are found in this category, most oxygen-demanding wastes are organic compounds. Typically they come from such sources as sewage, both domestic and animal; industrial wastes from food-processing plants; wastes from paper-mill activities; tanning operation by-products; and effluent from slaughterhouses and meat-packing plants. Their effects are a function of the amount of water available for dilution.

For this reason, it is not surprising to find that low-DO problems are especially common in late summer and early fall, when water levels are normally low.

Most compounds involved in this type of pollution contain carbon as their most abundant element. One reaction they undergo, with bacterial help, is the oxidation of carbon to CO_2:

$$C + O_2 \rightarrow CO_2 \qquad 22\text{–}1$$

In this reaction 32 grams of oxygen are required to oxidize 12 grams of carbon. The carbon can thus be thought of as demanding nearly three times its weight in oxygen. On this basis, 9 ppm of oxygen is needed to react with approximately 3 ppm of dissolved carbon. This amounts to a reaction between the dissolved oxygen from a gallon of water and a small drop of oil. It becomes easy to see how waters can quickly be depleted of dissolved oxygen.

Because oxygen-demanding wastes rapidly deplete the DO, it is important to be able to estimate the amount of these pollutants in a given body of water. The biochemical oxygen demand (BOD) of water is a quantity related to the amount of wastes present. In a water sample, the BOD indicates the amount of dissolved oxygen used up during the oxidation of oxygen-demanding wastes. It is measured by incubating a sample of water for five days at 20°C. The amount of oxygen consumed (BOD) is determined by a chemical analysis of the DO concentration of the water before and after incubation.

A BOD of 1 ppm is characteristic of nearly pure water. Water is considered fairly pure with a BOD of 3 ppm, and of doubtful purity when the BOD value reaches 5 ppm. Public health authorities object to runoffs entering streams if the BOD of the runoff exceeds 20 ppm. A comparison of these BOD levels with the range of values characteristic of the sources given in Table 22–2 indicates the seriousness of the problem. Obviously, these pollutants must be highly diluted upon entering water if the dissolved oxygen is not to be rapidly and completely depleted. The problem is especially critical for bodies of water already low in dissolved oxygen.

Table 22–2. Characteristic BOD Levels

Source	*BOD range (ppm)*
Untreated municipal sewage	100–400
Runoff from barnyards and feed lots	100–10,000
Food-processing wastes	100–10,000

Data selected from *Wastes in Relation to Agriculture and Forestry*, U.S. Department of Agriculture, pp. 41–44.

The disappearance of plant and animal life is an obvious result of the oxygen depletion of water. The reason is either a direct killing or migration to other areas. A less obvious but important result is a shift in water conditions from those favoring aerobic activity (oxygen required) to those that support anaerobic activity (oxygen not needed). This occurs when the oxygen levels become so low that the aerobic microorganisms are destroyed or driven away and anaerobic ones take their place. The products of decomposition following these two pathways are quite different, as shown in Table 22–3.

Table 22–3. Comparison of Decomposition End Products under Differing Conditions

Aerobic conditions	*Anaerobic conditions*
$C \longrightarrow CO_2$	$C \longrightarrow CH_4$
$N \longrightarrow NH_3 + HNO_3$	$N \longrightarrow NH_3$ + amines
$S \longrightarrow H_2SO_4$	$S \longrightarrow H_2S$
$P \longrightarrow H_3PO_4$	$P \longrightarrow PH_3$ and phosphorus compounds

Redrawn, with permission, from L. Klein, *River Pollution*, vol. 2: *Causes and Effects* (London: Butterworth & Company, Ltd., 1962).

Methane (CH_4) is odorless and flammable; amines have a fishy smell; hydrogen sulfide is bad-smelling and toxic; and some phosphorus compounds have unpleasant odors. When these contributions are added to the odor of decaying fish or algae, it becomes apparent that a shift from aerobic to anaerobic conditions of decomposition is not favored by users of fresh air.

Disease-Causing Agents

Water is a potential carrier of pathogenic microorganisms which can endanger health and life. The pathogens most frequently transmitted through water are those responsible for infections of the intestinal tract (typhoid and paratyphoid fevers, dysentery, and cholera) and those responsible for polio and infectious hepatitis. Historically, the prevention of waterborne diseases was the primary reason for pollution control in water. Modern disinfection techniques have greatly reduced this danger in the United States. This is not true in other parts of the world, where, for example, cholera epidemics are still common. The decrease of the waterborne disease hazard in the United States is illustrated in Figure 22–2, where the typhoid-fever death rate is shown over a period between 1880 and 1970.

The fact that such disease-causing agents are under control must not result in a false sense of security. The occurrence of a polluted water supply leading to an outbreak of disease is always a possibility. The organisms responsible are present in the feces or urine of infected people and are ultimately discharged into a water supply.

Even though it might seem desirable, a direct check for these organisms is not routinely performed on water supplies. Indirect methods are used for the following reasons: Pathogens are likely to gain entrance into water only sporadically, and once in the water they do not survive for long periods. Consequently, their presence could easily be missed by routine sampling.

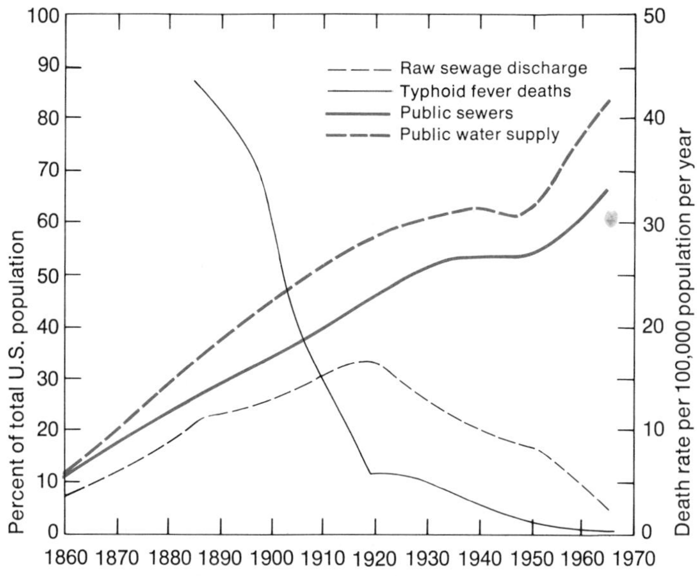

Redrawn from J. N. Pitts and R. L. Metcalf, *Advances in Environmental Sciences*, vol. 1 (New York: Wiley-Interscience, 1969), p. 48.

Figure 22–2.
Growth of Public Water Supply and Sewerage in the United States
A slow but steady decline of typhoid fever deaths during the first sanitary movement beginning in 1860 was followed by the crisis period of World War II and the beginning of the second sanitary movement of 1950.

Laboratory procedures are likely to fail to detect pathogens that are present in very small numbers. Also, it takes 24 hours or longer to obtain results from a laboratory examination. If pathogens were found in a water sample, it is likely that many people would have already used the water and would be subject to infection.

The coliform bacteria form the basis of the indirect method commonly used. These benign organisms live in the large intestine and absorb nutrients from their surroundings. They incite no diseases and are always present in feces. Their presence in water is an indication of fecal discharge into the water. They are present in large numbers, making their detection quite easy. It is estimated that billions are excreted by an average person per day. These natural inhabitants of the human bowel do not find environmental conditions in natural waters suitable for multiplication and, in fact, they begin to die rapidly. Their presence in water samples therefore permits a rough diagnosis of the time elapsed since fecal contamination took place. If fecal contamination is recent, it can be assumed that pathogenic organisms may be present along with the harmless coliform bacteria. The absence of coliform bacteria implies that recent intestinal discharges are not present in the water, and presumably the water is free of pathogens.

It should be mentioned that bacteria responsible for the decomposition of the organic constituents of sewage have no sanitary or public health significance. They are not found in the intestinal tracts of man or animals, and they are not pathogenic.

Plant Nutrients

Nutrients are an important limiting factor in the growth of all plants. With all other factors equal, the rate and profuseness of plant growth are proportional to the amount of nutrient available.

The enrichment of water with nutrients is a naturally occurring biological process called **eutrophication.** The term comes from two Greek words meaning *well nourished.* This enrichment leads to other slow processes collectively referred to as **natural aging of lakes.** Sometimes the term *eutrophication* is used to describe both nutrient-enrichment and lake-aging processes. The existence of peat and muck soils and of deposits of coal and oil show that eutrophication and aging have taken place in the past.

The following steps take place in the eutrophication and aging of a lake: (1) Streams from a drainage basin gradually bring soil and nutrients to a newly formed lake, increasing the fertility of the lake water. (2) The increased fertility gives rise to an accumulating growth of aquatic organisms, both plant and animal. (3) As living matter increases and organic deposits pile up on the bottom of the lake, it becomes more shallow, warmer, and richer in nutrients. (4) Plants take root at the bottom and gradually occupy more and more of the space. Their remains accelerate the filling of the basin. (5) The lake gradually becomes a marsh and finally a field or forest as it is overrun by vegetation.

This aging process is measured in thousands of years. The actual time required depends upon the size and mineral content of the basin and on the climate.

In a strict sense, eutrophication is not a matter of water pollution, because it takes place naturally and is essential to aquatic life. It does become a pollution problem when man, through his activities, accelerates the process and the resultant aging of lakes. The high concentration of natural nutrients and those contributed by man produces rapid plant growth, which first becomes apparent as algae "blooms." The term **bloom** is used when the concentration of an individual species exceeds 500 individuals per milliliter of water.

Algae blooms and the proliferation of other aquatic weeds create numerous problems. Excessive plant growth is often unsightly and interferes with recreational uses of water. The blooms also contribute unpleasant tastes and odors and, when they die and decay, become consumers of dissolved oxygen. This latter process leads to the deoxygenation effects previously discussed.

Most elements needed for growth are available to the plant in amounts well in excess of its needs. A few, however, are present in amounts very close to the minimum quantities required. These may be utilized by the plant almost to the point of exhaustion. Considerable interest is centered around these elements, because they appear to behave as natural controls in preventing excessive plant growth. The growth of a plant will cease when the least available element has been depleted. Man's role in eutrophication appears to be one of increasing the amounts of these growth-limiting elements.

Great changes in the availability of some elements in water have occurred because of man's activities.

Nutrient requirements of plants. Numerous studies are in progress to determine the relationships between nutrient concentrations present in water and the needs of plants. This research into the water chemistry of nutrient cycles has proved to be a very complex problem because of the variability in aquatic systems. Results obtained so far indicate that a single limiting reactant will not be found. The three elements studied in most detail, phosphorus, nitrogen, and carbon (the elements usually needed by plants in the largest amounts), have all been found to be limiting reactants, depending on the water conditions. Phosphorus appears to be limiting to algae growth in nutrient-poor lakes, but it is probably present in excess in nutrient-rich lakes. It is generally agreed that nitrogen is a limiting nutrient in some lakes and in many or perhaps most estuarine and coastal waters. Carbon appears to be a limiting factor only under restricted circumstances such as those present in extremely eutrophic soft-water lakes.

Although all three of these elements may cause excessive plant growth, discussions concerning plant nutrient problems usually center around phosphorus. Control of the phosphorus content of water appears to be the most feasible for two reasons: (1) the sources of phosphorus, mainly sewage, are not as varied as those of carbon or nitrogen; (2) the chemical technology for removing phosphorus from water is further advanced than that for either carbon or nitrogen. Even if carbon or nitrogen were the limiting reactant in a body of water, phosphorus theoretically could be made limiting if enough of it were removed.

Detergents and phosphates. Table 22–4 shows the phosphorus content of various types of water. Notice the large amount of phosphorus contained in domestic waste water (sewage). The primary source of phosphorus in sewage is synthetic detergents. It is estimated that up to 70 percent of the phosphorus in sewage comes from the use of detergents in the household. This fact has caused many arguments about removing phosphates from household detergents.

Table 22–4. Concentration of Phosphorus in Water (ppm)

Water source	*Total P content*
Lake waters	0.01–0.04
Forested streams	0.02–0.10
Agricultural drainage	0.05–1.00
Domestic waste water	3.50–9.00

Reprinted from *Journal, American Water Works Association*, vol. 61, p. 388, by the permission of the Association. Copyrighted 1969 by the American Water Works Association, Inc., 2 Park Avenue, New York, N.Y. 10016.

The chemical formulation of a detergent involves numerous compounds The two main ingredients are a surfactant and a builder. The **surfactant** functions as a wetting agent. It lowers the surface tension of wash water and permits it to penetrate fabrics better. The most common surfactants used are linear alkyl sulfonates (see Chapter 18). The primary role of the **builder** in a detergent formulation is to act as a sequestering agent, tying up hard-water ions in the form of large water-soluble ions (see Chapter 10). Builders also react with wash water and cause it to become alkaline; the alkalinity aids in the effective removal of dirt. The most commonly used builders are polyphosphates, such as sodium tripolyphosphate, $Na_5P_3O_{10}$. The active sequestering agent, the tripolyphosphate ion, has the chemical structure given below:

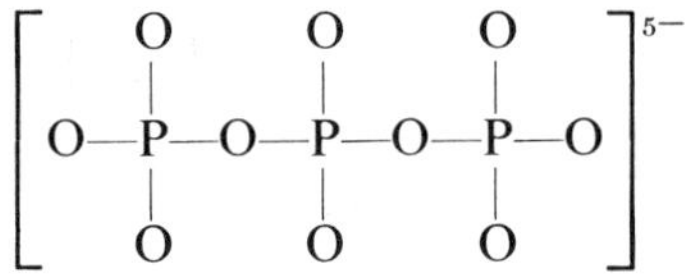

$$\left[\begin{array}{ccccccccccc} & & O & & & & O & & & & O & \\ & & | & & & & | & & & & | & \\ O & - & P & - & O & - & P & - & O & - & P & - & O \\ & & | & & & & | & & & & | & \\ & & O & & & & O & & & & O & \end{array}\right]^{5-}$$

Polyphosphates are useful builders because they are readily broken down into simple phosphates, which are nontoxic to aquatic life and pose no health hazards to man. In addition, polyphosphates are safe to use on colors, fibers, and fabrics, and are noncorrosive to metals (washing machines). Their only drawback is that they happen to be plant nutrients.

There is a great deal of conflict among scientific opinions and no absolute proof that the elimination of phosphates from laundry detergents will solve the eutrophication problem. Nonetheless, detergent manufacturers are taking steps to reduce or eliminate phosphates from their products. The most promising compound to emerge as a possible phosphate substitute was trisodium-nitriloacetate (NTA) with the formula $N(C_2H_2O_2Na)_3$. It was used temporarily during 1970 in some detergent formulations but was withdrawn pending studies of its possible hazards. The development of detergents containing nonsequestering builders (nonphosphate detergents) has proved to be difficult. The substitute builders are carbonates and silicates which react with water, producing a *p*H high enough to represent a health hazard if the wash water is ingested or splashed onto the skin or into the eyes.

Synthetic Organic Compounds

The production of synthetic organic chemicals in the United States has increased by a factor of nearly 14 since the end of World War II (Table 22–5). Much of the increase was caused by the development of new compounds for use in new consumer products rather than the increased use of a few known compounds. Synthetic organic compounds are commonly used as fuels, plastics, plasticizers, fibers, elastomers, solvents, paints, and insecticides. The relationship of these compounds to the environment is only slowly becoming apparent, because many of them have only recently been developed and put to use in commercial products. The areas of particular concern are their toxicities and resistance to natural degradation.

Table 22–5. Production of Synthetic Organic Compounds

Year	*Production (billions of pounds)*
1943	10
1953	24
1963	61
1967	102
1968	114
1969	126
1970	138

Data compiled from *Chemical and Engineering News*, 47:77A (September 1, 1969), and J. N. Pitts and R. L Metcalf, eds., *Advances in Environmental Science*, vol. 1, p.7 (New York: Wiley-Interscience, 1969).

DDT insecticides. The compound known as DDT illustrates the problems that can result from the use of synthetic organic compounds. DDT is a chlorinated hydrocarbon with the following molecular structure:

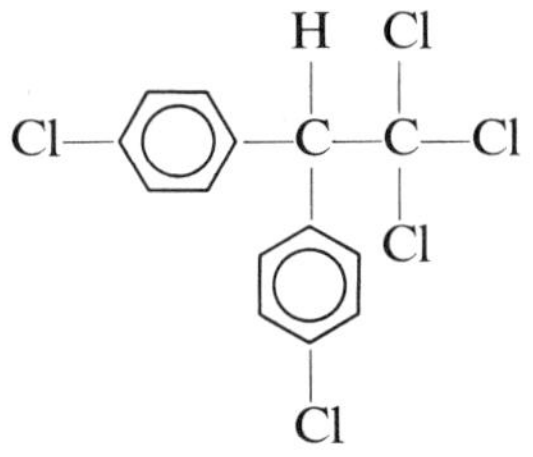

```
                 H   Cl
                 |   |
Cl—⟨◯⟩—C—C—Cl
                 |   |
                 |   Cl
               ⟨◯⟩
                 |
                 Cl
```

The story of this well-known compound is one of rags to riches to rags. It has been in the news almost continually since the late 1940s, when it was first widely used as an insecticide. At first all comments about DDT seemed to be favorable, as the short-term results of its use against agricultural and health pests proved to be spectacular. Today, most statements concerning DDT are negative. The change has taken place because of recent discoveries concerning its long-term, sometimes detrimental, effects on the environment.

In 1874 Othmar Zeidler, a German chemist, synthesized the compound dichlorodiphenyltrichloroethane (DDT) as part of his doctoral research. The compound was not particularly unusual, having been synthesized by simple substitutions on the common ethane molecule. He published his work as a short note in a journal and the compound lay forgotten for many years.

Sixty-five years later this same compound was rediscovered by a Swiss entomologist, Paul Mueller, who was studying the usefulness of various compounds as insecticides. He found Zeidler's compound to be extremely toxic to insects. The Swiss firm for which he worked began immediately to manufacture the substance and in 1942 delivered 6 pounds to the U.S. for testing.

The onset of World War II had deprived the U.S. of its supplier of pyrethrum, a louse control agent, and the military quickly found the new compound a useful substitute. Other successful applications followed rapidly. DDT was heralded for halting a typhus epidemic in Italy during 1943–1944.

As a result of the use of DDT, World War II was the first war to claim more dead by combat wounds than by insect-spread communicable diseases.

After the war, DDT was effectively used to combat the insect carriers of malaria, yellow fever, and typhus. Millions of houses and people were dusted and sprayed with the compound in all-out campaigns against fleas, houseflies, and mosquitoes. It saved literally millions of lives. Statistics show that DDT campaigns in Ceylon reduced human mortality by 34 percent in a single year. Such crop pests as the cotton boll worm were brought under control, and the increased crop yields provided economic benefits for the farmer.

In 1948 Paul Mueller was awarded the Nobel Prize in chemistry for his discovery of the insecticidal properties of DDT. The compound was at the zenith of its popularity.

Problems with DDT. Because of the tremendous success of DDT in controlling pests, its use in large quantities was continued through the 1950s and into the early 1960s. This large-scale use, and the fact that DDT is not easily broken down into harmless compounds, created two problems that only slowly became apparent and eventually led to a ban against its use in the United States. (1) Certain insects developed a resistance or immunity to the compound. (2) Natural processes were found to spread DDT from target areas to nontarget areas and ultimately throughout the environment.

Insect resistance to DDT led to the development of many new insecticides, a number of which are more toxic than DDT itself. The primary advantage of the new compounds is their rapid degradation in nature. Some persist for only two to three weeks after use, while DDT remains unchanged in the environment for years.

A 1971 report of the National Research Council of the National Academy of Sciences suggested that the oceans are the ultimate accumulation site for as much as 25 percent of all DDT used to date. This accumulation results because DDT very easily reaches ponds, lakes, and rivers in a variety of ways. It may be applied directly to water in the form of aerial sprays to control waterborne insects, or it may accidentally fall on water when it is sprayed from airplanes to control forest or agricultural pests. It may be carried down from the atmosphere in rain; rainwater samples have been found that contained up to 0.34 ppb DDT. DDT residues also may reach water in surface runoff from soil. The pesticide is relatively insoluble in water but it strongly adsorbs to organic matter like that present in many soils. When streams or rivers become contaminated with relatively high levels of DDT, it is usually through wind or water erosion of treated soils.

Even though large amounts of DDT have reached the waterways, and ultimately the oceans, the actual concentration in the oceans is quite low because of the dilution effect exerted by the huge volume of ocean water involved. Concentrations of DDT in the parts-per-billion range for fresh water and parts-per-trillion for ocean water are typical.

Biological amplification of DDT. A source of great concern regarding DDT

Figure 22–3.
A Typical Food Chain Concentration of DDT

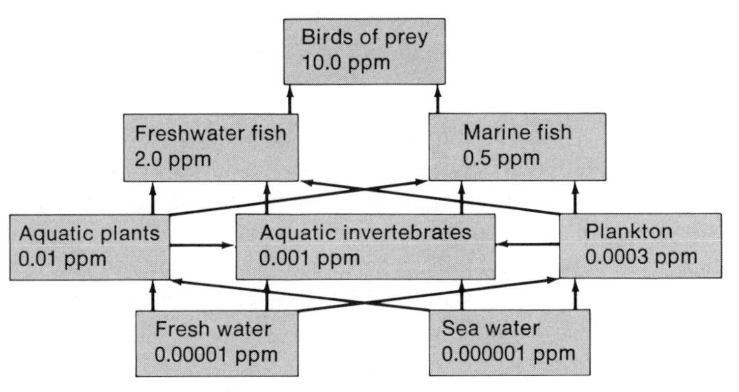

Redrawn, with permission, from part of a chart in Clive A. Edwards, *Persistent Pesticides in the Environment* (Cleveland: CRC Press, 1971).

is the ability of many plants and animals to concentrate it (and other similar compounds) in their body tissues. When the process of **biological amplification** is repeated through several links in a food chain, very high concentrations of DDT residues can be deposited in species at the top. Man is found at the top of every food chain in which he is involved. Typical values for a food-chain concentration pattern are given in Figure 22–3.

Laboratory and field studies of fish and birds indicate possible problems related directly and indirectly to the presence of DDT in marine and other aquatic environments. In fish DDT can (1) induce behavioral changes leading to reproduction problems, (2) affect the mortality rate of the young, (3) be directly toxic to certain species, and (4) accumulate in body tissues to the extent that human consumption of the fish might possibly be dangerous. In birds DDT may accumulate to lethal levels through the consumption of DDT-laden fish or earthworms. Reproductive failure in birds may be caused by sublethal DDT doses. A correlation has been found between DDT residues in eggs and eggshell thickness for a number of birds of prey. The resulting thin-shelled eggs cannot withstand the rigors of incubation.

Oil

During recent years the public has become increasingly aware of the presence of oil pollutants in the sea. Photographs and descriptions of oil-soaked birds floundering on oil-covered beaches have become almost routine presentations of the visual news media.

Oil pollution is an almost inevitable consequence of the dependence of a rapidly growing population on oil-based technology. The use of natural resources such as oil on a grand scale, without losses, is almost impossible. The extent of such losses, intentional or accidental, is steadily increasing and is causing great concern. It is estimated that the total oil influx into the ocean is between 5 and 10 million tons annually. Less than 10 percent of this comes from accidental spillage; the other 90 percent comes from routine activities involving drilling rigs, oil tankers, refineries, and filling stations.

Oil is not a pure substance but a complex mixture of many aromatic and aliphatic hydrocarbons (see Chapter 19). This variety of components must be considered when the fate and effects of oil on the environment are discussed.

Fate of spilled oil. According to an old adage, oil and water don't mix, and it might be expected that spilled oil would just float around until it was washed ashore. This, however, is not the case. One gallon of oil will cover an area of four acres of water when completely spread out. While it is spreading, many changes take place. All oils, regardless of type, contain volatile components that readily evaporate. Within a few days, 25 percent of the volume of a typical oil spill is lost through evaporation. The remaining oil is subjected to emulsification processes which, in a sense, do cause oil and water to mix.

Two types of emulsion are formed, oil-in-water and water-in-oil. The oil-in-water type is often formed in turbulent seas, and the resulting oil droplets become dispersed throughout the water. Water-in-oil emulsions, on the other hand, appear in calmer water as a sticky layer floating on the surface.

Most oil that survives emulsification is degraded by spontaneous photo-oxidation and oxidation by microorganisms. The microorganisms are the most powerful source of oil decomposition found in the sea.

By the end of three months at sea only about 15 percent of the original volume of an oil spill remains. This is in the form of black, tarry lumps of a dense asphaltic substance, and it is these lumps that frequently wash up on beaches.

When massive spills occur close to shore, sufficient time is not available for the process described above to effect the total amount, and a thick, sticky film of oil is deposited on any solid that comes in contact with the spill.

Effects of oil pollution. The problem of oil pollution may be thought of in terms of short-term and long-term effects. Short-term effects are immediately obvious and have received the most publicity. Long-term effects are only slowly becoming apparent and are currently the subject of much study.

Short-term effects fall into two categories, those caused by coating and asphyxiation, and those that result from oil's toxicity. The effects caused by coating and asphyxiation are:

Reduction of light transmission. Measurements showed that ambient light intensity 2 meters below an oil slick was 90 percent lower than at the same depth in clear water. Photosynthesis by marine plant life may be hampered by extended periods of such conditions.

Dissolved oxygen reduction. Oil films retard the rate of oxygen uptake by water. Significantly lower dissolved oxygen levels were found in water beneath a slick than in clear water. The final effects of this have not been determined.

Damage to water birds. Swimming and diving birds become covered with oil which causes their feathers to mat together, reducing their buoyancy and preventing flight. The insulative value of the feathers is also lost, and some birds quickly die of exposure in cold water.

Smothering. The effects of oil on plant life along shorelines have become a matter of concern. Intertidal algae and lichens have been killed by smothering coats of oil which washed ashore in some areas.

The toxic effects of oil result from some of the many compounds that make up this very complex mixture. These compounds vary widely in their molecular weights and structures.

Until recently, low-boiling saturated hydrocarbons were considered to be harmless to a marine environment. It has now been demonstrated that these substances, in low water concentrations, produce anaesthesia and narcosis in a wide variety of lower animals. Cell damage and death result in animals exposed to high concentrations. It is indicated that larvae and other young forms of marine life may be especially affected.

Low-boiling aromatic hydrocarbons are abundant in oil and represent its most dangerous fraction. Such compounds as benzene, toluene, and xylene are found in this fraction. These substances are acute poisons for man as well as for all other living things. Even more toxic to fish are naphthalene and phenanthrene, also present in oil.

The aromatic compounds are more water-soluble than the saturated hydrocarbons. The aromatics can kill marine organisms by direct contact (with oil) or through contact with dilute solutions created when the compounds dissolve into the water from the oil. The effect of these compounds diminishes somewhat with time, because they are volatile and escape by evaporation.

The long-term and especially low-concentration effects of oil components on living systems are not as apparent as short-term effects. Some possible areas of concern have been proposed and are being studied.

Many biological processes of importance to the survival of marine life, and occupying key positions in their life processes, are mediated by extremely low concentrations of chemical messengers in sea water. Some marine predators, for example, are attracted to their prey by organic compounds present in the sea water at concentrations in the ppb range. Similar chemical attractions and repulsions play important roles in such processes as finding food, escaping from predators, locating habitats, and sexual attraction.

There is reason to believe that some compounds of polluting oil interfere with these processes by blocking the taste receptors of organisms or by mimicking the natural stimuli, inducing false responses by the organisms. High-boiling saturated and aromatic compounds are the ones likely to interfere in this way. Such interference could have disastrous effects on the survival of some marine species.

Biological amplification of hydrocarbons. The fate of organic compounds which enter the marine food chain has been studied. Results indicate that hydrocarbons, once incorporated into a particular organism, are quite stable, regardless of their structure. There is also evidence that these compounds can pass through many members of the marine food chain without becoming altered. Such compounds are thus subject to food-chain amplification.

The food chain eventually reaches marine organisms that are harvested for human consumption. This presents the possibility that the flavor of seafood might be negatively affected, or, worse still, that potential long-term poisons might accumulate in food used for human consumption. This latter point is especially unnerving when it is considered that some of the high-boiling aromatic hydrocarbons involved are known or thought to be carcinogenic.

Inorganic Chemicals and Mineral Substances

This category of water pollutants includes inorganic salts, mineral acids, and finely divided metals or metal compounds. These substances enter natural waters as a result of activities in various smelting, metallurgical, and chemical industries; mine drainage; and various natural processes. Their presence in water may increase the water's acidity, salinity, or toxicity.

Acidity. *Acid mine drainage* is the primary source of pollutants that increase water acidity. This drainage adversely affects thousands of miles of streams, and is one of the most significant causes of water-quality degradation in coal-producing areas. The actual pollutants present in mine drainage are sulfuric acid (H_2SO_4) and soluble iron compounds. These substances are formed as the result of a reaction between air, water, and the pyrite (FeS_2) present in coal seams. Certain types of bacteria are involved in the reaction, but their role is not completely understood. This reaction can take place in both underground and surface mines.

During mining operations in deep mines, the strata between the coal seam and the surface are invariably disturbed. Fissures appear, through which water drains into the mine from many surface areas. This water, containing the harmful pollutants, is eventually discharged into surface streams, either naturally or through man-made processes. Damaging mine drainage is formed in surface mines if the surface runoff water comes in contact with pyrite-containing coal.

The total unneutralized acid drainage from coal mines in the United States is estimated at more than 4 million tons of H_2SO_4 equivalent per year. About 60 percent of the mine-drainage pollution problems originate in mines once worked but now abandoned.

Coal mining is not the only source of acidic water pollutants. Other types of mining make their contributions, as do various other industries. Large amounts of acid are used to clean oxides and grease from metals. The used acids from these pickling operations become serious problems in some localities.

The presence of carbonate (CO_3^{2-}) and bicarbonate (HCO_3^-) ions in fresh natural waters causes their *p*H to be generally between 6.5 and 8.5. Large influxes of strong acid can cause significant lowering of these values. The effects depend upon the magnitude of the *p*H drop involved. Some of the known effects are summarized in Table 22–6.

Table 22–6. Adverse Effects of Decreased Water pH

Water pH	*Effects*
Below 6.0	Excessive corrosion of plumbing systems, boats, piers, and related structures takes place.
Below 4.5	Agricultural crop damage occurs.
Below 4.0	Aquatic life is destroyed—all vertebrates, most invertebrates, many microorganisms, and most higher plants are destroyed.

Salinity. Salinity of water is not uncommon. About 97 percent of the total water in the world is found in oceans and seas in the form of salt water. It is common knowledge that such water is not suitable for consumption by man. The remaining 3 percent is classified as fresh water, but it can and does acquire salinity. The sources of salinity are varied and include:

Industrial effluents—Inorganic salts are a major constituent of many industrial effluents. Salts are the products of acid-base neutralizations, many of which are used in various smelting, metallurgical, and chemical industries. Acid mine drainage can also cause salt formation.

Irrigation—Water used in irrigation dissolves large amounts of minerals as it percolates down through soil.

Salt brines—Occasionally, salt brines from mines or oil wells are released into normally fresh water.

Ocean salt—Large rivers, by their continuous outflow, normally prevent salt water of the ocean from backing up. During times of low runoff, river currents may be overcome by tidal flow from the sea, and salty water may move many miles upstream.

Highway use—The use of salt on highways to melt winter ice and snow is creating serious problems in many rural areas. Salt from the highways has killed nearby trees and shrubberies of frontage homesteads. Worse still, the salt has penetrated into the ground waters and polluted rural wells.

Large amounts of salinity in water cause other problems besides those related to human consumption. Dissolved inorganic and mineral substances exert adverse effects on aquatic animal and plant life, leading to irrigation problems in the agricultural industry. Damage to aquatic life is primarily related to the osmosis process (assuming the dissolved substances are nontoxic). Generally, the concentration of dissolved materials in body fluids is the maximum that an aquatic organism can tolerate. When these organisms are in contact with water containing higher concentrations, there is a tendency for water to move out of the living cells into the surrounding water. The resulting increase in solute concentration within the cells of the organism can lead to death. Many freshwater species disappear when waters become brackish.

One of the most serious long-term effects of increased salinity of waters involves the use and reuse of water in irrigation. It has been estimated that about 25 percent of the irrigated land of the U.S. is now affected to some degree by water salinity. Irrigation water brought onto a field always contains some dissolved salts. The concentration of these salts is in the range of 25–8000 mg/l. Plants extract water from the irrigated field, but most of the dissolved salts are excluded by the roots. Water that evaporates from the soil surface leaves dissolved salts behind. These two processes cause residual salts to accumulate in the soil.

In order to preserve the salt balance of the soil and avoid damage to crops, the excess salt accumulation must be leached from the soil with excess irrigation water. Thus, drainage water from the soil contains an increased concentration of salts, which it carries back to the general water supply. Irrigation does not actually produce a pollutant in the form of dissolved salts but merely returns the salts to the general water supply in a more concentrated form.

In addition to total salt content, the nature of the individual salt components affects the soil. These components can be altered by previous use of the water. For example, water used in municipal and some industrial applications is softened before use. The ions Ca^{2+} and Mg^{2+} are replaced by Na^+, resulting in water enriched with Na^+ ions. This high Na^+ concentration adversely affects soils and plants by breaking up soil aggregates and causing the soil to become highly impermeable.

Toxicity. The toxic properties of numerous inorganic compounds, particularly those of some of the heavier metallic elements, have been known for years. Some of these compounds have desirable properties and are routinely manufactured. Appropriate precautions are taken to insure the safety of individuals involved in the manufacturing process. The use of these compounds has led to their introduction into the environment either directly or indirectly, intentionally or unintentionally. The recent detection of these metallic substances in air and water in concentrations approaching toxic levels has created a great amount of concern. The most toxic, persistent, and abundant of these compounds in the environment appear to be those of the metals mercury (Hg), lead (Pb), arsenic (As), cadmium (Cd), chromium (Cr), and nickel (Ni). These metals are known to accumulate in the bodies of organisms, remain for long periods, and behave as cumulative poisons. We shall use mercury to illustrate the problems that can occur with heavy metals.

Mercury is used in a wide variety of ways, as shown by Table 22–7. The largest amounts are used in the chlor-alkali industry for the production of chlorine (Cl_2) and caustic soda (NaOH); the process involves the electrolysis of NaCl solutions, with mercury serving as the cathode. The second greatest consumption of mercury occurs in the manufacturing of electrical apparatus.

Table 22–7. Mercury Consumption in the United States (1969)

Area of use	*Annual consumption (pounds)*
Chlor-alkali industry	1,572,000
Electrical apparatus	1,382,000
Paint	739,000
Instrumentation	391,000
Catalysts	221,000
Dental preparations	209,000
Agriculture	204,000
General laboratory use	126,000
Pharmaceuticals	52,000
Pulp and paper making	42,000
Amalgamation	15,000
Others	1,082,000
Total	6,035,000

Adapted from U.S. Senate Subcommittee on Energy, Natural Resources, and the Environment Hearings, *Effects of Mercury on Man and the Environment* (1970), part 2, p. 116.

The mercury-containing products of this industry include mercury vapor lamps, silent electrical switches, and mercury batteries. The use of mercury and its compounds as fungicides makes up the third largest category. Mercury compounds are used to destroy fungi in the paint, pulp and paper, and agricultural industries. Mercury compounds are added to paint to act as a mildewcide and latex preservative. Paint for use on ships contains mercurial **antifouling** agents to prevent barnacle growth. Agricultural industries use organomercury compounds as **seed dressings** to inhibit the growth of fungi on seeds. Table 22–8 illustrates some typical mercury-containing compounds and their uses.

Table 22–8. Typical Mercurials and Their Uses

Compound	*Formula*	*Uses*
Mercury oxide	HgO	Paint industry—antifouling agent
Phenylmercury acetate	$C_6H_5{-}Hg{-}C(=O){-}OCH_3$	Paint, paper and pulp industries—mildewcide
Methylmercurydicyandiamide (Panogen)	$CH_3{-}Hg{-}N(H){-}C(=NH){-}NHCN$	Seed dressing
Methylmercury acetate	$CH_3{-}Hg{-}C(=O){-}OCH_3$	Seed dressing

A major concern relating to mercury pollution is its appearance in food chains. The contamination of natural food by mercury came to the attention of the public in 1970 through announcements that abnormally high mercury concentrations had been found in both fresh- and saltwater fish. Mercury is thought to enter the food chain by two routes, water and seed dressings. Mercury present in water is concentrated in living tissues through food chains; small mercury-containing organisms are ingested in quantity by larger organisms which, in turn, are eaten by still larger ones and so forth. In the case of fish, mercury levels in the 1–3 ppm range can be easily reached in highly polluted water. Abnormally high levels are considered to be those that exceed the FDA-established maximum of 0.5 ppm for food. Mercury enters food chains through seed dressings when translocation occurs and the mercurial seed dressing is transferred from the seed to the crop and is ultimately consumed.

The occurrence of mercury poisonings is well documented. Five major instances are given in Table 22–9. These incidents resulted from the eating of mercury-contaminated fish or mercury-treated seed grains.

Table 22–9. Major Mercury Poisonings

Location	*Year*	*Results*
Minamata Bay, Japan	1953–60	111 dead or seriously injured
Iraq	1961	35 dead, 321 injured
West Pakistan	1963	4 dead, 34 injured
Guatemala	1966	20 dead, 45 injured
Nigata, Japan	1968	5 dead, 25 injured

Data selected from U.S. Senate Subcommittee on Energy, Natural Resources, and the Environment Hearings, *Effects of Mercury on Man and the Environment* (1970), part 2, p. 110.

The toxic action of mercury in the body is not completely understood, but certain facts are known. All mercury compounds are toxic when present in sufficient quantity. Different mercury compounds exhibit somewhat different characteristics of toxicity, distribution, accumulation, and retention times in the body, and these compounds may be changed from one type to another by biological processes in the environment or the body. Mercury, through various processes, inhibits normal cellular activities in the body, and this damage is usually permanent; there is no known effective treatment.

Sediments

Sediments are almost taken for granted because of the naturally occurring process of erosion. Sediments produced by that process represent the most extensive pollutants of surface waters. It is estimated that suspended solid loadings reaching natural waters are at least 700 times as large as the solid loadings from sewage discharge.

Less commonly known are the effects man has had on erosion rates, and

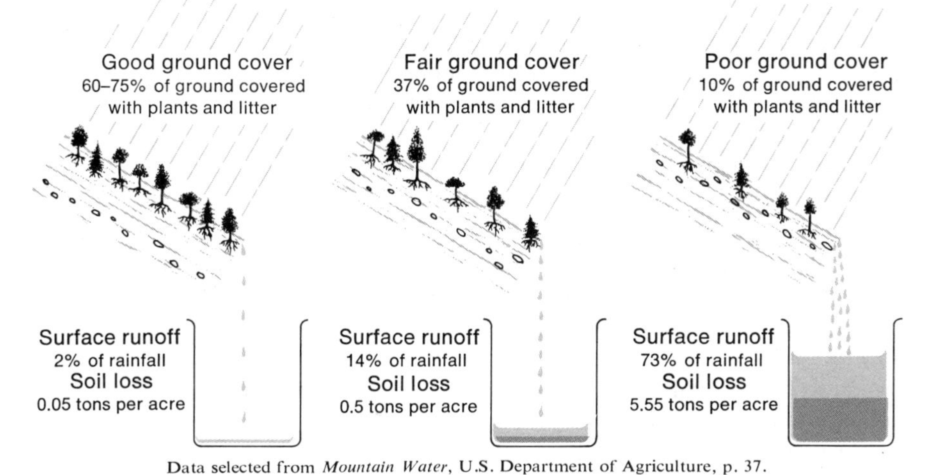

Data selected from *Mountain Water*, U.S. Department of Agriculture, p. 37.

Figure 22–4.
Effect of Ground Cover on Erosion

the far-reaching effects of sedimentation. Erosion rates of land are increased four to nine times by agricultural development and may be increased by a factor of 100 as a result of construction activities. Strip-mining activities greatly influence the rate of erosion in an area. Sediment yields from strip-mined areas average nearly 30,000 tons per square mile annually, an amount ten to sixty times as great as yields from agricultural lands. The amount of sediment washed from an area depends very much upon the condition of the land (Figure 22–4).

Sediment production is a greater problem in some parts of the United States than in others. This variation is not all caused by man's activities, but also involves the type of soils, geology, topography, precipitation, and vegetation cover.

The detrimental effects of sediment in water are the filling of stream channels, harbors, and reservoirs; destruction of aquatic animals; reduction of light penetration into water, reducing the rate of photosynthesis; and clouding of the water.

Radioactive Materials

Uranium and its decay products are examples of elements which possess highly unstable nuclei, whose radioactive emissions are highly injurious or even lethal to living organisms. Four activities are potential sources of radioactive pollutants, and each is known to have been involved in environmental pollution: (1) the mining and processing of ores to produce usable radioactive substances; (2) the use of radioactive materials in nuclear weapons; (3) the use of radioactive materials in nuclear power plants; (4) the use of radioactive materials in medical, industrial, and research applications.

Ore processing. Typically, uranium ore contains about 2–5 lb of U_2O_3 per ton. Large amounts of ore must be processed to produce the usable material; in the processing operations, radioactive wastes are produced. Perhaps the greatest problem is caused by **uranium tailings,** a finely divided solid material that remains after useful materials have been removed. Huge piles of these tailings are found in uranium-producing areas. An estimated 12 million tons are piled up in the Colorado River Basin, an active center of uranium production. Often, no one seems interested in the tailings once mining operations are halted.

A radiation pollution problem is created by these tailings because they contain radioactive decay products of uranium. Two of these are thorium-230 ($^{230}_{90}Th$) and radium-226 ($^{226}_{88}Ra$). Substances such as these can be dissolved or eroded from piles of tailings by rainfall, and in this form they find their way into the general water supply. Radium and thorium chemically resemble calcium and so tend to be absorbed by the bones when taken into the body. Some waters in the Colorado River Basin have had the concentration of $^{226}_{88}Ra$ raised to twice the maximum level permissible for human consumption.

Weapons testing. A well-known source of radioisotopes is nuclear weapons testing. The quantity and variety of radioactive materials produced depends on the type of weapon tested. Some of the radioisotopes are very short-lived, lasting only a few seconds or minutes, while others may have a half-life of several hundred years. These materials reach the earth as radioactive fallout. Concern about fallout led to the signing of the Limited Nuclear Test Ban

Figure 22–5.
Radiation Levels in Nature (all plots are not on the same scale)

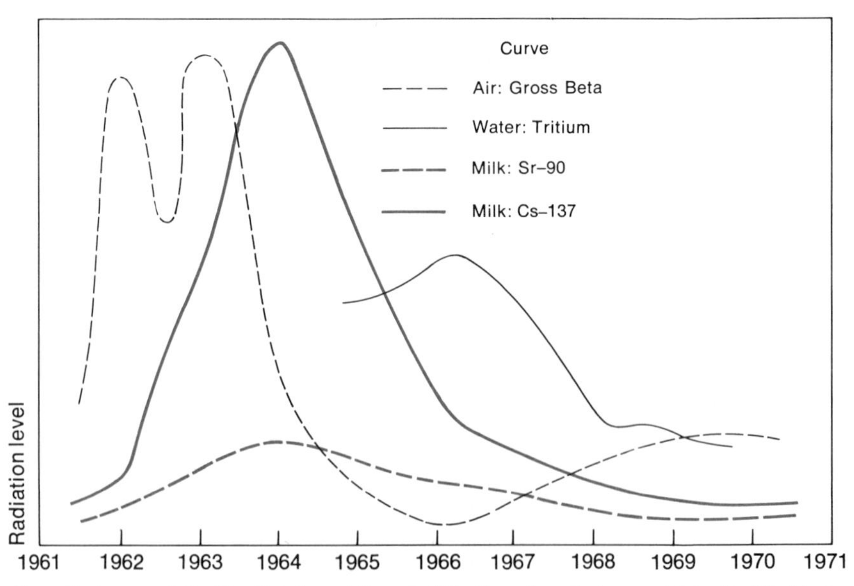

From *Environmental Quality*—2nd Annual Report of the Council on Environmental Quality (Washington, D.C.: U.S. Government Printing Office, 1971), p. 222.

Treaty in 1963. According to the terms of the treaty, the U.S. and U.S.S.R. will limit testing of nuclear weapons to underground detonations. The effects of this ban on radiation levels can be seen from Figure 22–5.

Underground detonations, performed in the U.S. at the Nevada test site, are designed to limit the amount of atmospheric fallout, and in most cases do so. Occasionally, accidental releases of radiation reach the atmosphere because of **venting** at the explosion site. When this happens, radioactive gases are forced up through the hole used to place the explosives. The leakage may occur through channels left for control and monitoring cables, natural cracks or fissures in the rock, fractures created by the blast, or combinations of these. Sometimes venting takes place when the materials used to fill the hole are simply blown out. Leakage may occur immediately after the explosion or later, when rock and earth fall into the cavity created by the explosion and leave a channel for gas venting.

Ventings were reported in 12 of 190 tests conducted at the Nevada site from 1961 through 1969. Others are known to have occurred since 1969. The amount of radiation released is, of course, variable. It is released near ground level and thus poses no worldwide fallout problems, but it may create very serious local hazards.

Atmospheric fallout, regardless of the source, can have far-reaching effects and can be conveyed to man in a variety of ways, as shown in Figure 22–6.

An example of the effects of radioisotopes on man serves to illustrate the hazard of radioactive pollutants. Strontium-90 ($^{90}_{38}Sr$), a component of radio-

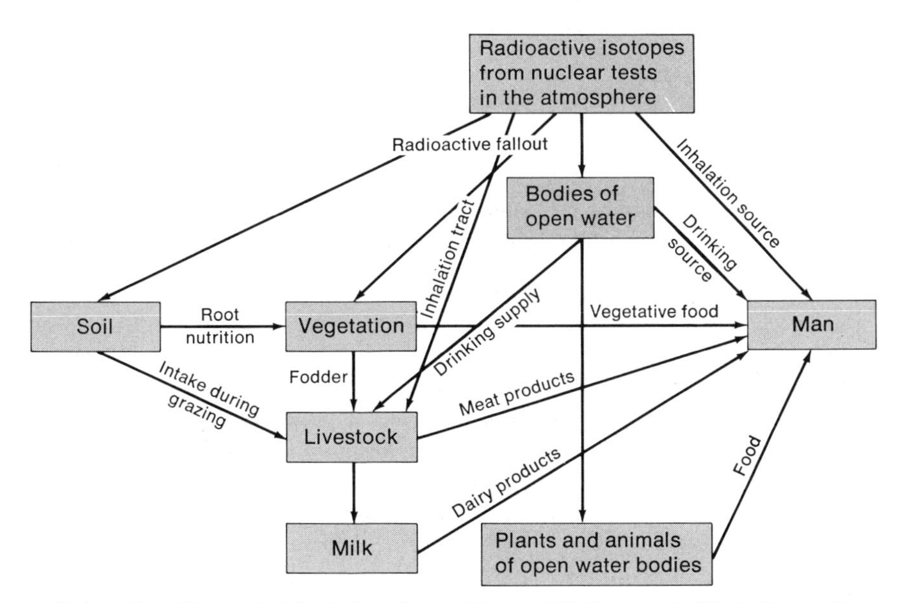

Redrawn from *Wastes in Relation to Agriculture and Forestry*, U.S. Department of Agriculture, p. 18.

Figure 22–6.
Transmission of Radioactive Fallout to Man

active fallout, has a half-life of 28 years and is chemically similar to calcium. Calcium is absorbed from the soil by plants and passed on to feeding animals, where it is used in the formation of bones and teeth. Man obtains calcium from both plant and animal sources, such as milk, vegetables, and cereal grains. Strontium is obtained from these same sources and, because of chemical similarities to calcium, is also deposited in bones and teeth. The marrow of bones is the main site of red blood cell formation. The presence of radioactive strontium-90 in the surrounding bone tissue seriously curtails this production, leading to anemia or more serious disorders.

Nuclear power generation. More and more, man is turning to nuclear generation in attempts to satisfy an increasing demand for electrical power. Nuclear generators have some advantages over more commonly used types. In very large generating facilities, nuclear energy is cheaper to use than that from conventional sources such as fossil fuels or falling water. Also, nuclear generators do not produce the SO_x or particulate pollutants commonly associated with fossil-fuel burning generators. By the year 2000, nuclear power is expected to be the main source of electricity. The statistics of Table 22–10 illustrate this fact.

Table 22–10. Sources of Electrical Energy Generation (in trillions of kilowatt-hours)

	Source of electrical energy					
Year	*Coal*	*Gas*	*Petroleum*	*Hydro*	*Nuclear*	*Total*
1968	2.1	1.0	0.4	0.7	0.04	4.24
2000 (est.)	7.3	1.2	0.3	1.5	11.4	21.7

The operating principles of nuclear power plants were discussed in Chapter 16. These operations produce four types of radioactive pollutants:

Low-level radioactive liquid wastes. Radioactive isotopes are formed when impurities in the primary coolant water and corrosion products from coolant pipes are bombarded with neutrons. This can be controlled to some degree by the use of demineralized coolant water. Much of this type of waste is now disposed of by sealing it in containers which are subsequently dropped into the ocean. There are unknowns in this disposal technique, such as the lifetime of the containers used, and the directions and speeds with which the containers or their contents move when subjected to underwater currents.

Liquid and gaseous wastes from fuel elements. Complete sealing of the fuel in steel or zirconium containers is apparently impossible to attain or sustain. Minute cracks allow fission products to escape into the primary coolant. This adds to the low-level-waste disposal problem discussed above.

Fission products. In from one to three years, fission products (the ashes of nuclear fuels) accumulate to the point that they absorb sufficient neutrons to slow or stop the chain reaction. At this point, the extremely radioactive fuel elements are removed and shipped in special containers to a fuel-reprocessing plant. Here the fission products are separated from the remaining usable fuel. The fuel is returned to the reactor for further use and the waste fission products are stored, in liquid form, in huge underground stainless steel tanks. Hundreds of millions of gallons of these high-level wastes are now in storage.

Heat. The secondary cooling system carries huge amounts of heat away from the reactors and puts it into natural water supplies. The effect of this is discussed next.

Heat

Heat is not ordinarily thought of as a pollutant, at least not in the same sense as a corrosive chemical. However, the addition of excess heat to a body of water causes numerous adverse effects. This serious problem of **thermal pollution** originates primarily with the practice of using water as a coolant in many industrial processes. Most water used for this purpose is returned, with heat added, to the original sources. At present, about 70 percent of the water diverted to industrial use serves as a cooling medium. It is estimated that by 1980 approximately 20 percent of all freshwater runoff in the United States will be used for cooling purposes.

Used coolant water frequently has a temperature 20°F higher than the river or stream to which it is returned. This added heat raises the temperature of the natural waters, with the results that: (1) the amount of dissolved oxygen in the water is decreased; (2) the rates of chemical reactions are increased; (3) false temperature cues are given to aquatic life; and (4) lethal temperature limits may be exceeded.

The decreasing ability of water to contain dissolved oxygen as the temperature increases is shown graphically in Figure 22–7. Some of the effects of low

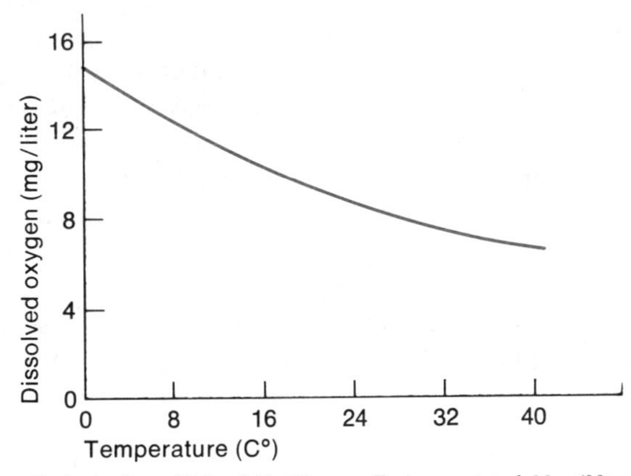

Redrawn from Richard H. Wagner, *Environment and Man* (New York: W. W. Norton and Company, Inc., 1971), p. 137.

Figure 22–7.
The Quantity of Oxygen Dissolved in Water
The higher the temperature, the lower the oxygen content of the water. Thermal pollution can lower the oxygen content below the point necessary to sustain many animals.

dissolved-oxygen levels on bodies of water were discussed earlier under "oxygen-demanding wastes."

The addition of heated water to a cooler body of water may accelerate the lowering of DO levels because of density differences between the two. The less dense warm water tends to form a layer on top of the cooler, denser water, particularly when the body of cool receiving water is deep. The resulting blanket of "hot" water cannot dissolve as much atmospheric oxygen as the underlying cold water, which is denied contact with the atmosphere. Normal biological reduction of the DO level of the atmospherically unreplenished lower layer may lead to anaerobic conditions.

Another effect of this stratification may show up downstream from a dam, when the oxygen-deficient lower level is discharged through the lower gates of a dam. Serious effects on downstream fish life may result. Also, the ability of the stream below the dam to assimilate oxygen-demanding wastes will be curtailed.

The effects of heat in water sometimes show up in nature without man's interference. On hot summer days the temperature of shallow waters sometimes reaches a point that critically reduces DO level; under these conditions, suffocated fish are often found on the surface.

A rough rule of thumb often used by chemists is that the rate of any chemical reaction, including those of respiration and oxidation, approximately doubles with every 10°C (18°F) increase in temperature. In thermally polluted water, fish require more oxygen because of an increased respiration rate. However, the available oxygen in such water has been decreased. Thus, thermal pollution affects fish in a double-barreled fashion.

Other reactions are also influenced. Trout eggs hatch in 165 days when incubated at 37°F. When the water temperatures are 54°F, only 32 days are required, and no hatching occurs at water temperatures in excess of 59°F. Such a result can be disastrous to fish populations. If the fish hatch early and find no natural food organisms available, they do not survive. The natural food of such hatchlings depends on a food chain originating with plants whose abundance is a function of day length as well as temperature.

The life cycle and natural processes of many aquatic organisms are closely and delicately geared to water temperature. Fish often migrate, spawn, and are otherwise distributed in response to water-temperature cues. Shellfish such as oysters spawn within a few hours after their environment reaches a critical temperature. These normal life patterns of aquatic organisms can be completely disrupted by artificial changes in water temperatures.

A problem of thermal pollution not related to aquatic life is the reduced cooling capability of warmed waters. This is important because nuclear reactor power sources require about 50 percent more cooling water for a given temperature increase than similar power plants using fossil fuels, and it appears that nuclear plants will be used extensively in the future.

Review Questions

1. What do the words *polluted* and *pure* mean when used with reference to natural waters?
2. Use the data of Table 22–1 and estimate the extent of water pollution in your region of the country. How does your region compare to the country as a whole?
3. Why would cold water in high mountain lakes contain less dissolved oxygen at saturation than similar lakes at sea level?
4. List the common types of water pollutants that are classified as oxygen-demanding wastes.
5. Suppose wastes such as those of Table 22–2 are diluted by a factor of 2000. Classify the resulting water as nearly pure, fairly pure, or of doubtful purity if the original wastes had BOD levels (a) equal to the low value of the range and (b) equal to the high value of the range.
6. Why are water tests not carried out to directly detect the presence of disease-causing organisms? What indirect method is used?
7. Criticize the following statement: Eutrophication of bodies of water is an undesirable and unnecessary process.
8. What three elements have been identified as limiting growth factors for plants in water? Which one is most often the topic when plant nutrient problems are discussed? Why is it the topic?
9. What single consumer product has been most affected by attempts to solve the plant nutrient problem? Discuss the changes that have been made in this product.
10. The pesticide DDT is an example of a synthetic organic compound that has become a pollution problem. The effectiveness of DDT as a pesticide is partly the result of its persistence—it doesn't change rapidly into non-toxic compounds. How is this property related to the pollution problems associated with the compound?
11. What is meant by the term *biological amplification?* What pollutants are known to undergo this process. Why must man be concerned about the process?
12. In what ways does DDT reach natural water supplies?
13. During the three months following an oil spill far out at sea, about 85 percent of the oil apparently disappears. What happens to the components that make up this 85 percent?
14. What are the short-term effects of oil pollution?
15. What are some long-term effects of oil pollution?
16. List the three general effects on water attributed to inorganic chemicals and mineral substances.
17. Give one or more examples of the substances and their sources that contribute to the effects given in question 16.
18. What is known about the toxic activity of mercury in the body?
19. What are the main sources of mercury pollution?
20. What are the results of water pollution by sediments?

21. What are the main sources of radioactive pollutants?
22. List the four categories of pollutants that are produced by the nuclear generation of electricity.
23. Explain why Figure 22–6 makes you concerned about atmospheric testing of nuclear devices—or the accidental introduction of radioactive materials into the atmosphere.
24. Why is heat considered to be a water pollutant?
25. Discuss the adverse effects caused by releasing excess heat into a body of water.

Suggestions for Further Reading

Aaronson, T., "Mercury in the Environment," *Environment*, May 1971.

Bazell, R. J., "Water Pollution," *Science*, Jan. 22, 1971.

Carter, L. J., "DDT: The Critics Attempt to Ban Its Use in Wisconsin," *Science*, Feb. 7, 1969.

Clark, J. R., "Thermal Pollution and Aquatic Life," *Scientific American*, March 1969.

Goldwater, L. J., "Mercury in the Environment," *Scientific American*, May 1971.

Grant, N., "Mercury in Man," *Environment*, May 1971.

Keller, E., "Fish Kills," *Chemistry*, Sept. 1968.

Keller, E., "The DDT Story," *Chemistry*, Feb. 1970.

Newman, F., "A Water Pollution Study," *Chemistry*, Jan. 1969.

Peakall, D. B., "Pesticides and the Reproduction of Birds," *Scientific American*, April 1970.

Pratt, C. J., "Chemical Fertilizers," *Scientific American*, June 1965.

Stoker, H. S., and S. L. Seager, *Environmental Chemistry: Air and Water Pollution*, Scott Foresman and Co., Glenview, Illinois, 1972.

Williams, C. M., "Third Generation Pesticides," *Scientific American*, July 1967.

Woodwell, G. N., "Toxic Substances and Ecological Cycles," *Scientific American*, March 1967.

23 Introduction to Biochemistry I: Carbohydrates and Lipids

Biochemistry, the science of the chemical substances and reactions common to living things, is an interesting and challenging area for study: interesting because we are all living organisms curious about ourselves, and challenging because of the variety and complexity of the subject.

This chapter presents some basic concepts of biochemistry with the aim of generating some appreciation for this area as well as providing necessary background material for the topics of the chapters to follow. Carbohydrates and lipids, two important types of biochemical substances, will be the main topics.

Carbohydrates

This group of compounds consists of polyhydroxy aldehydes or ketones or substances that yield them when subjected to hydrolysis reactions. The group name comes from the early observation that most compounds of the group had empirical formulas of $C_x(H_2O)_y$, suggesting that they were hydrates of carbon, or carbohydrates. The idea proved incorrect, but the group name has been retained.

The compounds included in this group constitute about 70 percent by weight of the average human diet, and in the diet of Americans provide between 50 and 60 percent of the energy needed by the body. In less prosperous parts of the world as much as 80 percent of the body energy is obtained from these foods, because they are the least expensive.

Monosaccharides

The simplest carbohydrates are the **monosaccharides** or simple sugars. Many are known, but we shall limit our discussion to eight specific examples: four **pentoses**—sugars containing five carbon atoms, and four **hexoses**—sugars containing six carbon atoms.

Each of the four pentoses is represented in Example 23–1 by two molecular structures. Either structure is acceptable, but the cyclic will be used because it is the predominant form shown by most saccharides in the solid and dissolved states.

None of the pentoses is found free in nature. A number do occur as components of **polysaccharides**—carbohydrates containing many monosaccharide units—and may be obtained from these higher saccharides by hydrolysis reactions.

Arabinose is found as a component of the polysaccharide gums of some plants such as cherry trees (cherry gum) or acacia plants (gum arabic). It is one of the main components of commercial mucilage.

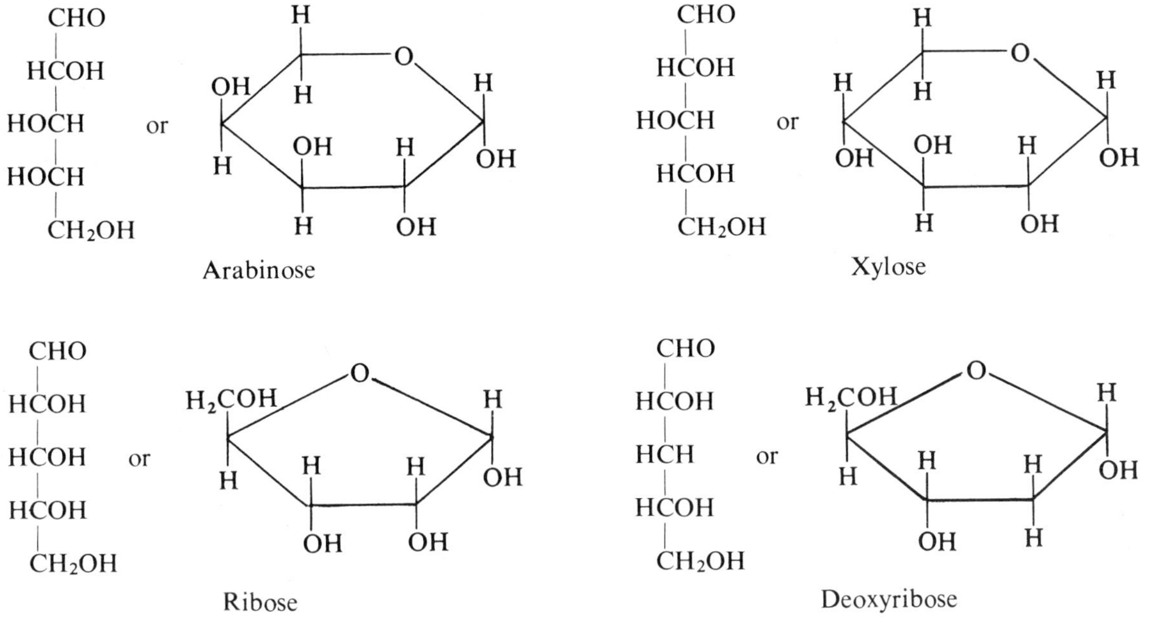

CHO
HCOH
HOCH
HOCH
CH_2OH

or

Arabinose

CHO
HCOH
HOCH
HCOH
CH_2OH

or

Xylose

CHO
HCOH
HCOH
HCOH
CH_2OH

or

Ribose

CHO
HCOH
HCH
HCOH
CH_2OH

or

Deoxyribose

Example 23–1. Pentose Monosaccharides

Xylose is found in the polysaccharides of seed hulls, straw, corncobs, and wood and is obtained from these sources by hydrolysis in hot acid. It is used in the dyeing and leather tanning industry and also as a food for diabetics.

Ribose and deoxyribose (ribose minus an oxygen atom) are biologically very important. They are found in every cell and are involved in enzyme systems, genes and the transfer of genetic information, and protein synthesis.

A number of the hexoses occur in the free state in nature. Of the four represented in Example 23–2, glucose and fructose are found free while mannose and galactose are found only in a combined state.

Glucose, sometimes called dextrose, is widely found in the free state. Its sources are numerous, including honey and fruits such as grapes, figs, and dates. It is a component of starch, a polysaccharide, and is obtained from this source by hydrolysis. Glucose is sometimes called blood sugar because it is the main sugar transported by the blood to body tissues. Most other sugars taken into the body are converted to glucose before being absorbed into the blood. The normal concentration of glucose in the blood is between 80 and 90 mg per 100 ml of blood (.08–.09 percent). The sugar is commonly used as a sweetener in various confections and other foods—especially baby foods.

Fructose, also known as levulose or fruit sugar, is the sweetest of the common sugars. Table 23–1 lists the relative sweetness of some common sugars based on sucrose, which has been arbitrarily given a value of 100. Fructose

Table 23–1. Relative Sweetness of Sugars (Sucrose = 100)

Sugar	*Relative sweetness*
Lactose	16
Galactose	22
Maltose	32
Xylose	40
Glucose	74
Sucrose	100
Invert sugar	130
Fructose	173

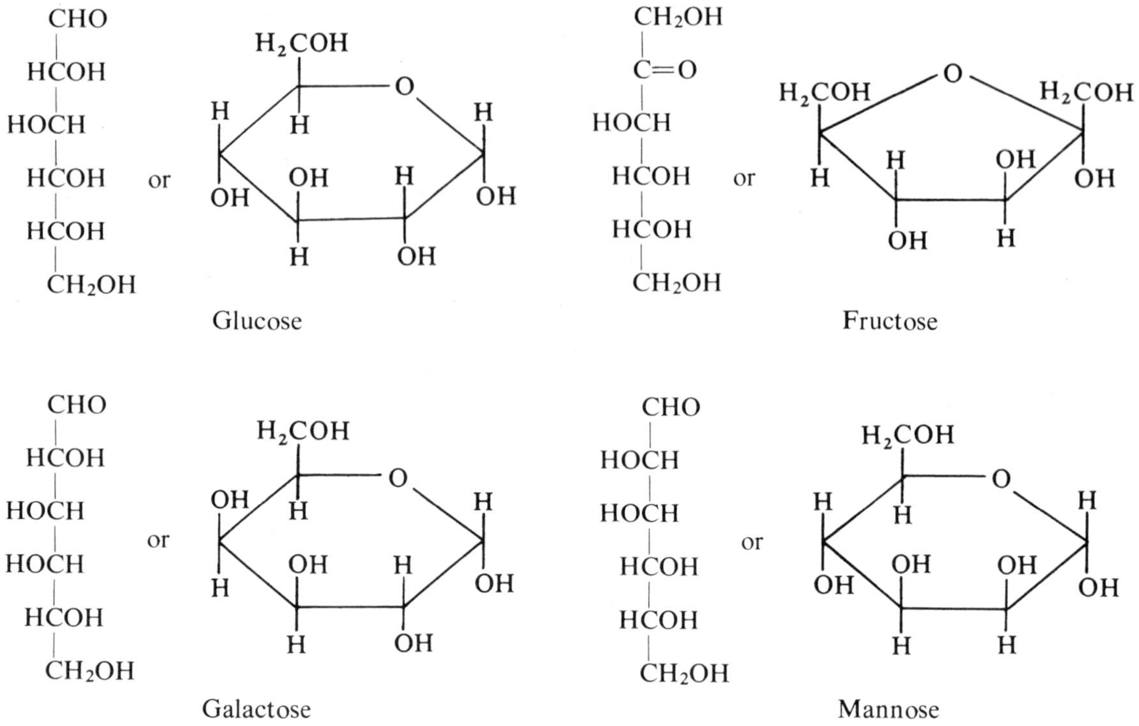

Glucose

Fructose

Galactose

Mannose

Example 23–2. Hexose Monosaccharides

is found in the free state in many fruits and is present in honey in a 1:1 ratio with glucose. It is also found as a component of some higher saccharides and is commonly used as a food sweetener.

Galactose is not found free in nature but only in combination with other sugars in the form of higher saccharides. It is usually obtained by hydrolysis of the disaccharide lactose. Galactose is sometimes called a cerebroside or brain sugar because of its occurrence in a combined state in brain and nerve tissue. It is used medicinally and as a reactant in synthetic organic chemistry.

Mannose is found in the polysaccharides of some plants. It is a component of the polysaccharide mannan found in berries and in vegetable ivory, a hard, white, ivorylike material occurring in the endosperm of palm nuts.

Disaccharides

The next carbohydrates to be considered are the **disaccharides,** sugars composed of two monosaccharide units. A typical disaccharide can be thought of as the product of a condensation reaction between two monosaccharides. A reaction of this type between two glucose molecules to form maltose is:

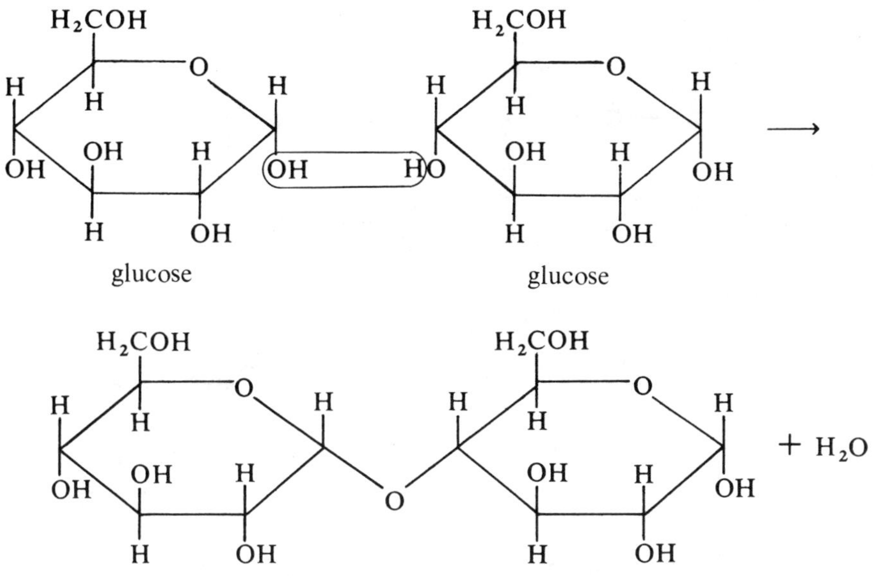

glucose + glucose $\longrightarrow$

maltose $+ H_2O$ 23–1

The resulting etherlike *acetal* linkage between the monosaccharide units is formed by the elimination of water from two OH groups. The reverse reaction, hydrolysis, in which water is added to the acetal linkage, generates the original monosaccharides. Hydrolysis reactions, often involving an enzyme, are common in the higher carbohydrates and, as already mentioned, provide a source for some of the simple sugars.

Example 23–3 gives the formulas of four disaccharides. Each of these sugars has the monosaccharide glucose as one of the components. The glucose unit is represented in Example 23–3 in two different forms, called alpha (α) and beta (β), which are shown in Example 23–4. When the circled OH and CH_2OH groups are on the same side of the plane of the ring, the β form is given. The α form has the two groups on opposite sides of the plane. Notice that the reference OH and CH_2OH groups are attached to carbon atoms of the ring, which are in turn attached to the ring oxygen. The other monosaccharides may be classified in a similar manner.

The best known and most widely used disaccharide is **sucrose**—the common sugar found on the dining table. It occurs free in the juice of many plants, especially sugar cane and sugar beets, the commercial sources. Obtained from these sources in a very pure form, sucrose is probably produced in greater quantities than any other pure organic chemical.

Sucrose is composed of the monosaccharides glucose and fructose. It is 30 percent sweeter than glucose but only about 58 percent as sweet as fructose (see Table 23–1). When it is cooked with acid-containing foods such as fruit, jam, or jelly, partial hydrolysis takes place, and the resulting mixture of glucose, fructose, and sucrose is sweeter than the original sucrose. Foods

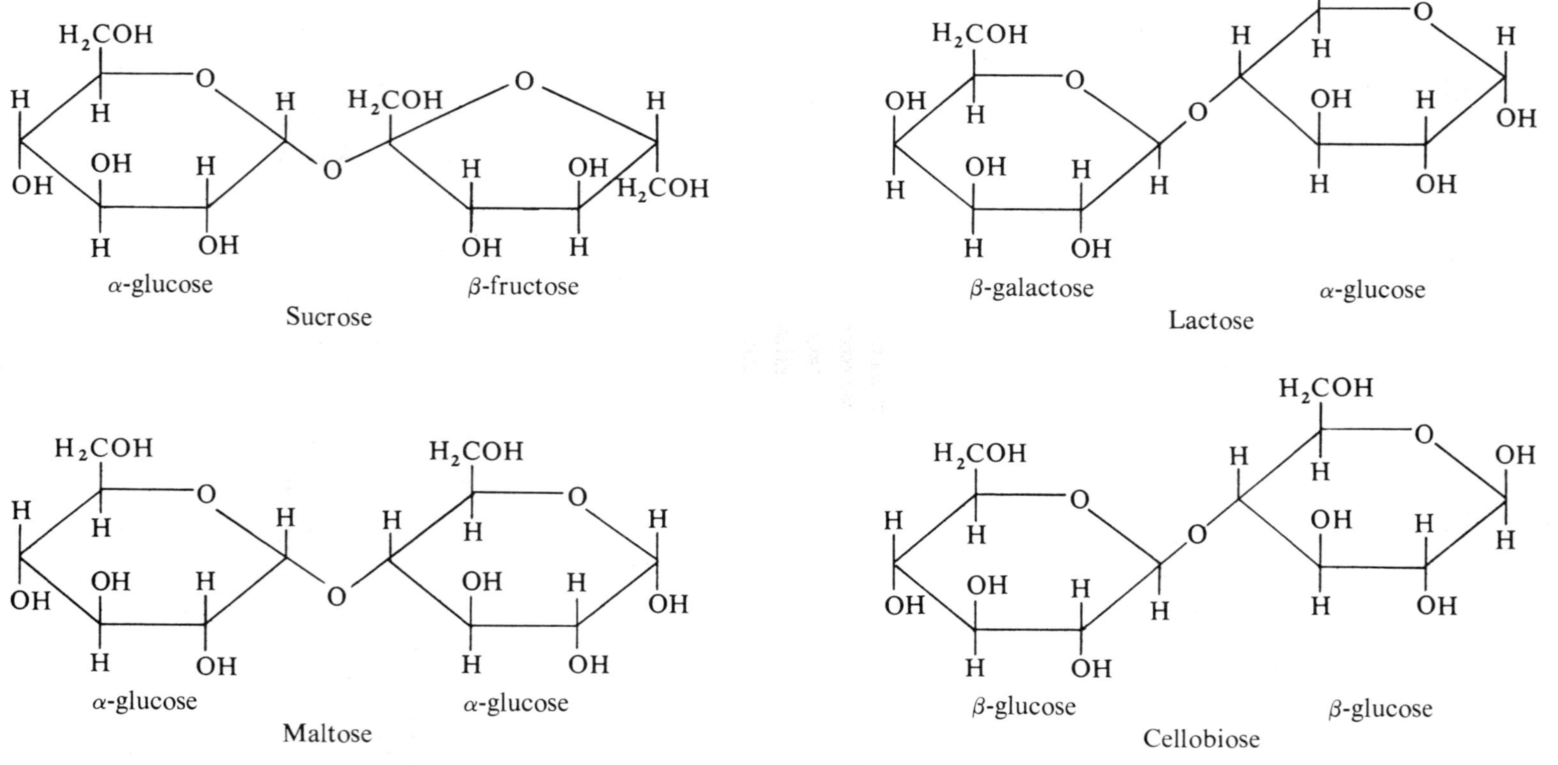

Example 23–3. Disaccharides

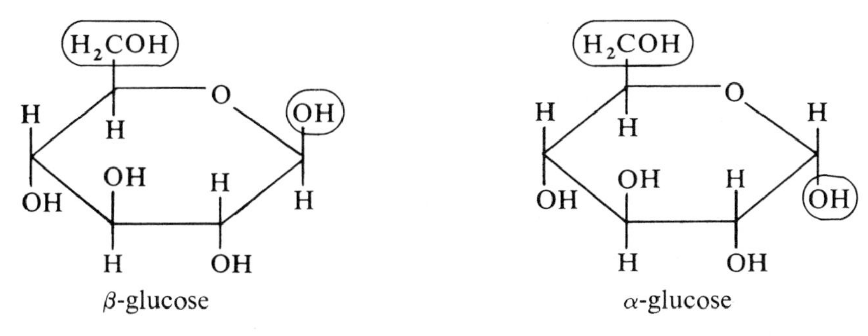

Example 23–4. Alpha and Beta Forms of Glucose

prepared in this way are actually sweeter than the pure sugar originally put in them.

This same hydrolysis reaction of sucrose is used in the preparation of candy cordials such as cherry chocolates. The solubilities of the sugars involved are also important; glucose and fructose are much more soluble in water than is sucrose. In the candy-making process the fruit is thoroughly moistened and then coated with sucrose. The sucrose-covered fruit is then dipped into chocolate. Upon aging, the moist sucrose inside the chocolate shell hydrolyzes and forms glucose and fructose—which, because of high solubilities, dissolve in the available water and form the sweet syrup characteristic of these delicacies. An enzyme may be added to the sugar in order to speed up the hydrolysis reaction. The syrup resulting from such reactions, **invert sugar** (Table 23–1), is used in numerous other applications as a sweetener.

Caramel, another popular candy product, is also the result of reactions involving sucrose. Upon heating, sucrose melts, loses water, and apparently undergoes further condensation to form higher carbohydrates. With continued heating the familiar brown color appears. The reactions involved in this caramelization process are not completely understood, but the presence of nitrogen-containing compounds increases their rates, so milk is often added to the sugar before the attempt to caramelize it.

Lactose, or milk sugar, occurs only in milk. Human milk contains about 6 percent and cow's milk about 4.8 percent. It is synthesized in the mammary glands of animals and consists of one unit each of the monosaccharides α-glucose and β-galactose. The sugar is obtained from the whey of milk; although not very sweet (Table 23–1), it is used to some extent as a sweetener in bakery and confectionary goods. It is an important ingredient in the commercially produced infant foods that simulate mother's milk.

Maltose, composed of two α-glucose units, does not occur in the free state naturally but is an intermediate in the enzymatic hydrolysis of starch to glucose. The name is derived from the fact that during the germination or malting of barley, starch is hydrolyzed to the sugar. It is used as a sweetener in some foods and as a nutrient for growing bacterial cultures.

Cellobiose is also a disaccharide made up of two glucose units. However, the glucose units in this case are both in the β form. This sugar can be thought of as the primary unit of the polysaccharide cellulose. It is produced upon the partial hydrolysis of cellulose in the same way that maltose is produced from starch.

Polysaccharides

Polysaccharides are condensation polymers of the monosaccharides and thus contain many monosaccharide units. The molecular weights of these compounds are quite high. In our discussion we shall only deal with starch, glycogen, and cellulose, although numerous other polysaccharides are known.

Starch is a substance well known to those who cook and do laundry. It occurs naturally in many plants, where it serves as a food storage system. Natural starches contain two distinct types of molecules; a type called *amylose* makes up 10 to 20 percent, and the remainder is a type called *amylopectin*.

Amylose is composed of unbranched chains of α-glucose (Example 23–5). The chain takes on the configuration of a helix with an open core running its entire length. The molecular weight of amylose is about 50,000 amu. It is insoluble in water but will form a suspension of very small particles. A product called *soluble starch* is produced when amylose is partially hydrolyzed by acid into units small enough to be soluble in water.

An interesting reaction takes place between amylose and iodine, yielding an intensely blue substance, which is thought to be produced when iodine molecules fill the hollow core of the starch helix. As the starch is hydrolyzed into smaller and smaller units, the color with iodine lightens and finally disappears—apparently because of the breakdown of the starch helix.

Amylopectin is also made up of α-glucose chains, but the chains are cross-linked by a condensation reaction between the terminal OH group of one chain and the CH_2OH group of another chain. This type of linkage is shown in Example 23–6.

The molecular weight of amylopectin is greater than that of amylose, with values of up to 1 million amu.

Glycogen is a carbohydrate storage material for animals analogous to the starch of plants. Much of the excess glucose an animal takes in is stored in the liver for future use in the form of this so-called **animal starch.** Structurally, glycogen is very similar to amylopectin; however, the molecular weight is much higher, often exceeding 3 million amu.

Cellulose, a polysaccharide of glucose, is found mainly in the structural parts of plants. It is a linear polymer of glucose with the individual glucose units in the β form. The resulting linkage bonds are much more stable toward hydrolysis than those formed by the α-glucose of starch and glycogen. The structure of cellulose is illustrated in Example 23–7.

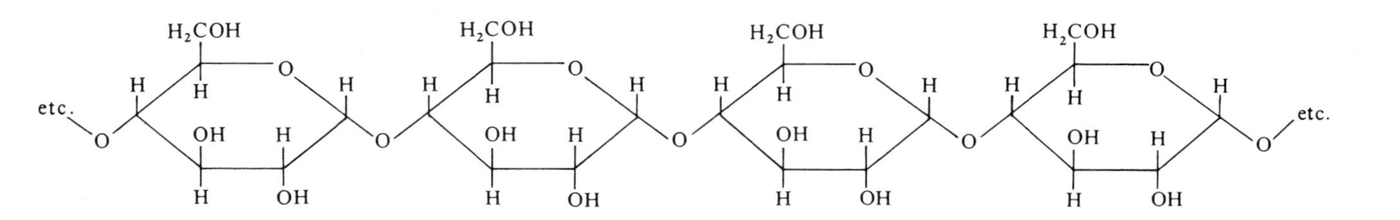

Example 23–5. Section of Amylose Chain

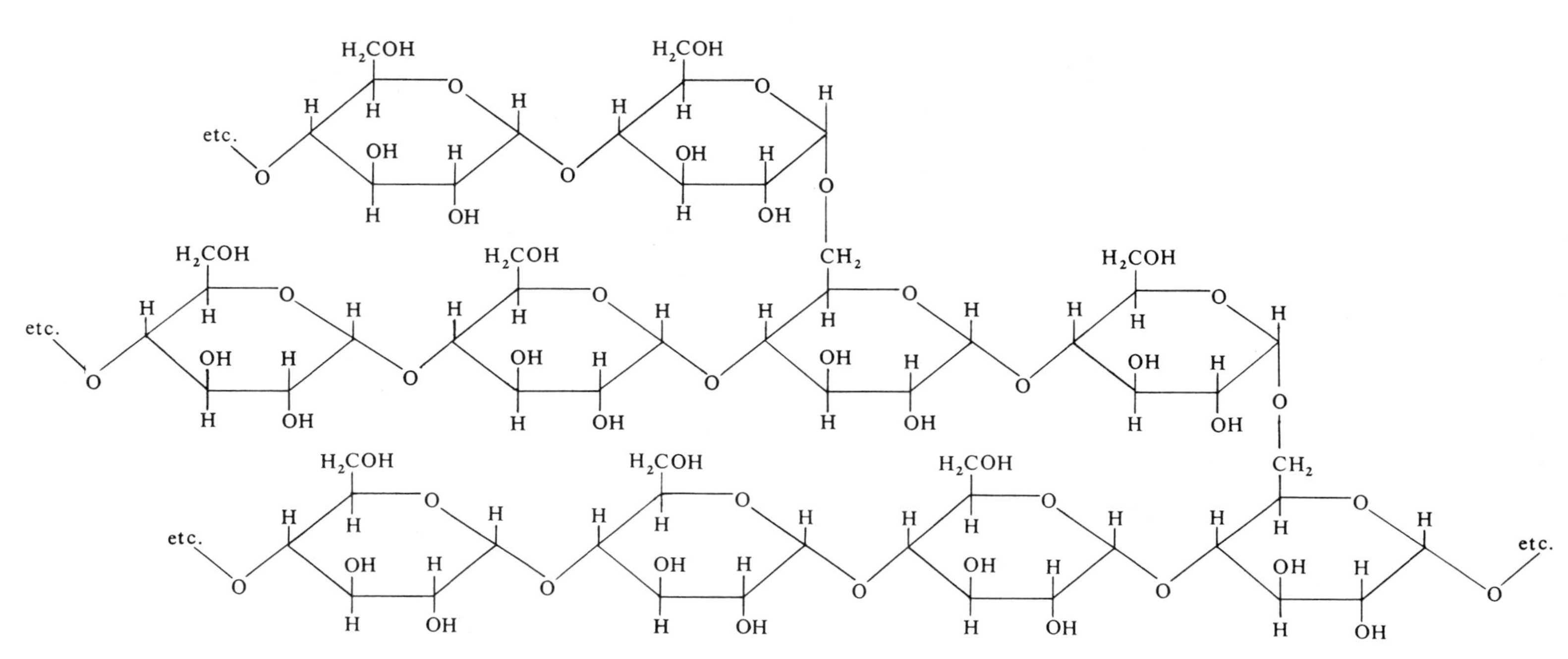

Example 23–6. Amylopectin Chain Section Showing Cross-Linking

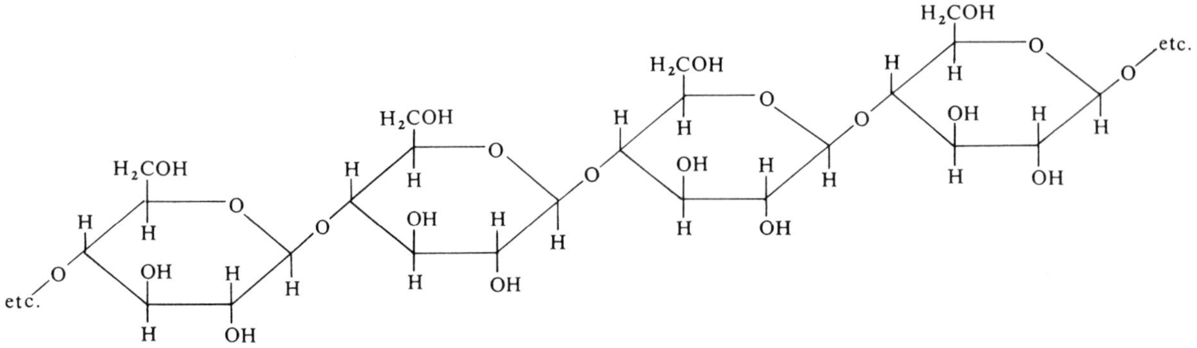

Example 23–7. Section of Cellulose Chain

The α linkages of starch can be hydrolyzed in the body of man and other animals, which leads to the wide usage of starch as a food. Man and most other animals, however, lack the necessary enzymes to hydrolyze the β linkages of cellulose, thus limiting its use as a food. Some simple bacteria possess the necessary enzymes for this hydrolysis and are capable of producing simpler carbohydrates from cellulose. Higher animals which use cellulose as a food do so only with the assistance of such bacteria. Herbivores such as cows, sheep, horses, and even the termite are animals in this category. Each has a colony of bacteria somewhere in the digestive system and utilizes the simpler carbohydrate resulting from bacterial efforts.

Cellulose is used by man in a variety of ways—for example, we build our shelters with it (wood), write and print on it (paper), wear it (cotton, linen, and rayon), package with it (paper and cellophane), and use it as an explosive (guncotton and nitrocellulose).

Flax and cotton are natural forms of cellulose used as textile fibers. Another fiber, rayon, and its close relative, cellophane, are known as *regenerated cellulose.* These two useful forms of cellulose are manufactured by the **viscous rayon process,** in which cellulose in the form of wood pulp is dissolved in a concentrated sodium hydroxide solution and then treated with carbon disulfide. The result is a syrupy, viscous solution called *cellulose xanthate.* This solution is forced into an acid bath through small holes to give rayon fibers or through a narrow slit to give cellophane sheets. The acid bath changes the cellulose xanthate back into cellulose. Rayon produced in 1892 by this process was one of the earliest man-made textile fibers. The reactions involved in the process are given below:

$$(C_6H_{10}O_5)_x + NaOH + CS_2 \rightarrow \left(C_6H_9O_4\!-\!O\!-\!C \begin{matrix} \diagup\!\!\diagup S \\ \\ \diagdown S\!-\!Na \end{matrix} \right)_x \quad 23\text{–}2$$

cellulose — cellulose xanthate

$$\left(C_6H_9O_4\!-\!O\!-\!C \begin{matrix} \diagup\!\!\diagup S \\ \\ \diagdown S\!-\!Na \end{matrix} \right)_x + H_2SO_4 \rightarrow (C_6H_{10}O_5)_x + CS_2 + Na_2SO_4 \quad 23\text{–}3$$

cellulose xanthate — rayon or cellophane

Lipids

The group of compounds known as lipids is made up of substances with widely different compositions and structures. For this reason it is difficult to define or classify them in a single concise fashion. The most commonly used single definition is based upon solubility behavior. If a biological material is crushed and well mixed with a nonpolar solvent such as ether, chloroform, or carbon tetrachloride, the substances which dissolve in the solvent are classified as **lipids.**

These compounds are important in the diet of man for several reasons. They serve as sources of energy and, sometimes to our chagrin, excess carbohydrates and other energy-yielding foods are stored in the body as the lipids known as fat. They also serve as insulation for some vital organs, providing protection from mechanical shock and an aid in maintaining correct body temperature. According to some estimates, as much as 40 percent of the average American diet consists of edible lipids. Certain lipids must be provided by the diet because the body is unable to synthesize them from other available materials.

In this discussion lipids will be divided into three classifications: fats and oils, waxes, and steroids.

Fats and Oils

Fats and **oils** are the most abundant lipids in nature. Chemically, both are esters in which the alcohol is glycerol, a trihydroxy alcohol, and the acids are all of a variety called *fatty acids.* The main distinction between a fat and oil is the melting point; fats are solids at room temperature while oils are liquids. A general reaction between glycerol and three different fatty acids to form a fat or oil is given below (equation 23–4).

Notice that water is removed during this reaction. The reverse reaction, hydrolysis, in which water is added, results in the regeneration of the fatty acids and glycerol. This is one of the most important reactions of fats and oils, just as it was for carbohydrates.

Since each fat or oil contains glycerol, any differences between them are probably caused by differences in the fatty-acid components. Table 23–2 gives

$$
\begin{array}{llll}
R{-}\overset{O}{\overset{\|}{C}}{-}OH & HO{-}CH_2 & & R{-}\overset{O}{\overset{\|}{C}}{-}O{-}CH_2 \\
 & \quad\ | & & \qquad\qquad\ | \\
R'{-}\overset{O}{\overset{\|}{C}}{-}OH \;+ & HO{-}CH & \rightarrow & R'{-}\overset{O}{\overset{\|}{C}}{-}O{-}CH + 3H_2O \\
 & \quad\ | & & \qquad\qquad\ | \\
R''{-}\overset{O}{\overset{\|}{C}}{-}OH & HO{-}CH_2 & & R''{-}\overset{O}{\overset{\|}{C}}{-}O{-}CH_2 \\
\text{fatty acids} & \text{glycerol} & & \text{fat or oil (an ester sometimes called a } \textit{triglyceride}\text{)}
\end{array}
\qquad 23\text{–}4
$$

the names, formulas, melting points, and common sources of some fatty acids.

These fatty acids have a number of characteristics in common: (1) They are usually monocarboxylic, containing only one $-C\begin{smallmatrix} \nearrow O \\ \searrow OH \end{smallmatrix}$ group per molecule. (2) The hydrocarbon chain is unbranched. (3) The total number of carbon atoms is even. (4) The hydrocarbon chain may be saturated, or it may contain as many as three or four double bonds.

The melting points of the acids depend upon two factors, molecular weight and degree of unsaturation (or number of double bonds per molecule). It is easily seen in Table 23–2 that for the saturated fats an increase in melting point accompanies an increase in molecular weight. A comparison of the melting points of stearic, oleic, and linoleic acids illustrates the effect of unsaturation on melting point. All three contain 18 carbon atoms and have nearly identical molecular weights of 284, 282, and 280 amu, respectively, and yet the melting points are 70°, 16°, and −5°.

The acids contained in a fat or oil largely determine the melting point. Thus, it is not surprising to find higher percentages of unsaturated acids in the oils as compared to the solid fats. The fats obtained from animal sources (lard, beef fat, butter, etc.) contain high percentages of saturated fatty acids and are solids at room temperature. The corresponding products from vegetable sources contain much lower percentages and are liquids at room temperature. Most of the commercial liquid oils used for cooking, salad dressings, and so forth are of vegetable origin.

Fats and oils undergo numerous reactions; we shall consider four that are of commercial importance. Three are useful and the fourth undesirable.

Hydrolysis is the reverse of the esterification reaction which produced the fat or oil; its products are glycerol and the fatty acids. An example is given below for an oil derived from cottonseed (equation 23–5).

Table 23–2. Fatty Acids

Name	*Formula*	*Melting point (°C)*	*Common sources*
Saturated			
Butyric acid	$CH_3(CH_2)_2{-}C(=O){-}OH$	−7	Butterfat
Caproic acid	$CH_3(CH_2)_4{-}C(=O){-}OH$	−3	Butterfat
Caprylic acid	$CH_3(CH_2)_6{-}C(=O){-}OH$	16	Butterfat
Capric acid	$CH_3(CH_2)_8{-}C(=O){-}OH$	31	Butterfat, coconut oil
Lauric acid	$CH_3(CH_2)_{10}{-}C(=O){-}OH$	44	Coconut oil
Myristic acid	$CH_3(CH_2)_{12}{-}C(=O){-}OH$	54	Butterfat, coconut oil, nutmeg oil
Palmitic acid	$CH_3(CH_2)_{14}{-}C(=O){-}OH$	63	Lard, beef fat, butterfat, cottonseed oil
Stearic acid	$CH_3(CH_2)_{16}{-}C(=O){-}OH$	70	Lard, beef fat, butterfat, cottonseed oil
Cerotic acid	$CH_3(CH_2)_{24}{-}C(=O){-}OH$	97	Beeswax, wool fat
Unsaturated			
Palmitoleic acid	$CH_3(CH_2)_5CH{=}CH(CH_2)_7{-}C(=O){-}OH$	32	Cod liver oil, butterfat
Oleic acid	$CH_3(CH_2)_7CH{=}CH(CH_2)_7{-}C(=O){-}OH$	16	Lard, beef fat, olive oil, peanut oil
Linoleic acid	$CH_3(CH_2)_4CH{=}CHCH_2CH{=}CH(CH_2)_7{-}C(=O){-}OH$	−5	Cottonseed oil, soybean oil, corn oil
Linolenic acid	$CH_3CH_2CH{=}CHCH_2CH{=}CHCH_2CH{=}CH(CH_2)_7{-}C(=O){-}OH$	−11	Linseed oil

$$
\begin{array}{l}
CH_3(CH_2)_{14}-\overset{\overset{\displaystyle O}{\|}}{C}-O-CH_2 \\
CH_3(CH_2)_7CH{=}CH(CH_2)_7-\overset{\overset{\displaystyle O}{\|}}{C}-O-CH \quad + 3H_2O \rightarrow \\
CH_3(CH_2)_4CH{=}CHCH_2CH{=}CH(CH_2)_7-\overset{\overset{\displaystyle O}{\|}}{C}-O-CH_2 \\
\text{oil (from cottonseed)}
\end{array}
$$

$$
\rightarrow \left\{ \begin{array}{ll}
CH_3(CH_2)_{14}-\overset{\overset{\displaystyle O}{\|}}{C}-OH & HOCH_2 \\
\text{palmitic acid} & \\
CH_3(CH_2)_7CH{=}CH(CH_2)_7-\overset{\overset{\displaystyle O}{\|}}{C}-OH & + HOCH \\
\text{oleic acid} & \\
CH_3(CH_2)_4CH{=}CHCH_2CH{=}CH(CH_2)_7-\overset{\overset{\displaystyle O}{\|}}{C}-OH & HOCH_2 \\
\text{linoleic acid} & \text{glycerol}
\end{array} \right. \qquad 23\text{–}5
$$

When the hydrolysis reaction is carried out in the presence of a strong base, the process is called **saponification.** In this commercially important reaction the products are glycerol and the salts of the fatty acids or soaps. An example is given in equation 23–6, using the same cottonseed oil.

The salts obtained depend upon the base used. Sodium salts, known as **hard soaps,** are found in most cake soap used in the home. Potassium salts, **soft soaps,** are used in some shaving creams and liquid soap preparations. Insoluble soaps are formed with calcium and other heavy metals, and were mentioned earlier in the discussion of water softening. Heavy metal soaps, besides being the primary ingredients in bathtub ring and soap scum, are used in facial powders and heavy lubricating greases.

The third commercially important reaction is **hydrogenation,** in which some of the double bonds of the fatty acids are reacted with hydrogen. The result is a decrease in the fatty-acid degree of unsaturation and a corresponding increase in melting point of the fat or oil. The reaction is most often employed in the production of solid cooking shortenings or margarines from liquid vegetable or animal oils. It is important not to complete the reaction and totally saturate all double bonds. If this is done the product is hard and waxy—not the smooth, creamy product desired by the consumer. The reaction is given in equation 23–7, again using the same cottonseed oil. A metal catalyst (finely divided nickel) is used to speed up the reaction.

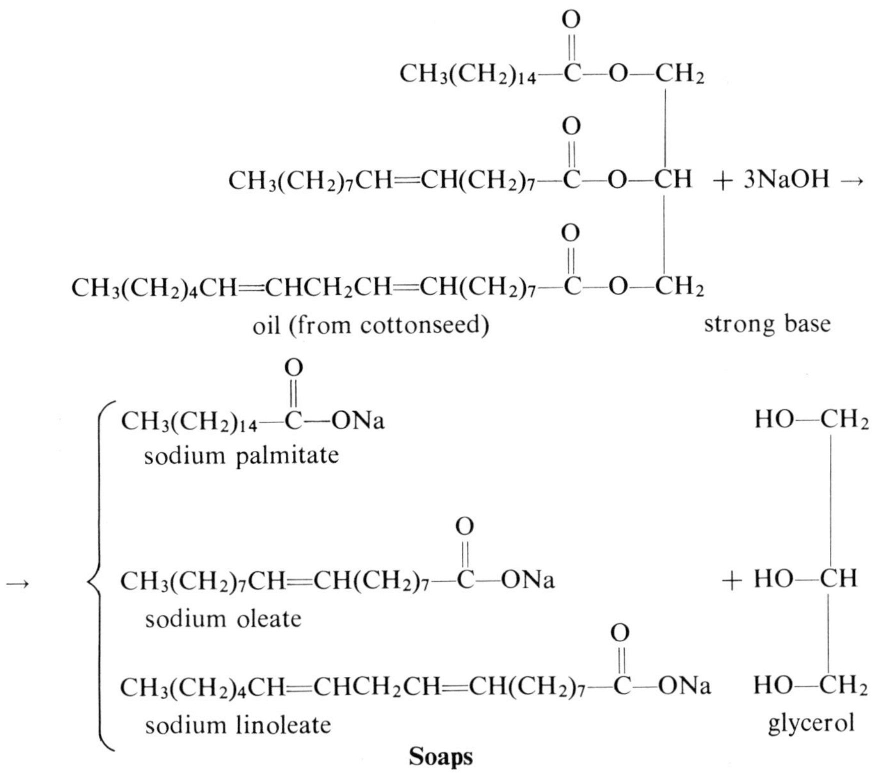

$$CH_3(CH_2)_{14}-\overset{O}{\overset{\|}{C}}-O-CH_2$$

$$CH_3(CH_2)_7CH{=}CH(CH_2)_7-\overset{O}{\overset{\|}{C}}-O-CH \quad + 3NaOH \rightarrow$$

$$CH_3(CH_2)_4CH{=}CHCH_2CH{=}CH(CH_2)_7-\overset{O}{\overset{\|}{C}}-O-CH_2$$

oil (from cottonseed) strong base

$$\rightarrow \left\{ \begin{array}{l} CH_3(CH_2)_{14}-\overset{O}{\overset{\|}{C}}-ONa \\ \text{sodium palmitate} \\ CH_3(CH_2)_7CH{=}CH(CH_2)_7-\overset{O}{\overset{\|}{C}}-ONa \\ \text{sodium oleate} \\ CH_3(CH_2)_4CH{=}CHCH_2CH{=}CH(CH_2)_7-\overset{O}{\overset{\|}{C}}-ONa \\ \text{sodium linoleate} \end{array} \right. + \begin{array}{l} HO-CH_2 \\ HO-CH \\ HO-CH_2 \\ \text{glycerol} \end{array} \qquad 23\text{–}6$$

Soaps

$$CH_3(CH_2)_{14}-\overset{O}{\overset{\|}{C}}-O-CH_2$$

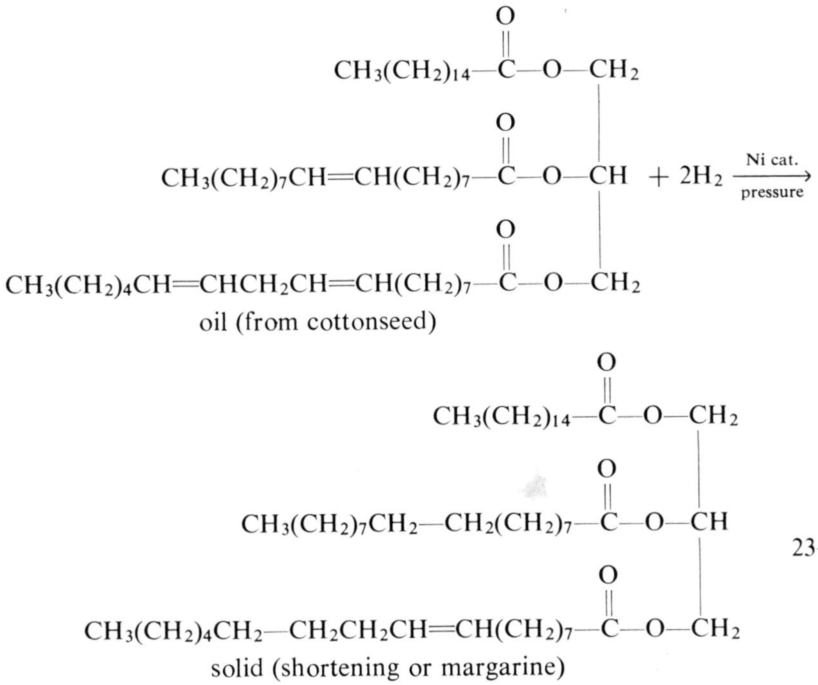

$$CH_3(CH_2)_7CH{=}CH(CH_2)_7-\overset{O}{\overset{\|}{C}}-O-CH \quad + 2H_2 \xrightarrow[\text{pressure}]{\text{Ni cat.}}$$

$$CH_3(CH_2)_4CH{=}CHCH_2CH{=}CH(CH_2)_7-\overset{O}{\overset{\|}{C}}-O-CH_2$$

oil (from cottonseed)

$$CH_3(CH_2)_{14}-\overset{O}{\overset{\|}{C}}-O-CH_2$$

$$CH_3(CH_2)_7CH_2-CH_2(CH_2)_7-\overset{O}{\overset{\|}{C}}-O-CH \qquad 23\text{–}7$$

$$CH_3(CH_2)_4CH_2-CH_2CH_2CH{=}CH(CH_2)_7-\overset{O}{\overset{\|}{C}}-O-CH_2$$

solid (shortening or margarine)

Under rather extreme hydrogenation conditions, involving high pressures and catalysts, not only are the double bonds saturated but the acids are removed from the glycerol and reduced to alcohols. Some of these alcohols are useful materials from which detergents are made. An example is given below.

$$\begin{array}{r}
CH_3(CH_2)_{14}-\overset{\overset{\displaystyle O}{\|}}{C}-O-CH_2 \\
CH_3(CH_2)_7CH{=}CH(CH_2)_7-\overset{\overset{\displaystyle O}{\|}}{C}-O-CH \\
CH_3(CH_2)_4CH{=}CHCH_2CH{=}CH(CH_2)_7-\overset{\overset{\displaystyle O}{\|}}{C}-O-CH_2
\end{array} + 9H_2 \xrightarrow[\text{high pressure}]{\text{cat.}}$$

oil (from cottonseed)

$$\rightarrow \underbrace{\left\{\begin{array}{l}
CH_3(CH_2)_{14}-CH_2OH \\
CH_3(CH_2)_7CH_2CH_2(CH_2)_7-CH_2OH \\
CH_3(CH_2)CH_2CH_2CH_2CH_2CH_2(CH_2)_7-CH_2OH
\end{array}\right.}_{\text{alcohols}} + \underbrace{\begin{array}{l} HOCH_2 \\ HOCH \\ HOCH_2 \end{array}}_{\text{glycerol}} \qquad 23\text{–}8$$

The fourth reaction we shall consider (the undesirable one) is actually a combination of reactions which leads to rancid fats and oils. **Rancidity** occurs when fats and oils are left exposed to warm, moist air for a period of time and changes take place which produce disagreeable odors and flavors. The changes primarily involve two kinds of reactions, hydrolysis of ester linkages and oxidation of carbon-chain double bonds. The hydrolysis of ester linkages releases the fatty acids, many of which have disagreeable odors, especially those of lower molecular weight. The odor of butyric acid, for example, is the characteristic odor of rancid butter—and unwashed socks. The oxidation of double bonds also produces fatty acids and, in addition, some aldehydes which also have disagreeable odors and flavors.

Waxes

The lipids classified as **waxes** are also esters of fatty acids but with high-molecular-weight alcohols other than glycerol. They are not as easily hydrolyzed as the fats and oils and so are sometimes seen in nature as protective coatings on plants and animals. The fatty acids involved are generally the same as those of the fats and oils. The alcohols usually contain only one hydroxide group per molecule and from 12 to 32 carbon atoms. Some naturally occurring waxes are given in Table 23–3.

Table 23–3. Some Naturally Occurring Waxes

Name	*Primary alcohol*	*Primary acid*	*Source*	*Uses*
Beeswax	Myricyl, $CH_3(CH_2)_{29}OH$	Palmitic, $CH_3(CH_2)_{14}C(=O)—OH$	Honeycomb	Ointments, cosmetics, candles
Lanolin	Mixture	Mixture	Wool	Base for ointments, salves, and creams
Spermaceti	Cetyl, $CH_3(CH_2)_{15}OH$	Palmitic, $CH_3(CH_2)_{14}C(=O)—OH$	Head of sperm whale	Ointments, cosmetics, candles
Carnauba	Myricyl, $CH_3(CH_2)_{29}OH$	Cerotic, $CH_3(CH_2)_{24}C(=O)—OH$	Palm leaves	Floor and auto wax

It is obvious from Table 23–3 that waxes are commonly used in cosmetics Soaps and other lipids, including fats and oils, are also very often included. Table 23–4 contains four typical formulas or recipes for common cosmetics.

Table 23–4. Typical Cosmetic Formulations

Face powder		*Vanishing cream*	
Calcium carbonate, $CaCO_3$	25%	Glycerol monostearate (a partially hydrolyzed fat)	10%
Calcium stearate (a hard soap)	5%	Spermaceti wax	5%
Kaolin (clay—$Al_2O_3 \cdot 2SiO_2 \cdot 2H_2O$) (opaque, absorbs oils)	10%	Paraffin wax (a hydrocarbon)	5%
Talc, $MgSi_4O_{10}(OH)_2$ (opaque material, covers and provides "slip")	50%	Stearic acid	2%
Titanium dioxide, TiO_2 (opaque covering material)	5%	Titanium dioxide, TiO_2	0.5%
Zinc stearate (a soap)	5%	Potassium hydroxide, KOH	0.1%
Coloring agents	trace	Glycerol	3%
Perfumes	trace	Water	74.2%
		Perfumes	0.2%
Cleansing cream		*Lipstick*	
Beeswax	8%	Paraffin (a hydrocarbon)	45%
Soft paraffin (a hydrocarbon)	6%	Cerosine (a hydrocarbon)	15%
Hard paraffin (a hydrocarbon)	8%	Beeswax	15%
Liquid paraffin (a hydrocarbon)	62%	Lanolin	24.5%
Spermaceti wax	5%	Intense coloring agents	0.5%
Borax, $Na_2B_4O_7 \cdot 10H_2O$	0.5%	Perfumes or flavoring agents	trace
Water	10.5%		

All ingredients in cosmetics are very well mixed together to form a homogeneous powder, cream, or solution. Don't look for lists of contents on cosmetic containers. According to law, anything classified as a cosmetic may be sold without a list of ingredients.

Steroids

The lipids classified as **steroids** are compounds with formulas which generally contain the specific carbon-atom skeleton illustrated in Example 23–8. The

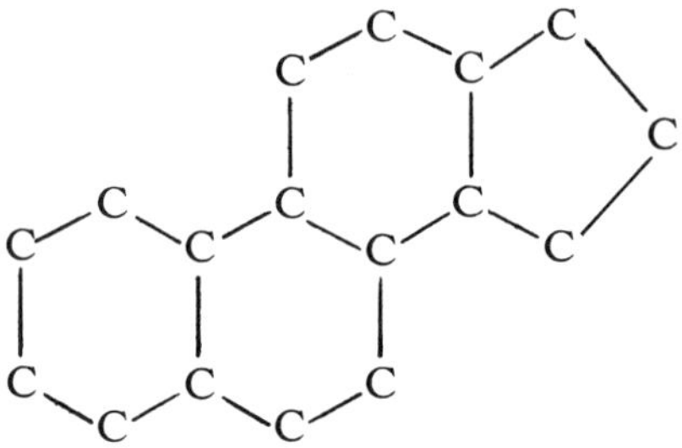

Characteristic steroid carbon skeleton

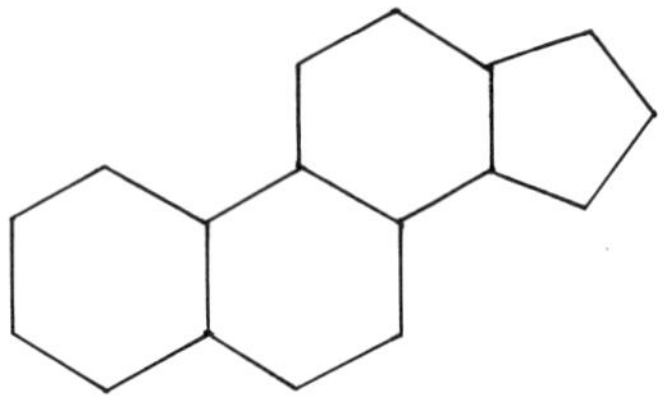

Commonly used form, called the steroid nucleus

Example 23–8. Steroid Structure

alcohol group, ketone group, and double bond are often found in steroids. However, the ester linkage, so characteristic of fats, oils, and waxes, is seldom present. Important compounds in this group are cholesterol, many types of hormones, and bile salts.

Cholesterol is the best known and most abundant steroid in the body. It is found in the brain and nerve tissue as well as being the main component of gallstones. It has received widespread attention because of an apparent correlation between cholesterol concentrations in the blood and the disease known as hardening of the arteries or atherosclerosis. The structure of this steroid alcohol is:

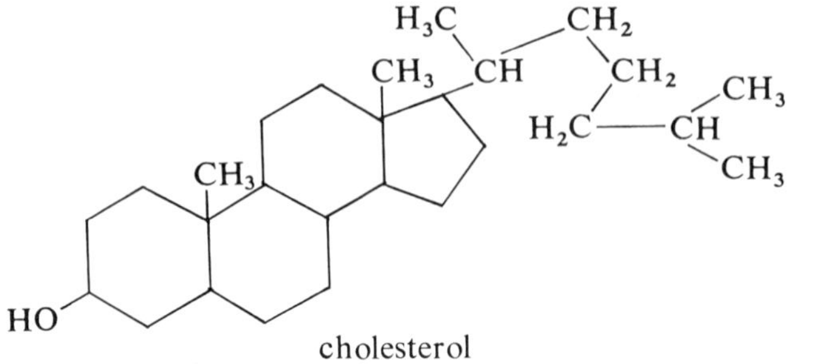

cholesterol

The steroids known as **sex hormones** control the development of secondary male and female sexual characteristics at puberty, and they are important in the normal reproductive processes. The formulas of some interesting ones are given in Example 23–9.

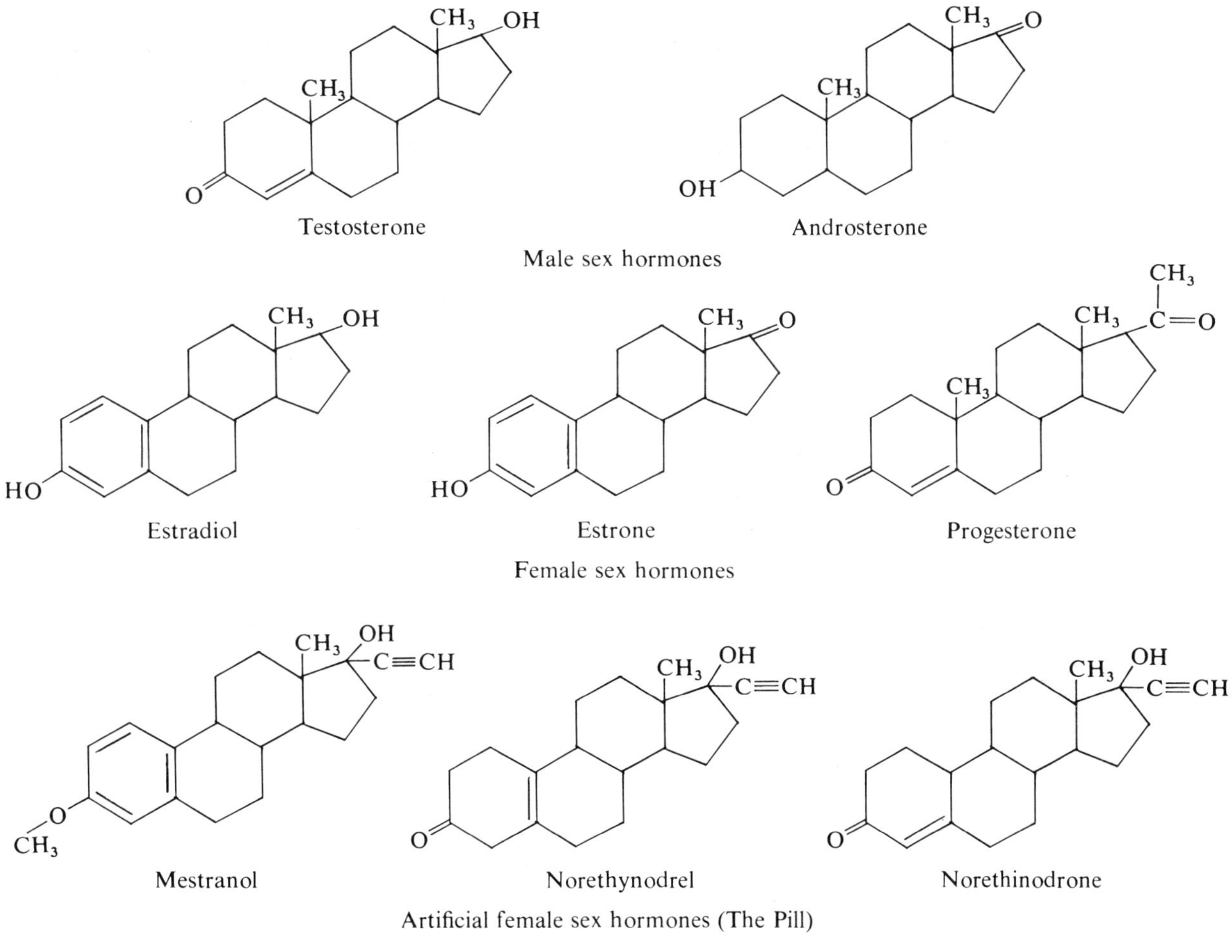

Testosterone Androsterone

Male sex hormones

Estradiol Estrone Progesterone

Female sex hormones

Mestranol Norethynodrel Norethinodrone

Artificial female sex hormones (The Pill)

Example 23–9. Structures of Some Sex Hormones

Testosterone and **androsterone,** examples of the male sex hormones or **androgens,** control the development of masculine secondary sexual characteristics such as deepening of the voice, distribution of body hair at puberty, and beard growth. They also control the function of the glands involved in reproduction. These hormones are produced by the testes.

The ovaries of the female produce estradiol and estrone. **Estrone** is involved in the development of secondary female sexual characteristics and in general functions much like the testosterone of the male. **Estradiol** is an important factor in the control of the ovulation cycle in the female. The main function of progesterone is to prepare the inner wall of the uterus for implantation of a fertilized ovum. When pregnancy occurs, this hormone also prevents further ovulation, retains the embryo in the uterus, and develops the mammary glands prior to lactation.

Synthetic derivatives of the female sex hormones have been developed and have attracted a good deal of attention. When taken regularly at appropriate intervals, these compounds can function very effectively in preventing ovulation and pregnancy. A mixture of progesterone and estradiol produces the same results when administered, but they must be injected into the body to be effective. The synthetic compounds have been produced with structural changes which allow them to be taken orally. These **oral contraceptives** are generally mixtures of two or more analogs with structures and functions similar to those of the natural hormones. Enovid, for example, is a mixture of mestranol and norethynodrel.

Another important group of hormones is produced by the outer part of the adrenal gland. These hormones affect the metabolism of food taken into the body, maintain the proper balance of electrolytes (ions) in the various body fluids, and help control inflammation and allergic reactions in the body. A typical member of this group of 28 compounds is **cortisone,** which has the structure given below.

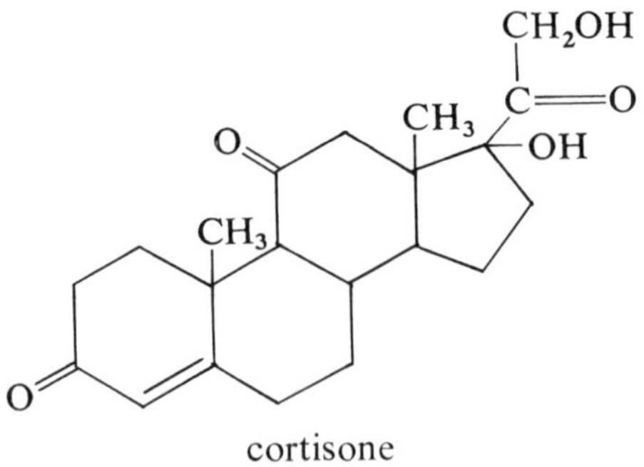

cortisone

This hormone is important in the control of carbohydrate metabolism and has also been found effective in relieving symptoms of rheumatoid arthritis.

The last group of steroids we shall consider is made up of the **bile salts,** materials which act as emulsifying agents for water-insoluble lipids during the digestive process. They are found in the bile, a digestive juice manufactured by the liver and stored in the gall bladder until needed. The structures of two of the acids from which the salts are derived are given below. A basic substance reacts with the carboxylic acid group to form the salt.

cholic acid

deoxycholic acid

Review Questions

1. Which of the pentoses of Example 23–1 and the hexoses of Example 23–2 are found free in nature?
2. The disaccharides maltose and cellobiose are both made up of two glucose units. In what way are the two disaccharides different? (See Example 23–3.)
3. Why is invert sugar sweeter than sucrose?
4. Describe and compare the following polysaccharides: amylose, soluble starch, amylopectin, and cellulose.
5. Discuss the chemical similarities and differences of the following pairs of lipids:
   a) Solid fat and liquid oil
   b) Solid fat and wax
6. List five general products that could be produced by subjecting a triglyceride oil to various reactions.
7. Draw the structure of a triglyceride containing palmitic, myristic, and butyric acids (Table 23–2). What common food might contain such a triglyceride?
8. Some toilet soaps, according to advertisements, are obtained from coconut oil. Refer to Table 23–2 and write the formulas for three soaps that might be obtained by the saponification of coconut oil.
9. Draw the structure of the main ester found in carnauba wax (see Table 23–3).
10. How are steroids different from the other lipids discussed in this chapter?
11. Compare the structure of a bile salt to that of a soap. Describe how a bile salt might behave as an emulsifying agent for water insoluble lipids (see Figure 10–4).

Suggestions for Further Reading

Battista, O. A., "Sugar—The Chemical with a Thousand Uses," *Chemistry*, April 1965.
Dole, V. P., "Body Fat," *Scientific American*, Dec. 1959.
Fieser, L. F., "Steroids," *Scientific American*, Jan. 1955.
Green, D. E., "The Synthesis of Fats," *Scientific American*, Feb. 1960.
"New Artificial Sweeteners," *Chemistry*, June 1970.
Nordsiek, F. W., "The Sweet Tooth," *American Scientist*, Jan./Feb. 1972.
Patton, S., "Milk," *Scientific American*, July 1969.

24 Introduction to Biochemistry II: Proteins and Enzymes

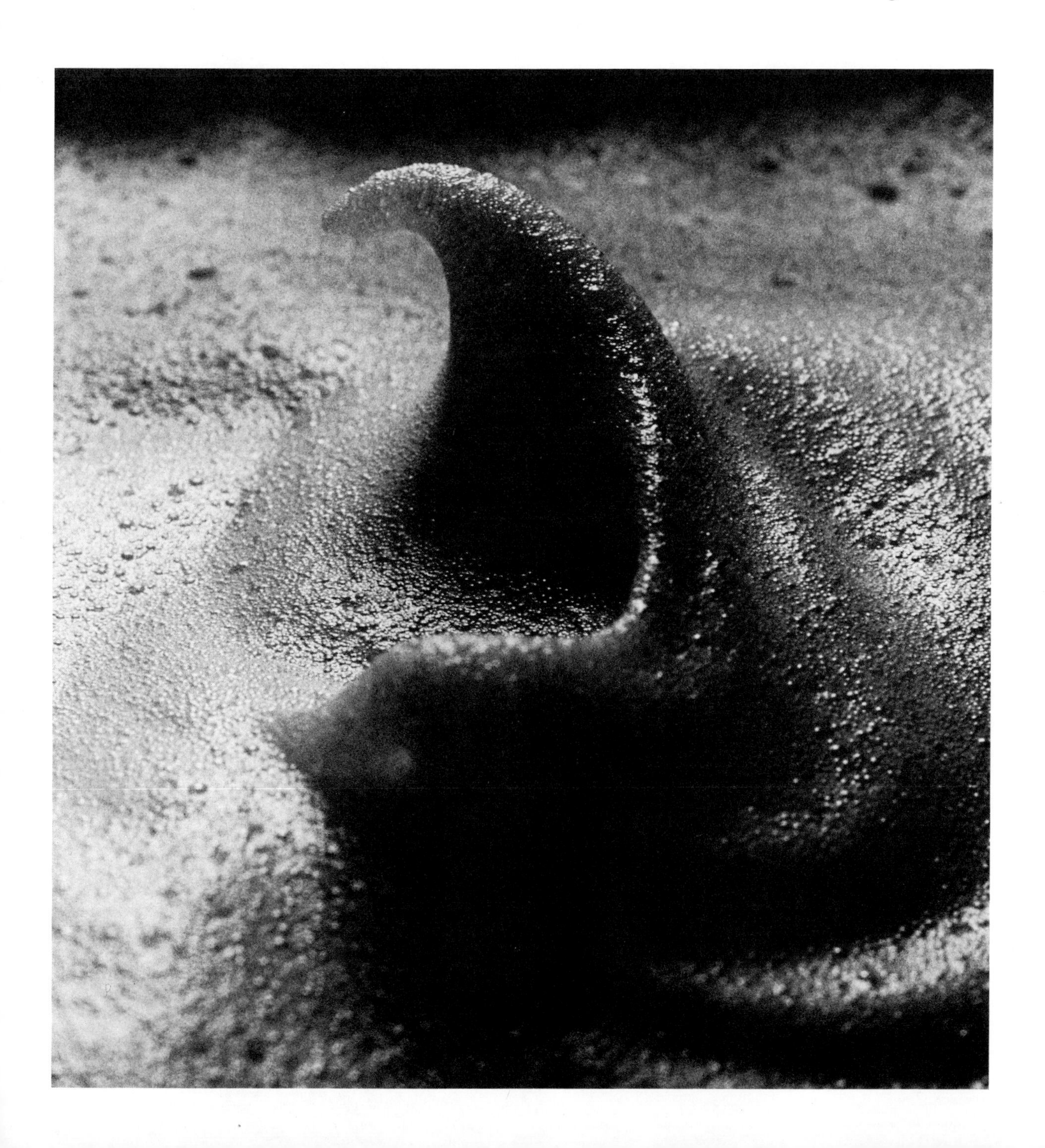

This chapter continues our presentation of the basic concepts of biochemistry. Proteins, the third important type of biochemical substance, and enzymes, the biochemical catalysts, are the topics.

Proteins

Seldom do we see, hear, or read a commercial advertisement extolling the amount of carbohydrates or lipids contained in a food product. This is not true for proteins. We are told that food product X, Y, or Z contains large amounts of protein which, the commercial emphasizes, are the building blocks of the body.

This latter statement is correct, for the compounds classified as proteins are found in all cells and, in fact, constitute at least one-half of the total dry weight of the body. The Dutch chemist G. J. Mulder is credited with first recognizing the importance of these substances, and it was he who named them **proteins,** from the Greek word *proteis* which means *preeminence* or *of first importance*. In a book published in 1844, he described these compounds as *one* substance occurring in plants and animals—one that was without doubt the most important of the known components of living matter, and without which life would be impossible.

Proteins are essential constituents in the diet of all animals, including man. They are needed if the body is to synthesize body tissues, enzymes, certain hormones, and certain blood components. In addition, they are used in the maintenance and repair of existing tissues and may serve as a source of energy for the body.

In contrast to carbohydrates and lipids, which are stored in the body as energy reserves, proteins are not stored to any appreciable extent and must be continually taken in as part of the diet. Good sources of dietary protein from plants are beans, wheat, corn, and seeds. Animals provide excellent sources of protein in the form of such products as milk, eggs, and various meats. In general, a good diet will include more than just one of the products given above in order to insure a supply of all the types of protein needed by the body.

Amino Acids

Chemically, protein is a condensation polymer of monomer units called **amino acids,** which are bifunctional molecules containing both a carboxylic acid and an amine functional group. The general formula for an amino acid is given below, where R represents the side chain, which is the main distinguishing difference between amino acids.

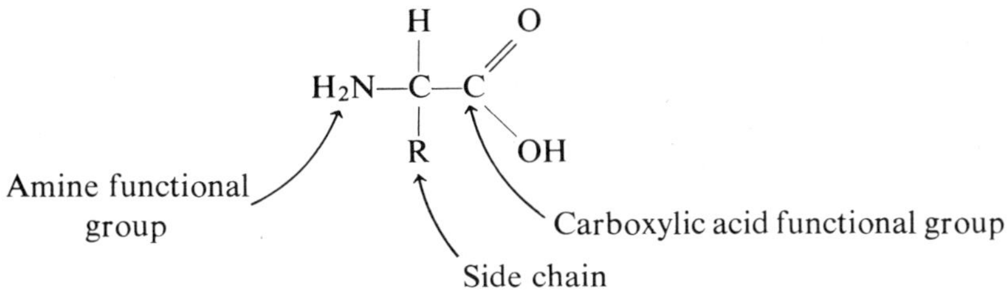

$$H_2N-\underset{\underset{R}{|}}{\overset{\overset{H}{|}}{C}}-C\begin{matrix}\nearrow O \\ \searrow OH\end{matrix}$$

Amine functional group
Side chain
Carboxylic acid functional group

This general acid and all others we shall consider are called alpha (α) amino acids, because the amine group is attached to the carbon chain at the alpha position, the first carbon after the acid group.

The R side chain may contain a number of different elements, such as carbon, hydrogen, oxygen, sulfur, and nitrogen, and may also include such structural features as aromatic rings, —OH groups, $-NH_2$ groups, and $-C\begin{matrix}\nearrow O \\ \searrow OH\end{matrix}$ groups. This variety in side chains causes differences in the properties of the amino acids and resulting proteins. The differences in properties depend primarily upon whether the side chains are polar, nonpolar, acidic, or basic.

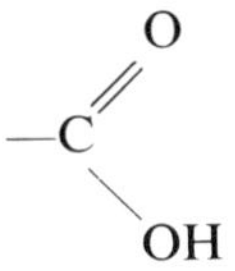

Structural formulas of the twenty amino acids commonly found in proteins are given in Table 24–1 together with common names and abbreviations used when describing proteins. The R side chains are circled in each structure. An asterisk indicates essential amino acids which cannot be synthesized in the body and must be obtained directly from food.

The amino acids are white, crystalline solids, most of which are quite water soluble. This solubility, together with their relatively high melting point, suggests that they resemble inorganic salts and exist in the form of polar molecules. Studies of amino acids in solution confirm that they are charged molecules. It has also been found that the charge changes with the *p*H of the solution. In acid solution, a hydrogen ion becomes attached to the amine group. In a basic solution a hydrogen ion is given up by the acid group, and in neutral solutions both of the above occur to give a very polar but uncharged molecule. This is illustrated in Example 24–1.

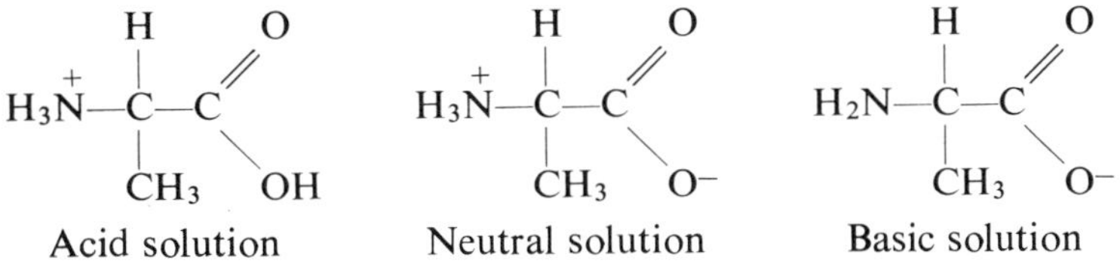

$$H_3\overset{+}{N}-\underset{\underset{CH_3}{|}}{\overset{\overset{H}{|}}{C}}-C\begin{matrix}\nearrow O \\ \searrow OH\end{matrix} \qquad H_3\overset{+}{N}-\underset{\underset{CH_3}{|}}{\overset{\overset{H}{|}}{C}}-C\begin{matrix}\nearrow O \\ \searrow O^-\end{matrix} \qquad H_2N-\underset{\underset{CH_3}{|}}{\overset{\overset{H}{|}}{C}}-C\begin{matrix}\nearrow O \\ \searrow O^-\end{matrix}$$

Acid solution — Neutral solution — Basic solution

Example 24–1. Ionic Forms of Alanine in Solution

Table 24–1. Amino Acids of Proteins (abbreviations are in parentheses)

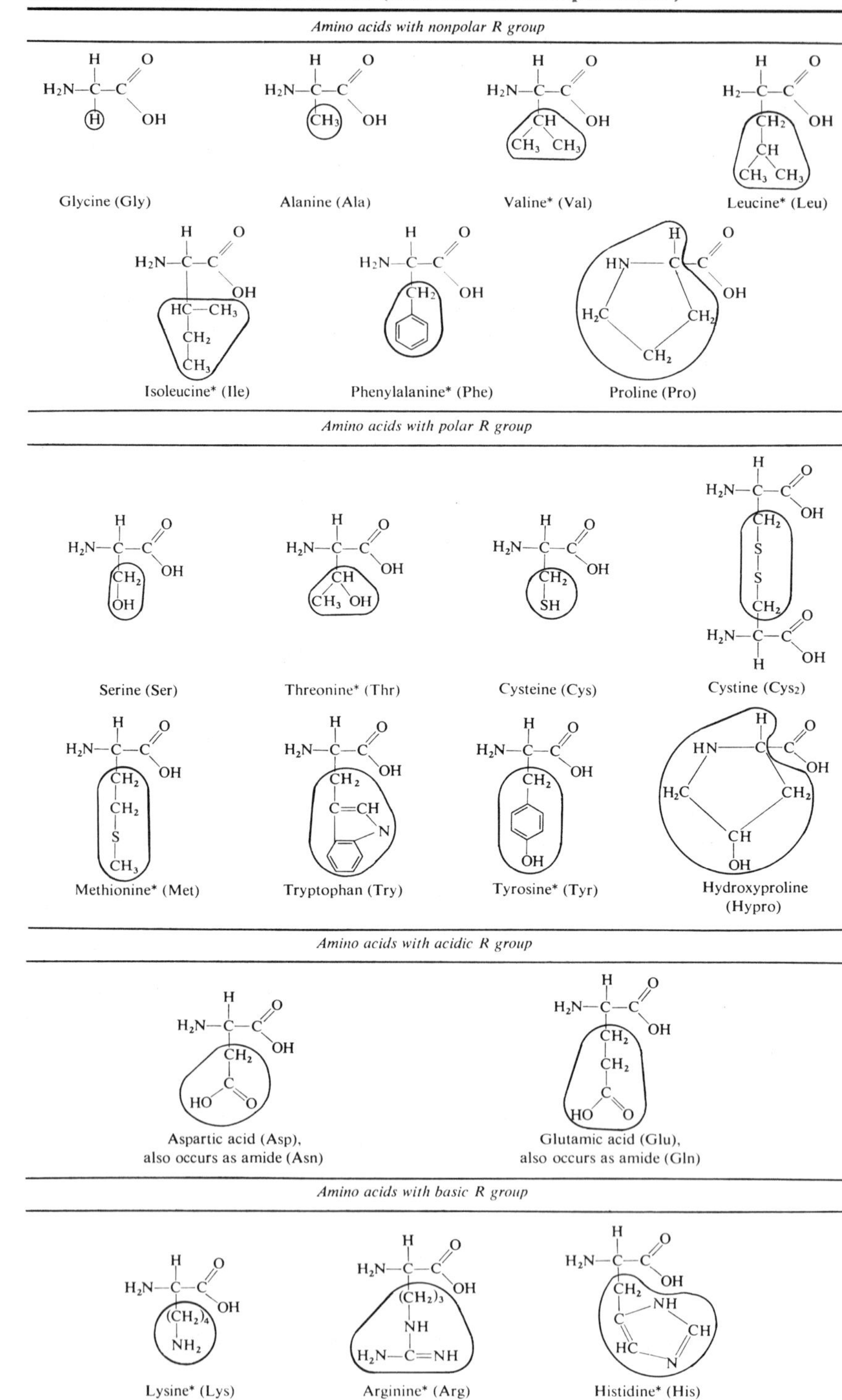

Individual amino acids can undergo numerous reactions characteristic of the amine, carboxylic acid, and any other groups in the molecular structure. By means of such reactions, amino acids in the body may be converted into many relatively simple but vital compounds. Phenylalanine is converted into adrenalin, for example, by the pathway given below.

phenylalanine → tyrosine → 3,4-dihydroxyphenylalanine →

3,4-dihydroxyphenylethylamine → noradrenaline (norepinephrine) → adrenaline (epinephrine)   24-1

In spite of such reactions, the major function of amino acids is to serve as building blocks for protein synthesis—our opening commercial again.

Structure of Proteins

The reaction responsible for the polymerization of amino acids into protein was discussed in Chapter 20. It is the reaction of a carboxylic acid with an amine to give an amide. Each amino acid contains both of the functional groups necessary for such reactions to take place. When the reaction occurs between the acid group of one amino acid and the amine group of a second, a bond or amide linkage is formed between the two. A general reaction of this type is:

$$H_2N-\underset{R}{\overset{H}{C}}-\overset{O}{\overset{\|}{C}}-OH + H-\overset{H}{N}-\underset{R'}{\overset{H}{C}}-\overset{O}{\overset{\|}{C}}-OH \rightarrow$$

$$H_2N-\underset{R}{\overset{H}{C}}-\overset{O}{\overset{\|}{C}}-\overset{H}{N}-\underset{R'}{\overset{H}{C}}-\overset{O}{\overset{\|}{C}}-OH + H_2O \qquad 24\text{–}2$$

amide linkage

The result of this reaction between two amino acid monomers is a new dimer molecule, called a dipeptide, which still contains an amine group plus an acid group. In protein chemistry, the amide linkage is called a **peptide linkage.** The dipeptide can continue to react with other amino acids to give tripeptides, tetrapeptides, and so forth.

$$\underbrace{H_2N-\underset{R}{\overset{H}{C}}-\overset{O}{\overset{\|}{C}}-\overset{H}{N}-\underset{R'}{\overset{H}{C}}-\overset{O}{\overset{\|}{C}}-OH}_{\text{dipeptide}} + \underbrace{H-\overset{H}{N}-\underset{R''}{\overset{H}{C}}-\overset{O}{\overset{\|}{C}}-OH}_{\text{amino acid}} \rightarrow$$

$$H_2N-\underset{R}{\overset{H}{C}}-\overset{O}{\overset{\|}{C}}-\overset{H}{N}-\underset{R'}{\overset{H}{C}}-\overset{O}{\overset{\|}{C}}-\overset{H}{N}-\underset{R''}{\overset{H}{C}}-\overset{O}{\overset{\|}{C}}-OH + H_2O \qquad 24\text{–}3$$

peptide linkages

tripeptide

Molecules made up of from 5 to 50 amino acid units are called **polypeptides.** The term **protein** is usually reserved for structures containing more than 50 amino acid units, although in some instances *polypeptide* and *protein* are used interchangeably.

The sequence of amino acids in polypeptide or protein chains is called the **primary structure.** These sequences are difficult to determine, especially for very large proteins, and as a result only relatively few have been completely obtained. Three such polypeptide structures are shown in Figure 24–1.

All three examples in Figure 24–1 are human hormones. Oxytocin and vasopressin are produced by the pituitary gland. Oxytocin stimulates the contraction of the smooth muscle of the uterus, and vasopressin causes increased blood pressure and an increase in water retention by the kidneys. The dark band between the two cystine units denotes a disulfide link. This link, which is not a part of the primary structure, will be discussed later.

Figure 24–1.
Primary Structure of Some Polypeptides

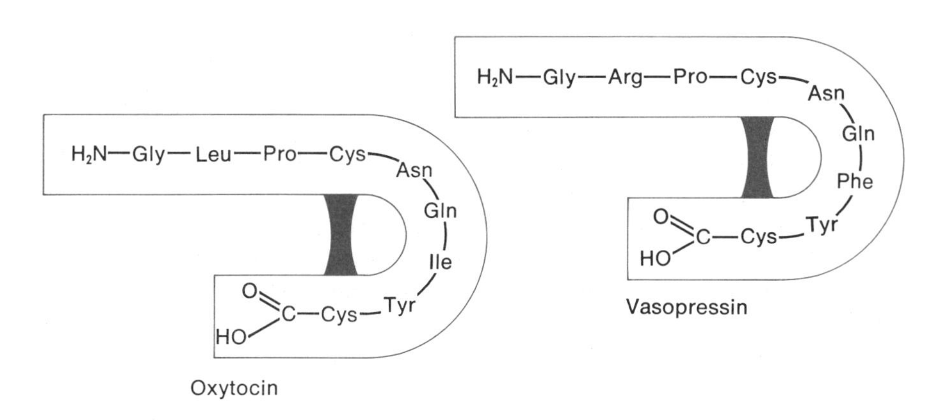

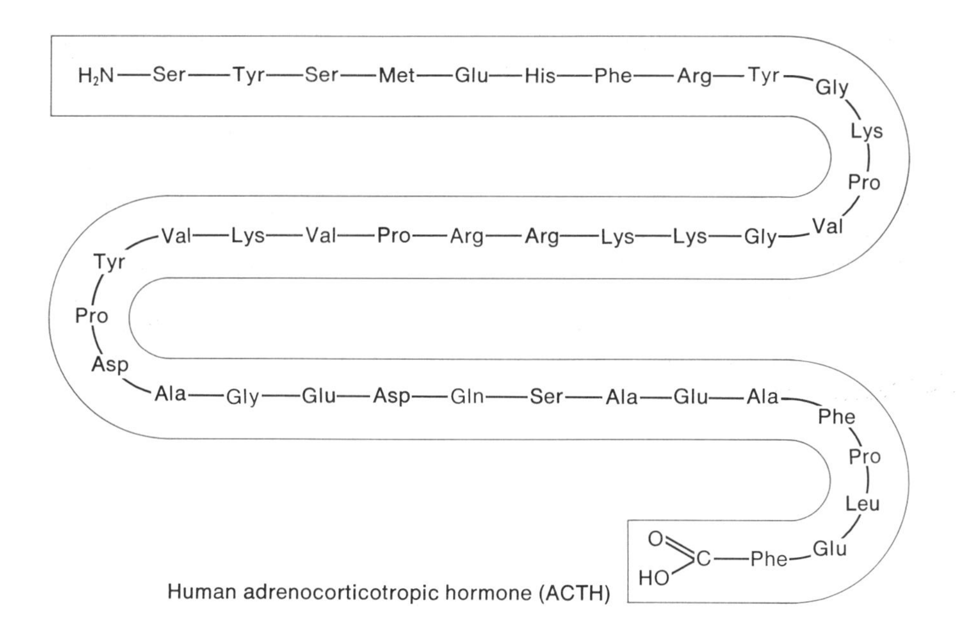

ACTH is also a hormone of the pituitary gland. Only the amino acid sequence is given in Figure 24–1. The structure is folded to fit the page and in no way represents the actual shape of the molecule. This hormone stimulates the adrenal gland to produce and secrete its steroid hormones, some of which were discussed in Chapter 23.

The primary structure or amino acid sequence is very important if protein molecules are to function normally. A case in point is the hereditary blood disease, sickle-cell anemia. The hemoglobin (a protein) of an afflicted individual contains a valine unit in place of a glutamic acid unit. This substitution, involving only one of the approximately 300 amino acid units of the hemoglobin, causes a drastic change in the properties of the molecule. The abnormal hemoglobin tends to precipitate in the red blood cells containing it.

This distorts the red cells into a characteristic shape that gives the disease its name. The distorted cells clump together and sometimes block veins and arteries; in some cases the cells themselves break open. Children who inherit the trait from both parents seldom live beyond the age of two.

If the only structural characteristic of proteins and polypeptides were the amino acid sequence, all such substances would be made up of random arrangements of polypeptide chains. Experimental evidence shows that this is not the case. At least two characteristic **secondary structural features** of proteins have been found. One of these, the **alpha helix,** results when the peptide chain making up the primary structure is coiled into the shape of a helix. Hydrogen bonds between the —C=O of some peptides of the chain and the —NH of others hold the chain in this configuration.

The other type of secondary structure is the **pleated sheet.** In this structure hydrogen bonding takes place between —C=O and —NH groups of different chains and holds the chains in a sheetlike arrangement. The pleats are the result of the natural bond angles of the polypeptide chains. Both of these secondary structures are shown in Figure 24–2.

A third or **tertiary structure** of proteins has also been found to exist. This feature involves the interaction of groups in the side chains of the peptides making up the primary structure. The disulfide linkages of oxytocin and vasopressin mentioned earlier illustrate the interactions involved. Besides the disulfide bond, the interactions include the salt bond, caused by attractions between charged saltlike groups; hydrogen bonds; and hydrophobic bonds, which result from the tendency of nonpolar groups to clump together when placed in an aqueous medium. These various interactions are shown in Figure 24–3.

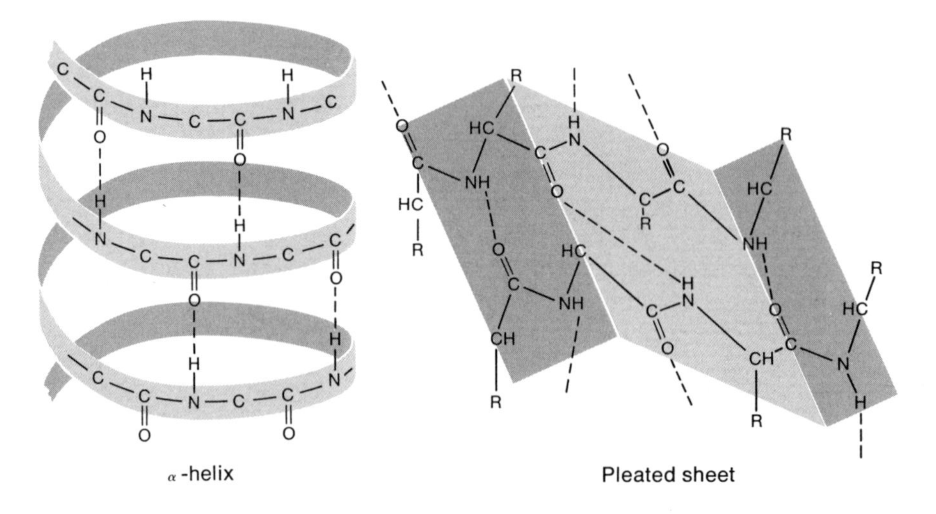

Figure 24–2.
Secondary Structure of Proteins

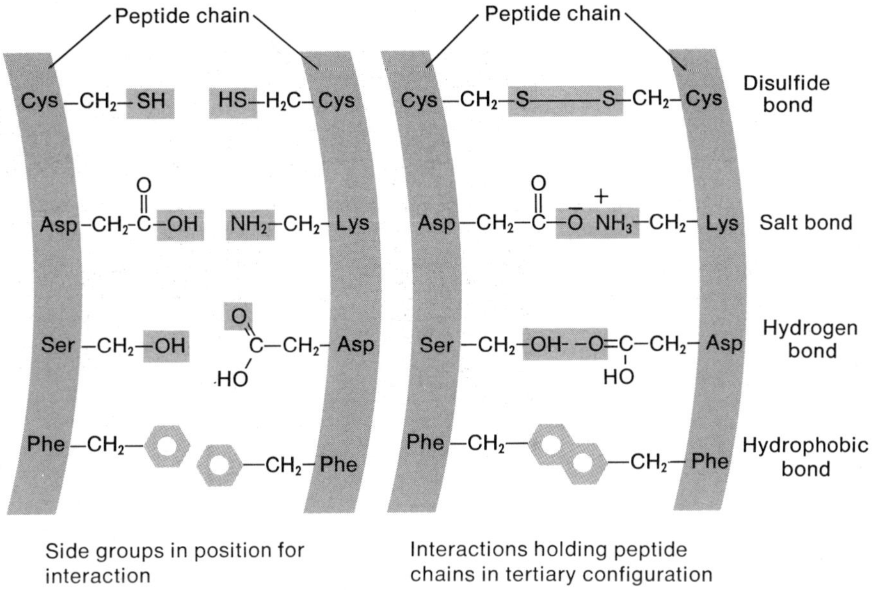

Peptide chain

Cys $-CH_2-SH$ $HS-H_2C-$ Cys

Asp $-CH_2-C(=O)-OH$ NH_2-CH_2- Lys

Ser $-CH_2-OH$ $O=C(HO)-CH_2-$ Asp

Phe $-CH_2-C_6H_5$ $C_6H_5-CH_2-$ Phe

Side groups in position for interaction

Peptide chain

Cys $-CH_2-S$——$S-CH_2-$ Cys   Disulfide bond

Asp $-CH_2-C(=O)-\bar{O}$ $\overset{+}{N}H_3-CH_2-$ Lys   Salt bond

Ser $-CH_2-OH$- -$O=C(HO)-CH_2-$ Asp   Hydrogen bond

Phe $-CH_2-C_6H_5$ $C_6H_5-CH_2-$ Phe   Hydrophobic bond

Interactions holding peptide chains in tertiary configuration

Figure 24–3.
Side Group Interactions Leading to Tertiary Protein Structures

The tertiary structures resulting from these interactions can be visualized, for example, as a folding or clumping of the alpha helix into a bent or ball-like arrangement. The curvature in the chains of oxytocin and vasopressin are tertiary structural features.

Reactions of Proteins

Proteins undergo numerous reactions. We shall discuss three of these: hydrolysis, denaturation, and precipitation.

Hydrolysis reactions, as seen before, involve the addition of water groups. The site for hydrolysis in proteins and polypeptides is the peptide linkage originally formed in the reaction between the acid and amine functional groups of amino acids. Hydrolysis causes the bonds to rupture and regenerates the original functional groups. Large proteins are broken into smaller proteins, peptides, or possibly individual amino acids. The general reaction is given below.

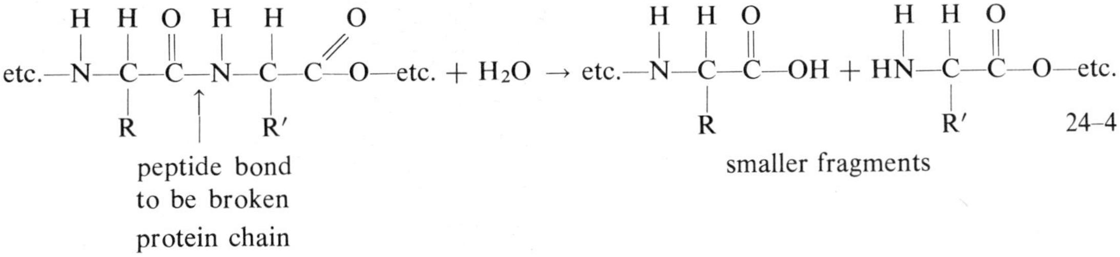

$$\text{etc.}-\underset{|}{\overset{H}{N}}-\underset{R}{\overset{H}{C}}-\overset{O}{\overset{\|}{C}}-\underset{\uparrow}{\overset{H}{N}}-\underset{R'}{\overset{H}{C}}-\overset{O}{\overset{\|}{C}}-O-\text{etc.} + H_2O \rightarrow \text{etc.}-\overset{H}{N}-\underset{R}{\overset{H}{C}}-\overset{O}{\overset{\|}{C}}-OH + H\overset{H}{N}-\underset{R'}{\overset{H}{C}}-\overset{O}{\overset{\|}{C}}-O-\text{etc.} \qquad 24\text{–}4$$

peptide bond to be broken
protein chain

smaller fragments

Hydrolysis of protein accounts for a number of the changes that occur when protein-rich food is cooked. In meats, the protein collagen, which constitutes half of the protein of the body, is partially hydrolyzed to the familiar material gelatin.

Denaturation reactions include all those in which any bonds are ruptured in the protein with the exception of the peptide linkages of the primary structure. Denaturation destroys the natural conformation of a protein, as represented by the secondary and tertiary structures, and is often accompanied by precipitation or coagulation of the protein. Denaturation is usually a nonreversible reaction, though in a few cases reversal is possible. A variety of reagents and conditions can cause denaturation; some of these are listed in Table 24–2.

Table 24–2. Protein Denaturation

Denaturing reagent or condition	*Example*
Heat	Cooking of egg white, sterilization of surgical instruments
Agitation	Beating of egg white
Reducing agent	Home permanent
Organic solvents	Alcohol used to disinfect skin
Acids or bases	Clouding of eye cornea from accidental contact
Heavy-metal ions (Ag^+, Pb^{2+}, Hg^{2+})	Disinfectant solutions

Denaturation of bacterial protein takes place when alcohol is used as a disinfectant. However, the commonly used 70 percent solution is more effective than pure alcohol. Pure alcohol quickly denatures and coagulates the surface of the bacteria. The coagulated surface acts as a barrier to further penetration by the alcohol and protects the bacteria. A 70 percent solution denatures more slowly and allows complete penetration before coagulation takes place.

Heavy-metal salts are toxic when taken internally because they denature and precipitate cell protein with which they come in contact. Egg whites and milk are used as antidotes because they are quickly denatured and combine with the metal ions to form insoluble solids. An emetic is used to remove the resulting solid from the stomach before the precipitated egg and milk proteins can be digested and the toxic metal ions released.

The permanent wave for hair is possible because of a reversible denaturation of the hair. The protein of hair (keratin) contains a high proportion of sulfur-containing amino acids such as cysteine or methionine. Disulfide linkages between these acids are largely responsible for the shape of the hair. In the permanent waving process a substance (waving lotion) is put on the hair which reduces and ruptures the disulfide bonds. A typical waving lotion ingredient is ammonium thioglycolate,

$$H{-}S{-}CH_2{-}\overset{\displaystyle O}{\overset{\|}{C}}{-}ONH_4$$

Figure 24–4.
Permanent Wave Process

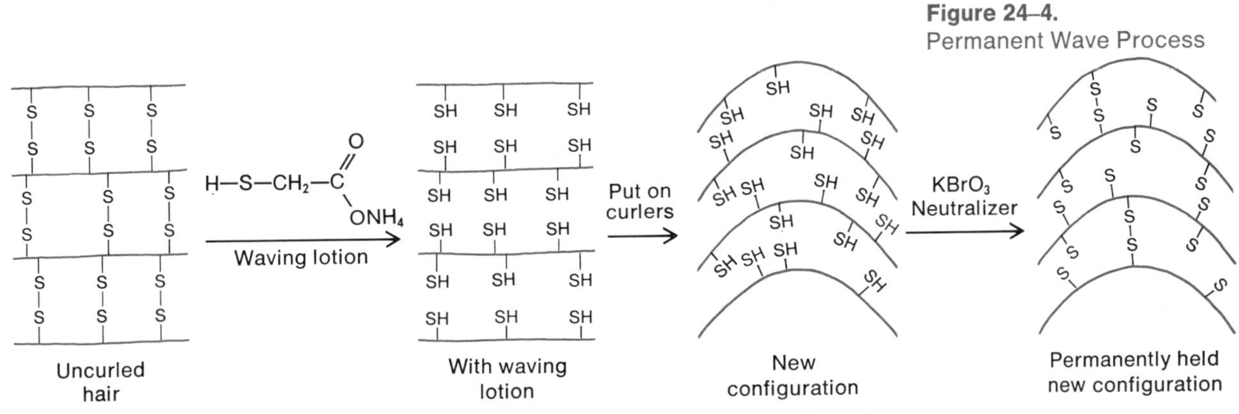

The hair is then put on rollers and an oxidizing substance (neutralizer) is added which causes bonds to reform. The new bonds hold the hair in the desired configuration. A typical neutralizer is potassium bromate, $KBrO_3$. The process is illustrated in Figure 24–4.

Enzymes

In a typical soap-making operation, the saponification (hydrolysis) of fats takes place in large vats at a temperature near the boiling point of water, 212°F. The same hydrolysis reaction takes place in the body at the much more comfortable temperature of 98.6°F. The presence and activity of marvelous biological catalysts called **enzymes** make such reactions possible at the acceptable temperature. Life as we know it would not be possible without these enzymes to aid and control the processes of metabolism.

The Chemical Nature of Enzymes

Most enzymes consist of a large polypeptide or protein coupled with some other substance called a cofactor or **coenzyme.** The protein or polypeptide portion is called the **apoenzyme.** The apoenzyme shows no enzyme activity until it is combined with a proper coenzyme. In some enzymes the coenzyme is a simple divalent metal ion such as Mg^{2+}, Ca^{2+}, Co^{2+}, Mn^{2+}, or Zn^{2+}; in others it is a nonprotein organic molecule. In some enzymes both types of coenzymes are necessary.

The Names of Enzymes

The names of most enzymes end in *-ase*. Prefixes to the ending indicate either the type of reaction catalyzed or the type of substance (substrate) upon which the enzyme acts. Maltase, for example, is an enzyme which catalyzes the hydrolysis of maltose. Maltase would also be called by the group name

hydrolase, because it is active in a hydrolysis reaction. Table 24–3 gives a few examples of enzyme nomenclature.

Table 24–3. Examples of Enzyme Nomenclature

Group name	*Specific name*	*Type of reaction catalyzed*
Hydrolase	Amylase	Hydrolysis of amylose
	Sucrase	Hydrolysis of sucrose
	Lipase	Hydrolysis of lipids
Oxidoreductase	Dehydrogenase	Transfer of H from one molecule to another
	Oxidase	Transfer of H from one molecule to oxygen
Transferase	Transaminase	Transfer of amine group enabling synthesis of amino acids from nonprotein sources

The names of some common enzymes are exceptions to these nomenclature rules. Three enzymes of the digestive system are known by their common names pepsin, trypsin, and chymotrypsin.

Coenzymes and Vitamins

In 1912 Casimir Funk, a 26-year-old scientist, proposed his vitamin (vital amine) hypothesis. This hypothesis was based upon the work of others as well as his own. He postulated that there were four vitamins which protected the body from four specific diseases.

Today numerous vitamins are known, and many have been found to be coenzymes or their **precursors**—substances from which another substance is easily produced. This is especially true of the vitamins of the B group. Table 24–4 gives names and structures of some typical vitamins which behave as coenzymes or coenzyme precursors. Notice that not all of these vitamins are amines.

Mode of Enzyme Action

A theory to account for catalytic activity was proposed by the Swedish chemist Svante Arrhenius in 1888. He suggested that all catalytic behavior was based upon an intermediate compound formed between the catalyst and a reacting substrate molecule. In biochemical terms, the substrate, S, and enzyme, E, form an intermediate compound, S-E, which then breaks apart yielding the products, P, and the enzyme, E:

$$S + E \rightarrow S\text{-}E \rightarrow P + E \qquad 24\text{–}5$$

or specifically:

$$\underset{(S)}{\text{sucrose}} + \underset{(E)}{\text{sucrase}} \rightarrow \underset{(S\text{-}E)}{\text{intermediate}} \xrightarrow{\text{add } H_2O} \underbrace{\text{glucose} + \text{fructose}}_{(P)} + \underset{(E)}{\text{sucrase}} \qquad 24\text{–}6$$

Table 24–4. Examples of Vitamins

Name	*Structural formula*	*Coenzyme function*
Thiamine (B_1)	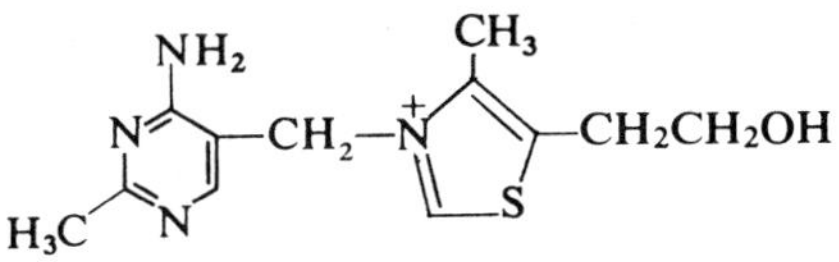 NH_2, CH_3, N, $—CH_2—\overset{+}{N}$, $—CH_2CH_2OH$, H_3C, N, S	Precursor
Riboflavin (B_2)	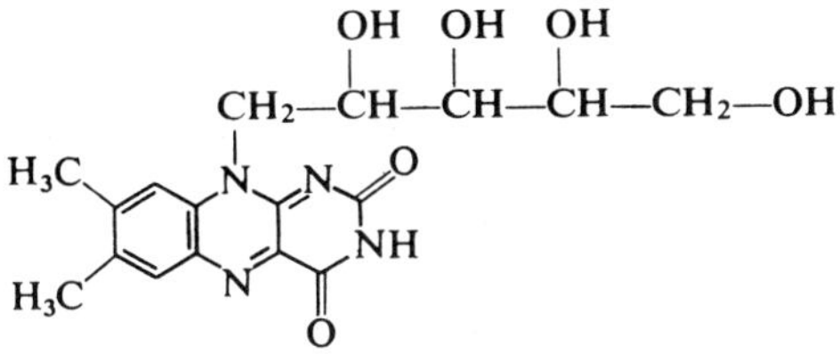 OH OH OH; $CH_2—CH—CH—CH—CH_2—OH$; H_3C, H_3C, N, N, O, NH, N, O	Precursor
Pyridoxine (B_6)	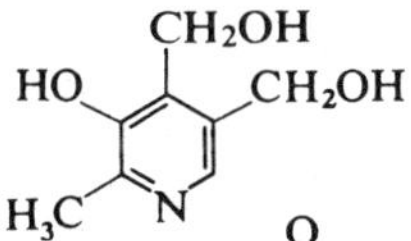 CH_2OH, HO, CH_2OH, H_3C, N	Precursor
Nicotinic acid (niacin)	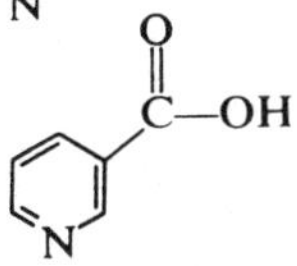 O, C—OH, N	Precursor
Biotin	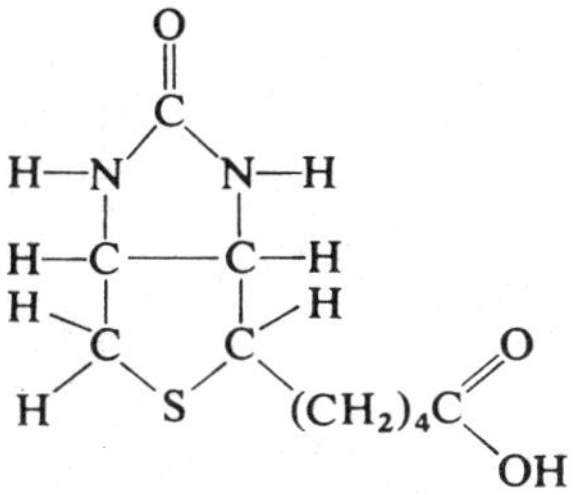 O, C, H—N, N—H, H—C, C—H, H, H, C, C, H, S, $(CH_2)_4C$, O, OH	Coenzyme
Lipoic acid	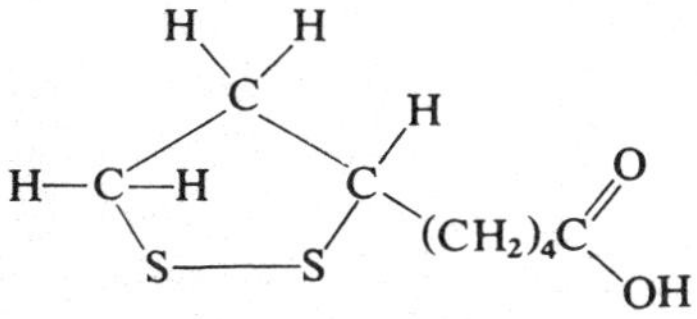 H H, C, H, H—C—H, C, S—S, $(CH_2)_4C$, O, OH	Coenzyme

The existence of intermediate compounds has been verified experimentally. The bonds holding the intermediate together are probably the same as those responsible for tertiary protein structure.

It has also been shown that the characteristic functional groups or other structural features essential to the formation of intermediates occur at a

Figure 24–5.
Lock-and-Key Theory of Substrate-Enzyme Interaction

Substrate

Enzyme

Intermediate compound

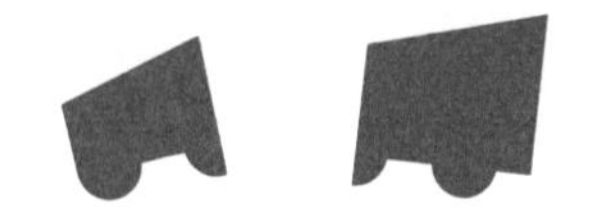

Products

Enzyme

specific location or **active site** on the enzyme surface. The substrate combines with the enzyme only at the active site to form the intermediate compound. The active site has a unique conformation which fits only a particular substrate in a manner analogous to a lock and key. This fact accounts, in part, for the specificity of enzymes. The lock-and-key theory is illustrated in Figure 24–5.

Review Questions

1. What functional groups are present in all amino acids?
2. What are essential amino acids?
3. Write an equation representing the reaction of two molecules of serine (Table 24–1).
4. How many different tripeptides could be formed by reacting together three different amino acids?
5. Discuss the meaning of the following terms and give an illustrative example for each one:
   a) Primary protein structure
   b) Secondary protein structure
   c) Tertiary protein structure
6. Describe at least four examples of protein denaturation. Include two that you consider to be desirable and two that are not desirable.
7. Discuss the role of vitamins in the function of enzymes.
8. What is a precursor?
9. Numerous enzymes are known to act upon very specific substrates. Is this observation consistent with the lock-and-key theory of enzyme activity? If it is, explain how.
10. Some enzymes are known as isomerases. What kind of reactions would they catalyze?

Suggestions for Further Reading

Doty, P., "Proteins," *Scientific American*, Sept. 1957.
"Enzymes in Detergents," *Chemistry*, Feb. 1970.
Kendrew, J. C., "The Three Dimensional Structure of a Protein Molecule," *Scientific American*, Dec. 1961.
Pauling, L., R. B. Corey, and R. Haywood, "The Structure of Protein Molecules," *Scientific American*, July 1954.
Phillips, D. C., "The Three Dimensional Structure of an Enzyme Molecule," *Scientific American*, Nov. 1966.
Raw, I., "Enzymes, How They Operate," *Chemistry*, June 1967.
Stein, W. H., and S. Moore, "The Chemical Structure of Proteins," *Scientific American*, Feb. 1961.

25 Chemical Processes in the Body

Each of us associates physical well-being with certain activities of the body. It is considered normal to eat and drink food and water, breathe, eliminate waste materials, and possess sufficient energy to move about vigorously.

When a hand is cut we expect to see a little bleeding, followed by clotting of the blood and eventual healing of the wound. We would be surprised—and seriously injured—if a hand were not quickly and involuntarily pulled back upon touching a hot stove. We expect and look forward to producing children who will exhibit many of our characteristics.

Such activities of the body are based on chemical reactions and processes. In this chapter we shall investigate a few of these activities from a chemical point of view.

Digestion

The pleasurable activity of eating is followed by the important process of digestion, in which large, complex molecules of food are chemically changed into relatively small, simple ones. The smaller molecules are absorbed into the body by other processes and used as sources of energy and as raw building materials.

The most common reaction of digestion is hydrolysis, which takes place at the amide or peptide linkage of proteins; the ester linkage of fats, oils, and waxes; and the acetal linkage of di- and higher saccharides.

The digestive tract, in which these processes take place, consists primarily of a long tube; its main parts and organs are illustrated in Figure 25–1.

In order to understand the processes and reactions that occur in each area of the digestive tract, we shall follow a food sample through. The sample is assumed to contain all three of the major foods—carbohydrates, lipids, and proteins.

In the mouth the food is torn and broken into small pieces by chewing. The resulting small particles provide a large surface area upon which the reactions of digestion can take place. As the food is chewed it is mixed with saliva, the first digestive juice. Saliva consists mostly of water (99.5 percent), but it also contains mucin (a lubricative glycoprotein made up of carbohydrate and protein units), the enzyme amylase (ptyalin), and the inorganic ions Ca^{2+}, K^+, Na^+, Mg^{2+}, $H_2PO_4^-$, Cl^-, and HCO_3^-. The *p*H of saliva is about 6.8. Saliva lubricates the food for easy passage into the stomach and, through the action of amylase, initiates the digestive hydrolysis of the starch amylose.

The food next passes into the stomach, where it is mixed with gastric juice, a mixture of several secretions originating in various glands or areas of the stomach wall. The components of gastric juice are water (99 percent), mucin, hydrochloric acid (HCl, 0.5 percent), the enzymes pepsin and gastric lipase, and the inorganic ions Na^+, K^+, Cl^-, and $H_2PO_4^-$. The *p*H of the juice is 1.6–1.8.

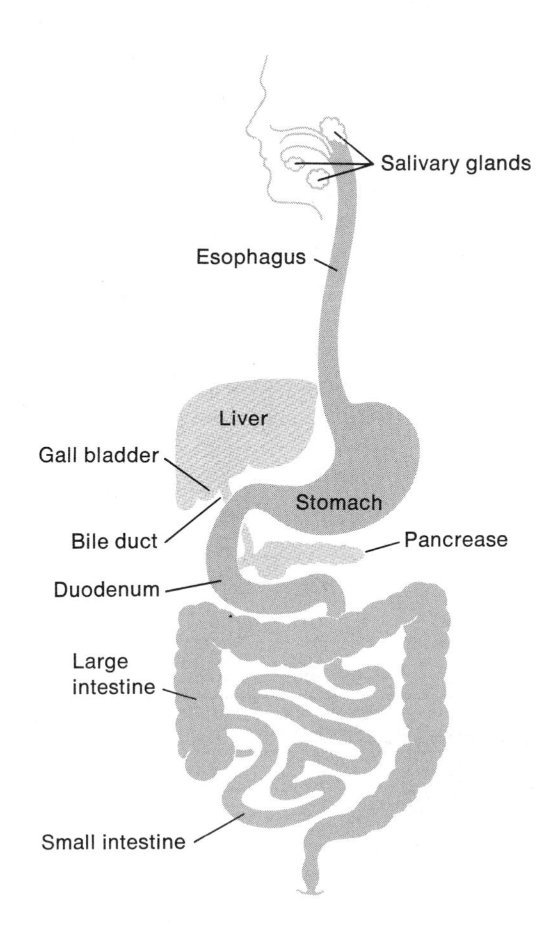

Figure 25–1.
The Digestive Tract

The enzyme pepsin catalyzes the hydrolysis of peptide bonds in proteins. It is especially effective on interior bonds of the protein chain and produces intermediate-sized peptides of varying chain length. Pepsin is most effective at a *p*H of 2 and is inactivated through denaturation at a *p*H near 6.

The enzyme lipase is not active in the acidic environment provided by the stomach; its exact purpose in gastric juice is not known. Some scientists have proposed that it is not really produced in the stomach but leaks in from the lower parts of the digestive tract. Another opinion is that lipase production in the stomach is a carryover from infancy when the stomach was much less acidic and lipids—which lipase causes to hydrolyze—were an important part of the diet.

Nature has provided an interesting protective system to prevent the stomach from being digested along with food proteins. Special mucous cells in the stomach wall continuously secrete a thick, viscous fluid which contains a mucoprotein made up of a polysaccharide and protein. The fluid coats the stomach wall and is slowly hydrolyzed, but because it is continually replenished the stomach wall remains protected. A similar mucous system protects the intestines. Some disorders in the body result in high stomach acidity—hyperacidity. Under such conditions the mucous coating is hydrolyzed more rapidly and may fail to adequately protect the stomach wall. The resulting damaged—hydrolyzed or digested—area of the stomach wall is called an **ulcer.** In extreme cases the attack on the wall results in perforation, which can be very serious, especially in cases involving a blood vessel where severe bleeding may result.

The mixture of partially digested food, called **chyme,** is slowly released from the stomach into the duodenum, where it is mixed with several enzyme-rich secretions which complete the process of digestion. The important digestive secretions added in the duodenum come from the pancreas (pancreatic juice), special cells in the duodenal wall (intestinal juice), and the liver (bile). Tables 25–1 and 25–2 summarize the digestive action of some typical components of pancreatic and intestinal juice. The materials produced by each component's activity are also given.

Table 25–1. Composition of Pancreatic Juice

Component	*Role in digestion*	*Products of activity*
Ions		
Na^+, K^+, Ca^{2+}, Cl^-, HCO_3^-, HPO_4^{2-}, *p*H: 7–8		
Enzymes		
Pancreatic lipase (steapsin)	Catalyzes hydrolysis of triglycerides.	Fatty acids and monoglycerides
Amylase	Catalyzes hydrolysis of amylose.	Maltose
Maltase	Catalyzes hydrolysis of maltose.	Glucose
Zymogens (inactive enzymes)		
Trypsinogen	Active form (trypsin) catalyzes hydrolysis of nonterminal peptide linkage near charged side chain.	Small peptides
	Trypsin also activates chymotrypsinogen and procarboxypeptidase.	Chymotrypsin and carboxypeptidase
Chymotrypsinogen	Active form (chymotrypsin) catalyzes hydrolysis of peptide linkages located near aromatic side groups.	Small peptides
Procarboxypeptidase	Active form (carboxypeptidase) catalyzes hydrolysis of peptide linkages next to terminal amino acids.	Small peptides and amino acids

Table 25–2. Composition of Intestinal Juice

Component	*Role in digestion*	*Products of activity*
Water		
Ions		
Na^+, K^+, Ca^{2+}, Cl^-, HCO_3^-, HPO_4^{2-}		
Enzymes		
Aminopeptidase	Catalyzes hydrolysis of peptide linkages next to terminal amino acids containing an unreacted amine group.	Small peptides and amino acids
Dipeptidase	Catalyzes hydrolysis of dipeptides.	Amino acids
Maltase	Catalyzes hydrolysis of maltose.	Glucose
Sucrase	Catalyzes hydrolysis of sucrose.	Glucose and fructose
Lactase	Catalyzes hydrolysis of lactose.	Glucose and galactose
Intestinal lipase	Catalyzes hydrolysis of lipids.	Fatty acids and monoglycerides or glycerin
Enterokinase	Activates trypsinogen of pancreatic juice.	Trypsin

The bile of the liver does not catalyze any hydrolysis reactions but plays an important role in lipid digestion. Lipids are not soluble in water, and in the watery medium of the digestive system they tend to form large globules. These large globules present a limited surface area for attack by the digestive juices. Bile contains bile salts which look and behave very much like soaps (see Chapter 23). These salts emulsify the lipids and break the large globules into many smaller droplets. The total amount of lipid is still the same, but a much larger surface area is available on which the hydrolysis reactions can take place. Some researchers feel that bile salts also behave as soaps to remove fatty coatings from particles of other types of food.

Upon completion of digestion, the four main carbohydrates of food—starch, lactose, maltose, and sucrose—have been converted into galactose, glucose, and fructose. Any cellulose, from the woody structure of fruits and vegetables, remains essentially unchanged and will pass through the tract and be eliminated from the body. The lipids have been changed to glycerol, fatty acids, and monoglycerides (one acid molecule attached to glycerol). The proteins, at the end of digestion, are in the form of amino acids and some simple water-soluble di- and tripeptides.

The digested food passes on into the small intestine, where most of the useful products of digestion are absorbed into the bloodstream for redistribution to other parts of the body. The remaining moist material passes into the large intestine, where excess water is reabsorbed into the blood. The resultant waste material is eliminated from the body.

Chemical Transport

In order to carry out its complex function, a living cell requires a steady supply of reactants such as nutrients and oxygen as well as a reliable system for removing the resulting waste products, such as carbon dioxide and water.

A single cell isolated in liquid is provided these two necessities by the simple process of diffusion. Reactants can diffuse into the cell through the thin cell wall, and the products of cell function can diffuse out to be absorbed by the surrounding liquid.

However, a complex organism consisting of many closely packed cells cannot depend solely upon diffusion processes. Some circulating system must bring necessary materials to the cells and remove wastes; otherwise, any surrounding liquid would rapidly be depleted of needed reactants and become saturated with wastes.

The blood and associated circulatory system provide the most active transport system in the body. This system is assisted in the transport function by the interstitial fluid surrounding individual tissue cells and by the lymph system. This latter system consists of a slow-moving fluid in a network of vessels which are closed at one end and empty into the blood system at the other end. Materials diffuse into the closed vessels and eventually end up in the blood.

The importance of the blood and its circulatory system becomes apparent when we consider their major functions:

Nutrients are transported to the cells making up the tissues of the body.
Waste materials are transported from the tissue cells to organs of excretion.
Oxygen is carried from the lungs to the tissue cells.
Carbon dioxide, a waste material, is transported from the tissue cells to the lungs for excretion.
Regulatory substances such as hormones, vitamins, and enzymes are distributed throughout the body.
Structures in the blood help protect the body from infectious microorganisms.
The blood helps control body temperature.
The acid-base balance of the body is partially controlled.
The water balance of the body is partially controlled.

The Circulatory System

The circulatory system of the blood is represented schematically in Figure 25–2. Blood is moved through the system by the pumping action of the heart. The blood first enters large arteries, which, through much branching, are reduced in size to small capillaries. Through the thin walls of capillaries most exchange of chemicals with the tissue cells takes place. The blood, after passing through the capillaries, is collected in larger veins for return to the heart. From the heart it travels to the lungs, then back to the heart, and the cycle is repeated.

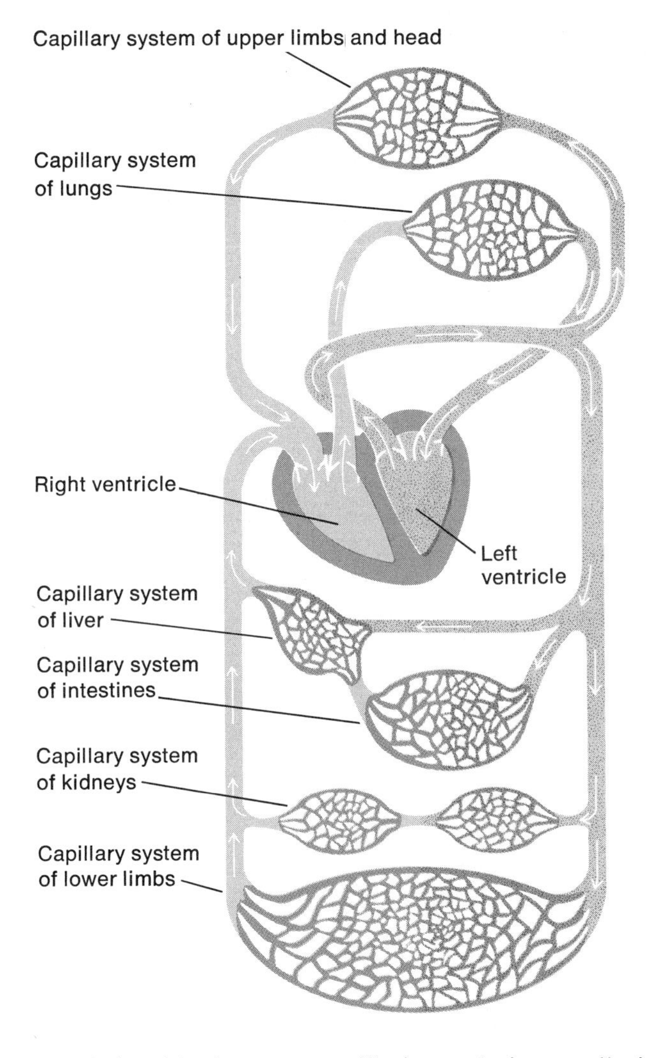

Figure 25–2.
The Circulatory System

The distance relationship between capillaries and tissue cells is shown schematically in Figure 25–3. Each cell is quite close to a capillary to ensure the rapid transport of materials from the blood, through the interstitial fluid, and into the cells.

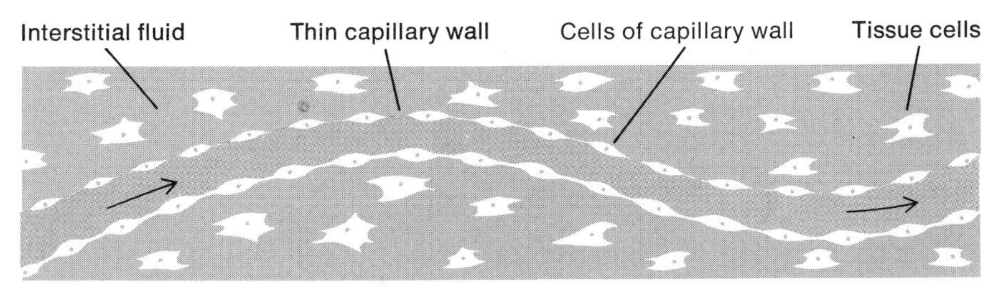

Figure 25–3.
Spatial Relationship Between Tissue Cells and Capillaries

From *Elements of General and Biological Chemistry* by John R. Holum (New York: John Wiley and Sons, Inc., 1972), p. 417.

Blood Composition

Blood is a suspension of solid cellular elements in a liquid medium. The liquid medium or **plasma** is a solution of various proteins and salts dissolved in water. The dissolved proteins include albumins, which are important in regulating the osmotic processes in blood and in transporting some relatively insoluble materials; globulins (alpha, beta, and gamma), which serve as aids in the transport of some metal ions (alpha and beta globulins are involved) and as a defense against infection (gamma globulin); and fibrinogen, which is the immediate precursor of fibrin, an insoluble protein that precipitates in a tangled mass during blood-clot formation. The dissolved salts in plasma consist mainly of the cations Na^+, K^+, Ca^{2+}, and Mg^{2+}, and the anions Cl^-, HCO_3^-, HPO_4^{2-}, and SO_4^{2-}. The ions are involved in the osmotic processes of blood and in this way aid in the maintenance of acid-base and water balance. Plasma from which the suspended cellular material and the protein fibrinogen have been removed is called **serum.**

The suspended cellular material of blood is made up of the so-called **formed elements.** These materials constitute 40 to 45 percent of the total volume of the blood. The formed elements include red blood cells (erythrocytes), white blood cells (leukocytes), and the blood platelets (thrombocytes).

Red blood cells contain a pigment, hemoglobin, which is extremely important in the respiration process. A single drop of blood holds about 100 million of these cells, which are formed in the bone marrow and have a useful lifespan of about 120 days.

White blood cells are larger than red cells. A drop of blood contains between 100,000 and 200,000 white cells. Several types are present, and each type has a useful function such as the destruction of infectious microorganisms, the production of antibodies, or the neutralization of poisons produced by microorganisms.

Blood platelets, smaller than red cells, number between 5 and 10 million per drop of blood. Their major function is to aid blood clotting. They contain a cephalin (a phosphorus-containing lipid) which is important in the early stages of clot formation.

Chemical Transport and Exchange

In order for chemical transport to take place, the substance transported must in some way become a part of the moving bloodstream. It may be dissolved in the water-based plasma (sugars, amino acids, ionic substances, and gases do this to some extent), it may become chemically bonded to some of the formed elements (oxygen and carbon dioxide do this with red cells), or it may form a suspension in the plasma of the blood (lipids do this).

Final delivery to the tissue cells requires the transported materials to move through the capillary wall into the interstitial fluid and through the cell wall into the cell itself. The reverse process must take place when waste products move from tissue cells to the blood. During this process, capillary walls behave as selectively permeable membranes or filters, allowing water and dissolved nutrients, including oxygen, to pass in one direction and water containing dissolved wastes to return in the other.

The movement of water and dissolved materials through the capillary walls is governed by two factors—the actual pressure of blood against the capillary wall and osmosis. The blood pressure results from the pumping action of the heart and is higher in arteries than in veins. The higher pressure on the arterial side of capillaries creates a tendency for water, with dissolved electrolytes and small molecules, to move out of the capillary into the interstitial fluid. Plasma, on the inside of a capillary wall, and interstitial fluid on the other side contain different concentrations of protein and set up an osmotic system. Since plasma is the more concentrated of the two solutions, water and dissolved materials tend to flow from the interstitial fluid into the capillary. The net direction of flow depends upon the difference in these two tendencies. As Figure 25–4 illustrates, blood pressure dominates on the arterial end and osmosis on the venous end of capillaries, with the net flow going in the indicated directions.

The mechanism described above allows the exchange of nutrients and wastes and also provides a means for maintaining fluid balance between the blood and interstitial space. Very large amounts of fluid are exchanged in this way through the capillary walls. It has been estimated that fluid exchange through the capillaries of a 160-lb man takes place at the rate of about 1500 liters (400 gallons) per minute.

The nutrients in the interstitial fluid come in contact with the tissue cells, which selectively absorb the needed materials and send out waste products. No net concentration change of dissolved solute occurs in the cell, but the solute composition varies, since nutrients are used up and wastes produced.

Respiration

As mentioned earlier, red blood cells contain a pigment called hemoglobin. The molecules of this material are composed of an iron-containing pigment,

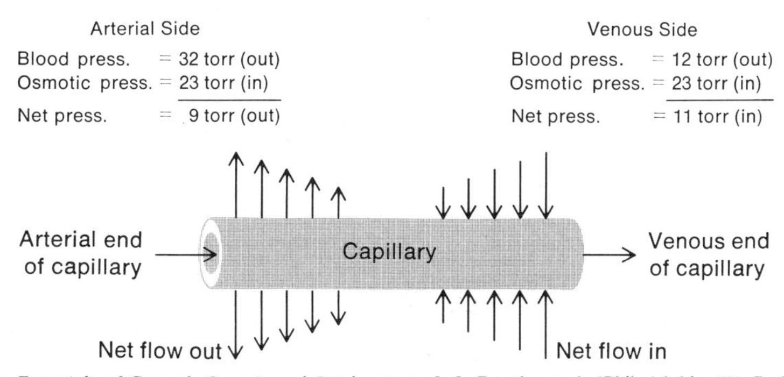

Figure 25–4.
Movement of Water Solutions Through Capillary Walls

From *Essentials of General, Organic and Biochemistry*, J. I. Routh, et al. (Philadelphia: W. B. Saunders Company, 1969), p. 672.

heme, and the protein called globin. Four heme units are attached to each globin unit to form one molecule of hemoglobin.

The most important function of hemoglobin occurs in respiration, where oxygen is transported from the lungs to the tissues, and carbon dioxide is carried back for excretion by the lungs. If it were not for the ability of hemoglobin to form a compound with oxygen, the blood could not possibly carry the amount of oxygen required by body tissues. An average adult has about 6 liters of blood, which can hold a maximum of about 12 ml of dissolved oxygen in solution without the aid of hemoglobin. If the body is to function normally, one liter of blood must contain 150 ml of dissolved oxygen. When the amount of dissolved oxygen drops to a level approaching 140 ml per liter of blood, the body begins to show some effects of oxygen deficiency. The ability of blood to dissolve the necessary amount of oxygen is due to a reversible reaction which takes place between hemoglobin (represented by HHb) and molecular oxygen:

$$O_2 + HHb \rightarrow HbO_2^- + H^+ \qquad 25\text{–}1$$

In the lungs, oxygen contained in the inhaled air comes in contact with about 1500 miles of exposed capillaries which present a surface area of 800 square feet. Oxygen diffuses through this large area and forms oxyhemoglobin, HbO_2^-, in the blood. At the same time, carbon dioxide, released from the blood, diffuses in the opposite direction into the lungs to be exhaled. The oxygenated blood travels back to the heart and then moves out through the arteries to the capillaries near the tissue cells. In the capillaries the oxygen is released from the hemoglobin and diffuses through the capillary walls into the interstitial fluid and then into the individual cells. Carbon dioxide produced by cell reactions moves in the opposite direction. Tables 25–3 and 25–4 show the reactions for the overall process.

Table 25–3. Respiration Reactions at the Lung

	Lung	*Plasma*	*Red blood cell*
1.	Inhaled O^2 —diffusion→	O_2 —diffusion→	O_2
2.			$O_2 + HHb \longrightarrow HbO_2^- + H^+$
3.		HCO_3^- —diffusion→	HCO_3^-
4.			$HCO_3^- + H^+ \longrightarrow H_2CO_3$
5.			$H_2CO_3 \xrightarrow[\text{anhydrase}]{\text{carbonic}} H_2O + CO_2$
6.			$HHb\text{-}CO_2 \longrightarrow HHb + CO_2$
7.	Exhaled CO_2 ←diffusion—	CO_2 ←diffusion—	CO_2
8.		Cl^- ←diffusion—	Cl^-

From *Elements of General and Biological Chemistry* by John R. Holum (New York: John Wiley and Sons, Inc., 1972), p. 419.

At the lungs, oxygen gas diffuses through the capillary membrane into the plasma, then through the red cell membrane into the cell (reaction 1). Inside the red cell, oxygen gas reacts with hemoglobin to form oxyhemoglobin and hydrogen ions (reaction 2). Bicarbonate ions diffuse from the plasma into the red cells (reaction 3), where they react with hydrogen ions generated in reaction 2 and rapidly produce carbonic acid (reaction 4), which decomposes into water and carbon dioxide (reaction 5). Carbamino hemoglobin ($HHb\text{-}CO_2$) breaks apart in the red cells to give hemoglobin and carbon dioxide (reaction 6). The carbon dioxide produced in reactions 5 and 6 diffuses from the red cells to the plasma and then to the lung for elimination (reaction 7). Chloride ions move from the red cells to the plasma (reaction 8) in sufficient numbers to maintain electrolytic balance as the bicarbonate ions move in the opposite direction in reaction 3. This movement of chloride ions is called the **chloride shift.**

Table 25–4. Respiration Reactions at the Tissue Cell

	Tissue cell	*Interstitial fluid*	*Plasma*	*Red blood cell*
1.	CO_2 —diffusion→	CO_2 —diffusion→	CO_2 —diffusion→	CO_2
2.				$CO_2 + H_2O \xrightarrow[\text{anhydrase}]{\text{carbonic}} H_2CO_3$
3.				$H_2CO_3 \rightarrow H^+ + HCO_3^-$
4.			HCO_3^- ←diffusion—	HCO_3^-
5.				$H^+ + HbO_2^- \longrightarrow HHb + O_2$
6.	O_2 ←diffusion—	O_2 ←diffusion—	O_2 ←diffusion—	O_2
7.				$CO_2 + HHb \longrightarrow HHb\text{-}CO_2$
8.			Cl^- —diffusion→	Cl^-

From *Elements of General and Biological Chemistry* by John R. Holum (New York: John Wiley and Sons, Inc., 1972), p. 419.

The reactions which take place at the tissue cells are essentially the reverse of those occurring at the lung. Carbon dioxide, produced during cell metabolism, diffuses from the tissue cell into the interstitial fluid; through the capillary walls into the plasma; and then into the red cells (reaction 1). Inside the red cells, carbon dioxide reacts with water, in the presence of carbonic anhydrase, to rapidly produce carbonic acid (reaction 2) which dissociates to give hydrogen ions and bicarbonate ions (reaction 3). The generated bicarbonate ions diffuse into the plasma (reaction 4). The bicarbonate ions in the plasma represent the form in which most carbon dioxide moves from tissue cells to the lungs. The hydrogen ions generated in reaction 3 react with oxyhemoglobin to produce hemoglobin and free oxygen (reaction 5). The free oxygen diffuses out of the red cells through the plasma and interstitial fluid and into the tissue cells (reaction 6). A small amount of the carbon dioxide in the red cells as a result of reaction 1 reacts with some hemoglobin produced in reaction 5 to form HHb-CO_2 (reaction 7). The chloride shift again takes place to maintain electrolytic balance (reaction 8).

Other materials have the ability to form compounds with hemoglobin. One well-known example is carbon monoxide, which combines with hemoglobin and gives a compound represented as HHb-CO. The bond between hemoglobin and CO is much stronger than the bond between hemoglobin and oxygen. As a result, oxygen cannot displace CO from hemoglobin, and the oxygen-transporting capacity of the blood is reduced. If sufficient HHb-CO is formed, the results prove fatal.

It is apparent from the mechanism given above that the normal reactions of respiration involving hemoglobin and its derivatives are also important in maintaining the normal acid-base balance of the body. The blood is made more acidic by the reactions between CO_2 and water to give H^+ and HCO_3^- ions. This corresponds to a retention of CO_2 in the blood. The reverse reaction, in which more CO_2 is excreted, tends to lower the acidity.

The Formation of Urine

Another very important process of chemical transport takes place in the kidneys. Arteries leading to each kidney branch into smaller and smaller vessels and eventually end up as small bundles of capillaries inside the kidney. Each bundle, called a **glomerulus,** is surrounded by a structure called a **Bowman's capsule.** Figure 25–5 illustrates these structures, together with other parts of the **nephron,** the tiny filtration-absorption unit of the kidney. Each kidney contains approximately a million nephrons.

As blood flows through the glomerulus, a liquid consisting of water, dissolved ions, and small molecules passes through the capillary walls into the interior of the Bowman's capsule. The formed elements of the blood, the plasma proteins, and any suspended lipids remain behind in the blood. The liquid inside the Bowman's capsule passes through a tube, the proximal convoluted tubule, where about 88 percent is reabsorbed into the blood through a capillary network surrounding the tubule. Nearly all electrolytes, all blood sugar, and most of the water are returned to the circulatory system in this manner. This type of reabsorption is called **obligatory,** since it is not variable and normally occurs without exception. The remaining 12 percent of the liquid may or may not be reabsorbed further along in the distal tubule. This **optional reabsorption** is under the control of the hormone vasopressin. When the body must conserve liquids, vasopressin is released and most of the liquid is reabsorbed. Under such conditions the volume of unabsorbed liquid (urine) is small. When vasopressin is absent, little optional reabsorption occurs, and the volume of urine produced increases. The urine passes through collecting tubules into the bladder, where it is stored until excreted.

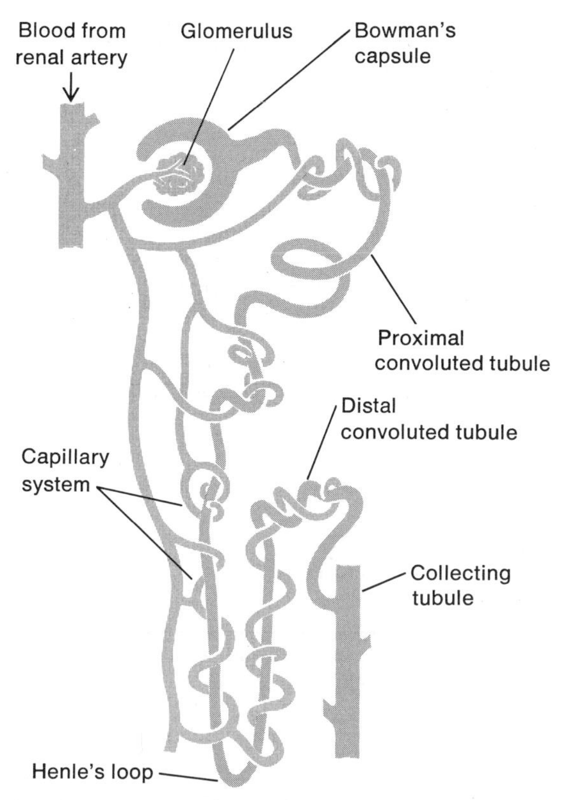

From *Elements of General and Biological Chemistry* by John R. Holum (New York: John Wiley and Sons, Inc., 1972), p. 410.

Figure 25–5.
The Kidney Nephron

The process taking place in the kidney is one of filtration, which produces a protein-free filtrate of plasma, followed by selective reabsorption of substances valuable to the body. The amount of material involved is rather amazing. In a normal adult, about 180 liters of water, 1000 grams of sodium chloride, 360 grams of sodium bicarbonate, 170 grams of glucose, and substantial amounts of wastes pass from the blood into the Bowman's capsules daily. Of this amount, only about 1.5 liters of water, 12 grams of sodium chloride, 30 grams of the waste product urea, and small amounts of other wastes are excreted as urine. The rest is reabsorbed into the blood.

In addition to waste removal, the kidneys play an important part in maintaining water, electrolytic, and acid-base balance in the body.

Water balance is maintained by the vasopressin hormone already mentioned. The water content of the body is reduced by the formation of greater volumes of urine as directed by the amount of hormone present.

Electrolytic balance is partially controlled by the water balance, since electrolytes containing ions such as Na^+, K^+, Cl^-, HPO_4^{2-}, and HCO_3^- are excreted or reabsorbed along with water. Some steroid hormones of the adrenal gland exert more specific control over electrolyte movement through membranes and, under appropriate conditions, can cause them to be retained or excreted as needed for balance.

A mechanism for the control of acid-base balance by the kidneys was proposed in 1950 by R. F. Pitts. Table 25–5 shows the reactions involved and their location in the nephron.

Table 25–5. Acid-Base Balance Reactions of the Kidney

	Capillary blood	*Wall of the distal tubule*	*Urine in the distal tubule*
1.		Nutrient + $O_2 \rightarrow H_2O + CO_2$	
2.		$H_2O + CO_2 \xrightarrow[\text{anhydrase}]{\text{carbonic}} H_2CO_3$	
3.		$H_2CO_3 \rightarrow H^+ + HCO_3^-$	
4.		H^+ —— diffusion →	H^+
5.			$H^+ + HPO_4^{2-} \rightarrow H_2PO_4^-$
6.		Na^+ ← diffusion ——	Na^+
7.	$Na^+ + HCO_3^-$ ← diffusion ——	$Na^+ + HCO_3^-$	
8.	$HCO_3^- + H^+ \rightarrow CO_2 + H_2O$		

From *Elements of General and Biological Chemistry* by John R. Holum (New York: John Wiley and Sons, Inc., 1972), p. 413.

According to Pitts, some nutrient in the wall of the distal tubule reacts with oxygen to give the usual products of carbon dioxide and water (reaction 1). The water and carbon dioxide react rapidly with each other, influenced by carbonic anhydrase, to give carbonic acid (reaction 2). The carbonic acid ionizes to give hydrogen ions and bicarbonate ions (reaction 3). The produced hydrogen ions diffuse into the urine contained in the distal tubule (reaction 4), where they react with hydrogen phosphate ions already there (reaction 5). Sodium ions migrate from the urine into the distal tubule wall (reaction 6) and then, together with bicarbonate ions from 3, they move into the capillary blood (reaction 7). The bicarbonate ions entering the blood replace any that are used up by reacting with hydrogen ions of the blood (reaction 8). The water and carbon dioxide produced in reaction 8 are eliminated through the kidneys and lungs as already discussed. In this way, sodium ions replace hydrogen ions in the blood as needed to control acidity.

Protein Synthesis

The ability of the body to repair itself by manufacturing replacement protein is truly amazing. Perhaps even more amazing is the transfer of this ability from parents to offspring. We recall from Chapter 24 that protein is a polymeric substance composed of many individual amino acids linked together in a specific sequence. We also remember that the sequence and type of amino acids in a protein chain are very important. Slight deviations from normal can cause serious malfunctions of the unit containing the protein. The sickle-cell anemia discussed in Chapter 24 is an example.

Let us consider the mechanism by which the body is able to properly synthesize proteins from available amino acids.

Nucleic Acids

Scientists have long known that the central body or nucleus of a cell is rich in a polymeric substance called deoxyribonucleic acid (DNA). All evidence available today indicates that DNA is the compound of which genes, the basic cellular units of heredity, are composed. Both DNA and its close relative ribonucleic acid (RNA) are very active in the process of protein synthesis. Both compounds are referred to by the general term, **nucleic acid.**

Polymeric nucleic acids, upon analysis, are found to be composed of monomeric units called **nucleotides,** which in turn are found to contain still simpler units. Each nucleotide is a single molecule containing one molecule of phosphoric acid (H_3PO_4), one molecule of a pentose (deoxyribose in DNA and ribose in RNA), and one molecule of a basic substance. The basic substance involved is one of five different nitrogen-containing heterocyclic compounds. Example 25–1 gives the structures and names of these bases along with the other components of nucleotides.

Phosphoric acid

Ribose

Deoxyribose

Pentoses

Adenine

Guanine

Thymine (only in DNA)

Cytosine

Uracil (only in RNA)

Purine bases

Pyrimidine bases

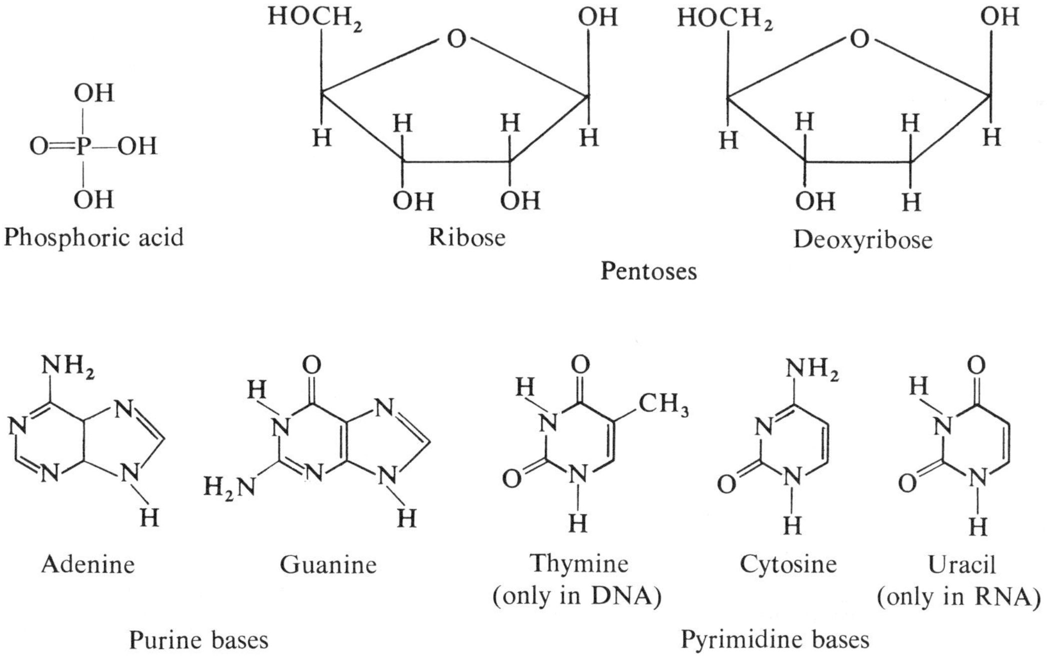

Example 25–1. Components of Nucleotides

The formation of nucleotides from the components given in Example 25–1 is represented in equation 25–2. Notice that water is removed during the bond formation. The reverse reaction, hydrolysis, generates the individual components; it is one type of reaction used to analyze for the composition of nucleotides. Since ribose was involved, the nucleotide formed in equation 25–2 is one of RNA.

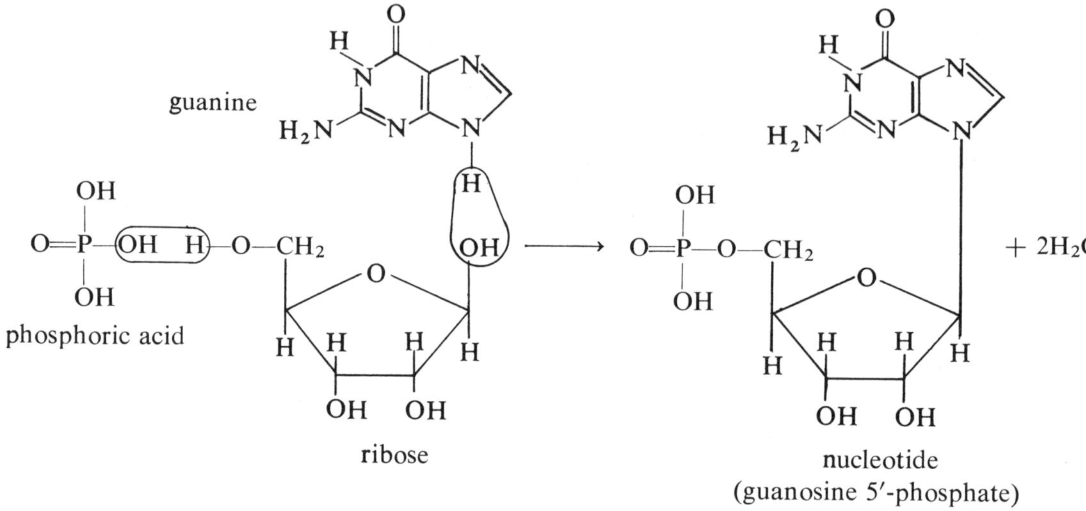

guanine

phosphoric acid

ribose

$\longrightarrow$

$+ 2H_2O$

nucleotide
(guanosine 5′-phosphate)

25–2

The base and phosphate are always attached to the pentose at the positions shown in equation 25–2. Each formed nucleotide has the potential to form ester linkages by further reaction. The two OH groups of both the pentose and the phosphoric acid are available for such reactions. Bonds involving these groups are formed when nucleotides polymerize to form nucleic acids. Such a process is shown in equation 25–3. In this polymeric nucleic acid chain the pentose is deoxyribose, so the resulting nucleic acid is DNA. One other difference between DNA and RNA is that the thymine of DNA is replaced by uracil in RNA.

In order to simplify future representations, a condensed formula will be used for nucleotides and nucleic acid chains, as illustrated by Figure 25–6. The condensed formulas make it quite apparent that all nucleic acids have one thing in common. The backbone, in every case, consists of alternating phosphate and pentose units. Of course, in DNA the pentose is deoxyribose and in RNA it is ribose; however, the backbone is fundamentally the same.

Nucleic acids differ in chain lengths and in the sequence of attached basic units. It has been found that uracil occurs only in RNA and thymine in DNA. Therefore, a given nucleic acid formula can represent DNA if the pentose is deoxyribose and thymine is present as a base. The same formula can represent RNA if ribose is substituted for deoxyribose and uracil for thymine.

thymine

guanine

cytosine

adenine

$+ 4H_2O$

25–3

Figure 25–6.
Condensed Structural Formula for Nucleic Acids

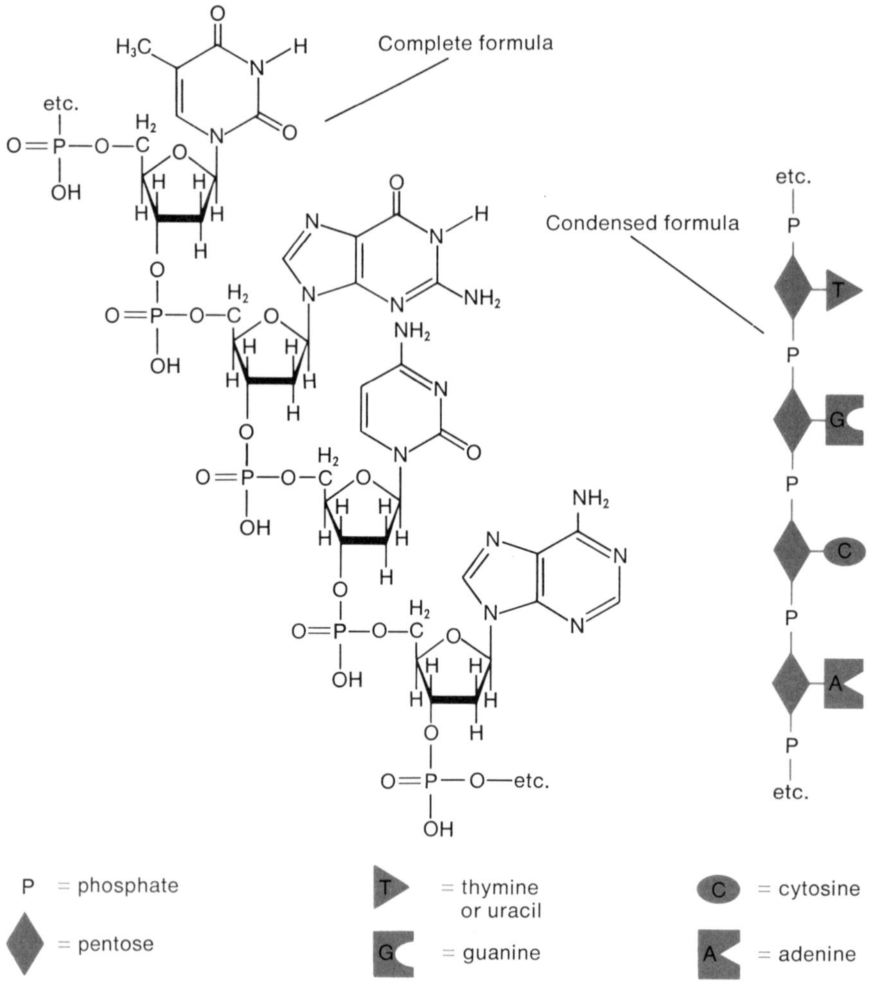

Complete formula

Condensed formula

P = phosphate

= pentose

= thymine or uracil

= guanine

= cytosine

= adenine

DNA

In 1953 J. D. Watson and F. H. C. Crick proposed a unique structure for the DNA molecule. They suggested that it consisted of two nucleic acid chains, each in the form of a helix coiled around the same axis. This proposed structure was not completely new, since a coiled structure had been proposed before; the unique feature was the manner in which the chains were held together in the helical configuration. They suggested that the configuration was the result of hydrogen bonding between base units of the chains. They further suggested that only certain pairs of bases had the correct size and structural properties to enable hydrogen bonds to form between chains. The pairs of bases able to hydrogen-bond with each other were adenine (a purine) with thymine (a pyrimidine), and guanine (a purine) with cytosine (a pyrimidine). Therefore, if adenine were in one chain, it would be hydrogen-bonded to thymine of the other chain. This type of structure is illustrated in Figure 25–7.

Figure 25–7.
The Double Helix Structure of DNA Illustrating Hydrogen Bonding Between Paired Bases

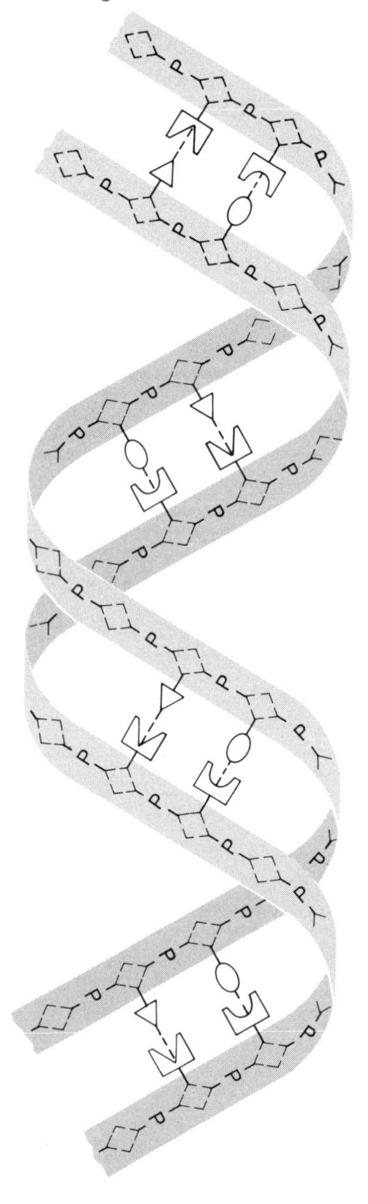

The double helix structure for DNA provided the first satisfactory explanation for the process by which DNA could replicate (duplicate) itself in a dividing cell. Replication was known to be necessary in order for each resulting cell to have a complete set of genes identical to those of the parent cell.

The mechanism of replication involves the unwinding of the helix, with each separated strand of nucleic acid then serving as a template or form on which a complementary strand, identical to the former partner, could be constructed. The complementary strand is forced to be identical to the former partner strand by the matching properties of the bases. The mononucleotides needed to build the new strand come from the liquid medium surrounding the original DNA in the cell nucleus. Such a source of mononucleotides in a surrounding medium is called a **mononucleotide pool.** This mechanism produces two double helices of DNA where only one existed before. The process is represented by Figure 25–8.

It is likely that replication takes place during the unwinding of the strands rather than after the strands have completely separated, as implied by Figure 25–8. Experimental evidence indicates that as soon as the strands begin to unwind, replication begins at the sites of the exposed bases, continuing to completion as the unwinding process is completed.

RNA

Ribonucleic acids are not found in genes and they differ somewhat from DNA. However, they play a very important role in the process of assembling proteins from free amino acids.

Three types are known. Each type is synthesized from mononucleotides under the direction of DNA, and each has a distinct function in protein synthesis.

Ribosomal RNA (*rRNA*). This type is found in the cell structures called ribosomes. Ribosomes, located in the cytoplasm of the cell outside the nucleus, are bodies which provide sites at which protein synthesis takes place. The role of ribosomes in the process is still the subject of intensive study and research. Ribosomal RNA does not direct the work of synthesis but apparently stabilizes some of the molecules involved in the process.

Messenger RNA (*mRNA*). Molecules of messenger RNA carry the code or directions for protein synthesis from the DNA of the cell nucleus out to the ribosomes. Molecules of mRNA are produced when a DNA molecule partially unwinds and exposes a chain segment with base units representing the particular message or code to be transmitted. Mononucleotides, from the mononucleotide pool, form a complementary strand of mRNA by pairing up

Figure 25–8.
DNA Replication

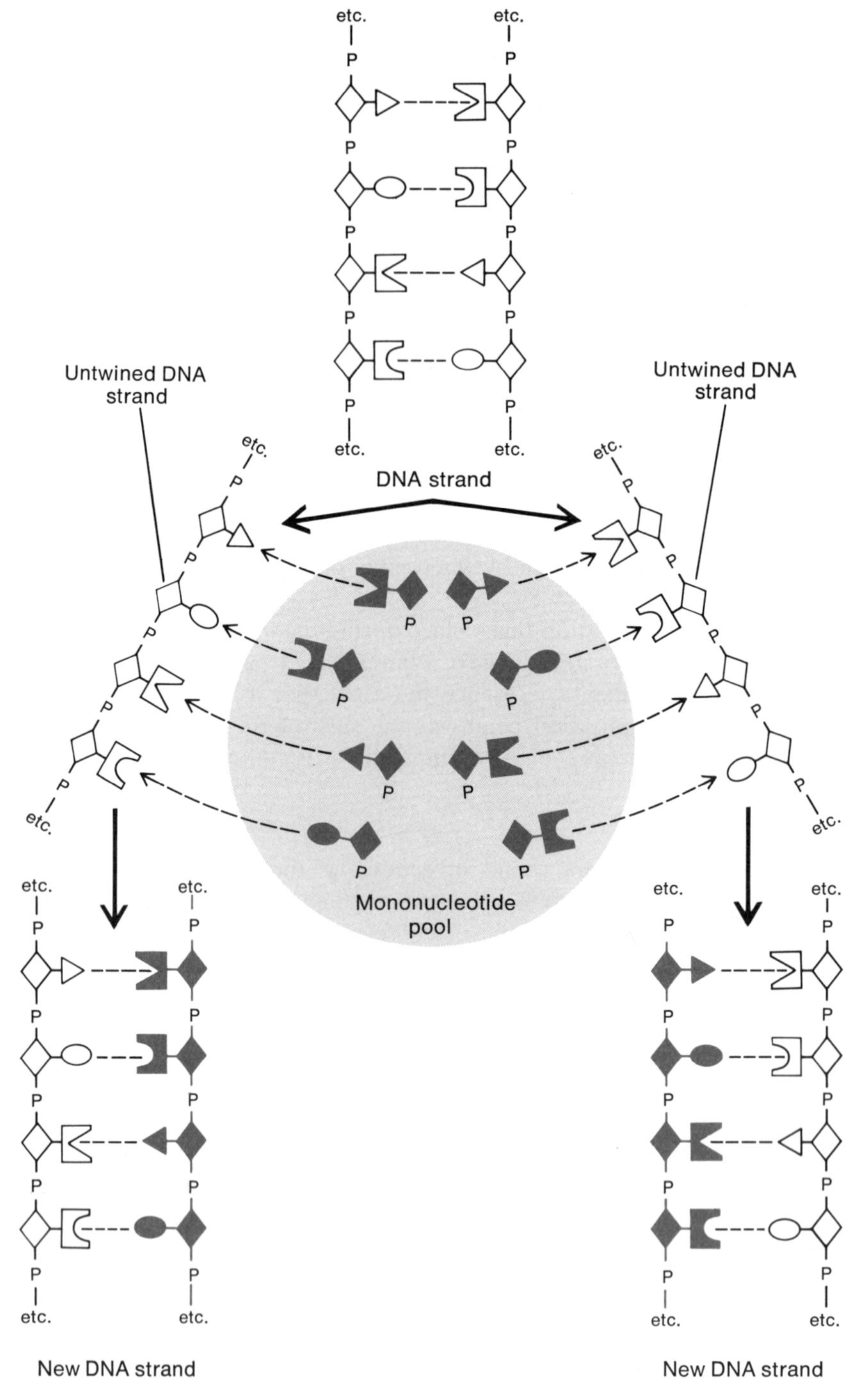

DNA strand

Untwined DNA strand

Untwined DNA strand

Mononucleotide pool

New DNA strand

New DNA strand

proper base units with those of DNA. The formed mRNA then moves away from the DNA and diffuses out of the nucleus to become attached to one or more ribosomes in preparation for the next step in protein synthesis.

Transfer RNA (tRNA). This type is also produced by the formation of complementary strands against a DNA template. It is of much lower molecular weight and tends to be more soluble than other types. For this reason it is sometimes called soluble RNA (sRNA).

The primary function of tRNA in protein synthesis is to deliver individual amino acids from an amino acid pool to the site of protein synthesis. In addition, the amino acids must be delivered in the sequence called for by the code of the mRNA. It is believed that at least twenty different molecules of tRNA exist, one for each of the twenty different amino acids used in protein synthesis.

In operation, tRNA becomes attached to the specific amino acid for which it is coded and then carries it to the protein synthesis site. Here the amino acid is delivered and attached to a growing polypeptide chain at the location called for by the mRNA code.

The Genetic Code

The production of a code to specify each of twenty different items (amino acids) is not difficult if twenty different letters or symbols can be used (one for each item). However, it appears that nucleic acids are limited to a code of four letters. The four letters are the four different bases that are attached to the nucleic acid backbone.

The solution to this problem is obvious: the code words will contain combinations of the four available letters. Only sixteen unique words result when each is made up of two letters. This is not enough to specify the twenty necessary amino acids. The use of three letters per code word results in 64 unique words, obviously more than enough to do the job. Experimental evidence indicates that three-letter words or **codons** are the type involved in the genetic code.

Table 25–6 gives some codons for messenger RNA. Remember that in RNA the base uracil replaces thymine of DNA. The letters C, G, U, and A represent the four bases cytosine, guanine, uracil, and adenine. The tRNA which carries the amino acids to the mRNA will contain code words that are complementary to those of the mRNA; this idea will be illustrated later.

The code of Table 25–6 is seen to be **degenerate.** This means that more than one codon can specify a particular amino acid. The code, however, is not overlapping, since no single codon specifies more than one amino acid. The significance of degeneracy in the code is not known.

The RNA codons of Table 25–6 are for mRNA extracted from a particular baccilus. Evidence is accumulating that a universal code may exist; the codon assignments of Table 25–6 might be valid for mRNA regardless of the source. We shall assume such a universal code in the following description.

Table 25–6. Codons for Messenger RNA

Amino acid specified	*Codons or code words*
Alanine	CCG, CUG, CAG
Arginine	GUC, CGC, AGA
Asparagine	CUA, UAA, CAA
Aspartic acid	GUA, GCA
Cysteine	UGU, GGU, UGC
Glutamic acid	GAA, GAG, AUG
Glutamine	ACA, UAC, CAG
Glycine	CGG, UGG, AGG
Histidine	CAC, AUC
Isoleucine	UAU, AAU, CAU
Leucine	CUC, UGU, UAU
Lysine	AAA, AAG, AUA
Methionine	UGA
Phenylalanine	UUU, UUC
Proline	CCC, CAC, CUC
Serine	UCG, CCU, ACG
Threonine	ACA, UCA, CAC
Tryptophan	UGG
Tyrosine	UAU, ACU
Valine	GUG, UUG, GUA

Synthesis of a Simple Peptide

In order to visualize the mechanism of protein formation, we shall diagrammatically represent the process for a part of the chain constituting the small peptide vasopressin. The structure of this hormone is:

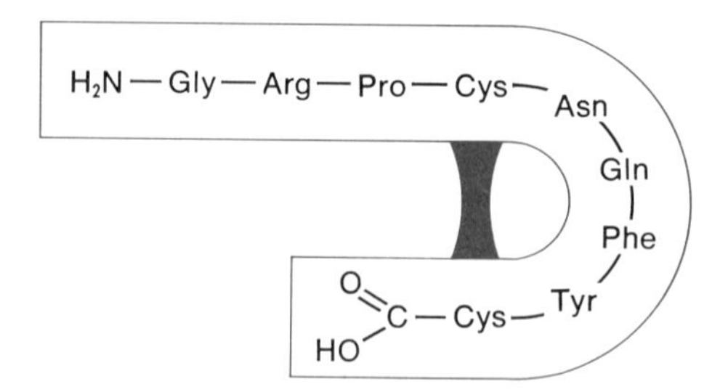

The process will be followed for the first four amino acids, beginning with glycine (Gly). The steps in the process are diagrammed in Figure 25–9.

According to Figure 25–9 a segment of a DNA helix unwinds to expose a sequence of bases representing the codons for the amino acids glycine, arginine, proline, cysteine, and so on (a). Mononucleotides from the mononucleotide pool move into proper sequence as determined by the base-pairing

Figure 25–9. Vasopressin Synthesis

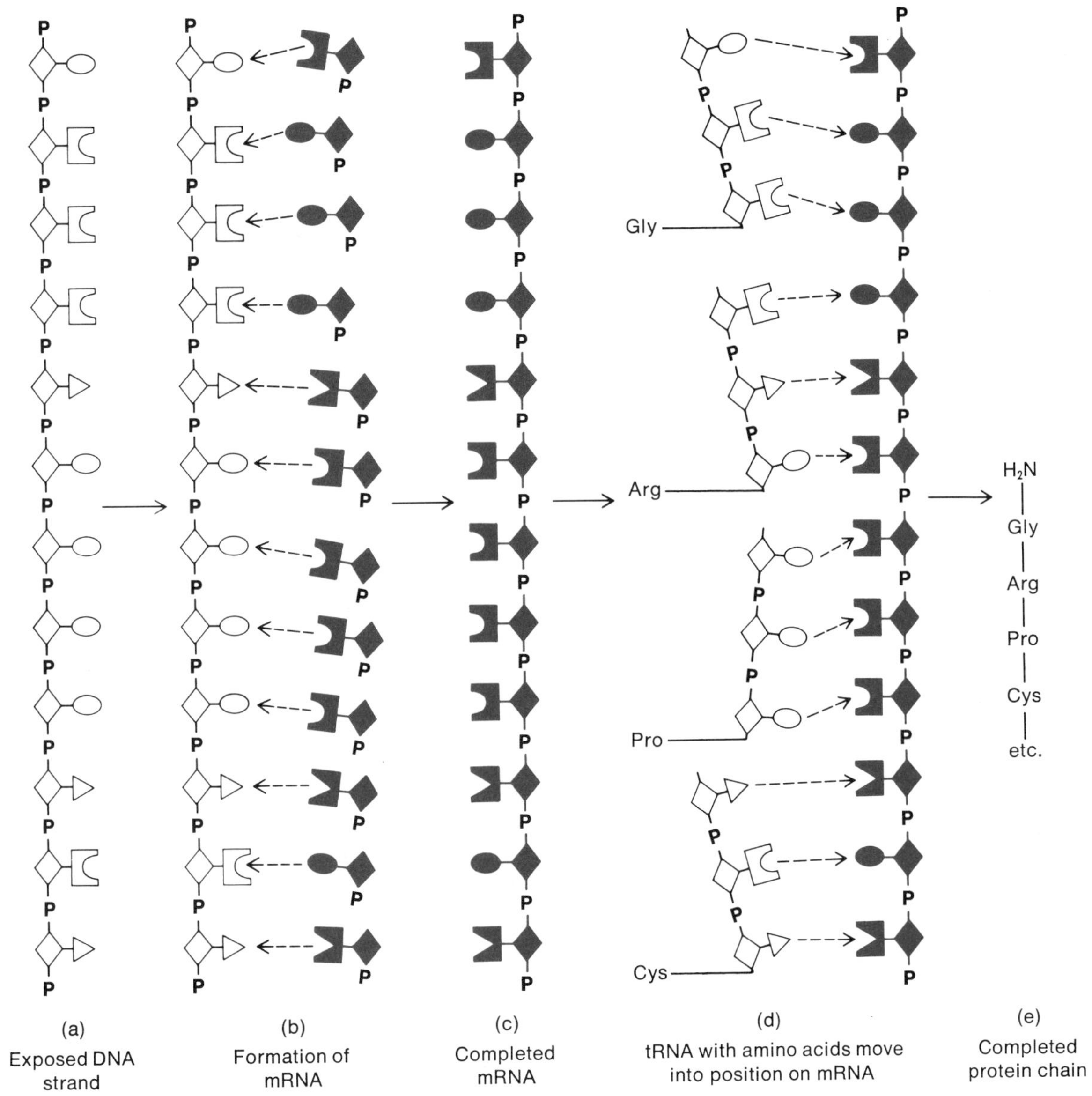

(a) Exposed DNA strand

(b) Formation of mRNA

(c) Completed mRNA

(d) tRNA with amino acids move into position on mRNA

(e) Completed protein chain

H_2N — Gly — Arg — Pro — Cys — etc.

possibilities with the exposed DNA chain (b). The completed mRNA moves away from the DNA, out of the nucleus and into the cytoplasm, where it becomes attached to a ribosome (c). Properly coded tRNA, carrying appropriate amino acids, moves into position on the mRNA (d). The sequence of amino acids is determined by the base-pairing possibilities between tRNA and mRNA. The amino acids react with each other to form peptide linkages, then move away from the tRNA in the form of a peptide with the primary structure of vasopressin (e). The peptide assumes characteristic secondary and tertiary configurations after detachment from the synthesis site.

Any tRNA involved in reactions such as those depicted in Figure 25–9 actually contain about 80 nucleotide units. For clarity, all but the ones representing codons were eliminated. The actual tRNA exists in the form of a long, folded chain, with the resulting segments coiled about one another in a manner similar to DNA. The codon nucleotides are located at one position of the resulting structure and the attached amino acid at another. The tRNA molecules with attached amino acids approach the mRNA in proper sequence separately, rather than all at once as implied in Figure 25–9.

This brief glimpse into the process of protein synthesis leaves many questions unanswered. Some answers that are known are beyond the scope of this treatment; others are still being sought.

The ideas involved in protein synthesis and the other processes of the body excite the imagination. Scientists have uncovered many of nature's secrets, but many remain to be discovered in this stimulating area.

Review Questions

1. Hydrolysis reactions are very important in the digestion of food in the body. Write a reaction to illustrate the hydrolysis of
   a) A disaccharide
   b) A dipeptide
   c) A fat
2. Discuss and compare the possible products that result when
   a) A dipeptide and a protein are hydrolyzed
   b) A disaccharide and a starch are hydrolyzed
3. Suppose food is eaten that contains the following carbohydrates: amylose (from starch), sucrose (table sugar), and lactose (from milk). Write reactions (including enzymes) to show how this mixture yields glucose during the digestion processes. See Tables 25–1 and 25–2.
4. List the major functions of the blood and its circulatory system.
5. Discuss the importance of osmosis and diffusion in the movement of various substances that takes place in the body.
6. What is the chloride shift and why is it important? See Tables 25–3 and 25–4.
7. When the following substances are transported by the blood, what form do they take and in what part of the blood (plasma, platelets, etc.) are they found?
   a) Oxygen
   b) Carbon dioxide
   c) Carbon monoxide

8. What four important functions are performed by healthy kidneys?
9. What are the structural differences between (a) ribose (found in RNA) and deoxyribose (found in DNA), and (b) uracil (found in RNA) and thymine (found in DNA)? See Example 25–1.
10. What bonds of a nucleic acid could be damaged by substances possessing the ability to catalyze hydrolysis reactions? See equations 25–2 and 25–3.
11. Suppose a DNA helix untwines and the following sequences of bases is exposed: GTG, TCG, AAA, CGG; where C, G, U, A, and T are cytosine, guanine, uracil, adenine, and thymine. What sequence of amino acids would be found in the tetrapeptide formed in the manner illustrated in Figure 25–9?

Suggestions for Further Reading

Chappell, G. S., *Through the Alimentary Canal with Gun and Camera*, Dover, New York, 1963.

Clark, B. F. C., and K. A. Marcker, "How Proteins Start," *Scientific American*, Jan. 1968.

Comroe, J. H., Jr., "The Lung," *Scientific American*, Feb. 1966.

Crick, F. H. C., "The Genetic Code," *Scientific American*, Oct. 1962.

Crick, F. H. C., "The Genetic Code III," *Scientific American*, Oct. 1966.

Davenport, H. W., "Why the Stomach Does Not Digest Itself," *Scientific American*, Jan. 1972.

Kornberg, A., "The Synthesis of DNA," *Scientific American*, Oct. 1968.

Mayerson, H. S., "The Lymphatic System," *Scientific American*, June 1963.

Neurath, H., "Protein Digesting Enzymes," *Scientific American*, Dec. 1964.

Nirenberg, M., "The Genetic Code II," *Scientific American*, March 1963.

Smith, H. W., "The Kidney," *Scientific American*, Jan. 1953.

Wood, J. E., "The Venous System," *Scientific American*, Jan. 1968.

26 Chemistry and Medicine

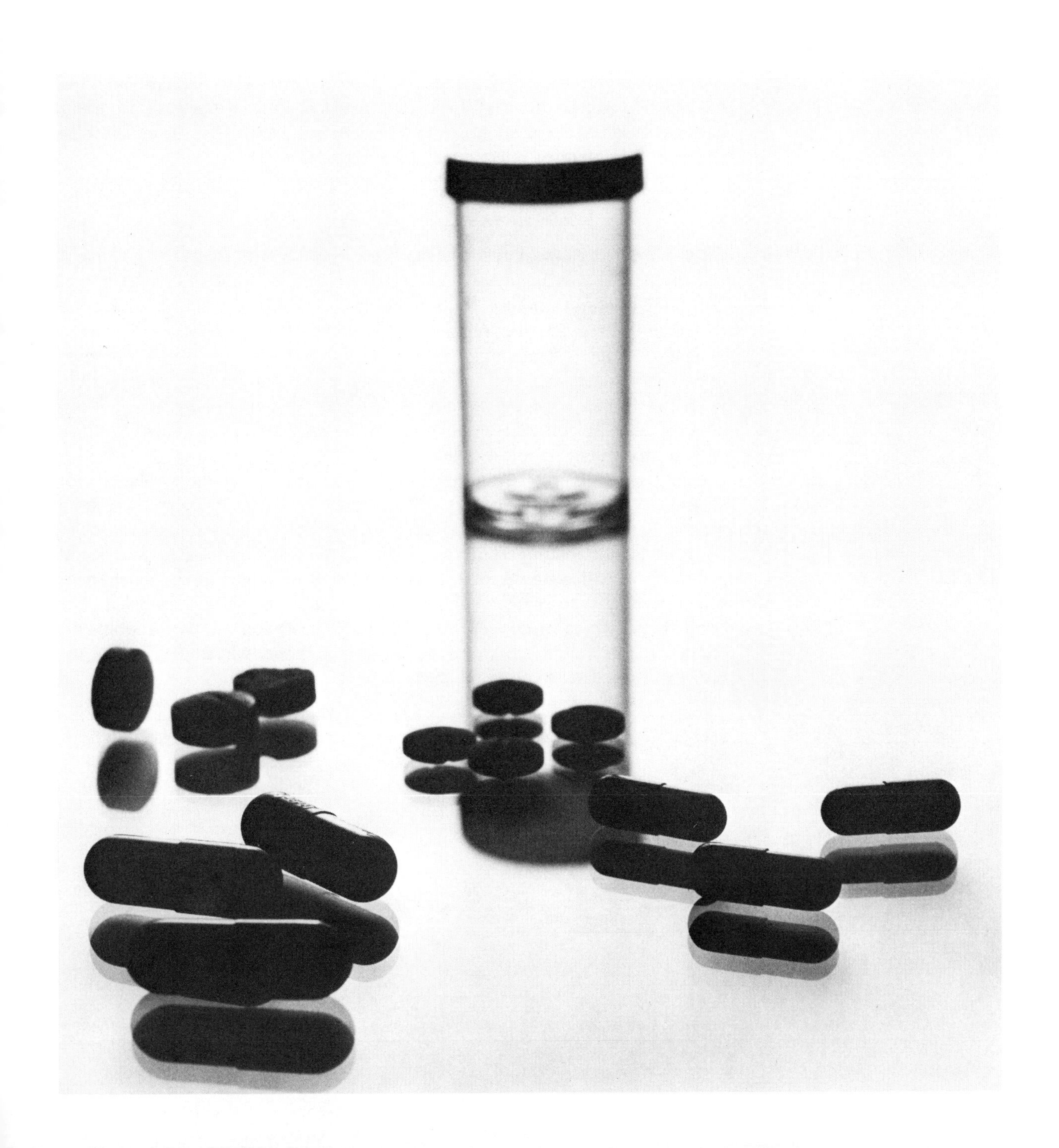

The use of chemical agents in the treatment of human ailments is not a recent innovation. The combination priest, pharmacist, and physician of ancient Babylonia (2600 B.C.) treated patients with various chemical-containing concoctions extracted primarily from plants. The Papyrus Ebers (1500 B.C.), a pharmaceutical record of ancient Egypt, contains a collection of 800 prescriptions in which 700 different drugs are mentioned. Mercury was used in the treatment of syphilis in western Europe as early as the fifteenth century.

In this chapter, rather than attempting to touch on all of the many areas of medicinal chemistry, we shall focus on four topics: the chemical nature of the nervous system, chemical poisons, the chemical control of pain, and the chemical treatment of infection.

The Chemical Nature of the Nervous System

The body contains about 10 billion (10^{10}) nerve cells or neurons, structured to transmit messages over relatively long distances (Figure 26–1). The axon of a neuron may be quite long—more than two feet in some cases—depending upon the distance between the source and the destination of the nerve impulses. A typical body nerve consists of a bundle containing many neurons held together by connective tissue and furnished with a blood supply to provide needed nutrients and oxygen.

Our discussion will be limited to the motor nerves of the **involuntary** or **autonomic** nervous system. This system carries messages between the **central** nervous system—the brain and spinal cord—and the organs and glands of the body that function involuntarily, such as the heart, stomach, and sweat glands. The **motor nerves** of this system carry messages from the brain or spinal cord to the organs or glands. Two kinds of nerves are involved—the **sympathetic,** which originate in the thoracic (chest) and lumbar (abdominal) regions of the spinal cord, and the **parasympathetic,** whose origins are the brain and sacral (lower) spinal cord.

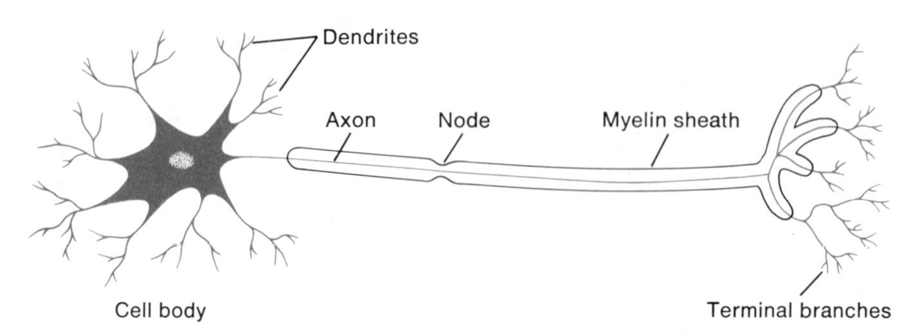

Figure 26–1.
A Single Nerve Cell or Neuron

The nerves do not form a continuous pathway from the brain or spinal cord to the intended organ or gland. An impulse, traveling along a nerve, must be transmitted across two or more very tiny gaps or **synapses** formed when one neuron terminates in close proximity to the beginning of another neuron or the acceptor cells of an organ or gland. This discontinuous pathway is sketched in Figure 26–2.

Transmission of Nerve Impulses

An impulse, originating in the brain or spinal cord, travels along the axon of a neuron in the form of an electrical charge generated by the selective flow of ions into and out of the axon through the cell membrane. When this impulse reaches the end of the axon, a chemical **mediator** (transmitter) is released and diffuses across the synapse, a distance of less than 5×10^{-6} cm or 2×10^{-6} inches. The mediator is accepted by the second neuron making up the synapse. This triggers an impulse in the accepting neuron, which travels as an electrical charge until it encounters another synapse between neurons or a neuron and an organ, and again a mediator is released. The acceptance of mediator by an acceptor cell of an organ or gland causes the cell to respond characteristically—a muscle cell contracts, for example. Two chemical mediators known to function in the involuntary nervous system are acetylcholine and norepinephrine (noradrenaline). Their chemical structures are given in Example 26–1.

$$CH_3-\overset{\overset{\displaystyle O}{\|}}{C}-O-CH_2-CH_2-\overset{+}{N}(CH_3)_3$$

Acetylcholine

$$(HO)_2C_6H_3-\underset{\displaystyle OH}{CH}-CH_2-NH_2$$

Norepinephrine

Example 26–1. Mediators of the Involuntary Nervous System

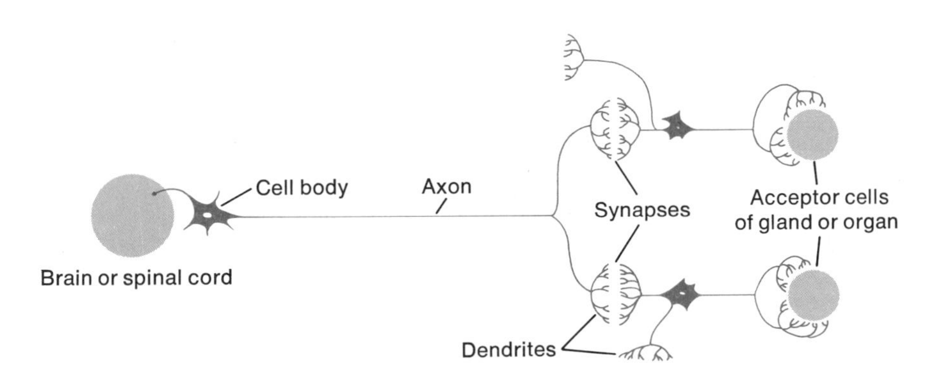

Figure 26–2. Pathway Followed by Nerve Impulses

Figure 26–3.
The Involuntary Nervous System
The mediators at the various synapses are indicated by A (acetylcholine) and N (norepinephrine).

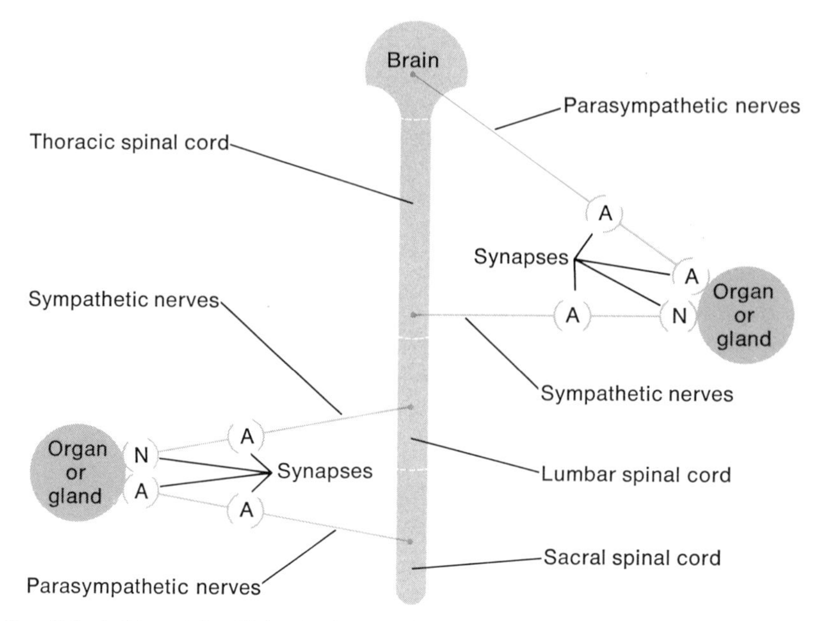

From *Take As Directed, Our Modern Medicines*, ed. by F. E. Shideman, Chemical Rubber Company, 1967, p. 107.

Figure 26–3 gives a schematic representation of the involuntary nervous system, indicating the mediator at work in the various synapses. Acetylcholine is the mediator at all synapses of the parasympathetic nerves and at the synapses between neurons of the sympathetic nerves. Norepinephrine is involved as the mediator at synapses between neurons of sympathetic nerves and acceptor cells of organs and glands.

A chemical serving as a mediator in nerve-impulse transmission must be stable enough to complete the journey across the synapse, but must be rapidly inactivated so the action on acceptor cells is brief. If the necessary concentration of mediator remains on acceptor cells, a stream of electrical impulses will be sent out, possibly causing undesirable spasmodic or convulsive behavior in the organs or glands involved.

Deactivation of mediators takes place either by metabolic (chemical) alteration, usually involving an enzyme, or by a process of binding to specific inactive sites. Acetylcholine is broken down into the inactive compounds acetic acid and choline by the action of the enzyme cholinesterase. Norepinephrine is metabolically destroyed by the action of two enzymes, monamine-oxidase and catechol-*o*-methyltransferase. In addition, significant amounts of norepinephrine are rapidly inactivated by bonding to selected sites on nerve terminals.

Chemical Control of the Nervous System

Many of the nerve messages of the body can be speeded up, slowed down, or even stopped by the action of different chemical substances. The mechanisms by which such effects are obtained become fairly obvious when we consider the steps involved in nerve-impulse transmission:

1. An electrical charge is conducted inside a neuron.
2. A chemical mediator is generated and relased by a neuron at each synapse.
3. The mediator diffuses across the synapse.
4. The mediator is accepted by another neuron, organ, or gland.
5. The mediator, after generating an impulse, is inactivated chemically or by binding.

A change takes place in nerve-impulse transmission when substances are present which:

Change (usually decrease) the ability of a neuron to contain or conduct an electrical charge.
Decrease the rate of generation or release of mediator at a synapse.
Decrease the rate of diffusion of mediator across a synapse.
Decrease the acceptance of mediator by acceptor cells.
Increase the rate of generation and release of mediator at a synapse.
Decrease the rate of deactivation of mediator on acceptor cells.
Mimic the behavior of mediator on acceptor cells.

The first four of these effects slow down or stop impulse transmission; the last three enhance it. The action of numerous poisons and medicinal chemicals derives from such effects.

The net result on the body depends upon which kind of nerves, sympathetic or parasympathetic, are acted upon. In essence, these two types of nerves work in opposition to each other to keep the body in a state of balance. The parasympathetic nerves regulate the body when it is relaxed or at rest. The heart and respiration rates, pupil size, and digestive processes are regulated by these nerves. The sympathetic nerves are important during times of stress when metabolic functions not absolutely needed are shut down or depressed and the body is made ready to defend itself.

Table 26–1 lists the effects of stimulation of the two types of nerves. Depression of nerve activity often causes the reverse effect.

Chemical Poisons

Almost any substance—including some necessary for good health, such as water—can prove harmful if ingested in large enough quantities. For this reason a **poison** will be defined as any substance which, through chemical action, can cause illness or death when incorporated into living organisms in relatively small quantities.

Table 26–1. Effects of Nerve Stimulation

Type of nerve stimulated	
Parasympathetic	*Sympathetic*
Pupils decrease in size	Pupils increase in size
Heart beat slows	Heart beat increases
Blood vessels relax	Blood vessels constrict
Blood pressure drops	Blood pressure increases
Bronchi constrict	Bronchi relax
Digestive secretions increase	Digestive secretions decrease
Bladder contracts	Mouth becomes dry
	Hands and fingers tremble

Many complex chemical processes take place in the body. Illness and death can result when poisons slow down, speed up, or in some other way change these processes. Because so many processes are susceptible, poisons can work in many ways.

Toxicity of Poisons

The toxicity (strength) of poisons and the related lethal dose vary over a wide range, as Table 26–2 reveals.

Table 26–2. Toxicity and Lethal Dose of Poisons

Toxicity rating	*Lethal dose (mg/kg of body wt.)*	*Lethal dose (g) for a 170-lb (77-kg) man*	*Examples*
Nontoxic	More than 15,000	More than 1200	Water
Slightly toxic	5000–15,000	400–1200	Ethyl alcohol, soap
Moderately toxic	500–5000	40–400	Ether, wood alcohol
Toxic	50–500	4–40	Aspirin, tobacco
Very toxic	5–50	0.4–4	Mercury salts, morphine
Extremely toxic	Less than 5	Less than 0.4 (less than 7 drops of liquid)	Heroin, potassium cyanide
Supertoxic	Less than .01	Less than 0.08 (less than 1 drop of liquid)	Nerve gas

Enzyme Inhibition

Enzymes are involved in many of the chemical processes of the body, and the inhibition of enzyme activity is a mechanism of some poisons. **Enzyme inhibition** usually takes place when the poison forms a complex (compound) with the enzyme and prevents it from reacting with a normal substrate. In some cases, called **reversible inhibition,** the poison-enzyme complex will break up and release the enzyme when the concentration of poison in the system drops below some specific level. In other instances the complex, once formed, does not break up and the process is called **irreversible inhibition.**

Reversible enzyme inhibition is shown by the compound neostigmine. This substance is similar to the chemical mediator acetylcholine (Example 26–2). Because of the similarity, neostigmine readily combines with the enzyme cholinesterase, which normally catalyzes the hydrolysis of acetylcholine. With the enzyme inhibited, acetylcholine, accepted by receptor cells of parasympathetic nerves, sends out impulse after impulse along the nerves to the organs. This results in overstimulation of the organs involved and causes convulsions, choking, heart irregularities, and eventual death.

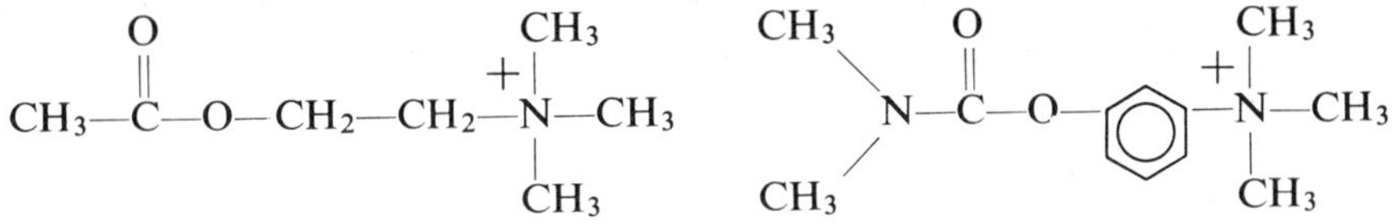

$$CH_3-\overset{\overset{O}{\|}}{C}-O-CH_2-CH_2-\overset{+}{N}(CH_3)_3$$

Acetylcholine

$$(CH_3)_2N-\overset{\overset{O}{\|}}{C}-O-C_6H_4-\overset{+}{N}(CH_3)_3$$

Neostigmine

Example 26–2. Structural Similarity of Acetylcholine and Neostigmine

In spite of its toxicity, appropriate amounts of neostigmine are used medicinally to slow a rapid heartbeat and to relieve the intestinal paralysis that often follows abdominal surgery.

Nerve gases and some insecticides (Example 26–3) react with cholinesterase and produce irreversible enzyme inhibition.

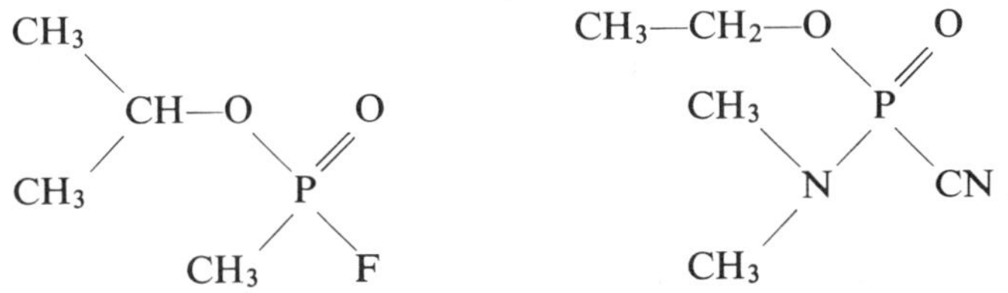

$$(CH_3)_2CH-O-P(=O)(CH_3)F$$

Sarin

$$CH_3-CH_2-O-P(=O)(CN)-N(CH_3)_2$$

Tabun

Nerve gases

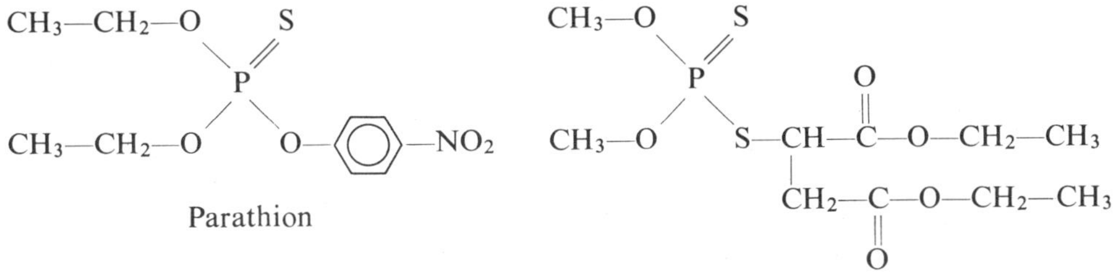

$$(CH_3-CH_2-O)_2P(=S)-O-C_6H_4-NO_2$$

Parathion

$$(CH_3-O)_2P(=S)-S-CH(-\overset{\overset{O}{\|}}{C}-O-CH_2-CH_3)-CH_2-\underset{\underset{O}{\|}}{C}-O-CH_2-CH_3$$

Malathion

Insecticides

Example 26–3. Substances Responsible for Irreversible Enzyme Inhibition—(note the structural similarities)

The symptoms of poisoning are the same as those given previously for neostigmine. The enzyme inhibition cannot be easily reversed by lowering the poison concentrations in the system; however, chemical antidotes are known.

Atropine, itself a powerful poison, is an effective antidote. This substance, found in the *Atropa belladona* or deadly nightshade plant, acts by occupying receptor sites of parasympathetic nerves without generating nerve impulses. The sites are made unavailable for reaction with impulse-producing acetylcholine, and the overstimulation of organs is controlled. In the absence of a nerve-stimulating poison, atropine is supertoxic and acts by depressing the stimulation of parasympathetic nerves. The symptoms are essentially those given in Table 26–1 for sympathetic-nerve stimulation.

Irreversible enzyme inhibition is among the mechanisms responsible for the effects of extremely toxic cyanide compounds. The cyanide ion, CN^-, combines irreversibly with cytochromoxidase, an enzyme which catalyzes the utilization of oxygen by body cells.

Lethal Synthesis

Some substances, not themselves toxic, are classified as poisons and act by providing the body with raw materials from which active poisons are produced. In the presence of fluoroacetic acid, which has the structure

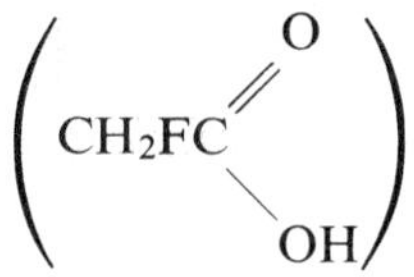

$$\left(CH_2FC\begin{matrix} \diagup\!\!\diagup O \\ \diagdown OH \end{matrix} \right)$$

the body generates fluorocitric acid rather than the normally produced citric acid. Fluorocitric acid inhibits the activity of an enzyme that plays a vital part in basic energy-producing processes of the body. These processes are upset and the body eventually dies. The term **lethal synthesis** is obviously appropriate for this mechanism.

Multiple Inhibition

The action of certain poisonous substances, including arsenic and heavy metals such as lead and mercury, is very general and may affect numerous body processes. These poisons apparently react with sulfhydryl groups, —SH, found on some proteins. Sulfhydryl groups are often involved in the enzymatic activities of proteins. Thus, many types of enzymes can be inhibited by these poisons. An antidote for this type of poisoning is any substance capable of forming a stable yet nontoxic compound with the poison. Complexing agents such as ethylenediaminetetraacetic acid, EDTA, have been used.

Nonenzymatic Nerve Interference

The synapses between neurons or between neurons and receptor cells of organs are the sites for other poison-related processes. Several mechanisms are known by which the proper transfer of nerve impulses is affected.

One mode of action is demonstrated by substances that react with receptor cells, prevent the normal chemical mediators from being accepted, but generate no impulses. Nerve impulses are thus stopped at the synapses and do not reach the organs. Atropine, mentioned earlier, is an example of such a poison:

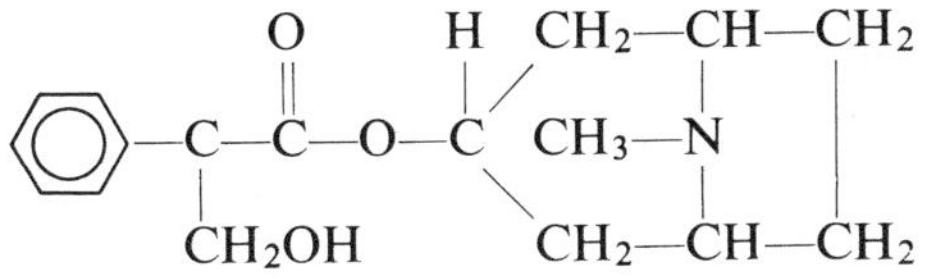

O H CH_2—CH—CH_2

—C—C—O—C CH_3—N

CH_2OH CH_2—CH—CH_2

atropine

Another mechanism involves substances with the ability to mimic the action of normal chemical mediators and cause nerve stimulation by reacting with receptor cells and generating nerve impulses. This is the action of epinephrine (adrenaline) and muscarine (Example 26–4), a toxic substance found in some mushrooms. These substances produce symptoms characteristic of overstimulation of the nervous system.

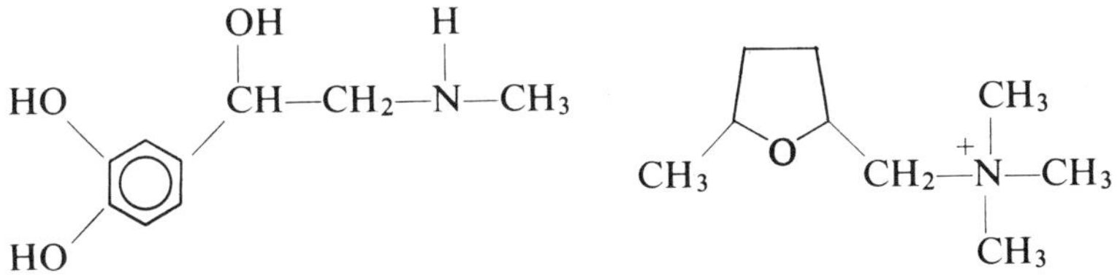

OH H

HO CH—CH_2—N—CH_3

HO

Epinephrine
(mimics norepinephrine)

CH_3

CH_3 O CH_2—$\overset{+}{N}$—CH_3

CH_3

Muscarine
(mimics acetylcholine)

Example 26–4. Poisons Which Mimic Chemical Mediators

Botulism, an extreme and fatal form of food poisoning, is caused by supertoxic proteins secreted from the botulinus bacilli (*Clostridium botulinum*) contained in some improperly processed canned food. This poison, more powerful than nerve gas, acts by preventing the release of acetylcholine mediator at nerve endings, thus halting nerve impulses at the synapses. Death results from respiratory failure.

Table 26–3 summarizes the poisons we have discussed and their mode of action. Obviously, we have not discussed every poison and mechanism for affecting the body. Entire books have been written on toxicology—the study of poisons and their mechanisms—and many scientists devote their lives to this study. Hopefully, enough examples have been given to illustrate the variety of substances and mechanisms involved.

Table 26–3. Poisons and Modes of Action

Poison	*Mode of action*
Neostigmine	Reacts with cholinesterase—enzyme inhib. (rev.)
Sarin	Reacts with cholinesterase—enzyme inhib. (irrev.)
Tabun	Reacts with cholinesterase—enzyme inhib. (irrev.)
Parathion	Reacts with cholinesterase—enzyme inhib. (irrev.)
Malathion	Reacts with cholinesterase—enzyme inhib. (irrev.)
Atropine	Reacts with receptor sites
Cyanide	Reacts with cytochromeoxidase—enzyme inhib. (irrev.)
Fluoroacetic acid	Provides material for lethal synthesis
Heavy metals	React with sulfhydryl groups—multiple enzyme inhib.
Epinephrine	Mimics chemical mediator norepinephrine
Muscarine	Mimics chemical mediator acetylcholine
Botulism toxin	Prevents release of acetylcholine from nerve endings

The Chemical Control of Pain

The relief of pain is one of the greatest needs in the practice of medicine. Whether it arises from injury, disease, or surgery, pain is of great and immediate importance to the patient and the physician or nurse.

Pain provides a natural protection mechanism for living organisms by indicating the presence of malfunctions or injuries. It is a cruel but essential guardian of life—one that man, from very early times, has sought to separate from necessary medical procedures. In the days before anesthetics, the sick sometimes chose to die rather than submit themselves to the surgeon's knife.

The Mechanism of Pain Control

The specific chemical mechanism involved in pain control is known for only a few substances. In general, however, it is achieved by depression of the activity of the central nervous system or localized nerves. It is reasonable to assume that in at least some instances this depression is caused by interference with nerve-impulse transmission in ways similar to those discussed earlier for the involuntary nervous system.

The brain is often the site of action for pain-controlling drugs. Mild depression of brain activity causes loss of interest in the surroundings and an inability to concentrate. Further depression slows the pulse and respiration and decreases the sensations of touch, heat, cold, sight, and pain, as well as motor (movement) and mental activities. In this way, **sedation** (calming without producing sleep), sleep, **analgesia** (reduction of pain), and **anesthesia** (unconsciousness) may be brought about by careful administration of drugs. Coma, respiratory and circulatory failure, and finally death can result if depression of vital brain centers, particularly the respiratory center, becomes too intense.

Some drugs have a selective action and work only on specific centers of the brain; others show a selective depression action which is dosage dependent. They affect one part of the brain when the usual dose is administered but extend the depression to other parts when the dose is increased. Fortunately, the brain centers controlling vital body processes are quite resistant to the action of most depressant drugs.

Mild Analgesics

Substances classified as **mild analgesics** will relieve muscle, joint, and minor headache pain, usually without producing drowsiness or interfering with mental alertness. They are not effective against deep-seated internal pain and are not addictive. Some commonly used substances in this category are aspirin and other salicylates, phenacetin, and acetanilide (Example 26–5).

Salicylic acid

Methyl salicylate (oil of wintergreen)

Acetylsalicylic acid (aspirin)

CH_3-CH_2O— Phenacetin

Acetanilide

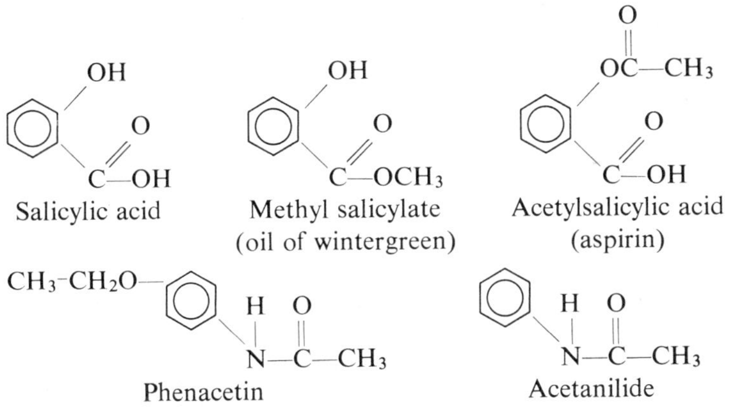

Example 26–5. Mild Analgesics

Salicylic acid was widely used in the 1870s as a fever depressant, but its acidic properties caused undesirable side effects. In 1899 the Bayer Company of Germany introduced the acetate ester of salicylic acid, naming it Aspirin. This name is no longer a trademark in the United States but retains that status in some countries. Aspirin was found to be an analgesic as well as a fever depressant and it is widely used today for both purposes.

Salicylic acid has properties similar to those of aspirin but less analgesic effectiveness. Methyl salicylate is not taken internally but is used in heat rubs to relieve muscle pain. It is rapidly absorbed when rubbed on the skin, especially when it is first dissolved in alcohol, oil, or lanolin.

Phenacetin and acetanilide are used as analgesics either alone or in combination with other drugs. A remedy known as APC tablets contains aspirin, phenacetin, and caffeine.

Tranquilizers

As the term implies, **tranquilizers** affect the emotional state of an individual by calming or quieting without decreasing the level of consciousness. These compounds are also used, in combination with other drugs, to relieve pain. The formulas of three compounds used as tranquilizers are given in Example 26–6.

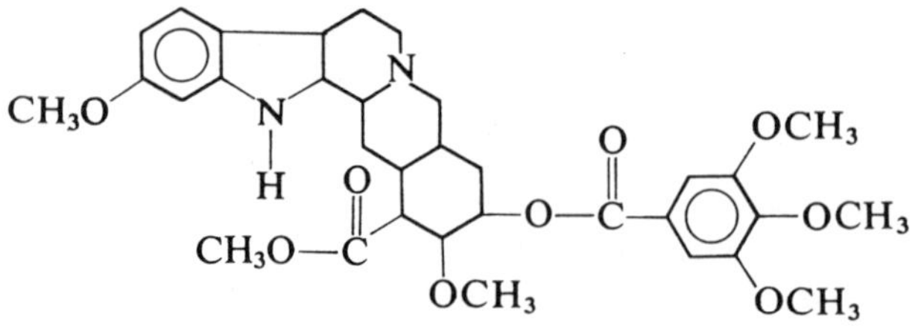

Reserpine

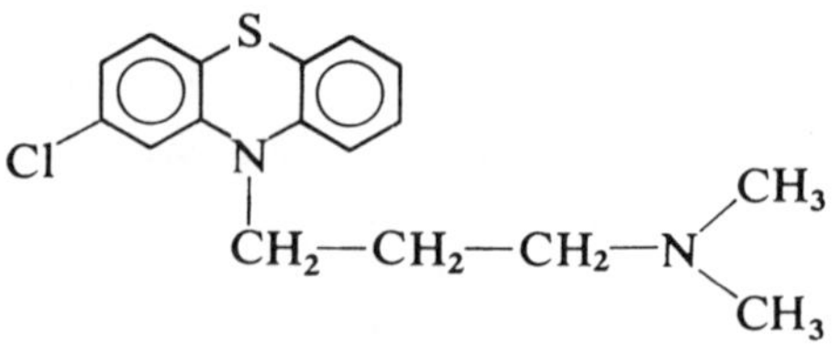

Chlorpromazine

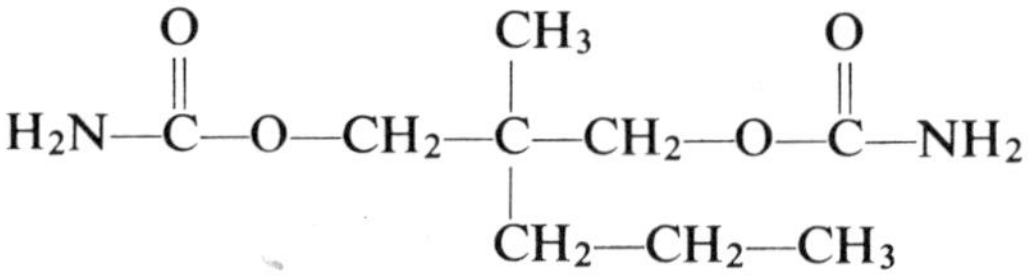

$$H_2N-\overset{O}{\overset{\|}{C}}-O-CH_2-\underset{CH_2-CH_2-CH_3}{\underset{|}{\overset{CH_3}{\overset{|}{C}}}}-CH_2-O-\overset{O}{\overset{\|}{C}}-NH_2$$

Meprobamate

Example 26–6. Tranquilizers

Reserpine depresses nerve activity by first causing nerve endings to release stored norepinephrine and then inhibiting the synthesis of any more. The lack of chemical mediator prevents impulse transmission across the synapse. Reserpine also acts on the brain as a central nervous system depressant.

Chlorpromazine depresses several centers of the brain and causes a variety of responses in the body. Blood pressure and body temperature drop, and nausea and vomiting are suppressed. Chlorpromazin also enhances the action of other central nervous system depressants.

Meprobamate has a depressant effect inside the spinal cord which causes a characteristic relaxation of skeletal muscles. The muscle relaxation is thought to provide better sleep for individuals undergoing treatment.

Sedatives and Hypnotics

Sedatives are drugs used to quiet and relax a patient without producing sleep. Drowsiness and sleep can result if the dosage is increased above that producing sedation. **Hypnotics** are drugs used to produce sleep. In many cases, the same drug in different dosages is used for both purposes. Examples 26–7 and 26–8 contain the formulas of some typical drugs used as sedatives and hypnotics.

With the exception of sodium bromide, which is used only as a sedative, all compounds in Example 26–7 are used as both sedatives and hypnotics. Chloral hydrate, the first synthetic hypnotic, was prepared in 1832 and was

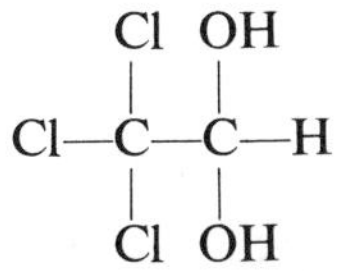

Chloral hydrate 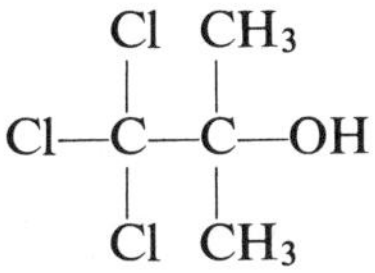 | Chlorobutanol 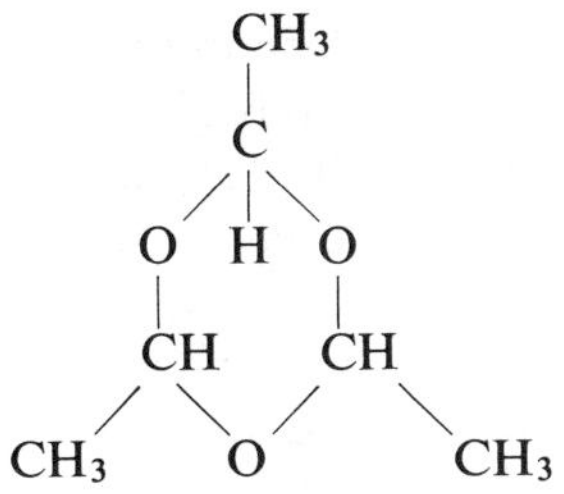 | Paraldehyde

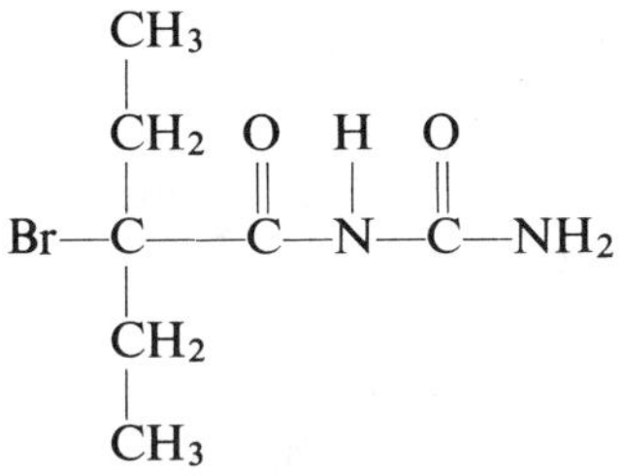

Carbromal | NaBr Sodium bromide

Example 26–7. Sedatives and Hypnotics

first used as a hypnotic in 1869. A mixture of this drug and alcohol is known as a Mickey Finn or knockout drops. The depressant effect of the two compounds is additive, but the mixture is not used medically.

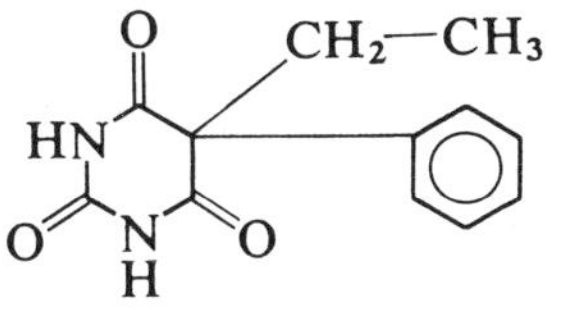

Phenobarbital 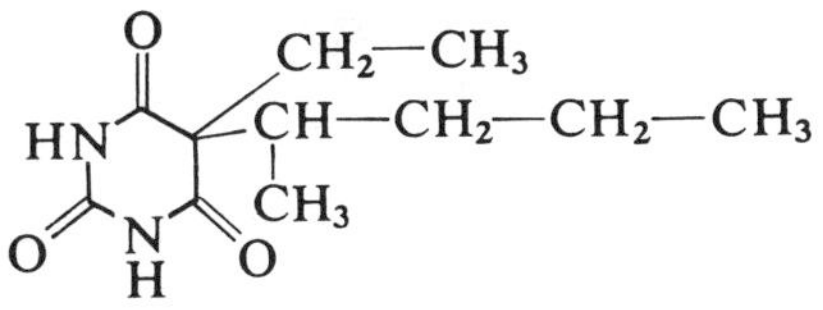 | Pentobarbital 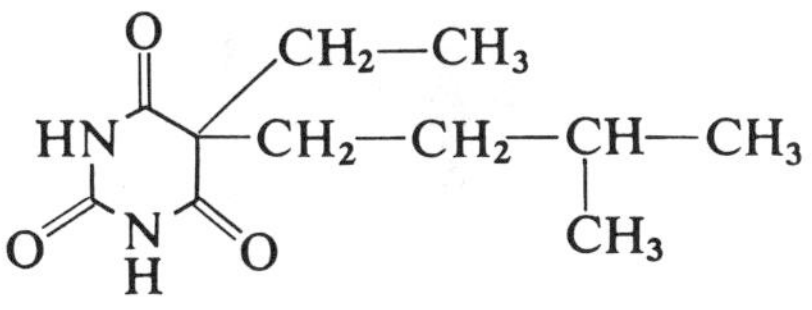 | Amobarbital

Example 26–8. Barbiturate Sedatives and Hypnotics

The representative compounds of Example 26–8, as well as other barbiturates, are derivatives of barbituric acid:

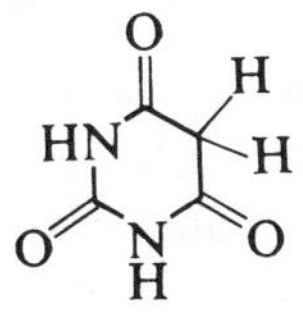

The compounds are frequently used as both sedatives and hypnotics. They are similar in their action but vary in the time required for a therapeutic

dose to take effect and in the duration of the effects. Phenobarbital takes an hour or longer to begin acting and is effective for 10–12 hours. Pentobarbital requires only 10–15 minutes for onset but is effective for only 3–4 hours. This type of variation allows the physician a great deal of latitude in prescribing to meet the needs of patients.

Individuals have become addicted to barbiturates through frequent use of abnormally large doses. The drugs present another hazard because they are central nervous system depressants; overdoses may cause respiratory failure and death.

Strong Analgesics

Through much of history the poppy has served as a source of pain-relieving materials. The hardened, dried juice from the unripe capsules of a particular poppy is called **opium,** a crude extract that formerly found wide use in medicine.

More than twenty different compounds have been extracted from crude opium. The most important one, morphine, makes up about 10 percent of the crude drug. Morphine is a good analgesic agent that is effective against major pain. Unfortunately, it is also very addictive and must be used with caution.

Codeine occurs naturally in opium and is also produced from morphine. Its analgesic activity, though fairly strong, is only about one-sixth that of morphine. It is also much less addictive. Codeine is used to relieve moderate pain and, because it depresses the cough center of the brain, as a cough suppressant.

Example 26–9 contains the structural formulas of morphine, codeine, and an illegal derivative, heroin.

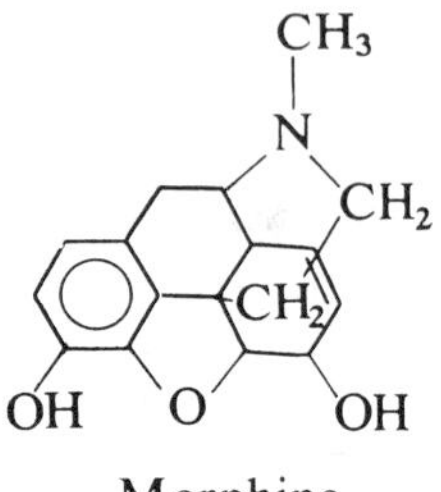

Morphine 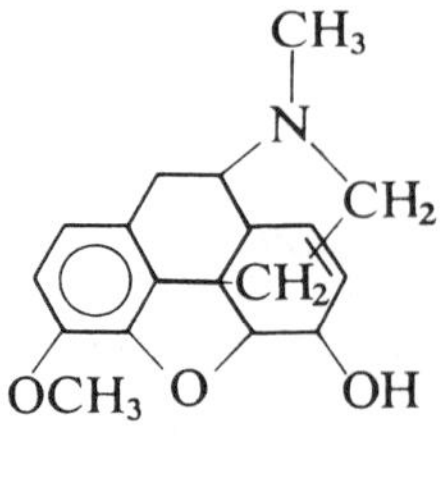 Codeine 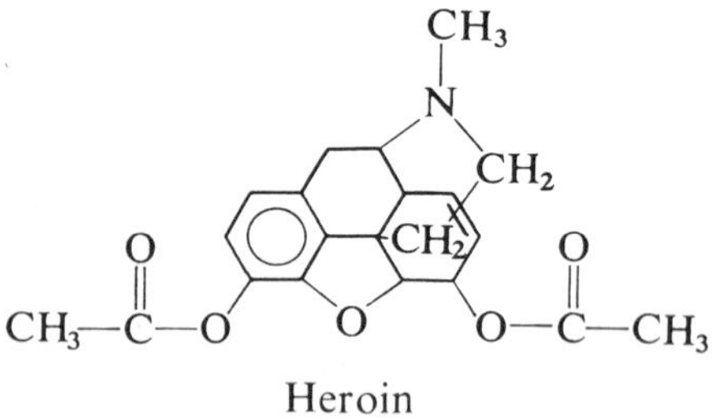 Heroin

Example 26–9. Strong Analgesics from Opium

Heroin is five to ten times as potent as morphine but is also many times more addictive. It cannot legally be manufactured or imported into the United States. The illegal traffic and use of this drug constitutes a great problem in the United States, Canada, and other western countries. Heroin has cough-suppressant properties and is used for this purpose in some countries.

Another strong analgesic is found in the leaves of the coca shrub that grows wild in the Andes Mountains. These leaves have been used for centuries by Inca Indians, who mixed them with lime and chewed them to lessen hunger and fatigue, especially before battle. The leaves were also applied directly to battle wounds for the relief of pain.

The active ingredient in the leaves is cocaine, an analgesic compound that blocks local nerve conduction when placed on or injected just under the skin. Cocaine cannot be injected directly into the bloodstream because it causes a general constriction of blood vessels. The resulting high blood pressure often leads to overwork and failure of the heart. This property of constricting blood vessels is useful when cocaine is applied to a small area. The blood supply to the area is decreased and the analgesic is not quickly washed away.

Cocaine was the first local analgesic to be successfully used in modern medicine. It proved to be the forerunner of numerous synthetic materials possessing similar properties. Example 26–10 gives the formulas of cocaine and some synthetic local analgesics; the trade name of the synthetic analgesics is given first, with the generic name in parentheses.

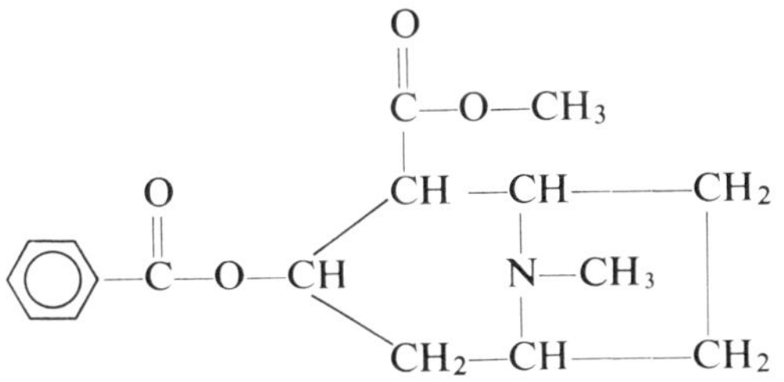

Cocaine

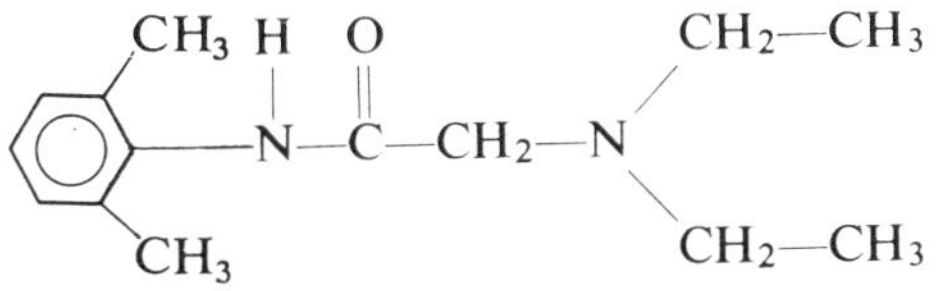

Xylocaine (lidocaine)

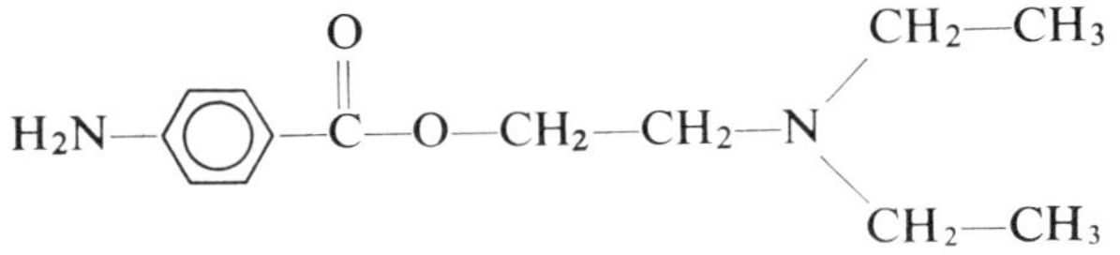

Novocain (procaine)

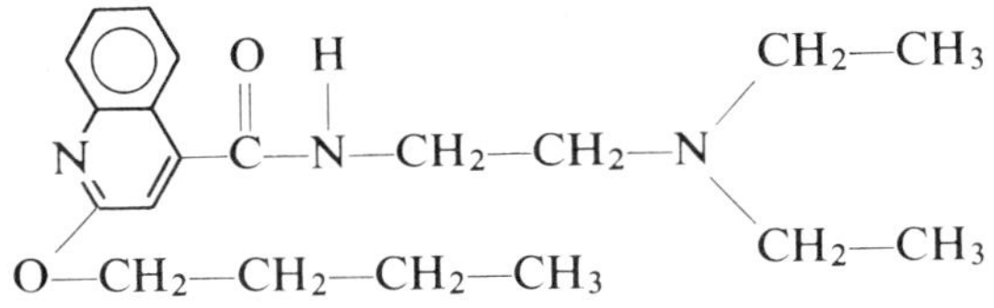

Nupercaine (dibucaine)

Example 26–10. Local Analgesics

Numerous commercial ointments and jellies contain a local analgesic that is absorbed by the skin and deadens local pain. In dental work, the analgesic is injected into the gum or inner lining of the cheek. A physician may inject analgesics under the skin to deaden the pain of minor surgery. In either case injection into blood vessels is avoided, so that the blood will not wash away the analgesics and thereby decrease the time available for the operation.

Local analgesics can be used to deaden large regions of the body. For example, operations may be carried out painlessly on the lower parts of the body after a local analgesic has been injected into the spinal column. The analgesic deadens all nerves it contacts and effectively, but temporarily, cuts off nerve impulses to the brain from lower parts of the body.

General Anesthetics

General anesthetics are substances that produce insensibility to pain and a readily reversible unconsciousness. They have a fairly wide safety margin (the difference between an anesthetic dose and a fatal dose).

Most general anesthetics are gases or volatile liquids. Before the introduction of general anesthetics a little over a century ago, opium and alcohol were used to blunt the pain of surgery. These substances could not be used safely because of the small difference between anesthetic and fatal doses. The patients were often given less than was needed for anesthesia, and much surgery was performed to a background of their agonized screams—the amount of pain depending on the speed of the surgeon.

In 1776 Joseph Priestly, the man who discovered oxygen, first prepared nitrous oxide gas. In 1800 Sir Humphrey Davy noted the analgesic properties of the gas and suggested that it be used in surgery. It was called **laughing gas** because of the exhilaration experienced by those who inhaled it. The first painless tooth extraction using nitrous oxide was performed in 1844 by Horace Wells. He unsuccessfully attempted a public demonstration of the technique in 1845. This failure held back for some time the development of nitrous oxide as an anesthetic.

Crawford Long, a Georgia physician, used ethyl ether as an anesthetic and performed a painless operation in 1842. He did not make his work public, and in 1846 another operation was successfully performed in the Massachusetts General Hospital by Dr. John Warren. The patient had been anesthetized with ethyl ether by a dentist named William Morton. Morton unsuccessfully attempted to keep the nature of the anesthetic a secret. In 1847 chloroform was first used as an anesthetic in England by Simpson and Keith, who used it for the relief of labor pains in patients.

During the past hundred years other inhalation agents have proved to be useful anesthetics. Today the usual practice is to provide balanced anesthesia by using several anesthetic substances rather than just one.

Typically, a patient facing surgery is first given a sedative such as those of Examples 26–7 and 26–8. Then, just before the operation, a drug such as a barbiturate is given to produce unconsciousness. Finally, an anesthetic is administered to produce analgesia, maintain unconsciousness, and often to relax muscles.

The names, formulas, uses, and characteristics of some general anesthetics are listed in Table 26–4.

Chemicals Against Infection

The human body possesses several natural defenses against infection. Skin, the tough outer body covering, is a major line of defense. When this line of defense is broached, other defenders take over and attempt to rid the body of

Table 26-4. General Anesthetics

Name	*Chemical formula*	*Uses and characteristics*
Ethyl ether	$CH_3—CH_2—O—CH_2—CH_3$	A volatile liquid administered by vapor inhalation. Safe for long operations. Vapors are flammable and irritating to mucous membranes.
Vinyl ether	$CH_2=CH—O—CH=CH_2$	A volatile liquid with properties similar to those of ethyl ether. Recovery from anesthesia is prompt.
Nitrous oxide	N_2O	A nonflammable gas that stimulates respiration. Often used in dentistry and obstetrics.
Ethylene	$CH_2=CH_2$	A nonirritating gas. Often used in obstetrics. Mixtures with oxygen are explosive.
Ethyl chloride	$CH_3—CH_2—Cl$	A volatile liquid. Powerful and fast-acting anesthetic. Rapid recovery. Highly flammable.
Chloroform	$CHCl_3$ (Cl—C(Cl)(Cl)—H)	A heavy volatile liquid. Rapid acting. More pleasant than ethyl ether. Nonflammable.
Halothane	H—C(Br)(Cl)—C(F)(F)—F	A nonflammable liquid. Safe and predictable. More potent than chloroform. Rapid recovery.
Cyclopropane	CH_2 / H_2C—CH_2 (three-membered ring)	Rapid-acting, pleasant gas. Rapid recovery. Very flammable and explosive with oxygen. No postanesthetic analgesia.

invading organisms. Certain types of white blood cells attack and destroy bacteria and other foreign bodies. Also, the body is able to produce antibodies that can destroy various organisms.

The process of antibody formation is not completely understood, but evidently each body has a system by which foreign substances are identified. Generated antibodies are primarily protein, although some contain carbohydrates in their structures. Vaccination is effective because the body is stimulated to produce antibodies against a particular organism. Rejection of transplanted organs is also caused in a large part by the generation of antibodies that attack the foreign tissue of the implanted organ.

Chemical substances may help the body defend itself against some infectious organisms. An antiseptic may disinfect a wound; an antibiotic may destroy infectious organisms already well established inside the body.

A perfect chemical for such uses would be completely effective against an infectious organism but have no adverse effects on the host. Many life processes are similar in most organisms, so agents poisonous to one life form are generally poisonous to other forms. Exceptions to this rule do exist, however, and it is these exceptions that make up most of the chemical agents used against infections.

Antiseptics

In the late nineteenth century an English surgeon, Joseph Lister, urged that phenol (carbolic acid) be used as a hospital antiseptic. In Lister's day surgery was performed, babies delivered, and wounds dressed by hands which had not been disinfected and often not even washed. Infection cost many patients their lives. Lister's ideas were eventually accepted along with the germ theory of infection, and the use of antiseptics to help combat infection is widespread today.

The phenol used by Lister is a very strong skin irritant. It has been replaced in most uses by substances with greater antiseptic strength and lower irritation properties. Formulas of a few antiseptic substances are given in Example 26–11.

Alcohols and phenollike compounds attack infectious organisms by denaturing their protein. Heavy-metal compounds denature and precipitate protein, but in some cases they also inactivate enzymes needed by the organism.

Sulfa Drugs

Compounds such as phenol and heavy-metal salts effectively combat infectious organisms outside the body but they usually prove to be quite toxic to the host when taken internally. A chemical that can kill invading organisms inside the body while causing minimal harm to the patient is called a **chemotherapeutic** agent; the use of such compounds to cure disease is **chemotherapy.**

Paul Ehrlich, a German scientist (1854–1915), is considered to be the father of chemotherapy. He observed that some kinds of bacteria could be stained with dyes more easily than normal tissue. From this observation he concluded that infectious organisms might have a different affinity for chemicals than normal tissue.

Ehrlich used this idea and attempted, successfully, to find a chemical cure for syphilis. He decided to use the toxicity of arsenic to kill the organism responsible for the disease. He attempted to attach arsenic to an organic molecule with a high affinity for the bacteria and a low affinity for the tissue of the patient. The arsenic compound, he reasoned, would concentrate in the bacteria and destroy them. The 606th modification of the organic structure attached to arsenic was effective. The compound, called salvarsan, and a relative, neosalvarsan, although quite toxic to man, were used for syphilis treatment until the advent of penicillin in the early 1940s. Dr. Ehrlich received a Nobel prize in 1908 for this pioneering work. The chemical structures of his active compounds are given in Example 26–12.

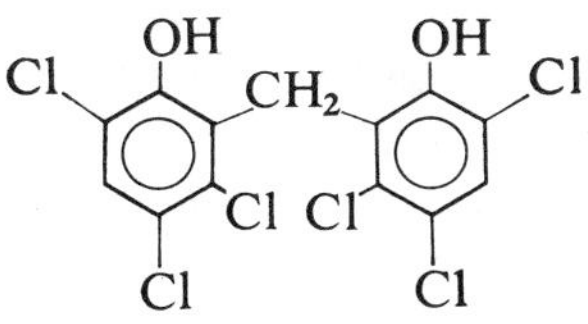

Cl, OH, CH_2, OH, Cl, Cl, Cl, Cl, Cl

Hexachlorophene

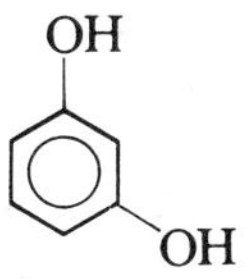

OH, OH

Resorcinol

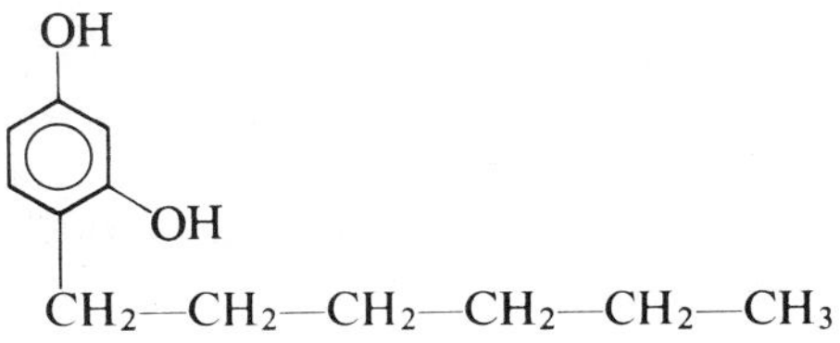

OH, OH, $CH_2—CH_2—CH_2—CH_2—CH_2—CH_3$

Hexylresorcinol

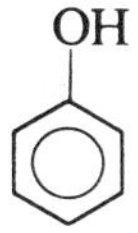

OH

Phenol

OH

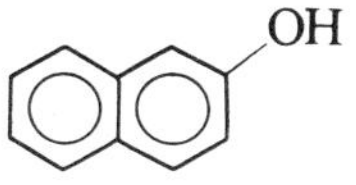

β-Napthol

Phenollike compounds

$CH_3—CH_2—OH$

Ethyl alcohol

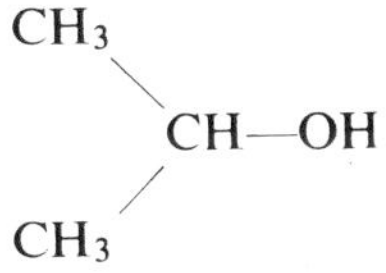

CH_3, CH—OH, CH_3

Isopropyl alcohol

Alcohols

$HgCl_2$

Mercury(II) chloride

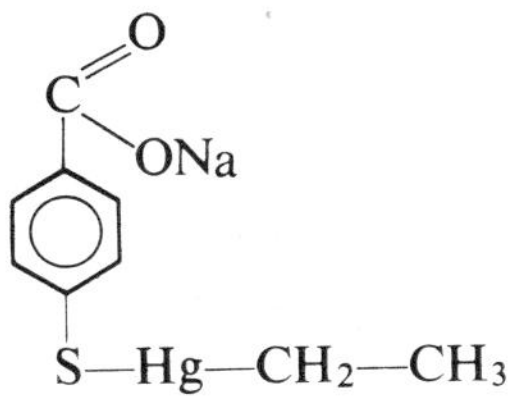

O, C, ONa, $S—Hg—CH_2—CH_3$

Merthiolate

$AgNO_3$

Silver nitrate

Heavy-metal compounds

Example 26–11. Some Common Antiseptics

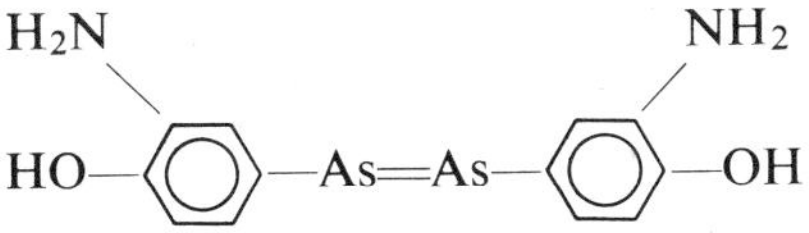

H_2N, NH_2, HO—, —As=As—, —OH

Salvarsan

H_2N, H, O, $N—CH_2—C—ONa$, HO—, —As=As—, —OH

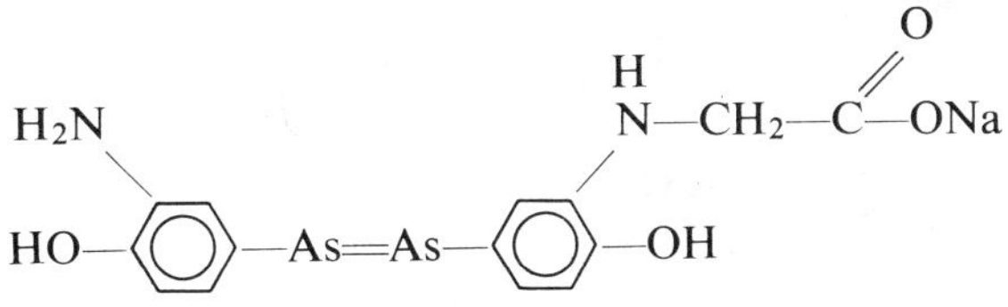

Neosalvarsan

Example 26–12. Chemotherapeutic Agents Used Against Syphilis

Scientists continued to study dyes in a search for other chemotherapeutic agents, and in 1935 an azo dye, prontosil, was announced to be a selective killer of certain types of bacteria called Gram-positive. These bacteria can be stained by a process developed by a man named Gram. Gram-negative bacteria, which cannot be stained by the process, were unaffected by prontosil. It was soon discovered that prontosil reacted in the body and gave sulfanilamide as one product.

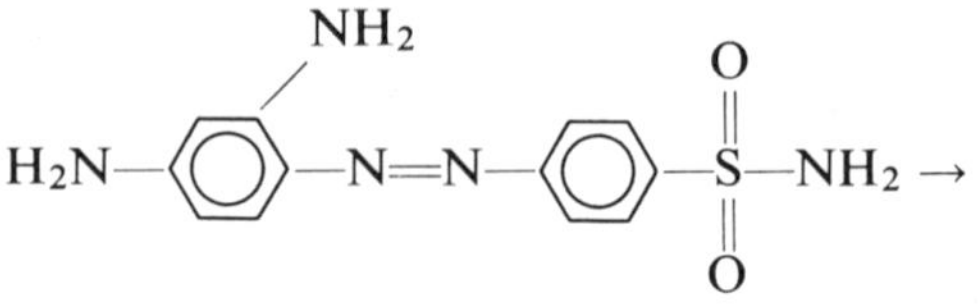

NH₂ — H₂N—⟨◯⟩—N=N—⟨◯⟩—S(=O)(=O)—NH₂ →

prontosil

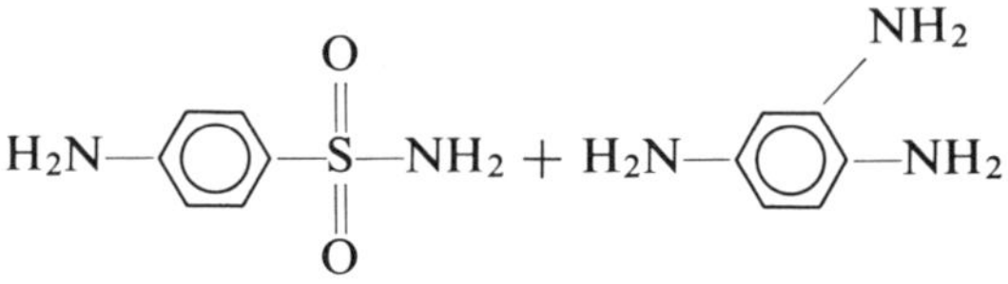

H₂N—⟨◯⟩—S(=O)(=O)—NH₂ + H₂N—⟨◯⟩(NH₂)—NH₂   26–1

sulfanilamide

Sulfanilamide proved to be the actual antibacterial agent. This compound and some of its derivatives make up the group of chemotherapeutic agents known as **sulfa drugs.** A few are structurally represented in Example 26–13.

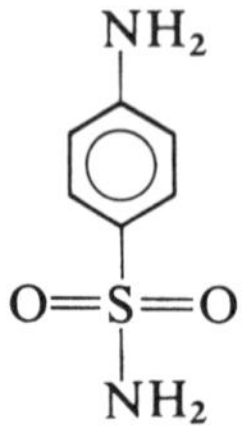

Sulfanilamide

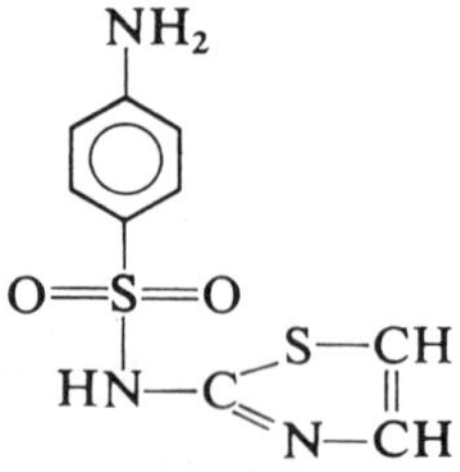

Sulfathiazole

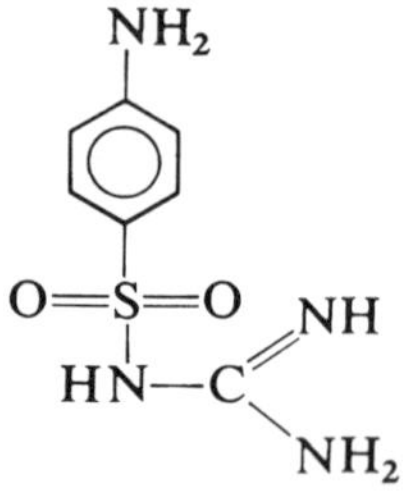

Sulfaguanidine

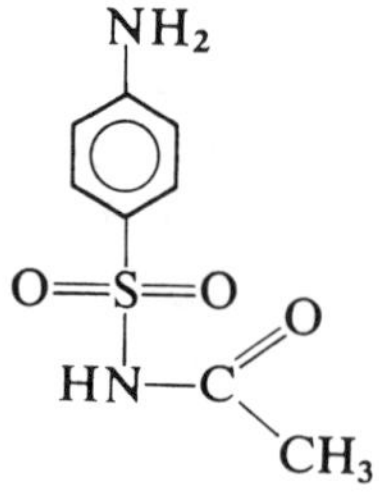

Sulfacetamide

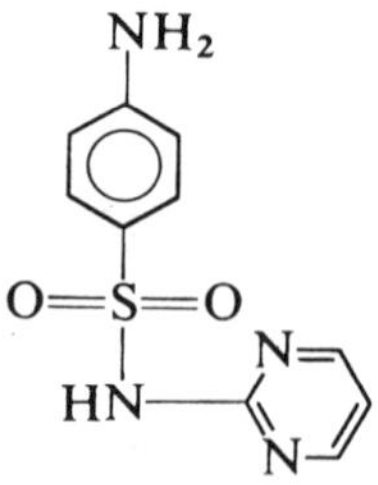

Sulfadiazine

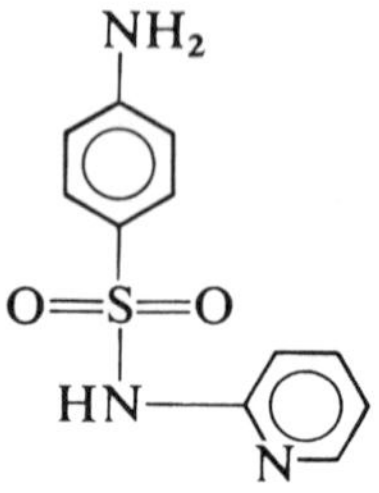

Sulfapyridine

Example 26–13. Some Sulfa Drugs

The success of these "wonder drugs" was almost unbelievable. Before sulfa drugs, the death rate from streptococcal meningitis was higher than 92 percent; after sulfa drugs were introduced, it dropped to less than 10 percent. During World War II many soldiers carried packets of "sulfa" which was sprinkled into wounds to prevent infection; this practice saved many lives.

Sulfa drugs, according to one theory, show bacteriocidal behavior by acting as **antimetabolites.** In this role, sulfa drugs take the place of compounds normally involved in vital metabolic processes carried out by bacteria. Once incorporated into the process, the antimetabolite fails to function like the normal component, and the bacteria die. It was thought that sulfanilamide took the place of para-aminobenzoic acid:

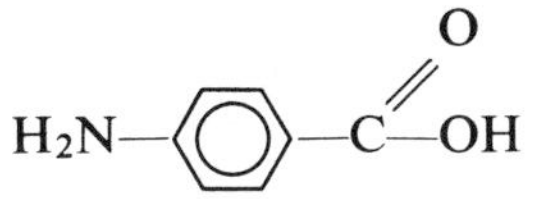

$$H_2N—C_6H_4—C(=O)—OH$$

At present there is some doubt that all sulfa drugs behave as true antimetabolites, but it is generally accepted that they do inhibit vital metabolic processes of bacteria, and slow down their reproduction and growth sufficiently to allow the natural defense mechanisms of the body to be effective.

Antibiotics

In 1877 Louis Pasteur, who had proved that microorganisms cause infection, observed the results of forcing two types of microorganisms to live together. He found that neither type could grow, even with a plentiful food supply. Pasteur summed up this observation with the statement, "Life hinders life."

Several years later a group of scientists found that a certain fungus, a penicillium, killed bacteria placed in contact with it. In 1897 mice were innoculated with disease-causing bacteria and a broth on which penicillium fungus had grown. The mice that received only bacteria died, while those that also received the broth lived. These results were ignored and soon forgotten.

Similar observations took place in 1928 when Alexander Fleming, a bacteriologist at London University, found mold growing on one of his culture plates. Before the plate was discarded, Fleming noticed that the staphylococcus bacteria on the plate were not growing in the vicinity of the mold. The mold apparently gave off some substance that inhibited the bacterial growth. A broth obtained by cultivating the mold was effective in curing pneumonia in mice. The active substance, subsequently named penicillin, proved very difficult to isolate and purify, and Fleming gave up on the project.

Ten years later this often-interrupted discovery process began again. Howard W. Flory and Ernest Chain headed an Oxford University group whose goal was to use penicillin against human infections. In 1940 they injected mice with deadly bacteria. Some mice also received penicillin. The results were the same as those observed and forgotten 43 years before. This time, however, the results were not forgotten, and a massive development program was undertaken in the United States. By 1943 penicillin was clinically available, and by 1945 enough was produced to supply everyone needing it.

Several types of penicillin have since been developed which vary slightly in chemical structure, offering useful differences in properties. The original penicillin was attacked by stomach acid and had to be given by injection. Acid-resistant types have been developed and can be taken orally, much to the relief of some patients. Certain bacteria developed the ability to generate penicillinase, an enzyme with the capacity to deactivate penicillin. New forms of penicillin were developed that are not affected by the enzyme and are capable of destroying such **resistant** bacteria. Example 26–14 gives the chemical structures of some penicillins. Notice the part of the structure common to all.

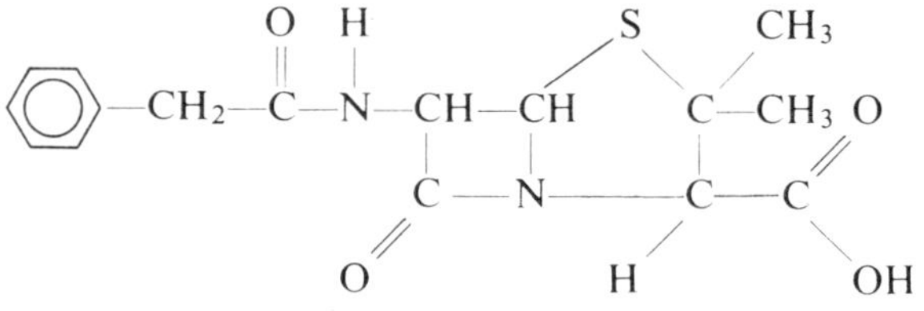

Benzyl penicillin
(original penicillin)

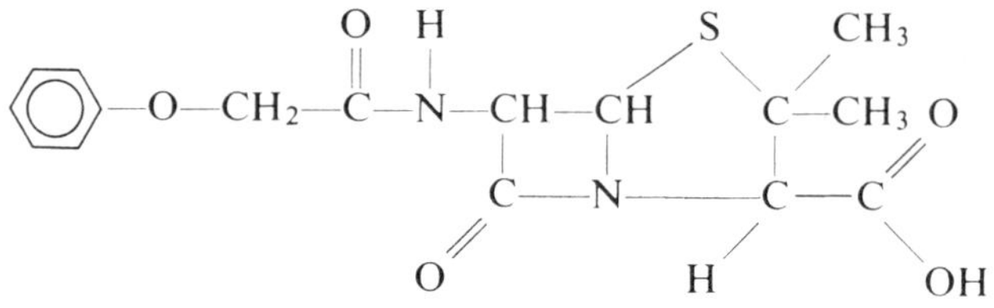

Penicillin V
(acid resistant)

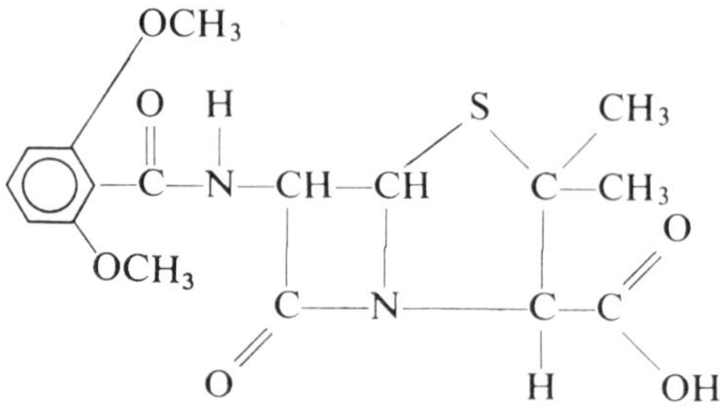

Methicillin
(penicillinase resistant)

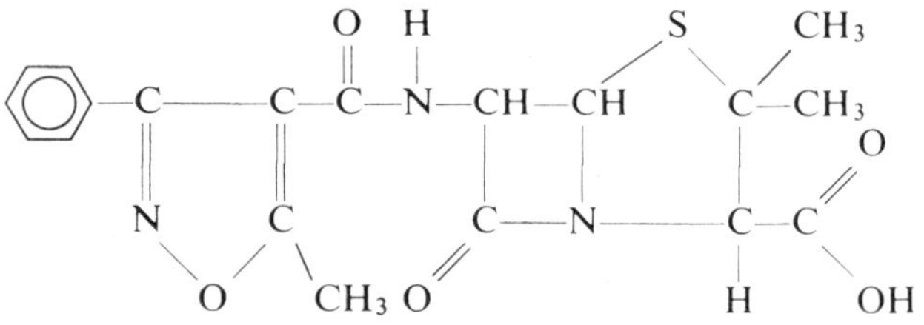

Cloxacillin
(resistant to acid and penicillinase)

Example 26–14. Structures of Some Penicillins

Penicillins are true **bacteriocides** that kill bacteria by interfering with the process of cell-wall construction. The bacteria are able to generate the molecular components of cell walls but cannot, in the presence of penicillin, put them together to form the wall. This inability to build walls allows the bacterial cell contents to spill out, and the organism dies.

About the time Fleming was making his observations concerning penicillin, another scientist was making a contribution to the search for antibiotics.

René J. Dubos wondered why dangerous microorganisms were not found in the soil. After all, animals were continually dying of infectious diseases and discharging the bacteria into the soil. Could it be, he wondered, that soil organisms were killing the bacteria? If so, perhaps these same soil organisms could produce antibiotics useful to man.

In 1937 Dubos launched a full-time search for such useful soil organisms. He found an agent that was effective against pneumonia and some other diseases, but it was also toxic to most animals. This limited success prompted others to perform systematic searches which led to the discovery of many important antibiotics.

Penicillin was effective primarily against Gram-positive bacteria, the same type treated with sulfa drugs. Using the ideas proposed by Dubos, Silman A. Waksman and his coworkers began a search in 1939 for useful soil organisms with wider antibiotic applications than sulfa or penicillin. During their five years of work they isolated and studied 10,000 cultures. Out of this work came one compound with low enough toxicity in man to be deemed useful. The carbohydratelike structure of this compound, streptomycin, is given in Example 26–15.

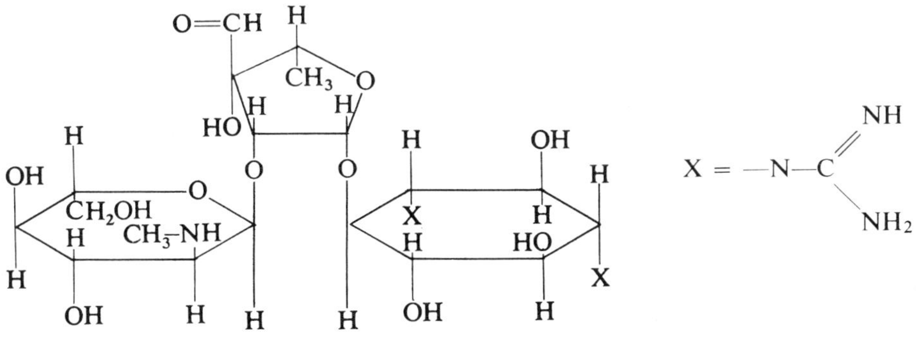

Example 26–15. Streptomycin

Streptomcyin is effective against some Gram-negative bacteria that are not affected by penicillin. Among the treatable diseases are tuberculosis, influenzal meningitis, bubonic plague, and tularemia. Streptomycin is another true bacteriocide. It destroys bacteria by binding to ribosomes, the sites of protein synthesis. Proteins produced by bacteria under these conditions are faulty, and the bacteria, deprived of vital proteins, die.

Further studies of soil organisms led to the discovery of other useful antibiotics, some of which are effective against many types of bacteria, both Gram-negative and Gram-positive. Compounds with this wide range of effectiveness are called **broad-spectrum antibiotics** (Example 26–16).

Tetracyclines are **bacteriostatic** agents without the ability to kill bacteria. Bacterial growth and reproduction are slowed down, and with these factors controlled, natural body defenses are able to destroy the bacteria. The tetracyclines exert this action by interfering with bacterial protein synthesis.

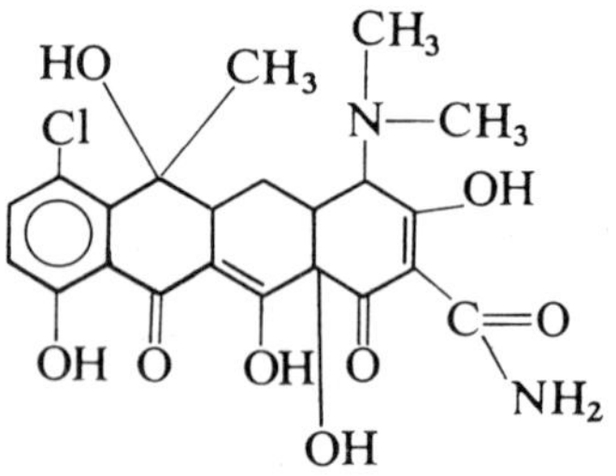

Aureomycin

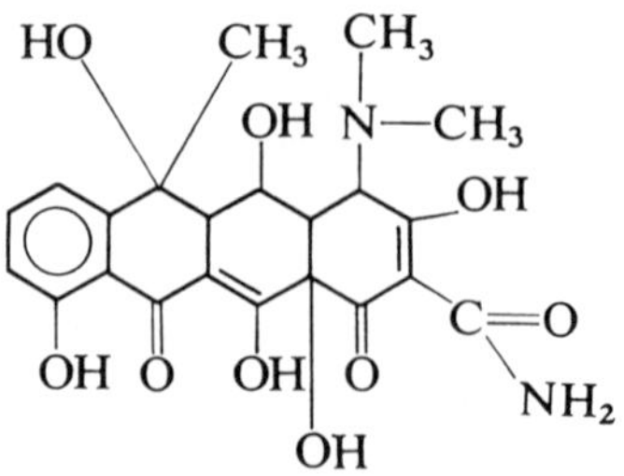

Terramycin

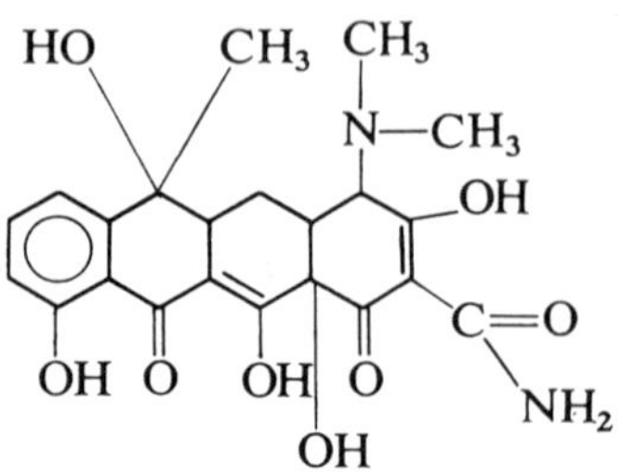

Achromycin

Tetracyclines

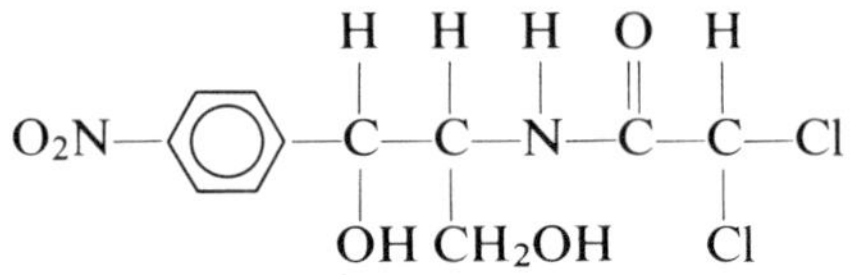

Chloromycetin

Example 26–16. Broad-Spectrum Antibiotics

Chloromycetin acts as a bacteriocide toward a few bacteria but is mainly a bacteriostatic agent. It modifies cellular RNA of bacteria and through this mechanism blocks protein synthesis. The drug is effective against diseases such as typhus, typhoid fever, spotted fever, and some forms of dysentery.

Review Questions

1. Describe three differences between sympathetic and parasympathetic nerves.
2. Compare the nerve gases and insecticides of Example 26–3 from a structural point of view.
3. List three substances that are classified as poisons but which can be used as medicines or poison antidotes when given in proper doses.
4. Compare the structures of morphine, codeine, and heroin (Example 26–9).
5. Chemically, what is opium?
6. Classify each of the following compounds as a mild analgesic, a tranquilizer, a nonbarbiturate sedative, a barbiturate sedative, a strong analgesic, a local analgesic, or a general anesthetic.
   a) Novocain
   b) Codeine
   c) Acetylsalicylic acid (aspirin)
   d) Halothane
   e) Chlorpromazine
   f) Phenobarbital
   g) Sodium bromide
   h) Paraldehyde
   i) Nitrous oxide

7. Describe the mode of action for each of the following substances that are used to combat infectious organisms:
   a) A phenolic antiseptic
   b) A sulfa drug
   c) A penicillin
   d) Streptomycin
   e) A tetracycline
8. What part of the molecular structure is common to
   a) Penicillins (Example 26–14)
   b) Sulfa drugs (Example 26–13)
   c) Tetracyclines (Example 26–16)
9. What contributions did the following individuals make toward the development of chemicals used against infection?
   a) Paul Ehrlich
   b) Joseph Lister
   c) Louis Pasteur
   d) Alexander Fleming
   e) René J. Dubos
10. What is meant by the term *resistant bacteria?*

Suggestions for Further Reading

Adams, E., "Barbiturates," *Scientific American*, Jan. 1958.

Adams, E., "Poisons," *Scientific American*, Nov. 1959.

Baker, P. F., "The Nerve Axon," *Scientific American*, March 1966.

Böttcher, H. M., *Wonder Drugs, A History of Antibiotics*, J. B. Lippincott Co., Phil. and New York, 1964.

Brookes, V. J., and M. B. Jacobs, *Poisons*, 2nd edition, D. van Nostrand Co., Inc., Princeton, N.J., 1958.

Chisolm, J. J., Jr., "Lead Poisoning," *Scientific American*, Feb. 1971.

Collier, H. O. J., "Aspirin," *Scientific American*, Nov. 1963.

DeKruif, P., *Microbe Hunters*, Harcourt Brace Jovanovich, Inc., New York, 1932.

Eccles, J., "The Synapse," *Scientific American*, Jan. 1965.

Gates, M., "Analgesic Drugs," *Scientific American*, Nov. 1966.

Gorini, L., "Antibiotics and the Genetic Code," *Scientific American*, April 1966.

Schubert, J., "Chelation in Medicine," *Scientific American*, May 1966.

Shideman, F. E., ed., *Take as Directed, Our Modern Medicines*, Chapters 7, 8, 9, and 10, Chemical Rubber Co., Cleveland, 1967.

Watanabe, T., "Infectious Drug Resistance," *Scientific American*, Dec. 1967.

Appendix I
Scientific Notation

In scientific work very large and very small numbers are often encountered. When these numbers are written in the usual way, many zeros must be used in order to locate the position of the decimal, and the result is cumbersome (for example, 602,000,000,000 and 0.000000000156).

Scientific notation provides a concise way of expressing such numbers as the product of a nonexponential and an exponential term. The general form is $M \times 10^n$, where M, the nonexponential term, is a number between 1 and 10, written with the decimal point located to the right of the first nonzero digit (standard position). The term 10^n represents the exponential term, where n is the power to which 10 is raised.

The number 10 raised to powers can be evaluated as given below.

$$10^1 = 10$$
$$10^2 = (10)(10) = 100$$
$$10^3 = (10)(10)(10) = (100)(10) = 1{,}000$$
$$10^4 = (10)(10)(10)(10) = (1000)(10) = 10{,}000$$

and

$$10^{12} = 1{,}000{,}000{,}000{,}000$$

Notice that the number of zeros following the number 1 and the number of places the decimal is moved to the right of 1 are identical to the n value. Therefore, a number such as 602,000,000,000 in which the decimal is located 11 places to the right of the first nonzero digit (6) can be written 6.02×10^{11}.

Negative exponents allow us to use the same general form ($M \times 10^n$) for numbers less than 1. Let us consider the significance of negative exponents. A rule of exponential operations states that when an exponential number is moved from the numerator to the denominator or vice-versa, its sign changes: Therefore, $10^{-1} = 1/10^1$, and we obtain the following values of 10 raised to negative exponents:

$$10^{-1} = \frac{1}{10^1} = \frac{1}{10} = 0.1$$
$$10^{-2} = \frac{1}{10^2} = \frac{1}{(10)(10)} = \frac{1}{100} = 0.01$$
$$10^{-3} = \frac{1}{10^3} = \frac{1}{(10)(10)(10)} = \frac{1}{(100)(10)} = \frac{1}{1000} = 0.001$$

and

$$10^{-8} = \frac{1}{10^8} = \frac{1}{100{,}000{,}000} = 0.00000001$$

Notice that the exponent indicates the number of places the decimal is located to the left of its standard position.

$$10^{-3} = .001 \leftarrow \text{standard position}$$

↑

decimal position

Therefore, a number less than 1, such as 0.000000000156, where the decimal is located 10 places to the left of the standard position, can be written 1.56×10^{-10}.

We see that in scientific notation, numbers with positive exponents represent numbers greater than 1 and numbers with negative exponents represent numbers less than 1.

Examples.

Express the following in scientific notation: (a) 527,600, (b) .00000146.

(a) 527,600. The decimal point is five places to the right of the standard position, so the exponent on 10 will be 5 when the decimal point in the non-exponential term is placed in the standard position:

$$5.276 \times 10^{5}$$

(b) .00000146. The decimal point is six places to the left of the standard position, so the exponent on 10 will be -6 when the decimal point in the non-exponential term is placed in the standard position:

$$1.46 \times 10^{-6}$$

Express the following in decimal form: (c) 3.04×10^{7}, (d) 6.22×10^{-4}.

(c) 3.04×10^{7}. The exponent 7 indicates the correct decimal-point position is seven places to the right of the standard position:

$$30{,}400{,}000$$

(d) 6.22×10^{-4}. The -4 exponent indicates a correct decimal-point location four places to the left of the standard position:

$$.000622$$

Appendix II Multiplication and Division of Exponential Numbers

Numbers written in scientific notation are often used in calculations. It is important that we be able to work with them in this form to avoid having to convert them back to decimal form. Two common mathematical operations in which we will use such numbers are multiplication and division.

Multiplication

Consider the product of $(a \times 10^b)(c \times 10^d)$. The two nonexponential terms are multiplied together in the usual way. The two exponential terms are multiplied by algebraic addition of the exponents. The sum of the exponents becomes the new exponent of 10. The resulting two products (nonexponential and exponential) are then multiplied together. Therefore,

$$(a \times 10^b)(c \times 10^d) = (a \times c)(10^{b+d})$$

or, using numbers,

$$(2 \times 10^3)(3 \times 10^4) = (2 \times 3)(10^{3+4}) = 6 \times 10^7$$

and

$$(2 \times 10^3)(3 \times 10^{-4}) = (2 \times 3)(10^{3+(-4)}) = (2 \times 3)(10^{3-4}) = 6 \times 10^{-1}$$

After such a multiplication, we usually move the decimal of the nonexponential part to standard position by changing the exponent.

For example:

$$(6.0 \times 10^5)(4.0 \times 10^3) = (6.0 \times 4.0)(10^{5+3})$$
$$= 24.0 \times 10^8 = 2.4 \times 10^9$$

or

$$(5.2 \times 10^3)(3.0 \times 10^{-7}) = (5.2 \times 3.0)(10^{3+(-7)})$$
$$= 15.6 \times 10^{-4} = 1.56 \times 10^{-3}$$

Division

The rules for division are similar; each type of number is dealt with separately, and the final answer is written as a product. The only change is that the result of dividing an exponential number is 10 raised to the quantity (numerator exponent − denominator exponent). Therefore

$$\frac{(a \times 10^b)}{(c \times 10^d)} = \left(\frac{a}{c}\right) \times \left(\frac{10^b}{10^d}\right) = \left(\frac{a}{c}\right) \times 10^{b-d}$$

or, using numbers,

$$\frac{(4 \times 10^6)}{(2 \times 10^3)} = \left(\frac{4}{2}\right) \times \left(\frac{10^6}{10^3}\right) = 2 \times 10^{6-3} = 2 \times 10^3$$

$$\frac{(4 \times 10^{-6})}{(2 \times 10^3)} = \left(\frac{4}{2}\right) \times 10^{-6-(3)} = 2 \times 10^{-9}$$

An adjustment of the decimal point may be necessary. For example,

$$\frac{(4.0 \times 10^{-3})}{(5.0 \times 10^{-2})} = \left(\frac{4.0}{5.0}\right) \times 10^{-3-(-2)} = 0.8 \times 10^{-3+2}$$

$$= 0.8 \times 10^{-1} = 8.0 \times 10^{-2}$$

Appendix III
The Metric System

In order to work quantitatively, we must use a system of units. If you were to ask someone to loan you four, the immediate response would be four what? You would then have to indicate that you wished to borrow four dollars, eggs, gallons of gasoline, or whatever it was that you wanted.

Scientists, as well as nonscientists, use units to express various quantities such as length and volume. The units used by scientists belong to the metric system. This system is also used by nonscientists in most parts of the world—the United States is still resisting its adoption.

In the metric system, prefixes are used to denote quantities that are larger or smaller than some reference quantity. Some of the commonly used prefixes are given in Table III–1.

Table III–1. Common Prefixes Used in the Metric System

Prefix	*Meaning*	*Numerical equivalent*
mega	million	10^6
kilo	thousand	10^3
deci	one-tenth	10^{-1}
centi	one-hundredth	10^{-2}
milli	one-thousandth	10^{-3}
micro	one-millionth	10^{-6}

Some useful relationships among various metric units, and between metric and English units, are given below.

LENGTH:

1 meter (m) = 100 centimeters (cm) = 1000 millimeters (mm)
1 meter = 39.37 inches = 3.28 feet
1 centimeter = 10 millimeters
1 inch = 2.54 centimeters

VOLUME:
1 liter (l) = 1000 milliliters (ml)
1 liter = 1.057 quarts (U.S.)
1 quart = 0.946 liters
1 milliliter = 1 cubic centimeter (cm^3 or cc) = 10^{-3} liters
1 milliliter = 0.061 cubic inches

MASS (WEIGHT):
1 kilogram (kg) = 1000 grams (g) = 1,000,000 milligrams (mg)
1 kilogram = 2.205 pounds avoir.
1 gram = 1000 milligrams = 10^{-3} kilograms
1 ounce = 28.35 grams
1 pound avoir. = 453.6 grams

Scientists also use a temperature scale different from the one commonly used in the United States. This scale, the centigrade or Celsius scale, is related to the familiar Fahrenheit scale by the following equations:

$$°F = \tfrac{9}{5}(°C) + 32, \qquad °C = \tfrac{5}{9}(°F - 32)$$

Thus, a centigrade reading of 40° can be converted to a Fahrenheit reading as follows:

$$°F = \tfrac{9}{5}(40) + 32 = (9)(8) + 32 = 72 + 32 = 104°F$$

Similarly, a Fahrenheit reading of 90° can be converted to a centigrade equivalent:

$$°C = \tfrac{5}{9}(90 - 32) = \tfrac{5}{9}(58) = 5(6.44) = 32.2°C$$

Index

2 3 4 5 6 7 8 9 10 11 12 13 14 15 16 17 18 19 20 21 22 23 24 25 80 79 78 77 76 75 74 73